Handbook of Experimental Pharmacology

Volume 135/I

Springer

Berlin
Heidelberg
New York
Barcelona
Hong Kong
London
Milan
Paris
Singapore
Tokyo

Estrogens and Antiestrogens I

Physiology and Mechanisms of Action of Estrogens and Antiestrogens

Contributors

S.E. Alves, E. Baral, I. Berczi, M. Birkhäuser, L. Cardozo,
O.L. Dewhurst, J.A. Dodge, J.M.W. Gee, F. Gomez, C. Hegele-
Hartung, A. Hextall, A.K. Hihi, C.D. Jones, K. Kauser,
C. Lauritzen, T.H. Lippert, B.S. McEwen, A.O. Mück, E. Nagy,
G. Neef, R.I. Nicholson, E. Nieschlag, V.K. Patchev, A. Purohit,
V.D. Ramirez, M.J. Reed, G.M. Rubanyi, H. Seeger,
L.T. Seery, M. Simoni, L. Sobek, E. von Angerer W. Wahli,
A. Weisz, J. Zheng

Editors

M. Oettel and E. Schillinger

 Springer

Professor Dr. med.vet.habil. MICHAEL OETTEL
Leiter Forschung und Entwicklung
Jenapharm GmbH & Co. KG
Otto-Schott-Str. 15
D-07745 Jena
GERMANY

Dr. rer.nat. EKKEHARD SCHILLINGER
Leiter Forschung
Fertilitätskontrolle und Hormontherapie
Schering Aktiengesellschaft
D-13342 Berlin
GERMANY

With 49 Figures and 18 Tables

ISBN 3-540-65016-4 Springer-Verlag Berlin Heidelberg New York

Library of Congress Cataloging-in-Publication Data

Estrogens and antiestrogens / editors, Michael Oettel and Ekkehard Schillinger. p. cm. – (Handbook of experimental pharmacology; 135) Includes bibliographical references and index. Contents: I. Physiology and mechanisms of estrogens and antiestrogens / contributors, S.E. Alves ... [et al.]. – II. Pharmacology and clinical application of estrogens and antiestrogens / contributors, E. Anderson ... [et al.]. ISBN 3-540-65016-4 (hardcover : set : alk. paper). – ISBN 3-540-65016-4 (alk. paper) 1. Estrogen–Physiological effect. 2. Estrogen–Therapeutic use. 3. Estrogen–Inhibitors–Physiological effect. 4. Estrogen–Inhibitors–Therapeutic use. I. Oettel, Michael, 1939– . II. Series. [DNLM: 1. Estrogens–pharmacology. 2. Estrogen Antagonists–pharmacology. 3. Estrogen Antagonists–therapeutic use. 4. Estrogens–metabolism. W1 HA51L v. 135 1999 / WP 522 E807 1999] QP905.H3 vol. 135 [QP572.E857] 615'.1 s–dc21 [615', 366] DNLM/DLC for Library of Congress 99-15299 CIP

© Springer-Verlag Berlin Heidelberg 1999
Printed in Germany

The use of general descriptive names, registered names, trademarks, etc. in this publication does not imply, even in the absence of a specific statement, that such names are exempt from the relevant protective laws and regulations and therefore free for general use.

Product liability: The publishers cannot guarantee the accuracy of any information about dosage and application contained in this book. In every individual case the user must check such information by consulting the relevant literature.

Cover design: design & production GmbH, Heidelberg

Typesetting: Best-set Typesetter Ltd., Hong Kong

Production Editor: Angélique Gcouta

SPIN: 10630946   27/3020 – 5 4 3 2 1 0 – Printed on acid-free paper

Preface

For many years, Springer has been publishing an impressive series of textbooks of pharmacology which have set standards in medical science. Surprisingly, an extensive overview of the current state of the art in research on estrogens and antiestrogens was still lacking. The present two volumes on estrogens and antiestrogens provide a comprehensive review of a field of research in which remarkable progress has been made over the past few years. New insights into the mechanisms of steroid hormone action resulted in a tremendous number of publications from which new principles of preventive and therapeutic applications of estrogens and antiestrogens emerged. Although various electronic data bases provide easy access to this copious information, there was a clear necessity for a monograph-style textbook which assesses and summarizes current knowledge in this rapidly expanding field of research. It should be noted, however, that, due to this dynamic development, it is barely possible to comprehensively update every aspect of basic and clinical knowledge on estrogens and antiestrogens. Thus, the intention of the editors was to provide the reader with an overview of the "classic" and most recently explored areas of research and stimulate future interests in basic and applied endocrinology.

Estrogens were among the first steroid hormones described in the scientific literature. Since they were first isolated, since the chemical, synthesic and pharmacological characterization of naturally occurring estrogens and, later on, of orally active derivatives, estrogen research has produced continuously hallmark results in reproductive endocrinology worldwide. Development of oral contraceptives consisting of estrogens and progestins has revolutionized not only fertility regulation but also society itself. The introduction of hormone replacement therapy and an emerging awareness of its benefits has dramatically improved the quality of life of postmenopausal women. The discovery of antiestrogens and their implication for the treatment of hormone-dependent cancers has saved millions of lives. At present, an explosion of knowledge in the field of molecular biology has a tremendous influence on basic research, which addresses cellular and molecular mechanisms of action of estrogens and antiestrogens. Estrogen research has gained substantial momentum from the discovery of different estrogen receptor isoforms (as exemplified in the chapter by Sven-Ake Gustafsson's group in Vol. I), identification of co-activators and co-repressors which interact with estrogen receptor-mediated transcription, and new insights into the physiological importance of estrogens gathered from studies on transgenic animals. Increasing understanding of basic mechanisms of estrogen action has paved the way for the discovery of a new class of selective ligands of the estrogen receptor, the so-called selective estrogen receptor modulators (SERMs). These compounds may open new avenues in the prevention and treatment of diseases. The growing importance of phytoestrogens has also been recognized. Recently, evidence has accumulated in support of the view that the biological role of estrogens is not confined to the female organism, but also has significant physiological and clinical effect on the male. The central nervous system and bone are constantly gaining importance as crucial targets of estrogen action. Finally, the issues of environmental pollution by estrogen-like compounds and metabolites and its biological consequences have become a source of increasing concern.

The contributors to this book are renowned experts in the corresponding fields of research. The editors thus focussed their effort on providing a balanced representation of basic and clinical research from academia and the pharmaceutical industry, while deliberately abstaining from any major interference with the content and style of individual chapters. This might have resulted in duplication of certain aspects; however, one should consider that even research can be contemplated from different points of view.

It was a rewarding endeavor to closely interact with the authors in the course of the editorial process, and we are indebted to them for their cooperation. We extend our thanks to Doris Walker from Springer for her patience and guidance through various stages of the development of this book. Many thanks also to Hans Herken – the editor-in-chief of the whole series – for his constant interest, his moderate but inevitable pressure on the editors and finally, his patience with us. The editorial assistance of Horst Wagner is gratefully acknowledged. We also appreciate the support of many staff members at Springer, Schering, and Jenapharm.

M. OETTEL and E. SCHILLINGER
Jena and Berlin, July 1999

List of Contributors

ALVES, S.E., Laboratory of Neuroendocrinology, The Rockefeller University, Box 165, 1230 York Avenue, New York, NY 10021, USA

BARAL, E., Department of Internal Medicine, Faculty of Medicine, The University of Manitoba, 795 McDermot Ave., Winnipeg, MB, Canada R3E 0W3

BERCZI, I., Department of Immunology, Faculty of Medicine, The University of Manitoba, 795 McDermot Ave., Winnipeg, MB, Canada R3E 0W3

BIRKHÄUSER, M., Abteilung für Gynäkologische Endokrinologie und Reproduktionsmedizin, Universitätsspital Bern, Frauenklinik, Schanzeneckstr. 1, CH-3012 Bern, Switzerland

CARDOZO, L., Department of Urogynaecology, King's College Hospital, Denmark Hill, London and 8 Devonshire Place, London W1N 1PB, United Kingdom

DEWHURST, O.L., Tenovus Cancer Research Centre, University of Wales College of Medicine, Tenovus Building, The Heath, Cardiff CF4 4XX, United Kingdom

DODGE, J.A., Endocrine Research, Lilly Research Laboratories, Lilly Corporate Center, Indianapolis, IN 46285, USA

GEE, J.M.W., Tenovus Cancer Research Centre, University of Wales College of Medicine, Tenovus Building, The Heath, Cardiff CF4 4XX, United Kingdom

Gomez, F., Division d'Endocrinologie, Diabétologie et Métabolisme, Départment de Médecine Interne, Centre Hospitalier Universitaire Vaudois (CHUV), CH-1011 Lausanne, Switzerland

HEGELE-HARTUNG, C., Fertility Control and Hormone Therapy, Research
  Laboratories, Schering AG, D-13342 Berlin, Germany

HEXTALL, A., Department of Urogynaecology, King's College Hospital,
  Denmark Hill, London and 38 Barnfield Avenue, Kingston upon Thames,
  Surrey KT2 5RE, United Kingdom

HIHI, A.K., Institut de Biologie Animale, Bâtiment de Biologie,
  Université de Lausanne, CH-1015 Lausanne, Switzerland

JONES, C.D., Endocrine Research, Lilly Research Laboratories,
  Lilly Corporate Center, Indianapolis, IN 46285, USA

KAUSER, K., Berlex Biosciences, Cardiovascular Department,
  15049 San Pablo Avenue, Richmond, CA 94804-0099, USA

LAURITZEN, C., Alpenstr. 49, D-89075 Ulm, Germany

LIPPERT, T.H., Sektion Klinische Pharmakologie in Gynäkologie und
  Geburtshilfe, Universitäts-Frauenklinik, Schleichstr. 4,
  D-72076 Tübingen, Germany

MCEWEN, B.S., Laboratory of Neuroendocrinology,
  The Rockefeller University, Box 165, 1230 York Avenue,
  New York, NY 10021, USA

MÜCK, A.O., Sektion Klinische Pharmakologie in Gynäkologie und
  Geburtshilfe, Universitäts-Frauenklinik, Schleichstr. 4,
  D-72076 Tübingen, Germany

NAGY, E., Department of Immunology, Faculty of Medicine,
  The University of Manitoba, 795 McDermot Ave., Winnipeg, MB,
  Canada R3E 0W3

NEEF, G., Preclinical Drug Research, Schering AG, Müllerstr. 170,
  D-13342 Berlin, Germany

NICHOLSON, R.I., Tenovus Cancer Research Centre,
  University of Wales College of Medicine, Tenovus Building, The Heath,
  Cardiff CF4 4XX, United Kingdom

NIESCHLAG, E., Institute of Reproductive Medicine of the University,
  Domagkstr. 11, D-48129 Münster, Germany

PATCHEV, V.K., Preclinical Research Division, Jenapharm GmbH & Co. KG, Otto-Schott-Str. 15, D-07745 Jena, Germany

PUROHIT, A., Endocrinology and Metabolic Medicine, Imperial College of Science, Technology and Medicine, St. Mary's Hospital, London W2 1NY, United Kingdom

RAMIREZ, V.D., Department of Molecular and Integrative Physiology, University of Illinois at Urbana-Champaign, 407 S. Goodwin Avenue, 524 Burrill Hall, Urbana, IL 61801, USA

REED, M.J., Endocrinology and Metabolic Medicine, Imperial College of Science, Technology and Medicine, St. Mary's Hospital, London W2 1NY, United Kingdom

RUBANYI, G.M., Berlex Biosciences, Cardiovascular Department, 15049 San Pablo Avenue, Richmond, CA 94804-0099, USA

SEEGER, H., Sektion Klinische Pharmakologie in Gynäkologie und Geburtshilfe, Universitäts-Frauenklinik, Schleichstr. 4, D-72076 Tübingen, Germany

SEERY, L.T., Tenovus Cancer Research Centre, University of Wales College of Medicine, Tenovus Building, The Heath, Cardiff CF4 4XX, United Kingdom

SIMONI, M., Institute of Reproductive Medicine of the University, Domagkstr. 11, D-48129 Münster, Germany

SOBEK, L., Preclinical Research Division, Jenapharm GmbH & Co. KG, Otto-Schott-Str. 15, D-07745 Jena, Germany

VON ANGERER, E., Universität Regensburg, Institut für Pharmazie, Universitätsstr. 31, D-93040 Regensburg, Germany

WAHLI, W., Institut de Biologie Animale, Bâtiment de Biologie, Université de Lausanne, CH-1015 Lausanne, Switzerland

WEISZ, A., Istituto di Patologia Generale e Oncologia, Facoltà di Medicina e Chirurgia, Seconda Università di Napoli, Larghetto S. Aniello a Caponapoli, 2, I-80138 Napoli, Italy

ZHENG, J., Department of Molecular and Integrative Physiology,
    University of Illinois at Urbana-Champaign, 407 S. Goodwin Avenue,
    524 Burrill Hall, Urbana, IL 61801, USA

Contents

CHAPTER 1

History of Estrogen Research
C. LAURITZEN 1

A. The Classical Experiments of Extirpation
   and Reimplantation 1
      I. Ovarian Extracts 1
B. Isolation of Estrogens 1
C. Analysis of the Steroid Structure; Nomenclature
   and Standardization 2
D. Partial and Total Synthesis of Estrogens 2
E. Localization of the Estrogen Production 2
F. Estrogens in Body Fluids and Tissues 3
G. Estrogens in Food 3
H. Estrogen Determination 4
I. Estradiol Effects at the Target Organs 4
      I. Vagina and Endometrium 4
      II. Breasts 4
      III. Hypothalamus and Pituitary 4
J. Biogenesis and Metabolism of Estrogens 5
K. Physiological Effects of Estrogens 5
      I. Genomic Actions 5
      II. Non-genomic Actions 6
L. Mechanism of Action of Estrogens 6
M. Effects on Lipids 6
N. Effects on Coagulation 7
O. Estrogens in Pregnancy 7
P. Fetal Endocrinology 7
Q. Estrogens for Treatment 7
      I. Preparations 7
      II. Treatment of Cycle Anomalies, Bleeding Disturbances
          and Dysmenorrhea 8
R. Side Effects of Estrogens 8
S. Treatment in Pregnancy 8
T. Treatment of Climacteric Complaints 8

U. Estrogens and Carcinogenesis   9
V. Estrogens and Life Expectancy   9
References   10

Part 1: Chemistry of Estrogens and Antiestrogens

CHAPTER 2

Steroidal Estrogens
G. NEEF   17

A. Introduction   17
B. Total Synthesis   18
C. Partial Synthesis   20
        I. Ring-A Substitution   21
       II. Ring-B Substitution   25
      III. Ring-C Substitution   28
       IV. Ring-D Substitution   30
        V. The Periphery   33
D. Labeling of Estradiol and its Derivatives   34
E. Bioconversions of Estrogens   35
References   35

CHAPTER 3

Non-steroidal Estrogens
J.A. DODGE and C.D. JONES   43

A. Introduction   43
B. Structural Classifications of Non-Steroidal Estrogens   44
        I. 1,2-Diarylethanes and Ethylenes   44
            1. Diethylstilbestrol (DES), hexestrol (HES) and
                Analogs   44
            2. Structure – Activity Relationships   45
            3. DES Metabolites   46
       II. Flavones and Isoflavones   47
      III. Macrolactones   47
       IV. Alkylphenols and Arylphenols........................   49
        V. Non-Aromatic Estrogens   49
       VI. Miscellaneous Non-Steriodal Templates   50
C. Conclusions   50
References   50

CHAPTER 4

Antiestrogens and Partial Agonists
E. VON ANGERER. With 17 Figures   55

A. Introduction 55
B. Triphenylethylene Derivatives 56
     I. Tamoxifen 56
     II. Triphenylethylene Derivatives Related to Tamoxifen 57
     III. Fixed-Ring Analogues of Tamoxifen 61
     IV. 1,2,3-Triarylpropenone-Derived Antiestrogens 63
C. Diphenylethylene Derivatives 66
     I. 1,1-Diphenylethylene-Derived Agents 66
     II. 1,2-Diphenylethylene Derivatives 67
     III. 2-Phenylindoles and Related Heterocycles 68
D. Steroidal Compounds 70
     I. 7α-Substituted Estradiol Derivatives 70
     II. 11β-Substituted Estradiol Derivatives 72
E. Conclusion 74
References 75

CHAPTER 5

Structure – Activity Relationships
E. von Angerer. With 2 Figures 81

A. Introduction 81
B. Estrogens and Antiestrogens with a Modified
     Steroid Structure 82
     I. Modifications of the A Ring 84
     II. Modifications of the D Ring 84
     III. Modifications of the B Ring 85
     IV. Modifications of the C Ring 87
     V. Conclusion 87
C. Non-Steroidal Estrogens and Antiestrogens 90
     I. Derivatives of Diethylstilbestrol 90
     II. 2-Phenylindole Derivatives 91
     III. 2-Phenylbenzo[b]thiophene Derivatives 94
     IV. Triphenylethylene Derivatives 94
     V. 2,3-Triphenyl-2H-1-Benzopyrans and
        Related Structures 98
     VI. Conclusion 100
References 101

Part 2: Molecular Biology of Estrogenic Action

CHAPTER 6

Structure and Function of the Estrogen Receptor
A.K. Hihi and W. Wahli. With 3 Figures 111

A. Introduction 111
B. Structure of the Estrogen Receptor 111
   I. The A/B Domain 112
   II. The DNA-Binding Domain 112
   III. The Ligand-Binding Domain 115
C. ER Functions and Transcription 116
   I. ER DNA-Binding Properties 116
   II. ER Ligand-Independent Activity 117
   III. Ligand-Dependent Transcriptional Activity 117
D. ER Isotype Diversity Generates Specificity 120
   I. ERα and ERβ Heterodimerization 120
   II. Tissue Distribution and Ligand Binding 120
   III. ERα and ERβ Differential Activity 121
E. ERα Knock-Out Mouse: A Functional Model 122
F. Future Directions 122
References 123

CHAPTER 7

Estrogen-Regulated Genes
A. WEISZ 127

A. Introduction 127
B. Transcriptional Control of Gene Activity by Estrogens 129
   I. Molecular and Cellular Determinants of
   Gene Responses to Estrogens 130
     1. The Estrogen Receptors 130
     2. Estrogen-Reponsive Gene Elements 132
       a) Estrogen Response Elements and
       Other Estrogen-Responsive DNA Elements 132
       b) Target Gene Promoters 135
     3. The Target Cell Environment 135
C. Non-transcriptional Control of Gene Activity by Estrogens 137
   I. Post-transcriptional Effects of Estrogens 138
   II. Extra-Genomic Effects of Estrogens 139
References 139

CHAPTER 8

**Regulation of Constitutive and Inducible Nitric Oxide Synthase
by Estrogen**
K. KAUSER and G.M. RUBANYI. With 6 Figures 153

A. Introduction 153
B. Isoenzymes of NOS 153
   I. Nitric Oxide Synthase-I 154

    II. Nitric Oxide Synthase-II 154
   III. Nitric Oxide Synthase-III 155
C. Regulation of NOS Isoenzymes by 17β-Estradiol 155
    I. Inhibition of NOS-II Activity by 17β-Estradiol 155
   II. Upregulation of Endothelial NOS-III Activity by
      17β-Estradiol 156
      1. Gender Difference and the Effect of 17β-Estradiol on
        Endothelial Function 158
      2. Potential Mechanism of Estrogen-Induced Increased
        Endothelial NO Production 159
        a) NOS-III Gene Expression 159
        b) NOS-III Enzyme Activity 161
        c) Bioactivity of NO 162
      3. Role of the Estrogen Receptor in the Regulation
        of NOS-III 162
      4. Role of Increased Endothelial NO in
        the Anti-Atherosclerotic Effect of 17β-Estradiol 164
D. Summary 165
References 166

CHAPTER 9

Non-Genomic Effects of Estrogens
V.D. RAMIREZ and J. ZHENG. With 5 Figures 171

List of Abbreviations 171
A. Introduction 172
B. Evidence for Fast Estradiol-Evoked Biological Responses 173
    I. Central Biological Responses 173
      1. Fast Estradiol Effects on the Catecholamine
        System 173
      2. Fast Estradiol Effects on the Hippocampus 174
      3. Fast Estradiol Effects on Other Areas of the
        Central Nervous System 176
   II. Peripheral Biological Responses 177
C. Evidence for Specific Estradiol Binding Sites in
    Cellular Membranes 180
    I. Central Sites 180
   II. Peripheral Sites 181
D. Evidence for Diverse Protein Estrogen Binders 182
    I. Estrogen Binders in Plasmalemmal Microsomal
      Fractions 182
   II. Estrogen Binders in Mitochondrial Lysosomal
      Fractions 184
   III. Estrogen Binders from Other Origins 184

E. Diverse Mechanisms in the Fast Actions of Estradiol 186
    I. Channel Regulator 187
       1. Ca^{2+} Channels 187
       2. Others 187
    II. Transduction Activator 188
       1. Cyclic Adenosine Monophosphate (cAMP) 188
       2. Phospholipase C (PLC) 188
       3. Mitogen-Activated Protein Kinase (MAPK) 188
    III. Metabolic Regulator 189
       1. ATPase/ATP Synthase 189
       2. Glyceraldehyde-3-Phosphate Dehydrogenase
          (G3PD) 190
       3. Others 191
F. Overview and Concluding Remarks 192
    I. The Continuum Theory 192
References 193

CHAPTER 10

Molecular Mechanisms of Antiestrogen Action
L.T. Seery, J.M.W. Gee, O.L. Dewhurst, and R.I. Nicholson 201

A. Introduction 201
B. Key Elements in Estrogen-Receptor Signalling Important
    for Antiestrogen Action 202
C. Molecular Actions of Antiestrogens 204
    I. Effects of Antiestrogens on Binding to the Receptor,
       Dimerisation and Nuclear Localisation 205
    II. Effects of Antiestrogens on Estrogen Receptor (ER) –
        Estrogen Response Element (ERE) Binding and
        Subsequent Transcriptional Activation 205
       1. Altered ER – ERE Binding Efficiency 205
       2. Changes in ER Conformation Influence
          AF-2 Activity 206
       3. Effects Enabled by AF-1 and the Phenomenon
          of AF-1/AF-2 Promoter Dependency 206
       4. Effects Enabled by the Cellular Levels of
          Co-Activators/Co-Repressors 207
       5. Effects on Ligand-Independent ERE
          Transactivation 207
       6. Effects Enabled by Promoter Elements and
          ERE Sequence 208
       7. Effects Enabled by ER Sub-Type 208
    III. Effects on Ap-1- and NF-κB-Mediated
         Transactivation 208

    IV. Effects on Antiestrogen-Specific Response
        Elements 209
     V. Effects on ER Degradation 209
    VI. Non-ER Action of Antiestrogens 209
  D. Modifying Effects of the Normal Cellular Phenotype on
     Antiestrogen Response 210
  E. Modifying Effects of the Cancer Cell Phenotype on
     Antiestrogen Response 212
  F. Conclusions 213
  References 213

Part 3: Biosynthesis and Metabolism of Endogenous Estrogens

CHAPTER 11

Estrogen Transforming Enzymes
M.J. REED and A. PUROHIT. With 2 Figures 223

A. Introduction 223
B. Metabolism of Estradiol by 17β-Hydroxysteroid
    Dehydrogenase Type 2 (HSD2) 223
      I. 17βHSD Superfamily 223
     II. Tissue Distribution 225
    III. Regulation of 17βHSD2 226
C. Cytochrome P_{450} Mono-Oxygenase Hydroxylation of
    Estrogens 227
      I. Cytochrome P_{450}s 227
     II. C2-and C16α-Hydroxylation of Estrogens 227
        1. Effect of 2- or 16α-Hydroxylation on
           Metabolite Estrogenicity 227
        2. Modulation by Thyroid Hormones, Body Weight and
           Diet 227
        3. Effect of Smoking on C2/C16α-Hydroxylation 228
        4. C2/C16α-Hydroxylation and Breast Cancer 229
        5. Induction of C2-Hydroxylation by Indole-3-
           Carbinol 229
        6. C2/C16α-Hydroxylation in Different Ethnic Groups 230
        7. Effect of Pesticides on C2/C16α-Estrogen
           Hydroxylation 231
        8. Inhibition of C2-Hydroxylation 231
        9. Estrogen-2-Hydroxylation by Placental Aromatase 231
D. Catechol Estrogens and Carcinogenesis 232
E. Estrogen Sulphates 233
      I. Estrogen Sulphate Formation 233
     II. Biological Role of Estrogen Sulphation 233
    III. Regulation of Estrogen Sulphation 234

F. Estrogen Glucuronides 235
G. Lipoidal Estrogens 235
H. Conclusions 237
References 237

CHAPTER 12

Metabolism of Endogenous Estrogens
T.H. Lippert, H. Seeger, and A.O. Mueck. With 3 Figures 243

A. Introduction 243
      I. Estradiol Breakdown Mechanisms 243
     II. Sex Differences in Estradiol Metabolism 246
    III. Catechol Estrogen Metabolism 248
B. Estrogen Metabolism and Disease 249
      I. Breast Cancer 251
     II. Endometrial Cancer 252
    III. Cervical Cancer 253
     IV. Prostate Cancer 253
      V. Papilloma of the Larynx 254
     VI. Liver Disease 254
    VII. Lupus Erythematosus 255
   VIII. Disease of the Thyroid 255
     IX. Weight Changes 256
      X. Depression 256
     XI. Osteoporosis 257
C. Drugs 257
D. Lifestyle 259
E. Pregnancy 261
      I. Changes in Metabolism During Pregnancy 264
References 265

Part 4: Physiology and Pathophysiology of Estrogens

CHAPTER 13

**Phylogeny of Estrogen Synthesis, Extragenital Distribution of
Estrogen Receptors and Their Development Role**
L. Sobek and V.K. Patchev 275

A. Phylogeny of Estrogen Synthesis 275
B. Extragenital Distribution of Estrogen Receptors and
      Their Functional Significance 276
      I. Breast 276

II. Bone 276
III. Gastrointestinal Tract 277
IV. Cardiovascular System 278
V. Immune System 278
VI. Pituitary 279
VII. Central Nervous System 280
VIII. Membrane-Bound ERs 281
C. Hormonal Regulation of ER Synthesis 282
I. Estrogens 282
II. Thyroid Hormones 283
III. Progesterone 283
IV. Androgens 283
V. Glucocorticoids 284
VI. Neurotransmitters 284
D. Developmental Role of ERs as Exemplified by Hormone-
Dependent Brain Differentiation 285
References 288

CHAPTER 14

Female Reproductive Tract

C. HEGELE-HARTUNG. With 4 Figures 299

A. Embryology of the Reproductive Tract 299
B. Sex Differentiation 300
I. Genetic Sex 300
II. Gonadal Sex 301
1. Male 301
2. Female 302
III. Somatic Sex 302
1. Male Sex Differentiation 302
2. Female Sex Differentiation 303
3. Estrogens and Sex Differentiation 304
C. Estrogens and Ovary 304
D. Fallopian Tubes 305
I. Steroidal Regulation 305
E. Uterus 306
I. Estrogen Regulation 307
II. Estrogens and Uterine Pathology 309
F. Cervix 309
I. Estrogen Regulation 309
G. Vagina 310
I. Estrogens and Vaginal Pathology 310
References 310

CHAPTER 15

**Estrogen and Brain Function: Implications for Aging
and Dementia**

S.E. Alves and B.S. McEwen 315

List of Abbreviations 315
A. Introduction 315
B. Current Views on Estrogen's Mechanism of Action 316
C. Non-Reproductive Brain Functions Influenced by Estrogen 317
    I. Cognition (Learning and Memory) 317
        1. Basal Forebrain Cholinergic System 317
        2. Hippocampal γ-Aminobutyric Acid (GABA)
           Interneurons 318
        3. Hippocampal CA1 Pyramidal Neurons 318
    II. Psychological Function 319
        1. Midbrain Serotonergic Neurons 319
    III. Sensorimotor Performance 321
        1. Nigrostriatal Dopaminergic System 321
    IV. Non-Neuronal Functions: "Brain Maintenance" 322
        1. Glial Cells 322
        2. Endothelial Cells of the Blood – Brain Barrier (BBB) .. 323
D. Summary 324
References 324

CHAPTER 16

Cardiovascular System

M. Birkhäuser. With 4 Figures 329

A. Introduction 329
B. Cardioprotective Mechanism of Estrogens 331
    I. Indirect Action on the Cardiovascular System by
       a Change of the Lipid Balance 331
        1. Atherogenic Lipoproteins 331
        2. High-Density Lipoprotein 332
        3. Lipoprotein (a) 332
        4. Influence of Estrogen Replacement Therapy and
           HRT on the Lipid Profile 333
    II. Direct Effect of Estrogens on the Vascular Wall 333
C. Influence of Progesterone and Progestagens 337
D. Conclusion 338
References 338

CHAPTER 17

Immune System

E. Nagy, E. Baral, and I. Berczi. With 1 Figure 343

A. Introduction 343
B. Effect on Primary Lymphoid Tissue 343
    I. The Effect of Estrogens 343
        1. The Bone Marrow 343
        2. The Thymus 344
C. Effect on Immune Function 344
    I. The Effect of Estrogens 344
        1. Receptors 344
        2. Cytokines 344
        3. Humoral Immunity 345
        4. Cell-Mediated Immunity 345
        5. Inflammation 345
    II. The Effect of Anti-Estrogens 346
        1. Lymphocytes 346
        2. Humoral Immunity 346
        3. Cell-Mediated Immunity 346
        4. Inflammation 347
D. Conclusions 347
References 348

CHAPTER 18

Male Reproductive Function
M. Simoni and E. Nieschlag 353

A. Introduction 353
B. Source of Estrogen in the Male 353
C. Localization of the Estrogen Receptor in the
   Male Genital Tract 353
D. Function of Estrogens in the Male 355
    I. Bone Maturation 355
    II. Testicular Function 355
    III. Prostate 357
    IV. Behavior 358
E. Conclusions 358
References 359

CHAPTER 19

The Effect of Estrogens and Antiestrogens on the Urogenital Tract
A. Hextall and L. Cardozo. With 2 Figures 363

A. Introduction 363
B. Pathophysiology 363
    I. Hormonal Influences on the Urogenital Tract 363
    II. The Effect of Ageing on the Bladder 364

C. Estrogen and Urinary Incontinence 365
        I. Epidemiological Studies 365
       II. Estrogen and the Continence Mechanism 367
      III. Estrogens for Stress Incontinence 368
       IV. Estrogens for Urge Incontinence 370
D. Estrogen for Recurrent Urinary-Tract Infections 370
E. Estrogen for Vaginal Atrophy 371
F. Antiestrogens and the Urogenital Tract 372
G. Conclusions 373
References 373

CHAPTER 20

Effects of Estrogens on Various Endocrine Regulations
F. GOMEZ 379

A. The Complexity of Estrogen's Action on Various Tissues 379
        I. Introduction 379
       II. Estrogen Receptor-Mediated Tissue Specificity 379
      III. Sex-Hormone-Binding Globulin-Mediated
           Estrogenic Action 380
B. Estrogens and the Thyroid 381
        I. Sexual Dimorphism in Thyroid Disease 381
       II. The Effect of Estrogens in Thyroid Enlargement 381
      III. Thyroid Dysfunction 382
       IV. Thyroid Hormone-Binding Globulin 382
        V. Thyrotropin Regulation 383
C. Estrogens and the Adrenal Glands 384
        I. Sexual Dimorphism in Adrenal Disease? 384
       II. Regulation of Adrenal Sex-Steroid Precursors 384
      III. Anti-toxic Properties of the Adrenal Glands 385
       IV. Corticosteroid-Binding Globulin 386
        V. Corticotropin Regulation 386
D. Estrogens and the Phosphocalcic Metabolism 387
        I. Bone Remodeling and Calcium Kinetics 387
       II. Parathyroid Hormone 389
      III. Vitamin D 390
References 391

Subject Index 397

Contents of Companion Volume 135/II

Part 5: Pharmacology of Estrogens and Antiestrogens

CHAPTER 21

**In Vitro and In Vivo Models to Characterise Estrogens
and Antiestrogens**
K.-H. Fritzemeier and C. Hegele-Hartung 3

CHAPTER 22

**Estrogen Receptor β in the Pharmacology of Estrogens
and Antiestrogens**
G. Kuiper, M. Warner, and J.-Å. Gustafsson 95

CHAPTER 23

**Interrelationship of Estrogens with Other Hormones or
Endocrine Systems**
N. Navizadeh and F.Z. Stanczyk 105

CHAPTER 24

Mammary Gland
G. Söderqvist and B. von Schoultz 113

CHAPTER 25

Cardiovascular System
G. Samsioe 129

CHAPTER 26

Bone
B. Winding, H. Jørgensen, and C. Christiansen 141

CHAPTER 27

Central Nervous System
R. Gallo, M. Stomati, A. Spinetti, F. Petraglia, and
A.R. Genazzani 151

CHAPTER 28

Liver Inclusive Protein, Lipid and Carbohydrate Metabolism
L. Sahlin and B. von Schoultz 163

CHAPTER 29

Pharmacology of Antiestrogens
A.E. Wakeling 179

CHAPTER 30

Oncology
E.V. Jensen 195

CHAPTER 31

Hormonal Resistance in Breast Cancer
S.R.D. Johnston, E. Anderson, M. Dowsett, and A. Howell 205

CHAPTER 32

Pharmacology of Inhibitors of Estrogen Biosynthesis
A.S. Bhatnagar and W.R. Miller 223

CHAPTER 33

Pharmacology of Inhibitors of Estrogen-Metabolizing Enzymes
M.J. Reed and A. Purohit 231

CHAPTER 34

**Pharmacology of Different Administration Routes – Oral vs
Transdermal**
R. Sitruk-Ware 247

Part 6: Kinetics and Toxicology of Estrogens and Antiestrogens

CHAPTER 35

**Pharmacokinetics of Exogenous Natural and Synthetic Estrogens
and Antiestrogens**
W. Kuhnz, H. Blode, and H. Zimmermann 261

CHAPTER 36

Toxicology of Estrogens and Antiestrogens
H. ZIMMERMANN 323

CHAPTER 37

Estrogens and Sexually Transmitted Diseases
M. DÖREN 353

**Part 7: Clinical Application and Potential of Estrogens
and Antiestrogens**

CHAPTER 38

Hormonal Contraception
H. KUHL 363

CHAPTER 39

Hormone Replacement Including Osteoporosis
H.L. JØRGENSEN and B. WINDING 409

CHAPTER 40

Gynaecological Disorders
G. SAMSIOE 423

CHAPTER 41

Oncology
R.Q. WHARTON and S.K. JONAS 431

CHAPTER 42

Cardiology
E.F. MAMMEN 447

CHAPTER 43

Urogenital Ageing and Dermatology
M. DÖREN 461

CHAPTER 44

Geriatric Neurology and Psychiatry
V.W. HENDERSON 473

CHAPTER 45

Estrogens and Antiestrogens in the Male
M. OETTEL   505

Part 8: Comparative Endocrinology

CHAPTER 46

**Comparative Aspects of Estrogen Biosynthesis and Metabolism and
the Endocrinological Consequences in Different Animal Species**
H.H.D. MEYER   575

CHAPTER 47

Therapeutic Use of Estrogens in Veterinary Medicine
M. OETTEL   603

Part 9: Estrogens, Antiestrogens, and the Environment

CHAPTER 48

Environmental Estrogens
S. MÄKELÄ, S.M. HYDER, and G.M. STANCEL   613

Subject Index   665

CHAPTER 1

History of Estrogen Research

C. LAURITZEN

A. The Classical Experiments of Extirpation and Reimplantation

That an ovarian secretion is necessary for the development and maintenance of both the organs of procreation and the secondary sexual organs, as well as for the change of genitals and breasts during the estrus cycle, was first demonstrated by KNAUER and HALBAN in Austria (1900), by MORRIS in the USA and by RUBINSTEIN in Dorpat (see SIMMER 1970) using castration and reimplantation experiments.

I. Ovarian Extracts

As early as 1896, MOND, from the Department of Gynecology and Obstetrics, University of Kiel, Germany, MAINZER, from the Landau Clinic, Berlin, and CHROBAK, from the I. Women Clinic, Vienna, treated the climacteric complaints of castrated women with ovarian extracts (Ovariin, Merck), which probably contained, if any, only small amounts of estrogens and progesterone.

The gynecologist HALBAN, from Vienna, first postulated an endocrine function of the ovaries and the placenta. He showed that the menstruation in primates is dependent on ovarian function. In 1912, ADLER (1876–1958), from the I. Women Clinic of the University of Vienna, and FELLNER, HERRMANN and ISCOVESCO, from France, used organic solvents to prepare a fairly pure extract from sow ovaries, which caused growth of the uterus in rabbits and guinea pigs (see SIMMER 1971).

B. Isolation of Estrogens

In 1923 and 1924, the zoologist and anatomist ALLEN (1892–1943) and the biochemist DOISY (1893–1986), from the Washington University School in St. Louis, developed a semi-quantitative test for the biological determination of estrogen activity, using vaginal cornification in spayed mice (the Stockard and Papanicolaou test) as the endpoint; 3 mg of the extract they had prepared corresponded to 1 rat unit. With the help of this test, ALLEN et al. (1929) isolated estrone (Theelin) from the follicular fluid of pigs, human placentae and urine of pregnant women. The isolation of estrone occurred at about the same time as Butenandt, from the institute of Windaus, Göttingen (1929) produced a raw oil extract (Progynon) from sow ovaries, in cooperation with Schoeller from Schering-Kahlbaum, Berlin. These results were confirmed in 1930 by means of similar investigations by Dingemanse in Amsterdam and D'Amour and Gustavson in Denver.

Estradiol was isolated from sow ovaries by McCorquodale et al. in 1935. Estriol was crystallized from the urine of pregnant women in 1929 by Marrian, London, and Doisy et al.

C. Analysis of the Steroid Structure; Nomenclature and Standardization

The structure of the estrogenic hormones was stated by Butenandt, Thayer, Marrian and Hazlewood in 1930 and 1931 (see Butenandt 1980). Following the proposition of the Marrian group, the estrogenic hormones were given the trivial names of estradiol, estrone and estriol. At the first meeting of the International Conference on the Standardization of Sex Hormones, in London (1932), a standard preparation of estrone was established.

D. Partial and Total Synthesis of Estrogens

The partial synthesis of estradiol and estrone from cholesterol and dehydroepiandrosterone was accomplished by Inhoffen and Hohlweg (Berlin 1940); the total synthesis was achieved by Anner and Miescher (Basel, 1948). Schwenk and Hildebrandt prepared estradiol from estrone (1933). The first non-steroidal estrogen, stilbestrol, was synthesized by Dodds et al. (1938).

E. Localization of the Estrogen Production

Zondek and Aschheim (1926) localized the production of the estrogenic hormones in the *theca folliculi*. In contrast, Allen et al. (1925) and Westmann (1934) found that the granulosa cells contained higher amounts of estrogen. Steinach and Holzknecht (1918) showed that interstitial cells of the ovary also produce estrogens. It is now established that the *theca folliculi* produces androgens, which, after diffusion to the granulosa are aromatized by these cells to estradiol (Channing et al. 1980).

The ovaries grow in size and weight from the age of 7 years. They loose follicles by ovulation and atresia, and shrink from the age of about 38 years to a minimum weight at age 60 (Watzka 1957). Follicle-stimulating hormone (FSH) stimulates the estrogen production in the granulosa layer, while luteinizing hormone (LH) promotes the production of androgens in the *theca folliculi* and advances aromatization (Bradburry 1961). The testes also produce estradiol (Dingemanse and Mühlbock 1938). The adrenals do not produce estrogens; their C19-steroids are aromatized in the fat tissue (SIITERI and McDONALD 1973).

F. Estrogens in Body Fluids and Tissues

ZONDEK and ASCHHEIM from the Charité's Women Clinic found, in 1926, huge amounts of estrone in urine from pregnant women. LOEWE (Dorpat 1925) and SIEBKE (Kiel, 1929) also detected estrogenic activity in the urine of cyclic females, with a peak in midcycle. Moreover, DINGEMANSE and MÜHLBOCK (Amsterdam, 1938) even found estrogens in the urine of males. ZONDEK's (1934) detection of estrogens in the urine of stallions and of pregnant mares was later used as a source of conjugated estrogens for substitution in climacteric women. The components were identified as the ring-B unsaturated sulfo-conjugated estrogens equlin, equilenin and their 17α-derivatives.

Estrogens are also produced in fat tissue from androgens (FRIEDMANN et al. 1985) by aromatization of androstendione, testosterone and dehydroepiandrosterone from ovaries and adrenals. The placenta is an incomplete endocrine organ and synthesizes mainly estriol from 16-hydroxylated fetal and maternal adrenal C-19 precursors (DICZFALUSY 1962; SIITERI and McDONALD 1973). As already pointed out, the testes also produce estrone and estradiol (HAEUSSLER 1934; ZONDEK 1934; DINGEMANSE and MÜHLBOCK 1938).

ZANDER et al. (1959) isolated estradiol and estrone from human ovarian venous blood, follicular fluid and luteal tissue (see MIKHAIL et al. 1963). The groups of MARRIAN (1950), LEVITZ et al. (1956) and BREUER et al. (1959) detected a number of qualitatively less important estrogens, mostly from the urine of pregnant women. Examples of these include 16-epiestriol, 16α- and 16β-hydroxyestrone, 16-oxoestradiol, 6- and 18-hydroxyestrone and some C-2 and C-4 hydroxylated or methylated estrogens (KRAICHY and GALLAGHER 1957; ENGEL et al. 1959).

The identification and determination of estrogens was greatly conveyed by the development of distribution methods between solvents (ENGEL and SLAUNWHITE 1953) and column-, paper- and thin-layer chromatography (MENINI and DICZFALUSY 1950).

G. Estrogens in Food

Estrogenic activity (stilbestrol) was found in sugar-turnips, potatoes, yeast, cat meal, rice, beer (hop), willow catkins and rhubarb (see DICZFALUSY and LAURITZEN 1961). Cabazo de negro, yam roots and soja beans are also sources of substances with estrogenic actitvity, e.g., Daidzein and Genistein. Conjugated estrogens (esterified estrogens) have been partly synthesized from soja beans (ADLERKREUTZ et al. 1991).

H. Estrogen Determination

The first chemical method making the determination of estradiol in urine possible was developed by Cohen and Marrian, in 1934, on the basis of the Koeber reaction. The method was improved by Bauld (1954), Brown (1955), and Ittrich (1960). Chromatographic, enzymatic, immunologic (Mayes and Nugent 1969) radioisotope methods (Abraham 1969; Korenman 1970; Tulchinsky and Korenman 1970) were elaborated, suitable for exact separation, hydrolysis and quantitative determination of substances in blood, the latter employing the immunological techniques of Berson and Yalow (1959), and Abraham (1969). Correlations of estrogen levels with diagnostic clinical problems were investigated by Paulsen (1965), Brown et al. (1953) and Hammerstein (1962).

I. Estradiol Effects at the Target Organs

I. Vagina and Endometrium

Stockard and Papanicolaou (1917) detected cyclical changes in the vaginal secretions of rodents. Papanicolaou extended these findings to cyclic women. However, Allen and Doisy (1923) were the first to show that removal of the ovaries and withdrawal of ovarian- and placental-extract injections provoked an estrogen withdrawal bleeding of the endometrium in rhesus monkeys. Hitschmann and Adler (1908), of the Schauta Clinic, Vienna, documented for the first time the dependency of the endometrial cycle on the function of the ovaries. Schröder, a gynecologist at the University of Rostock (later Kiel and Leipzig), definitely demonstrated the influence of follicular- and luteal-phase hormones on the structure of the functional layer of the endometrium, and described glandular-cystic hyperplasia as the consequence of a prolonged estrogenic proliferation when the luteal hormone is missing.

Using the classical experiment of administering hormone injections to a castrated woman, which produced proliferation, secretory transformation and menstruation, Kaufmann (1933) gained the first information concerning the quantitative relationship of estradiol and progesterone effects at the target organ endometrium.

II. Breasts

Bradburry (1932) originally described the effects of estrogens on the development and differentiation of the breasts.

III. Hypothalamus and Pituitary

Hohlweg (1934) first detected the positive feed-back of estrogens on the hypothalamus ("Sexualzentrum") and anterior pituitary ("Hohlweg effect").

BRADBURRY (1947, 1961) showed that estradiol exerts a biphasic action, when injected directly into the ovary. During the first 72h, the estrogen stimulates the follicular granulosa; thereafter, it activates the hypothalamic release of LH-releasing hormone (LHRH). Prolonged estrogen administration depletes the anterior lobe and suppresses the pituitary gonadotrophins. The pre-ovulatory surge of estradiol activates the hypothalamic surge of LHRH.

J. Biogenesis and Metabolism of Estrogens

Investigations concerning the biogenesis of estrogens from acetate and cholesterol were performed by LEVITZ et al. (1956) and PLOTZ (1957), using the administration of labeled compounds. In vitro experiments with tissue slices or homogenates (WOTIZ and LEMON 1956; RYAN 1958) resulted in the verification of the cascade of biogenesis and of biodegradation of estrogens as they are known today.

The entero-hepatic metabolism of estrogens was analyzed by ADLERKREUTZ (1974, 1976). BRUNELLI (1934) showed that estrogens are bound in the blood to sex hormone binding globulin (SHBG), while SAND-BERG and SLAUNWHITE (1956) demonstrated localization to fractions IV and V (COHN). The production of SHBG in the liver was proven by ROBERTS and SZEGO (1946). Androgens are metabolized to estrogens by fat tissue (LONGCOPE 1973).

K. Physiologic Effects of Estrogens

I. Genomic Actions

Estrogens act via specific nuclear receptors (see Sect. L.). The proliferative effects of estrogens on both vaginal endometrium and breasts have been dealt with in Sect. I. on target organs. HOHLWEG (1943) showed that castration increased the anterior pituitary content of gonadotrophins and that estradiol inhibits the development of castration cells in the anterior lobe following extirpation of the ovaries. He concluded that there is a feedback relationship between the ovarian estrogens on one side of the pituitary and a hypothalamic center ("Sexualzentrum") on the other side. He postulated a regulation center in the hypothalamus and described the negative and positive rebound effect ("Hohlweg effect"), which later explains the induction of ovulation by the ovulatory surge of LH and estradiol. Hohlweg's findings marked the birth of neuroendocrinology.

Estrogens are produced in the brain by aromatization from androgens. The fetal brain is imprinted as male by androgens from the human chorionic gonadotrophin (HCG)-stimulated fetal testes, which are aromatized in the brain during pregnancy (GOY and McEWEN 1980). Catecholestrogens seem to exert a specific effect in brain receptors and synapses, and stimulate the growth of neurons and their dendrites (Kraychy and Gallagher 1957; Rossmanith and Ulrich 1992). Estrogens increase transcription and biosynthesis of galanin (GABA), which is correlated to the regulation of dopamine and gonadotrophin-releasing hormone (GnRH) (Rossmanith and Ulrich 1992).

Estrogens stimulate osteoblast function and inhibit osteoclasts (Albright and Reifenstein 1948; Lindsay, et al. 1979, 1987) in an interplay with calcitonin, parathormone, vitamin-D hormone and interleukins. Estrogens activate mitosis and proliferation of the skin, connective tissue and mucous membranes (Rauramo and Punnonen 1976).

II. Non-genomic Actions

Quick, direct non-genomic actions are exerted without the binding of estrogen to the cellular estrogen receptors. An apparently non-genomic action is the acute vasodilating effect of estrogens on arteries and veins. This effect was first shown in rabbits by Reynolds (1939) and is also found in humans, not only in all sexual organs, but also in the coronary arteries, the carotids and the cerebral arteries. The vasodilation is mediated via production of nitric oxide (NO) (endothelium-derived relaxing factor) from arginine, and an inhibition of endothelin (Sarrel 1990). Estrogens also stimulate the production of the vasodilating prostacyclin and inhibit the formation of the vasoconstricting and thrombocyte-aggregating thromboxane. Acetylcholine only causes vasodilation in the presence of estrogens (Sarrel 1990).

L. Mechanism of Action of Estrogens

The investigation of Jensen and Jacobson (1962) showed that estrogens act via specific receptors in the nucleus of the target cells, finally producing specific proteins. Estrogens induce estrogen- and progesterone receptors. High doses of estrogens downregulate the estrogen receptors.

The different actions of different estrogens are caused by the avidity of binding to the same or to different estrogen receptors and by the duration of retention of the steroid–receptor complex (Korenman 1969). The estrogen receptor gene is localized on the long arm of the chromosome 6 (Gurpide et al. 1980; Bergink 1980).

The determination of estrogen receptors has gained importance mainly for the prognosis and treatment of mammary cancer. Estrogens influence a great number of enzymes (Astwood 1941) and estrogenic actions on growth and differentiation of target organs are mediated via several growth factors (Heldin and Westermark 1984).

M. Effects on Lipids

Estradiol and conjugated estrogens decrease the atherogenic low-density lipoprotein (LDL) and increase the antiatherogenic high-density lipoprotein

(HDL). They are radical-catchers, inhibit the atherogenic oxidation of LDL and decrease lipoprotein and homocystein, which also play a role in the development of atherosclerosis. Thus, atherogenesis and its sequelae can be partially inhibited by estrogens (OLIVER and BOYD 1956).

N. Effects on Coagulation

Significant effects of estrogens on coagulation are only found with oral medication, caused by the first liver passage of the hormone, which stimulates the enzymatic activity of the organ. Antithrombin III may be decreased in some patients, Factors VII and X may be dose dependently increased (ADAMS 1943; RATSCHOW 1944).

O. Estrogens in Pregnancy

The levels of estrogens greatly increase during pregnancy (ASCHHEIM 1931). Estradiol and estrone rise tenfold, estriol about a hundredfold compared with cyclic values. PHILLIPP (1929, 1942), from the Walter Stoeckel University Clinic, Berlin, identified the fetal part of the placenta as the locus of estrogen production. Histochemical investigation carried out by WISLOCKI et al. (1941) localized the estrogen production in the syncytium.

The placenta is an incomplete steroid-producing organ, as it is dependent on the supply of estrogen precursors, mainly C19-steroids as dehydroepiandrosterone from the fetal and maternal adrenals (DICZFALUSY 1961).

P. Fetal Endocrinology

Dehydroepiandrosterone and its sulfate are mainly produced in the highly hypertrophic fetal adrenals. Following hydoxylation at C-16, this compound is aromatized by the placenta to estriol (DICZFALUSY 1962). Estradiol and estrone are also produced from non-hydroxylated androgen precursors originating in the fetal and maternal adrenal cortices (DICZFALUSY 1962).

Q. Estrogens for Treatment

I. Preparations

The first trials using ovarian extracts were performed by MAINZER and MOND, (1896). In 1938, INHOFFEN and HOHLWEG synthesized ethinylestradiol, a highly active oral estrogen, not surpassed until today, which is mainly used for oral contraception and in the treatment of cycle disturbances. Conjugated estrogens were extracted from the natural source of pregnant mares' urine (ZONDEK 1934; GIRARD et al. 1932); they are called natural estrogens in contrast to the artificial synthetic estrogens ethinylestradiol and stilbestrol, which do not occur naturally in the human. At present, conjugated estrogens are also synthesized from soja beans. They are called "esterified estrogens".

In 1933, SCHWENK and HILDEBRANDT synthesized injectable estradiolbenzoate, which exerts a slightly protracted effect. LAQUEUR et al. (1948) synthesized estradiolpropionate and MIESCHER, a cristalline depot-preparation. Estradiolvalerate (JUNKMANN 1957) was prepared, which was orally active and injectable with protracted effectiveness. Lately micronized estriol was prepared with an improved enteral resorption rate.

II. Treatment of Cycle Anomalies, Bleeding Disturbances and Dysmenorrhea

Endometriotropic estrogens, such as estradiol, inhibit breakthrough bleedings caused by estrogen levels which are too low to maintain the endometrium. Withdrawal of estradiol will cause estrogen-withdrawal bleeding. Progesterone and progestogens will only exert their specific actions following a preceding estrogenic action, as for instance progesterone withdrawal bleeding or the stopping of dysfunctional bleedings.

R. Side Effects of Estrogens

Typical side effects of estrogens are mostly consequences of over-doses. They mimic physiologic effects in an exaggerated manner, like extracellular sodium and water retention producing edema, or the stimulation of pigmentation as an exaggerated stimulation of pigment hormone secretion. An increase in the risk of thrombosis and emboli by estrogen substitution is controversial.

The side effects of ethinylestradiol and natural estrogens are different and must not be intermixed (MÜHLBOCK et al. 1948).

S. Treatment in Pregnancy

There is no indication for treatment with estrogens in pregnancy. Treatment in pregnancy by exogenous application of ethinylestradiol and stilbestrol was, however, used in the fifties to treat imminent abortion and pregnancy diabetes. This therapy was unsuccessful and is no longer used, as administration of stilbestrol caused vaginal adenosis in the children and, in a small percentage, also vagino-cervical cancer (HERBST et al. 1972). However, teratogenic effects are not to be expected from natural hormones or ethinylestradiol, if taken inadvertently.

T. Treatment of Climacteric Complaints

Estrogens abolish the typical climacteric complaints caused by estrogen deprivation (ZONDEK 1935). They exert a psychotropic effect (CALDWELL and

WATSON 1956; BLEULER 1954) and improve the sleep pattern (THOMPSON et al. 1991). The effects on the quality and quantity of sleep and the improvement of depressive mood are mediated via an inhibition of the enzyme monoamine oxidase and an increase in free tryptophan levels, stimulating cerebral synthesis of serotonin (AYLWARD 1975). Estrogens prevent and remove atrophic urogenital complaints. Long-term estrogen substitution will prevent about 50% of myocardial and about 30% of apoplexia (PAGANINI-HILL 1994). Estrogens improve the cognitive cerebral functions and inhibit the development of Alzheimer's disease (OHKURA et al. 1994).

U. Estrogens and Carcinogenesis

The basic investigations concerning this important question were performed by KAUFMANN et al. (1949). This group stated that estrogens may have a "facultative syncarcinogenic action", especially in cases of prolonged use and with increased doses in genetically degenerated animals. LIPSCHÜTZ (1950) and MÜHLBÖCK et al. (1948) came to similar conclusions. The term "promoter function of estrogens" was coined. Estrogens have been shown not to be mutagenic or carcinogenic themselves (AMES). They act, however, as general and basic mitogenic and proliferative substances. Thus, estrogens will only act as promoters of existing cancer or pre-cancer cells, if the normal controlling functions of cell proliferations (proto-oncogenes, oncogenes, proliferation controlling and apoptotic substances) are compromised by a preceding genetic damage of DNA (PETO 1977).

Thus, estrogens only act as cancer promoters, if genetic damage has already occurred and cancer cells are present. Accordingly, the latency period between the origin of the first cancer cell and the clinical diagnosis of cancer is shortened. Because of this shortening of the latency period, there is less time for additional DNA genetic damage to occur. Consequently, the cancer is less malignant and has a better prognosis, as has been shown by clinical studies (COLLINS et al. 1980; BERGKVIST et al. 1989). The addition of a progestogen to the estrogen reduces the risk of occurrence of endometrial, ovarian and colon cancer.

Substitution of estrogens is possible after completed treatment of endometrial and ovarian cancer. In mammary cancer, tamoxifen is given for 5 years; thereafter, an estrogen substitution is possible, if an urgent indication is given (LAURITZEN 1993).

V. Estrogens and Life Expectancy

Long-term use of estrogens prolongs life expectancy for 2–3 years, by decreasing the incidence and the mortality of myocardial infarction, cerebrovascular accidents, osteoporosis and overall cancer morbidity and mortality (PAGANINI and HILL 1994; LAURITZEN 1993).

References

Abraham G (1969) Solid-phase radioimmunoassay of estradiol-17β J Clin Endocr Metab 29:866–870

Adam, W (1943) Östrogene erzeugen Hypertrombinämie. Zbl Gynäk 67:551–554

Adlercreutz H (1974) Hepatic metabolism of estrogens in health and disease. New Engl J Med 290:1081–1084

Adlercreutz H, Martin P, Järvenpää P (1976) Steroid absorption and enterohepatic recycling. Contraception 20:201–224

Adlercreutz H, Fotsis T, Bannwart C, Wöhälä K, Breunow G, Hase T (1991) Isotope dilution gas chromatographic-mass spectrometric method for the determination of lignans and isoflavonoids in human urine, including identification of genistein. Clin Chim Acta 199:263–278

Albright F, Smith PH, Rchardson AM (1941) Postmenopausal osteoporosis. JAMA 116:2465–2470

Allen E, Doisy EA (1923) An ovarian hormone. Preliminary report on its isolation, extraction, partial purification and action in test animals. JAMA 81:819–821

Allen E, Pratt J, Doisy EA (1925) The estrus hormone in the follicle. JAMA 83:399–404

Anner G, Miescher K (1948) Totalsynthese von Östradiol und Östron. Ann Chem 4:25–32

Aschheim S (1931) Experimentelle Grundlagen der Therapie mit Ovarialhormonen und Hypophysenvorderlappenhormonen in der Gynäkologie. In: Wolff-Eisner H (Hrsg) Handbuch der experimentellen Therapie. München

Aylward M (1976) Estrogens and plasma tryptophan levels in the perimenopausal patient. In: Campbell S (ed) The Management of the Menopause and Post-menopausal Years. MTP Press, Lancaster, pp 135–147

Bauld WS (1954) An improved method of estrogen determination. Biochem J 56:426–431

Bergink EW (1980) Oestriol receptor interactions, their biological importance and therapeutic implications. Acta Endocr (Kbh) 90 (Suppl):9–16

Berson SA, Yalow RS (1959) Assay of plasma insulin in human subjects by immuno-logical methods. Nature 184:1648–1649

Bleuler H (1948) Endokrinologische Psychatrie. Thieme, Stuttgart

Bradbury JT (1932) Study of endocrine factors in influencing mammary development and secretion in the mouse. Proc Soc Exper Biol Med 30:212–213

Bradbury JT (1961) Ovarian influence on the response of the anterior pituitary to estrogens. Endocrinology 41:501–513

Bradbury JT (1961) Direct action of estrogen on the ovary of the immature rat. Endocrinology 68:115–120

Breuer H, Knuppen R, Pangels G (1959) 2- and 4-hydroxylated estrogens. Acta Endocr (Kbh) 30:247–250

Brown JB (1955) A chemical method for the determination of oestriol, oestrone and oestradiol. J Endocr 60:185–193

Brown JB, Bradburry JT, Jungk EC (1953) The effect of estrogens and other steroids on the pituitary gonadotrophins in women. Am J Obstet Gynecol 65:733–737

Brunelli B (1934) Binding of steroids on serum globulin. Arch Int Pharmacodyn 49:262–265

Butenandt A (1929) über das Progynon, ein kristallisiertes weibliches Hormon. Natur-wiss 17:897–900

Butenandt A, Kaufmann C, Müller HA, Friedrich-Freksa H (1949) Experimentelle Beiträge zur Bedeutung des Follikelhormons für die Karzinomentstehung. Z Krebsforsch 46:482–488

Butenandt A (1980) Die Entdeckungsgeschichte des Östrons. Endokrinologie-Informationen. Demeter Verlag, Gräfelfing, 4:160–163

Caldwell J, Watson BB (1956) Psychotropic action of estrogens. J Geront 7:228–231

Channing CP, Schaerf FW, Anderson LD, Tsafriri A (1980) Ovarian follicular and luteal physiology. In: Reproductive Physiology II. Int Rev Physiol 22:117–129

Cohen H, Marrian GF (1934) A method for the determination of estrogens using the Kober reaction. Biochem J 28:1603–1607

Diczfalusy E, Lauritzen C (1961) Östrogene beim Menschen. Springer, Berlin

Dingemannse E, de Jongh DS, Kober SE, Laqueur E (1930) über kristallines Menformon. Dtsch Med Wschr 56:304–308

Dingemannse E, Mühlbock O (1938) Production of estrogens by the testes. Nature (London) 141:927–932

Dodds J, Goldberg L, Lawson W, Robins R (1938) Estrogenic activity of certain synthetic compounds. Nature 141:247–248

Doisy EA, Ralls JO, Allen E, Johnston CG (1924) The extraction and some properties of an ovarian hormone. J Biol Chem 61:711–715

Doisy EA, Veler CD, Thayer S (1929) Folliculin from the urine of pregnant women. Amer J Physiol 90:329–333

Engel LL (1959) New estrogens. Vitam Horm 17:205–210

Engel LL, Slaunwhite WR Jr, Carter P, Ekman G,Olmsted PC, Nathanson LT (1952) Identification of estrogens using solvent distribution. Ciba Found Coll on Endocrinology 2:123–129

Fishman J, Cox RJ, Gallagher TF (1960) 2-Hydroxyestrone: The new metabolite of estradiol in man. Arch Biochem Biophys 90:318–322

Frost HM (1963) Dynamics of bone remodelling. In: Bone Biodynamics. Thomas, Springfield

Girard A, Sandulesco G, Fridenson H, Ruttgers JJ (1932) Conjugated estrogens in the urine of pregnant mares. C R Soc Acad Sci 114:902–906

Goy RW, McEwen BS (1980) Sexual differentiation of the brain. Cambridge, MIT Press

Gurpide EH, Fleming C, Holinka F, Fridman D (1980) Estrogen metabolism. Pharmacotherapeutics 2 (Suppl):71–75

Haeussler EP (1934) Production of estrogens by the testes. Helv Chim Acta 17:531–535

Hammerstein J (1962) Hormonanalytische Untersuchungen zur Frage der endokrinen Korrelationen im biphasischen Menstruationszyklus der Frau. Arch Gynäkol 196:504–507

Heldin CH, Westermark B (1984) Growth Factors. Cell 37:9–14

Herbst AL, Kurman RJ, Scully RE (1972) Vaginal and cervical abnormalities after exposure to stilbestrol. Obstet Gynec 30:287–292

Hitschmann H, Adler L (1908) Der Bau der Uterusschleimhaut des geschlechtsreifen Weibes mit besonderer Berücksichtigung der Menstruation. Mschr Geburtsh Gynäkol 27:1–8

Hohlweg W (1943) Über das Sexualzentrum. Zbl Gynäk 67:1357–1387

Inhoffen HH, Hohlweg H (1938a) Partialsynthese von Östradiol und Östron aus Cholesterin. Angew Chem 53:471–476

Inhoffen HH, Hohlweg H (1938b) Neue per os wirksame weibliche Keimdrüsenhormonderivate. Naturwiss 26:96–100

Ittrich G (1960) Untersuchungen über die Extraktion des roten Kober-Farbstoffs durch organische Lösungsmittel zur Östrogenbestimmung im Harn. Acta Endocr (Kbh) 35:34–38

Jensen EV, Jacobson HJ (1962) Basic guides to the mechanism of estrogenic action. Rec Progr Hormone Res 18:387–404

Junkmann W (1953) Über protahiert wirksame Östrogene. Arch Exp Pathol 220:195–206

Kaufmann C (1933) Echte Menstruation bei einer kastrierten Frau durch Zufuhr von Ovarialhormonen. Zbl Gynäkol 57:42–46

Kober S (1931) Eine kolorimetrische Bestimung des Brunsthormons (Menformon). Biochem Z 239:209–212

Korenman SG (1969) Comparative binding affinity of estrogens and its relation to estrogenic potency. Steroids 13:163–177

Korenman SG, Perrin LF, McCallum TP (1969) A radiologic binding assay system for estradiol measurement in human plasma. J Clin Endocr 29:879–883

Korenman SG, Tulchinsky D, Eaton LW Jr (1970) Radio-ligand procedure for estrogen assay in normal and pregnancy plasma. Acta Endocr (Suppl) 147:291–304

Kraychy S, Gallagher TG (1957) Katecholestrogens. J Am Chem Soc 79:754–759

Laqueur E, De Jongh SE, Tausk M, Garenstrom JHH, Manus MBC (1948) Hormonologie, Physiologie en Pharmakologie van de Hormonen. Elsevier, Amsterdam

Lauritzen C (1961) Zur Geschichte der Östrogenforschung. Festschrift der Kali-Chemie

Lauritzen C (1986) Geschichte der gynäkologischen Endokrinologie des deutschen Sprachraums von 1935 bis zur Gegenwart. In: Beck L (Hrsg) Zur Geschichte der Gynäkologie und Geburtshilfe. Springer, Berlin, pp 221–265

Lauritzen C (1993) Östrogensubstitution vor und nach behandeltem Genital- und Mammakarzinom. In: Lauritzen C (Hrsg) Menopause. Hormonsubstitution heute. Aesopus, Basel, S 76–83

Levitz M, Condon CP, Dancis J (1958) New estrogens in pregnancy urine. Endocrinology 58:376–381

Lindsay R, Hart DM, Clark DM (1984) The minimum effective dose of estrogen for prevention of postmenopausal bone loss. Obstet Gynecol 63:750–754

Lindsay R, Hart DM, McLean A, Garwood A (1997) Bone loss during estriol therapy in postmenopausal women. Maturitas 1:279–282

Lipschütz A (1950) Steroid Hormones and Tumors. Wilkins, New York, pp 98–121

Loewe S (1925) Menformon im Harn von Frauen im Zyklus. Klin Wschr 4:1407–1410

Longcope C (1973) The significance of steroid production by peripheral tissues. In: Scholler R Endocrinology of the Ovary. Edition SEP, Paris

Mainzer F (1896) Zur Behandlung amenorrhoischer und klimakterischer Frauen mit Ovarialsubstanz. Dtsch Med Wschr 22:393–396

Marrian GF (1930a) The chemistry of oestrin. IV. The chemical nature of crystalline preparations. Biochem J 24:1021–1030

Marrian GF (1930b) IV. Internat Congr Biochem. Biochemistry of Steroids. Pergamon Press, London

Mayes D, Nugent CA (1965) Plasma estradiol determined with a competitive protein binding method. Steroids 15:389–403

Mc Corquodale DW, Thayer SA, Doisy EA (1935) The crystalline ovarian follicular hormone estrin. Soc Exp Biol 32:1182–1185

Menini E, Diczfalusy E (1961) Chromatographic distribution of estrogens. Endocrinology 58:485–490

Miescher K (1949) Östrogen-Depotpräparate. Experientia (Basel) 5:1–7

Mikhail G, Zander J, Allen WE (1963) Steroids in ovarian veine blood. J Clin Endocr 23:1267–1271

Mond H (1896) Weitere Mitteilung über die Einverleibung von Eierstocksubstanz bei natürlicher und anticipierter Climax. Münch Med Wschr 43:837–840

Mühlbock O, Knaus H, Tscherne E (1948) Die weiblichen Sexualhormone in der Pharmakotherapie. Huber, Bern

Neumann F (1985) History of the Endocrinological Research Schering AG. In: Kracht J, von zur Mühlen A, Scriba C Endocrinological Guide. Deutsche Gesellschaft für Endokrinologie. Brühlsche Universitätsdruckerei, Gießen

Okhura T, Isse K, Akazawa K, Hanamoto M, Yaoi Y, Hagino N (1994) Evaluation of estrogen treatment in female patients with dementia of the Alzheimer type. Endocr J 41:361–365

Oliver MF, Boyd GS (1956) The influence of sex hormones on circulating lipids in coronary sclerosis. Circulation 13:82–86

Paganini-Hill A (1994) Morbidity and Mortality Changes with Estrogen Replacemnt Therapy. In: Lobo E (ed) Treatment of the Postmenopausal Women. Raven Press, New York, pp 399–404

Papanicolaou GN (1948a) Detection of estrogen effect in vaginal smear of women. Am J Obstet Gynecol 31:316–321

Papanicolaou GN (1948b) The sexual cycle in the human female as revealed by vaginal smears. Amer J Anat 52:519–523

Paulsen AC (1965) (ed) Estrogen Assay in Clincal Medicine. University of Washington. Press Seattle/Wa

Philipp E (1942) Wie entstehen die Hormone der Plazenta? Geburtsh Frauenheilk 4:433–437

Plotz EJ (1957) Die Anwendung radioaktiver Isotope in der Erforschung des Östrogenstoffwechsels. Geburtsh Frauenheilk 17:595–599

Ratschow M (1944) Die Sexualhormone als Heilmittel innerer Krankheiten. Enke, Stuttgart

Reynolds SRM (1950) Determinants of uterine growth and activity. Rec Progr Horm Res 5:65–72

Richter E (1948) Zum 50. Geburtstag der gynäkologischen Hormonlehre. Wien Klin Wschr 61:667–671

Roberts S, Szego CM (1946) Formation of estroproteins in liver tissue. Endocrinology 39:138–142

Rossmanith WG, Ulrich U (1992) Neuroendokrine Aspekte der Hormonregulation in der Peri- und Postmenopause. In: Runnebaum B, Kiesel BL (eds) Hormonsubstitution in der Gynäkologie. Zuckschwerdt, München

Ryan K (1958) Metabolism of steroids in tissue slices and homogenates. Endocrinology 63:392–397

Sandberg AA, Slaunwhite WR Jr (1956) Binding of estrogens to plasma proteins. J Clin Endocr 16:923–927

Sarrel PM (1990) Ovarian hormones and the circulation. Maturitas 12:122:287–292

Schröder R (1914) über die zeitlichen Beziehungen der Ovulation zur Menstruation. Arch Gynäkol 101:1–35

Schwenk E, Hildebrandt F (1933) Die chemische Umwandlung von Östradiol zu Östron. Naturwiss 21:177–181

Siebke H (1929) Ergebnisse der Mengenbestimmung des Sexualhormons. Zbl Gynäk 53:2450–2455

Siiteri PK, McDonald PC (1973) Role of extraglandular estrogen in human endocrinology. In: Greep RO, Astwood E (eds) Handbook of Physiology. Endocrinology Vol 2(1), Washington

Simmer HH (1970) Robert Tuttle Morris (1875–1945), a pioneer of ovarian transplants. Obstet Gynec 35:314–318

Simmer HH (1971a) On the history of hormonal contraception. II. Ottfried Fellner (1873–19??) and estrogens as antifertility hormones. Contraception 31:1–7

Simmer HH (1971b) Josef Halban (1870–1937). Pionier der Endokrinologie der Fortpflanzung. Wien Klin Wschr 121:549–555

Simmer HH (1980) Die Entdeckung des Östriols. Endokrinologie-Informationen. Demeter Verlag, Gräfelfing, 4:252–262

Simmer HH (1986) Gynäkologische Endokrinologie in den Verhandlungen der Deutschen Gesellschaft für Gynäkologie von 1886 bis 1935. Beiträge deutschsprachiger Frauenärzte. In: Beck L (Hrsg) Zur Geschichte der Gynäkologie und Geburtshilfe. Springer, Berlin, pp 183–219

Slaunwhite WR Jr, Engel LL, Scott JF, Ham CL (1953) Distribution of steroids between solvents. J Biol Chem 201:615–619

Steinach E, Holzknecht G (1918) Zur Frage der Östrogenbildung in den interstitiellen Zellen des Ovars. Arch Entw Mech Org 42:307–311

Stockard CR, Papanicolaou GN (1917) Estrus in the vaginal cell smear of rodents. Amer J Anat 22:225–230

Tausk M (1981) Zur Geschichte der Östrogene. Endokrinologie-Informationen. Demeter Verlag, Gräfelfing, 5:199–204

Thijssen JHH, Veeman A (1986) A gas chromatographic method of measurement of small amounts of estrogens in urine. Steroids 11:369–387

Tulchinsky D, Korenman SG (1970) A radioligand assay for plasma estrone. J Clin Endocr 31:76–80

Wallach DS, Hennemann PH (1959) Prolonged estrogen therapy in postmenopausal women. JAMA 171:1637–1642

Watzka M (1957) Das Ovarium. In: Kaufmann E, Staemmler M (Hrsg) Bd VII/3, Springer, Berlin

Westman A (1934) über die Östrogenbildung in der Granulosazellschicht des Follikels. Arch Gynäk 158:476–481

Wislocky GB, Dempsey FW, Fawcett DW (1948) Placental Cytotrophoblast. In: Physiology of Pregnancy. Williams and Williams, Baltimore

Wotiz HH, Charransol G, Smith JN (1967) Gas chromatographic analysis of free and total plasma estrogen as heptafluorobutyrate. Steroids 10:127–154

Wotiz HH, Lemon HM (1959) Biochemistry of Steroids. Pergamon Press, London

Zander J, von Münstermann AM, Diczfalusy E, Martinsen B, Tillinger KG (1959) Identification and estimation of oestradiol-17β and oestrone in human ovaries. Acta Obstet Gynaekol Scand 38:724–736

Zander J (1981) Karl Kaufmann (1900–1980). Geburtsh Frauenheilk 2:81–86

Zander J (1986a) Steroids in the human ovary. J Biol Chem 232:117–122

Zander J (1986b) Meilensteine in der Gynäkologie und Geburtshilfe. In: Beck L (Hrsg) Zur Geschichte der Gynäkologie und Geburtshilfe. S 27–62 Springer, Berlin

Zondek B (1934) Estrogen production in the testes. Scand Arch Physiol 70:1334–1338

Zondek B, Aschheim S (1926) über die Funktion des Ovariums. Z Geburtsh Gynäkol 90:372–377

Part 1
Chemistry of Estrogens and Antiestrogens

CHAPTER 2
Steroidal Estrogens

G. NEEF

A. Introduction

The term estrogens denotes a subgroup of steroids with an aromatic A ring as a characteristic part of the tetracyclic molecular framework. The most prominent members of this class are the follicular hormone estradiol (**1**) and its main metabolites estrone (**2**), estriol (**3**), and 2-hydroxyestradiol (**4**) together with their sulfated and glucuronidated counterparts. A more extended definition includes the mare estrogens equilin (**5**) and equilenin (**6**), as well as some non-steroidal compounds derived from the stilbene series (FIESER and FIESER 1959).

The rapid oxidation of estradiol (**1**) with formation of estrone (**2**) observed after oral administration and first liver passage stimulated early synthetic variation of the natural hormone. As a result, 17α-ethynylestradiol (**7**) was synthesized and found to be highly effective by the oral route (INHOFFEN et al. 1938). With the advent of an estrogen/progestogen combination as a regimen for oral contraception in the early 1950s, 17α-ethynylestradiol (**7**) became the estrogenic component of contraceptive preparations and has maintained its

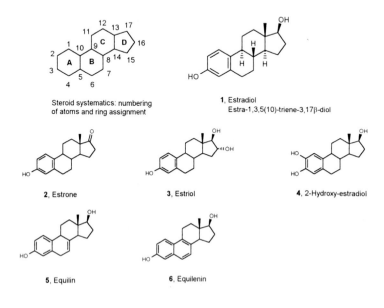

Steroid systematics: numbering
of atoms and ring assignment

1, Estradiol
Estra-1,3,5(10)-triene-3,17β-diol

2, Estrone

3, Estriol

4, 2-Hydroxy-estradiol

5, Equilin

6, Equilenin

dominant position, despite the synthesis and investigation of numerous analogues since then.

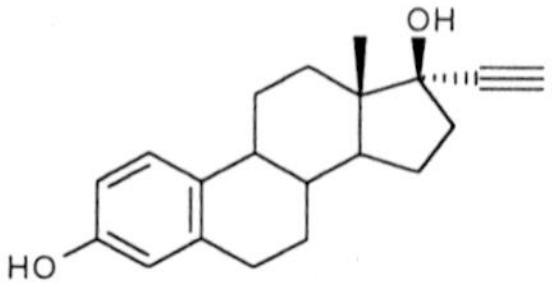

7, 17α-Ethynylestradiol

B. Total Synthesis

Apart from their biological interest estradiol and estrone represent valuable
starting materials for other pharmacologically important classes of steroids,
e.g., 19-nor-progestogens, antiprogestins and antiestrogens. The stereochemical challenges offered by the tetracyclic skeleton have been an additional stimulus for early chemical efforts directed towards replacing the natural sources
by chemical synthesis from readily available starting materials (Anner and
Miescher 1948; Cole et al. 1962). The different approaches to estrogen total
synthesis were reviewed comprehensively by Quinkert and Stark (1983).

Currently, the most economic process combines microbial degradation of
abundant cholestane derivatives (Sih and Benett 1962; Weber et al. 1975) with
pyrolytic methane extrusion from androsta-1,4-diene-3,17-dione (**9**) to form
estrone (Inhoffen and Zühlsdorff 1941). Scheme 1 exemplifies the process
for sitosterol (**8**), but cholesterol and mixtures of cholestanes can also be used
as starting materials for this process.

Scheme 1

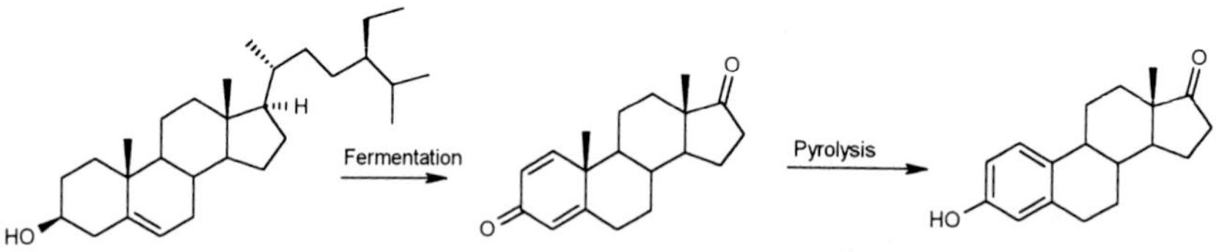

8, Sitosterol   **9**, Androsta-1,4-diene-3,17-dione   **2**, Estrone

Another strategy exploited technically is based on the Torgov synthesis
(Ananchenko and Torgov 1963) which starts from inexpensive 8-methoxy-
tetralone (**10**) and leads to formation of racemic estradiol in its original
version. The method gained additional interest when the pro-chiral dioxo-
intermediate (**12**) could be successfully subjected to enantioselective microbial reduction, thus avoiding the necessity of costly resolution at the final stage
of the synthesis (Kosmol et al. 1967; Rufer et al. 1967).

Scheme 2

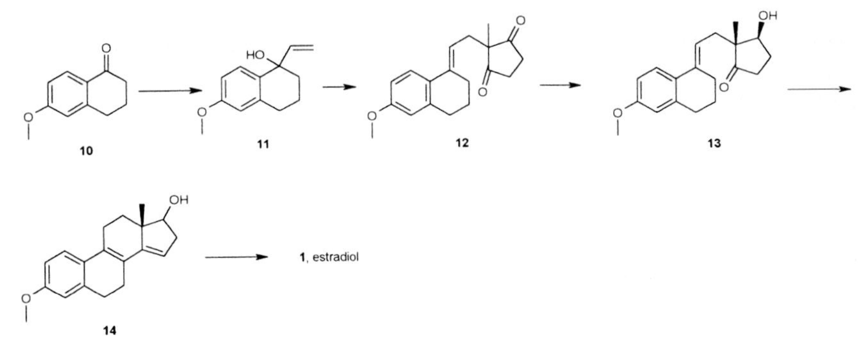

The elegant construction of an enantiomerically pure CD-ring building block by proline-catalyzed aldol condensation (**16 → 17**) does not only create the basis for a third approach to estrogen total synthesis, but represents an early outstanding example of catalytic asymmetric induction (EDER et al. 1971; HAJOS and PARRISH 1974).

Scheme 3

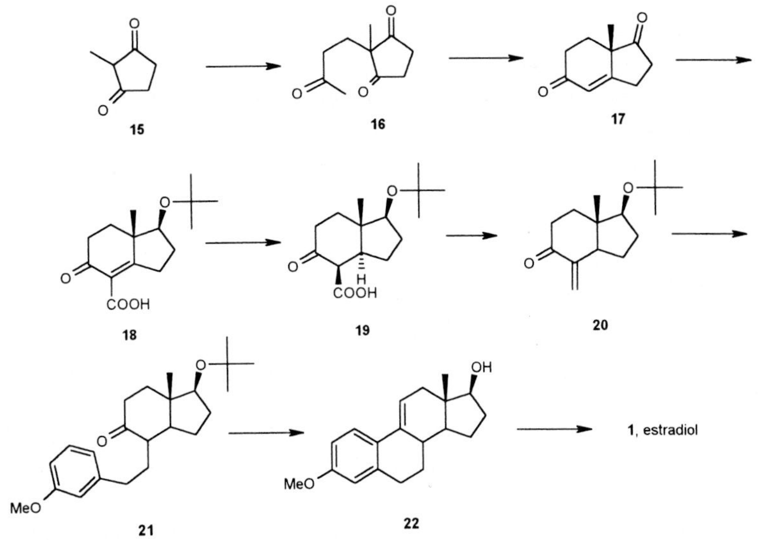

Diketone (**17**), commonly designated as Hajos-Wiechert-Eder ketone, was employed as a key intermediate in the synthesis of estradiol and various analogues (COHEN et al. 1975; EDER et al. 1976), as outlined in Scheme 3.

The shortcomings of the intermolecular Diels-Alder approaches (COLE et al. 1962) were elegantly overcome by the intramolecular versions of [4+2] cycloaddition, using appropriately substituted *o*-quinodimethane precursors (FUNK and VOLLHARDT 1980). Though hitherto unexploited for technical scale manufacture, the various approaches are intriguing examples of modern synthetic methodology applied to steroid framework synthesis (OPPOLZER 1980; KAMETANI 1978; QUINKERT et al. 1981).

Scheme 4

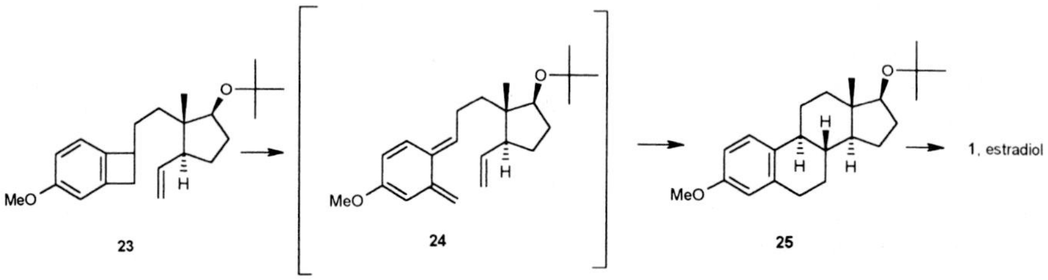

Conceptually closely related, the latter synthetic strategies essentially differ in their access to the kinetically unstable *o*-quinodimethane (**24**) as the central intermediate. Whereas the first successful approach led to formation of racemic estrone, subsequent attempts were aimed at getting the natural (+)-enantiomer (Scheme 4).

Efforts are ongoing to further elaborate the concepts of estrogen total synthesis by inter- or intramolecular cycloaddition (DOI et al. 1990; DANIEWSKI and KIEGILL 1988; TAKANO et al. 1992; BLAZEJEWSKI et al. 1992). Total synthesis was often helpful to replace unsatisfactory methodology of partial synthesis, offering convenient ways to e.g., substituted 8α-estradiols (EDER et al. 1978; BERMEJO-GONZALAEZ et al. 1982), 7α-methyl- (SAUER et al. 1982), 8α-methyl- (RUHLAND et al. 1995) and 12α,β-carboxy- (KUROSAWA et al. 1981) estradiol derivatives.

C. Partial Synthesis

Chemical activities aimed at structurally modified derivatives were started briefly after the structural elucidation of estrone and estradiol (BUTENANDT 1932); thousands of analogues have been synthesized and pharmacologically characterized in the meantime. The enormous number of conceivable variations as an inherent property of the steroid framework remains far from being exhausted. Increasingly, the search for new estrogens becomes guided by more detailed insight into the ligand–receptor interaction. Cloning and sequencing of the estrogen receptor and X-ray crystallographic data available for the ligand binding domain of the human estradiol receptor will help to focus further chemical efforts (GREEN et al. 1986; BOURGUET et al. 1995; LEWIS et al. 1995; BRZOZOWSKI et al. 1997).

Although increased estrogenic potency and oral bioavailability still rank among the objectives, structural modification is equally aimed at getting dissociated estrogens as well as using the hormone as a vehicle for photoaffinity labels, fluorescent groups, radioisotopes and cytotoxic moieties. Current knowledge on the estradiol pharmacophore was reviewed by Katzenellenbogen's group (ANSTEAD et al. 1997).

Chemical methodology has been developed to functionalize and substitute any of the skeleton-forming carbon atoms. In addition, techniques were made available to selectively invert the five stereogenic centers of the natural hormone which give rise to a theoretical number of 32 possible stereoisomers, most of which were synthesized and investigated.

From the chemist's point of view, the most challenging positions of the estradiol molecule are those not bearing a functional group or not being adjacent to a functionalized carbon atom.

I. Ring-A Substitution

The phenolic ring A of estradiol undergoes the large variety of electrophilic aromatic substitution reactions to introduce halogen (HORIUCHI et al. 1986; POIRIER and VOTTERO 1989; DIORAZIO et al. 1992), nitro- (SANTANIELLO et al. 1983), amino- (NUMAZAWA and KIMURA 1983), hydroxyalkyl- (LOVELY et al. 1996) and acyl- (MIKHAIL and DEMUTH 1989) groups into positions C-2 and C-4 of the molecule. Though C-2 attack is slightly favored, most procedures result in the formation of 2- and 4-substituted product mixtures, often along with the di-substituted parent derivative. Numerous attempts were described to achieve regioselectivity in favor of the 2-substituted product. Satisfactory yields of 2-hydroxy estrogens were obtained by deprotonation of the chromium tricarbonyl complexes (26) and oxidation with MoOPH [Oxodiperoxymolybdenum(pyridine)-(hexamethyl-phosphorictriamide)] (GILL et al. 1987).

Ortho-directed lithiation of *bis*(methoxymethyl)-protected estradiol (28) was found to occur at position C-2 exclusively. Subsequent silylation and N-bromo(iodo)-succinimide treatment gave 2-bromo(iodo) derivatives (29), albeit in moderate overall yield (PERET and RIDLEY 1987).

Much effort has been devoted to regioselective fluorination, since 2-fluoroestradiol could be expected to display a dissociated biological profile, being hormonally active but presumably resistant to metabolic attack at C-2. A reliable, though multistep procedure is based on fluorodeamination of

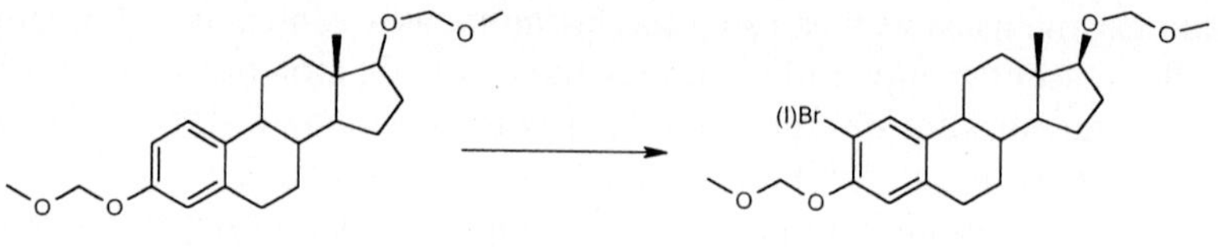

28   29

diazonium fluoroborates that require 2-amino-estradiol precursors (UTNE et al. 1968).

More recent methods take advantage of tricarbonylchromium arene complexes (30) (DIORAZIO et al. 1992) or, even more directly, make use of easily handled electrophilic fluorinating agents such as N-fluoropyridinium triflate (BULMAN PAGE et al. 1990) or N-fluorobis[(trifluoromethyl)sulfonyl]imide (PENNINGTON et al. 1992).

Scheme 5

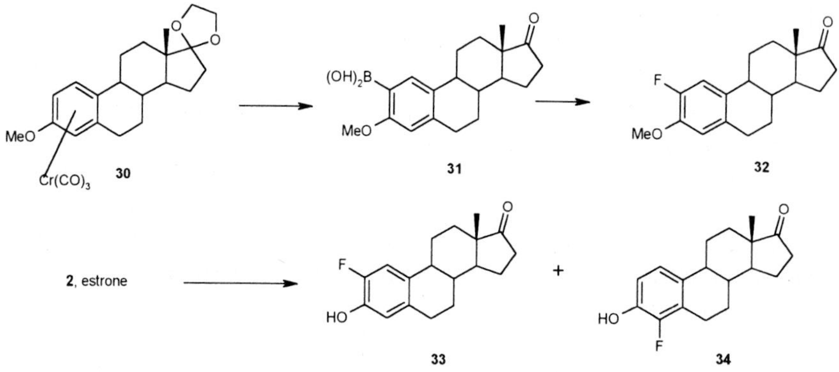

Due to electronic and steric reasons, carbon atom C-1 is far less accessible to substitution by commonly used procedures. Electrophilic substitution requires the transient introduction of an activating group at position C-4 (e.g., NH$_2$) in order to enable C-1 attack. After subsequent removal of the auxiliary C-4 substituent, C-1-halogenated estrogens are obtained (HYLARIDES et al. 1984) as shown in Scheme 6.

Nucleophilic aromatic substitution of estrogens is possible when the aromatic ring is complexed to a transition metal. KÜNZER and THIEL (1988a) succeeded in introducing a functional carbon nucleophile by reacting the chromium tricarbonyl complex (26) with the lithium anion of 1,3-dithiane to provide a C-1 formylated estradiol equivalent (40) in fairly good yield. The lithium anion of acetonitrile was found to react differently by replacing the methoxy group with formation of 3-cyanomethyl derivative (39) (KÜNZER and THIEL 1988b).

Scheme 6

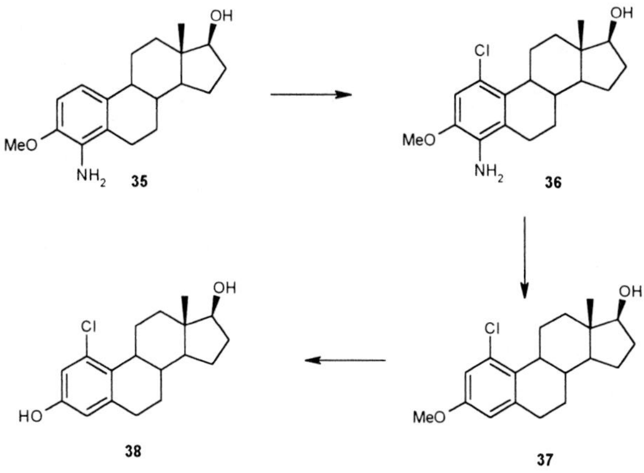

Scheme 7

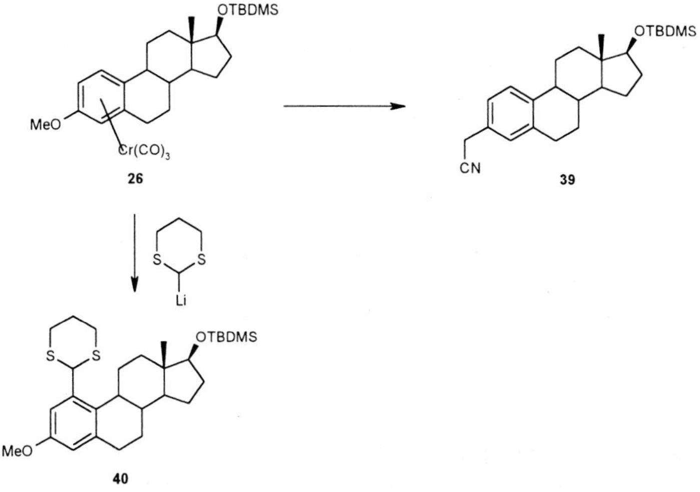

Rearrangement of 3-oxo-androsta-1,4-dienes does not result in the formation of 1-methyl substituted estrogens (DJERASSI and SCHOLZ 1948), but rather leads to inactive 1-methyl-4-hydroxy-derivatives (**41**). The undesired rearrangement can be blocked by starting with an androsta-1,4,6-triene precursor (**42**), the acid-catalyzed treatment of which yields 1-methyl-6-dehydroestrogens (**43**) which can be used to produce 1-substituted estradiol analogues (DJERASSI et al. 1950; LAING and SYKES 1968).

Lead tetra-acetate oxidation of estrone was shown to give moderate yields of a 1,3-dihydroxy derivative (GOLD and SCHWENK 1958).

Various procedures are available to replace the C-3 carbon–oxygen bond by other functionalities (SELCER and LI 1995; COUTTS and SOUTHCOTT 1990).

Scheme 8

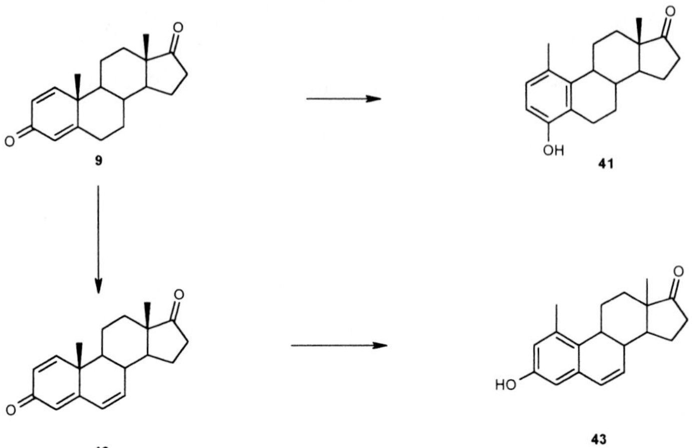

Nucleophilic exchange of the phenol triflate ester is possible in a limited number of cases (SPYRIOUNIS et al. 1990). Palladium(0)-complex catalysis has considerably broadened the repertoire of exchanging the phenolic OH group versus carbon functionality as exemplified by the transformation of estrone-3-triflate (**44**) to the 3-carboxy derivative (**45**) (CACCHI et al. 1986; CACCHI and LUPI 1992; KÜNZER and THIEL 1988b).

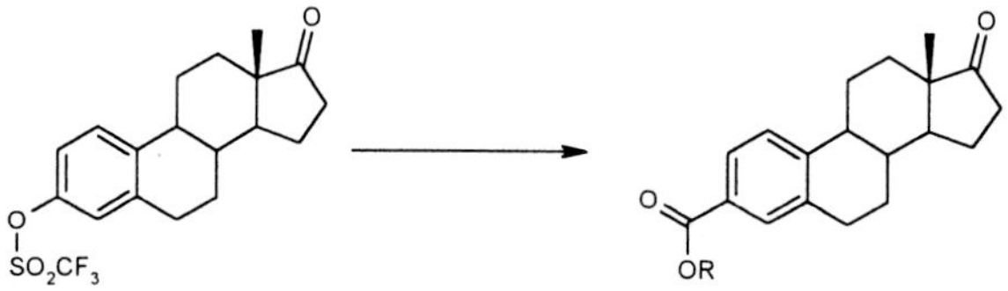

Being an essential element of the estradiol pharmacophore, the phenolic OH function has often been used to change pharmacodynamic characteristics or transport properties by creating oxygen linkages to form carboxylic esters, sulfates (LI et al. 1995), sulfamates (HOWARTH et al. 1994; SCHWARZ et al. 1996), phosphonates (HOWARTH et al. 1993), glycosides (HARREUS and KUNZ 1986) and glucuronides (OKHUBO et al. 1990). Recent findings indicate that estradiol-3-sulfamates may be useful as pro-drugs, capable of overcoming the afore-mentioned problems of oral estradiol administration (ELGER et al. 1995).

Of enormous value in the estrogen series is the reduction by alkali metals dissolved in ammonia, commonly known as Birch reduction (KAKIS 1963; DRYDEN 1971). The process does not only represent the key access to the whole class of 19-norsteroids, but has become indispensable for a large variety of subsequent chemical and microbial transformations, resulting in modified estrogens (Scheme 9).

Scheme 9

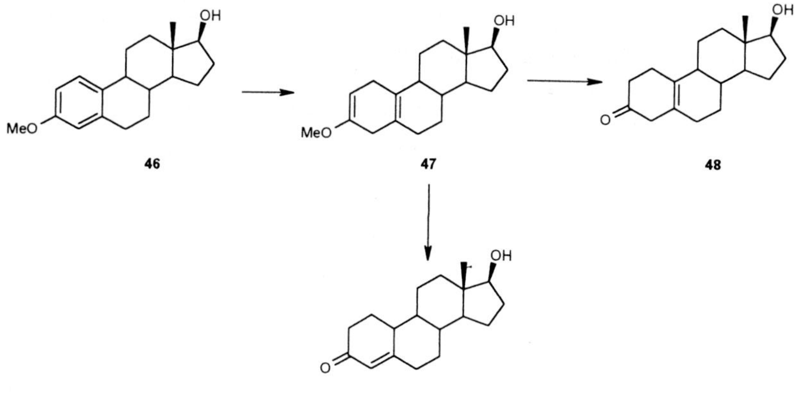

II. Ring-B Substitution

The introduction of substituents into the secondary positions C-6 and C-7 can be achieved using 6-keto-estradiol precursors (NUMAZAWA and YOSHIMURA 1996; KÜNZER et al. 1996). Chromium (VI) reagents have been shown to be capable of benzylic oxidation to form 6-keto estrogens (**51**) in one step, but yields are often moderate. Best results seem to be obtained with CrO$_3$/3,5-dimethypyrazole complex (AKANNI and MARPLES 1984; RATHORE et al. 1986). The oxidation with chromium reagents can be circumvented by starting with a 5,6-unsaturated 19-norsteroid precursor that requires A-ring aromatization as a final step. (BREVET et al. 1996).

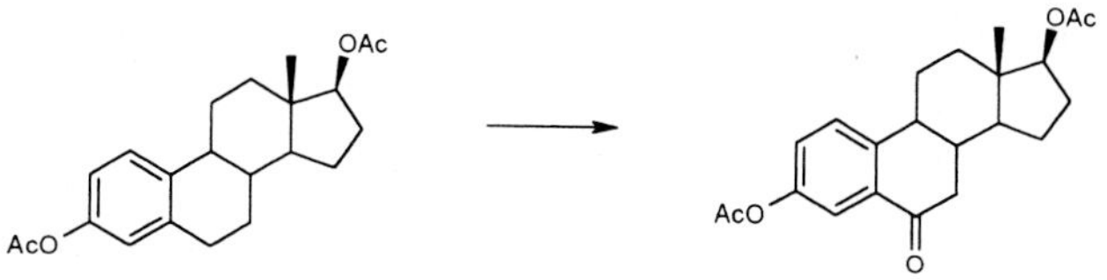

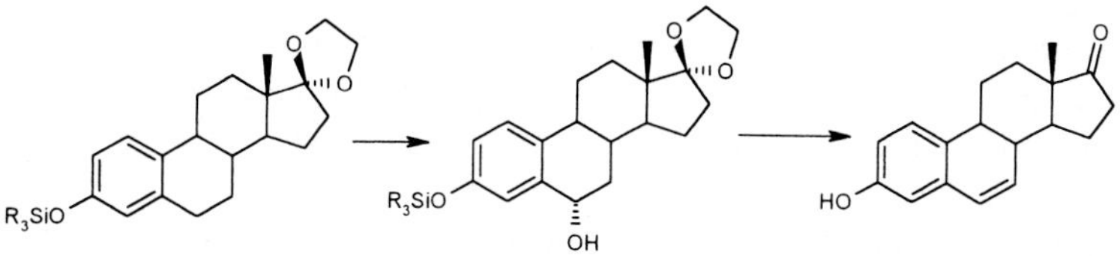

A direct stereoselective introduction of a 6α-hydroxy group was reported by RAO and WANG (1997), who deprotonated the protected estrone derivative (**52**) by lithium diisopropylamide and potassium 1,1-dimethylpropoxide, captured the resulting anion with trimethyl borate and oxidized it with 30% aqueous H_2O_2 to obtain a 76% yield of the 6α-hydroxy derivative (**53**) which was subsequently transformed into 6-dehydroestrone (**54**).

The previously mentioned tricarbonyl chromium complexes of protected estradiol derivatives can also be used to functionalize the benzylic site C-6 by deprotonation and subsequent reaction with an electrophile. The 6-hydroxymethyl compounds (**56** and **58**) were shown to be produced in a highly regio- and stereoselective manner, when the diastereomeric mixture of metal complexes (**55, 57**) was chromatographically separated and the individual isomers subjected to hydroxymethylation (JAOUEN et al. 1984).

Scheme 10

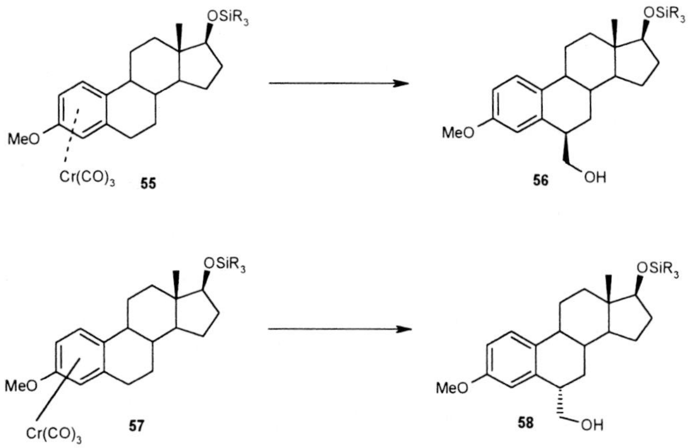

Since substitution at C-7 has turned out to result in highly active derivatives, versatile and stereoselective methods of forming 7α-alkylated and arylated estradiols are of particular interest. The most convenient route makes use of conjugate 1,6-addition to 3-oxo-estra-4,6-diene precursors (BUCOURT et al. 1978) and subsequent aromatization (RAO et al. 1994). The attachment of longer alkyl chains with additional nitrogen functionality resulted in the discovery of the first pure antiestrogen (**62**) (BOWLER et al. 1989) (Scheme 11).

The conjugate 1,6-addition to 4,6-estradienes often results in mixtures of C-7 epimers; stereoselectivity in favor of the 7α-isomer may be achieved by starting with an androsta-4,6-diene derivative (**63**) which, however, entails the inconvenience of additional steps necessary to remove the C-10-methyl group by reductive demethylation and aromatization (FRENCH et al. 1993) (Scheme 12).

Another highly stereoselective access to position C-7α applies the Prins reaction to equilin derivatives. Ene-reaction with formaldehyde leads to predominant formation of 7α-hydroxymethyl-product (**68**) which, by subsequent Birch reduction, tosylate formation and Zn/NaI reduction, is transformed into naturally configured 7α-methyl-estradiol (**70**) (KÜNZER et al. 1994) (Scheme 13).

Scheme 11

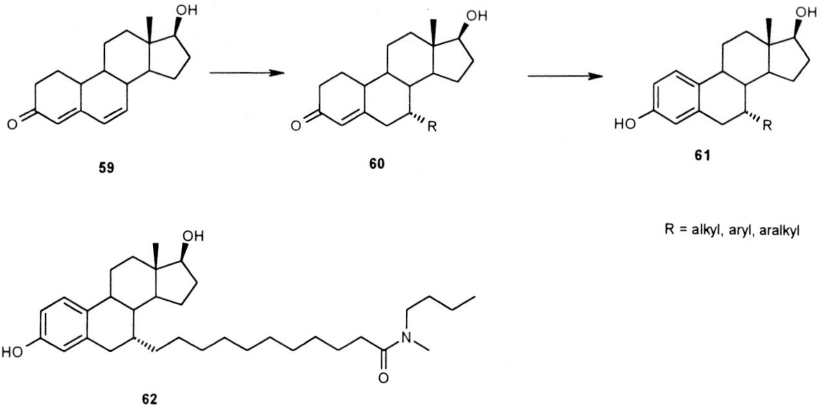

59　　　　　　　　　　**60**　　　　　　　　　　**61**

R = alkyl, aryl, aralkyl

62

Scheme 12

63　　　　　　　　　　**64**

65　　　　　　　　　　**66**

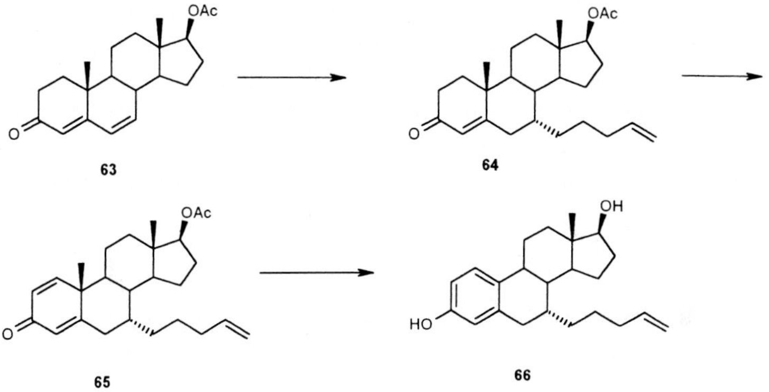

Scheme 13

67　　　　　　　　　　**68**　　　　　　　　　　**69**

70

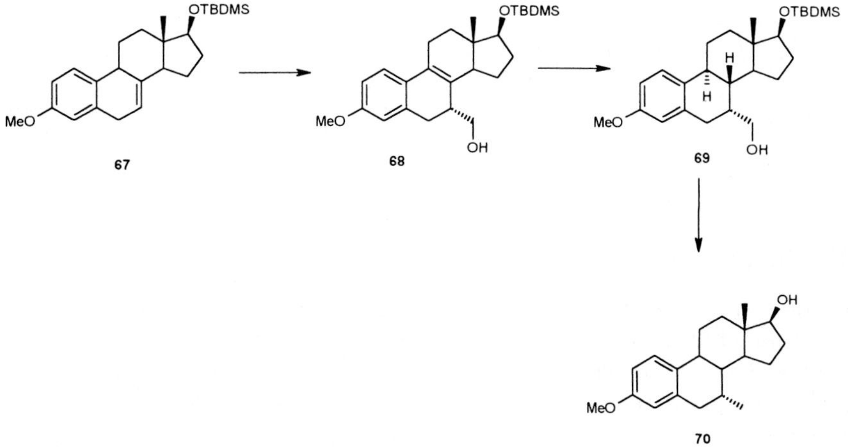

Less obvious are strategies to change stereochemistry or substitute the hydrogen atoms at tertiary positions C-8 and C-9 (Sykes and Rutherford 1971). Ene-type reactions of 8,9-unsaturated estradiol derivatives can be used to introduce substituents at position C-8α (Künzer et al. 1991).

Inversion of C-8 configuration is easily achieved by catalytic hydrogenation of 8(9)-unsaturated precursors that are obtained from the Torgov total synthesis. Acid-catalyzed rearrangement of a 8α,14α-methylene-bridged estradiol derivative is another strategy that gives way to 8α-methyl-estradiol (Künzer et al. 1991).

2,3-Dichloro-5,6-dicyano-1,4-benzoquinone (DDQ)-oxidation of estrone and estradiol derivatives leads to formation of a tertiary benzylic cation, which can either undergo proton loss to form a 9,11-double bond or can be trapped by trimethylsilyl cyanide to form 9α-nitrile (**74**) (Guy et al. 1992). A different mechanism must be assumed for the reaction with DDQ/hydrazoic acid that yields, preferentially, the 9β-azido derivative (**75**) (Guy et al. 1991).

III. Ring-C Substitution

Another site of particular biological interest is carbon atom C-11. Ceric ammonium nitrate oxidation of estrone 3-acetate or $\Delta^{9(11)}$-estrone 3-acetate is perhaps the most direct method of introducing oxygen functionality simultaneously into positions C-9α and C-11β. As the benzylic C-9 hydroxy group can be removed selectively by reduction, the process can be used to prepare 11β-

Scheme 14

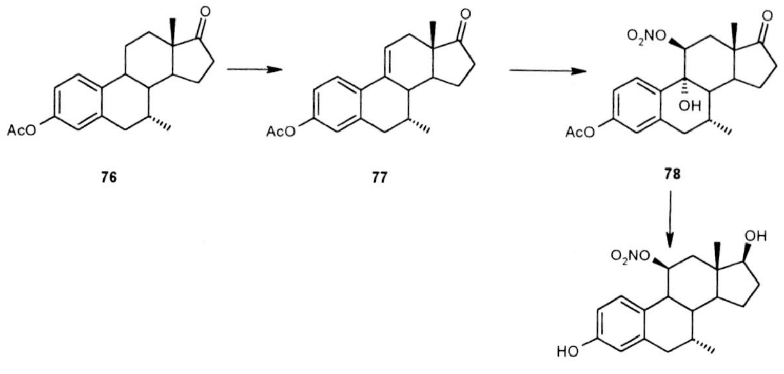

76   **77**   **78**

79

hydroxy-, alkoxy- and nitrato-substituted estrogens. The C-11-nitrate ester of
11β-hydroxy-7α-methylestradiol (**79**) was found to exceed the estrogenic
potency of 17α-ethynylestradiol (PETERS et al. 1989).

A convenient microbiological strategy is available for the synthesis of
11α-hydroxy-estrogens starting from 19-nor-estr-4-ene-3,17-dione (**80**). 11α-
Hydroxylation (CARNEY and HERZOG 1967) combined with subsequent micro-
bial aromatization results in a high overall yield of 11α-hydroxy-estrone (**82**)
(PETZOLDT et al. 1984).

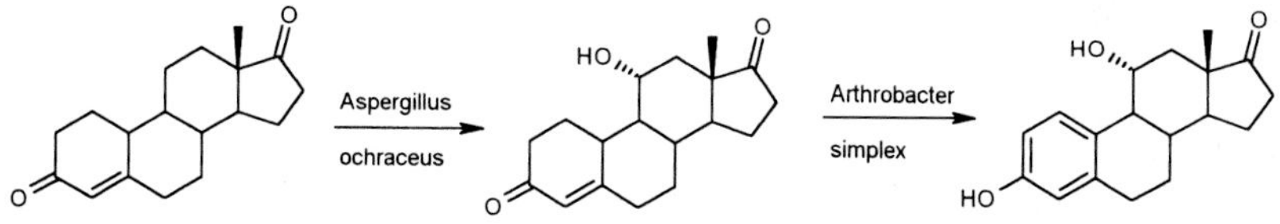

80   **81**   **82**

Estrogen and progesterone receptors tolerate bulky groups attached to
the axially oriented 11β-position. There is a certain similarity of effects
achieved by either 7α- or 11β-substituents, which can be viewed as roughly
interconvertible by rotation around a diagonal axis through positions C-3 and
C-16. A convenient strategy for the stereospecific access to C-11β consists of
a copper-catalyzed conjugate epoxide-opening and subsequent steps outlined
in Scheme 15 (BELANGER et al. 1981); as a result, another series of pure antiestrogens was found (NIQUE et al. 1989).

Little is known about the effects of substitution at position C-12. Besides
the previously mentioned access by total synthesis (see Sect. A), 9(11)-
unsaturated estrogens can be used as starting materials for 12-substituted
derivatives (such as **89**) (GUY et al. 1992).

Scheme 15

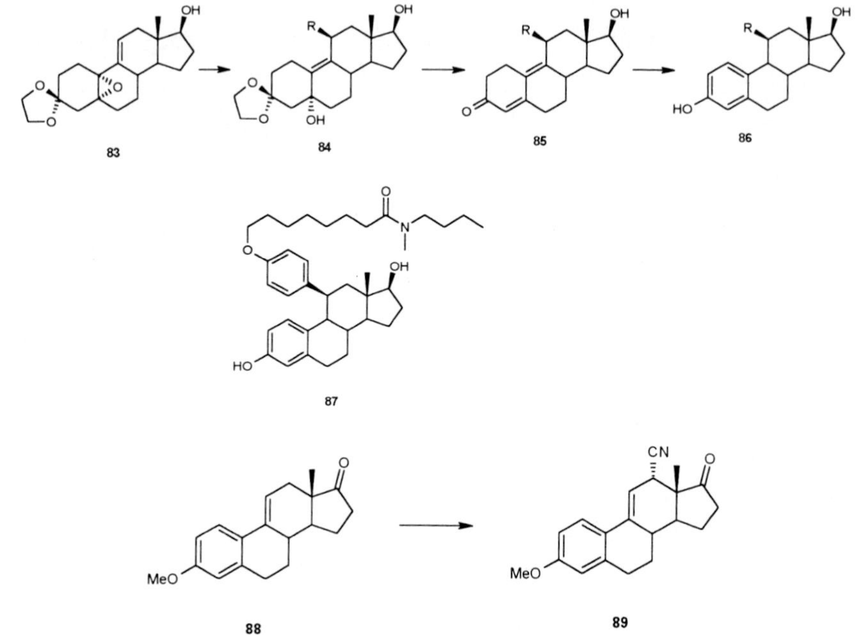

83   84   85   86

87

88   89

IV. Ring-D Substitution

Chemical modification of ring D is facilitated by the presence of the C-17 oxygen function, which as a C-17 carbonyl group can serve to introduce substituents at C-16, C-15 and C-14. The stereochemical course is often predictable due to the directing influence of the angular C-13 methyl group. Direct nucleophilic attack of a 17-oxo function almost exclusively occurs from the α-face of the molecule, resulting in 17α-substituted17β-hydroxy estrogens. Extensive use has been made of this most straightforward way to produce estradiol derivatives (ANSTEAD et al. 1996).

Various protocols exist to react 17-ketoenolates with stereoselective formation of 16-substituted estrogens (SCHÖNECKER et al. 1975; GOTO et al. 1977; FEVIG and KATZENELLENBOGEN 1978). Again, introduction of fluorine is of particular interest and led to the development of methodology for 16α,β-fluorination (PATRICK and MORTEZANIA 1988; SHIMIZU and ISHII 1989; UMEMOTO et al. 1990).

More complicated is the stereocontrolled substitution at carbon atom C-15. Michael-type addition to 17-keto-15-ene precursors usually results in epimeric mixtures with a more or less pronounced preference for β-substituted compounds of type **91** (CANTRALL et al. 1963; GROEN and ZEELEN 1979; POIRIER et al. 1991). Controlled formation of 15α-allyl substituted derivatives (**93**) can

Scheme 16

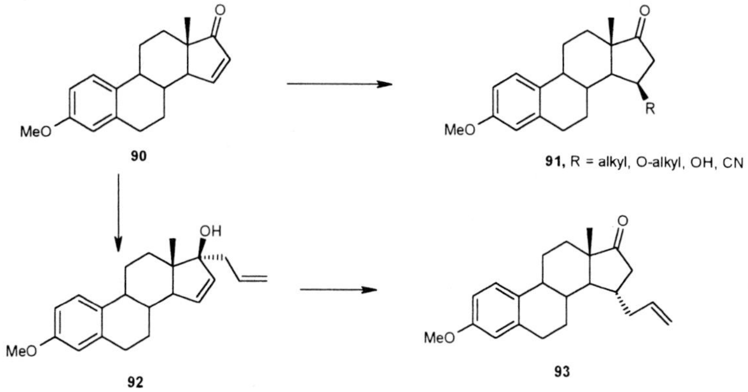

90

91, R = alkyl, O-alkyl, OH, CN

92

93

be achieved by Oxy-Cope rearrangement (BOJACK and KÜNZER 1994). 15,15-Dialkyl-substituted estrogens were prepared by BULL and LOEDOLFF (1996).

The Diels-Alder reaction of 14,16-dienes was extensively used to produce a large variety of estradiol derivatives, some of which were shown to exhibit high estrogenic potency (BULL and THOMSON 1986; BULL et al. 1994a, b).

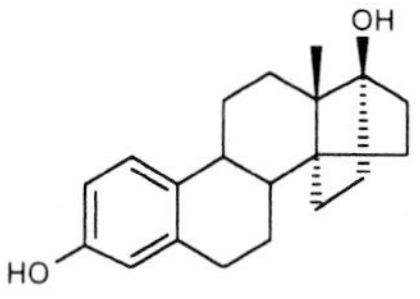

94

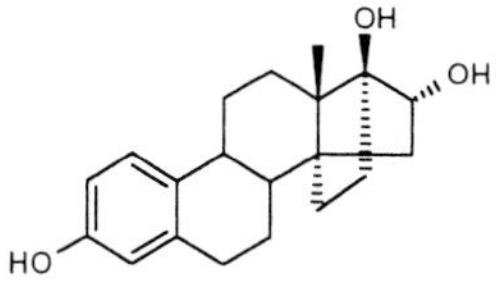

95

Dienolacetate (**96**) has served as the preferred diene component in the [4+2] cycloaddition reactions which proceed with high regio- and stereoselectivity resulting in the molecular geometry depicted below (SOLO et al. 1968; BULL and HOADLEY 1994).

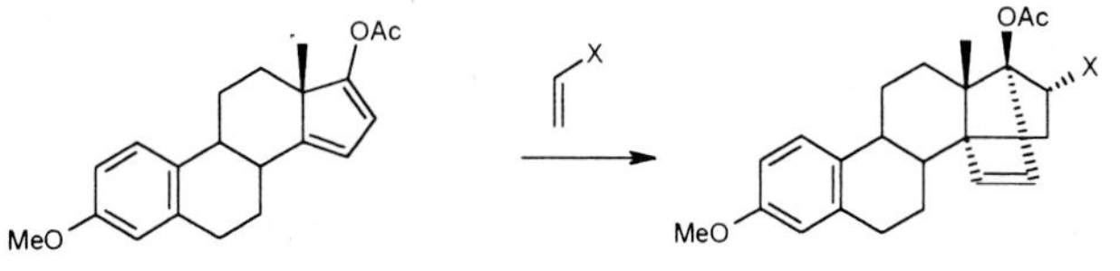

96

97, X = SO_2-phenyl, CHO, COOMe

The C-17-oxygen function can help to control substitution and stereochemical course in many ways. UV-irradiation of a 17-ketone was shown to invert the configuration at C-13 partially to form a 13α-estrone derivative (BUTENANDT et al. 1941). Considerably improved thermal variations of this transformation were developed based on the tendency of a C-17-imminium

Scheme 17

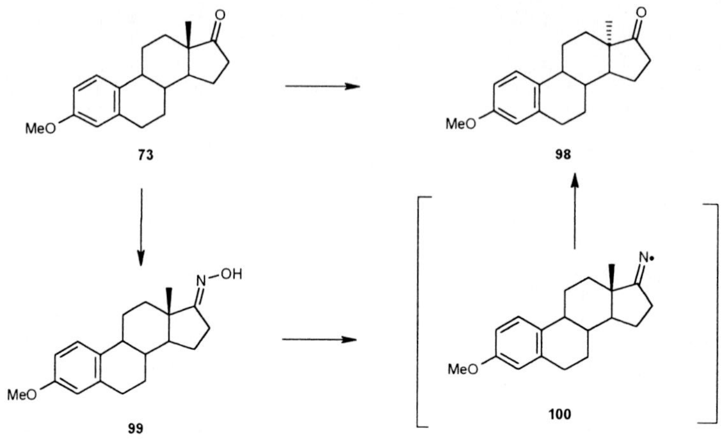

radical (100) to initiate homolytic cleavage of the C-17–C-13-carbon bond. Reverse attack of the resulting C-13 carbon radical resulted in predominant formation of 13α-configured estrone derivative (98) (BOAR et al. 1975; BOIVIN et al. 1992).

Much effort was devoted to stereochemical inversion at C-17 aiming at a convenient access to 17α-hydroxy estrogens. The problem is complicated by the sterically hindered neopentyl situation which directs hydride attack upon a 17-carbonyl group to occur from the α-face of the molecule. None of the numerous methods of transforming a chiral secondary carbinol into its stereoisomer by S_N2 displacement could be made to work satisfactorily in the

Scheme 18

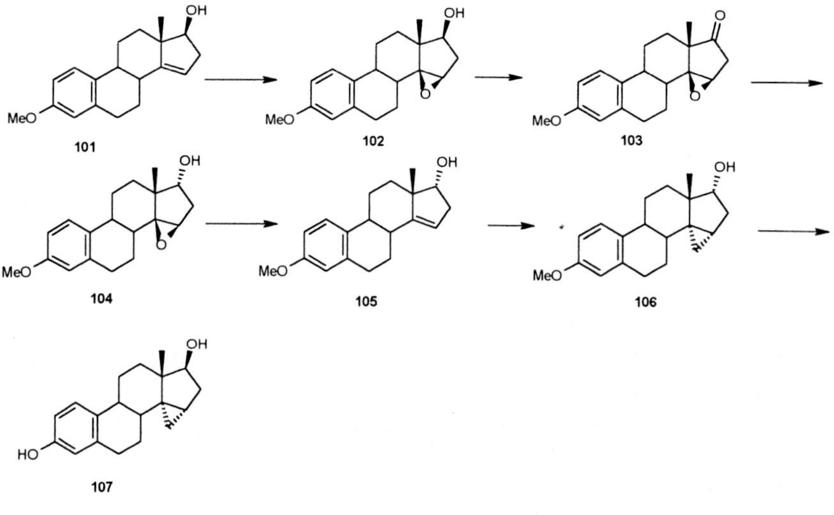

case of 17β-estradiol derivatives. The problem could be solved with the assistance of additional functionality, properly placed in ring D. The synthesis of 14,15α-methylene-estradiol, an equipotent derivative of the natural hormone, is a characteristic example how the directing effects of a 17β- or 17α-hydroxy group can be used (PROUSA et al. 1986; SIEMANN et al. 1995; KÜNZER and THIEL 1995) (Scheme 18).

V. The Periphery

Many attempts have been made to create structural hybrids between estradiol and other pharmacologically active principles in order to achieve site specific effects. A very early example is estramustin (**108**) which combines an alkylating mustard moiety with estradiol and has become widely used for the treatment of hormone-dependent cancers (FEX et al. 1967).

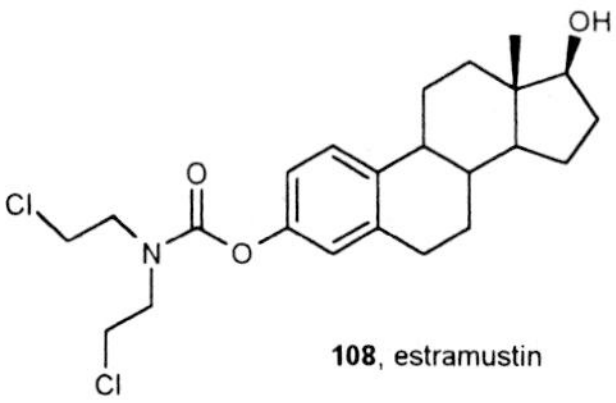

108, estramustin

Another intriguing example of a hybrid molecule (**109**) is the inclusion of estradiol in the ene-diyne framework, which is the core element of some potent antitumor antibiotics (MEERT et al. 1997). A third approach is aimed at site-selective photodynamic tumor therapy, by combining the estrogen with a porphyrin derivative (**110**) (MONTFORTS et al. 1992). The 6α-amino group was used to attach a fluorescence label to the estradiol skeleton (**111**) (ADAMCZYK et al. 1997).

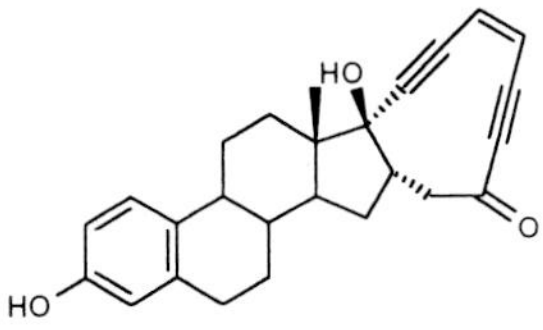

109

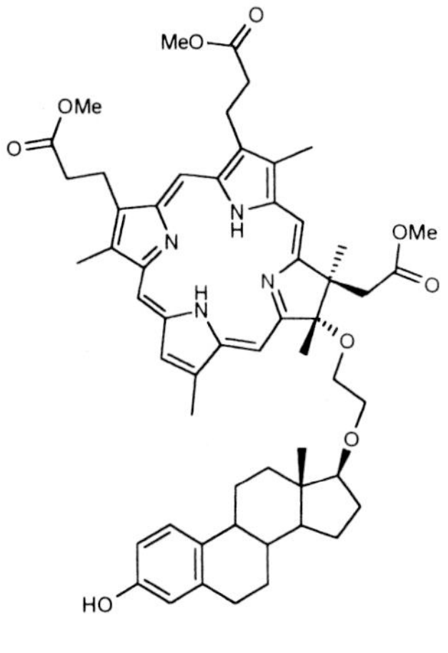

110

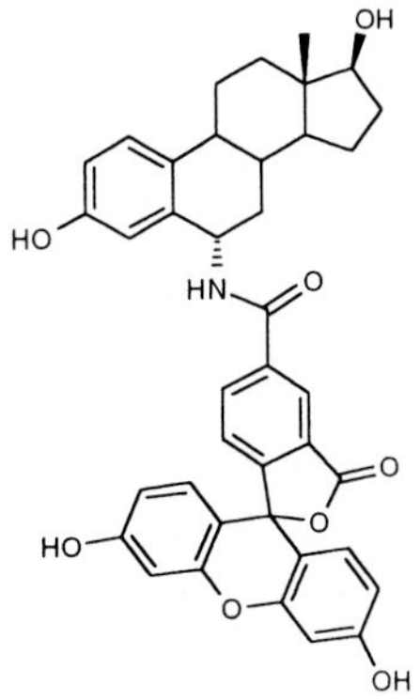

111

D. Labeling of Estradiol and its Derivatives

The availability of labeled material is indispensable for mechanistic, diagnostic and therapeutic purposes. A lot of effort was invested in getting stably labeled estradiol derivatives. The introduction of radioisotopes into the estrogen framework was reviewed by CUMMINS (1993).

The known tissue-selective distribution of the estrogen receptor makes labeled estradiol derivatives attractive candidates for in vivo imaging and diagnostics. An even more intriguing option lies in the site-selective treatment of estrogen receptor-positive mammary cancers. In spite of the promising perspectives for in vivo applications, the use of radiolabeled estrogens has remained restricted to in vitro diagnostics, mainly due to rapid disappearance and insufficiently selective tissue uptake of the compounds so far made available for clinical studies.

16α-^{125}Iodoestradiol (**112**) (HOCHBERG 1979), easily obtained by nucleophilic halogen exchange from 16β-bromoestradiol, was one of the first candidates to be considered for systemic use, seeming particularly suited due to its good binding affinity for the estrogen receptor. In principle, various animal models confirmed a selective retention in estrogen receptor-containing organs, but extremely rapid clearance of the radioisotope from both target and non-target tissues made the compound impractical. Subsequent attempts were made to enhance metabolic stability, receptor binding and lipophilicity by introducing additional substituents at positions C-2, C-4, C-7, C-11 and C-17 (ALI et al. 1993a; ALI et al. 1993b). No noticeable improvement of profile could be achieved.

Promising results were obtained with 16α-^{18}F-labeled estradiol (**113**) in clinical studies of positron-emission tomography imaging (McGUIRE et al. 1991; POMPER et al. 1990). Another candidate for positron emission tomography (PET) applications is ^{18}F-labeled 7α-pentylestradiol (**114**) (FRENCH et al. 1993).

112   113   114

The facile access to 17α-^{125}iodovinyl-estrogens (**115**), together with promising binding characteristics, stimulated renewed efforts and expectations to overcome the disadvantages of 16α-haloestrogens. Since few comparative studies are available, the anticipated improvement in biodistribution and label stability awaits confirmation (HANSON et al. 1984).

115

The application of 2- and 4-radiohalogenated estrogens for in vivo diag-nostics and therapeutics has also been under investigation (EAKINS et al. 1979). With the exception of fluorine substitution, a marked decrease in receptor binding affinity is observed after A-ring halogenation. Enhanced label stabil-ity and a higher rate of tissue uptake may outweigh the lower binding prop-erties. Tritiated (MAGGS et al. 1990), deuterated (RANJITH et al. 1989; BOUCHEAU et al. 1990) and 4-[^{14}C]-labeled (JACOBSOHN et al. 1991) estrogens were prepared for metabolic studies, as well as mechanistic and spectroscopic investigations.

E. Bioconversions of Estrogens

In contrast to the androstane and pregnane series of steroids, A-ring aromat-ics less readily undergo microbiological transformations that provide deriva-tives hydroxylated at positions otherwise difficult to access (KIESLICH 1981). Sometimes the problem can be overcome by microbial oxidation of precur-sors and subsequent chemical or microbiological aromatization of ring A, as demonstrated for 11α-hydroxy estrone (PETZOLDT et al. 1984). Direct trans-formations of estra-1,3,5(10)-trienes and estra-1,3,5(10),9(11) trienes were reported to hydroxylate positions C-2 and C-4, C-6α,β, C-7α, C-15α, C-15β and C-16 (FERRER et al. 1990).

Besides the normally used bacteria and fungi, microalgae were shown to produce A-ring aromatic steroids from androstane precursors (DELLAGRECA et al. 1996). Red blood cell cultures of various animal species and strains were used to demonstrate mono-O-methylation of 2- and 4-catechol estrogens (SUZUKI et al. 1993).

References

Adamczyk M, Mattingly PG, Reddy RE (1997) An efficient stereoselective synthesis of 6α-aminoestradiol: Preparation of estradiol fluorescent probes. Steroids 62:462–467

Akanni AO, Marples BA (1984) Improved preparation of 3β,17β-diacetoxyoestra-1,3,5(10)-trien-6-one. Synth Commun 14:713–715

Ali H, Rousseau J, van Lier JE (1993a) Synthesis of A-ring fluorinated derivatives of (17α,20 E/Z)-[^{125}I]Iodovinylestradiols: effect on receptor binding and receptor-mediated target tissue uptake. J Med Chem 36:3061–3072

Ali H, Rousseau J, van Lier JE (1993b) 7α-Methyl- and 11β-ethoxy-substitution of [^{125}I]-16α-Iodoestradiol: effect on estrogen receptor-mediated target tissue uptake. J Med Chem 36:264–271

Ananchenko SN, Torgov IV (1963) New Syntheses of estrone, d,l-8-iso-oestrone and d,l-19-nortestosterone. Tetrahedron Lett:1553–1558

Anner G, Miescher K (1948) Die Synthese des natürlichen Östrons. Totalsynthesen in der Oestronreihe III. Helv Chim Acta 31:2173–2183

Anstead GM, Carlson KE, Katzenellenbogen JA (1997) The estradiol pharmacophore: ligand structure-estrogen receptor binding affinity relationships and a model for the receptor binding site, Steroids 62:268–303

Bélanger A, Philibert D, Teutsch G (1981) Regio and stereospecific synthesis of 11β-substituted 19-norsteroids. Steroids 37:361–382

Bermejo-Gonzalez F, Neef G, Eder U, Wiechert R, Schillinger E, Nishino Y (1982) Synthesis and pharmacological evaluation of 8α-estradiol derivatives. Steroids 40: 171–187

Blazejewski J-C, Haddad M, Wakselman C (1992) An intramolecular Diels-Alder approach to angularly trifluoromethylated C/D cis estranes. Tetrahedron Lett 33: 1269–1272

Boar RB, Jetuah FK, McGhie JF, Robinson M, Barton DHR (1975) An improved synthesis of 13-epi-steroids. J Chem Soc Chem Commun:748

Boivin J, Schiano A-M, Zard SZ (1992) A novel and practical access to 13-epi-17-ketosteroids. Tetrahedron Lett 33:7849–7852

Bojack G, Künzer H (1994) An Oxy-Cope rearrangement approach to C(15)α-alkylated derivatives of estradiol. Tetrahedron Lett 35:9025–9026

Bojack G, Künzer H, Rölfing K, Thiel M (1996) A novel stereocontrolled approach to ring B alkylated estratetraenes. Tetrahedron Lett 34:6103–6104

Boucheau V, Renaud M, Rolland de Ravel M, Mappus E, Cuilleron CY (1990) Proton and carbon-13 nuclear magnetic resonance spectroscopy of diastereomeric 3- and 17ß-tetrahydropyranyl ether derivatives of estrone and estradiol. Steroids 55:209–221

Bourguet W, Ruff M, Chambon P, Gronemeyer H, Moras D (1995) Crystal structure of the ligand-binding domain of the human nuclear receptor RXR-α. Nature 375:377–382

Bowler L, Lilley TJ, Pittman JD, Wakeling AE (1989) Novel steroidal pure antiestrogens. Steroids 54:71–99

Brevet J-L, Fournet G, Gore J (1996) Improved syntheses of 3,17ß-diacetoxy-1,3,5(10)-trien-6-one. Synth Commun 26:4185–4193

Brzozowski AM, Pike ACW, Dauter Z, Hubbard RE, Bonn T, Engström O, Öhmann L, Greene GL, Gustafsson J-A, Carlquist M (1997) Molecular basis of agonism and antagonism in the estrogen receptor. Nature 389:753–758

Bucourt R, Vignow M, Torelli V, Richard-Foy H, Geynet C, Secco-Millet C, Redeuilh G,Baulieu EE (1978) New biospecific adsorbents for the purification of estradiol receptor. J Biol Chem 253:8221–8226

Bull JR, Thomson RI (1986) Cycloaddition route to 14α-formylestrone and derived 14α-substituted products. J Chem Soc Chem Commun:451–453

Bull JR, Grundler C, Laurent H, Bohlmann R, Müller-Fahrnow A (1994a) Cycloaddition mediated synthesis and rearrangement of 16-functional 14α,17α-etheno-19-norsteroids. Tetrahedron 50:6347–6362

Bull JR, Mountford PG, Kirsch G, Neef G, Müller-Fahrnow A, Wiechert R (1994b) Cycloaddition-fragmentation route to 14β-allylestrone and the derived 14α,17α-ethano analogue of estriol. Tetrahedron 50:6363–6376

Bull JR, Hoadley C (1994) Cycloaddition-oxidative cleavage pathways to 14β-formyl-19-norsteroids. Tetrahedron Lett 35:6171–6174

Bull JR, Loedolff MC (1996) Synthesis of 15,15-dialkylestradiols. J Chem Soc Perkin I:1269–1276

Bulman Page PC, Hussain F, Maggs JL, Morgan P, Park BK (1990) Efficient regioselective A-ring functionalization of estrogens. Tetrahedron 46: 2059–2068

Butenandt A (1932) Über die Chemie der Sexualhormone. Angew Chem 45: 655–656

Butenandt A, Wolff A, Karlson P (1941) über Lumi-oestron. Chem Ber 74:1308–1312

Cacchi S, Ciattini PG, Morera E, Ortar G (1986) Palladium-catalyzed carbonylation of aryl triflates. Synthesis of arenecarboxylic acid derivatives from phenols. Tetrahedron Lett 27:3931–3934

Cacchi S, Lupi A (1992) Palladium-catalysed hydroxycarbonylation of vinyl and aryl triflates: synthesis of α,β-unsaturated and aromatic carboxylic acids. Tetrahedron Lett. 33:3939–3942

Cantrall EW, Littell R, Bernstein S (1963) The synthesis of C-15β-substituted estra-1,3,5(10)-trienes. J Am Chem Soc 29:64–68

Charney W, Herzog HL (1967) Microbial transformations of steroids, Academic Press, New York

Cohen N, Banner BL, Eichel WF, Parrish DR, Saucy G (1975) Novel total syntheses of (+)-estrone 3-methyl ether, (+)-13β-ethyl-3-methoxygona-1,3,5(10)-trien-17-one, and (+)-equilenin 3-methyl ether. J Org Chem 40:681–685

Cole E, Johnson WS, Robin PA, Walker J (1962) A stereoselective total synthesis of oestrone, and related studies. J Chem Soc:244–278

Coutts IGC, Southcott MR (1990) The conversion of phenols to primary and secondary aromatic amines via a Smiles rearrangement. J Chem Soc Perkin Trans I:767–771

Cummins CH (1993) Radiolabeled steroidal estrogens in cancer research. Steroids 58:245–259

Daniewski AR, Kiegiel J (1987) A new route to a chiral synthon for the total synthesis of estrone. Synthesis:705–708

Daniewski AR, Kiegiel J (1988) A facile total synthesis of estrogens. J Org Chem 53:5535–5538

DellaGreca M, Previtara L (1996) Bioconversion of 17ß-hydroxy-17α-methyl-androsta-1,4-dien-3-one and androsta-1,4-diene-3,17-dione. Tetrahedron 52:13981–13990

Diorazio LJ, Widdowson DA, Clough JM (1992) Regiospecific synthesis of 2-fluoro-3-O-methylestrone using caesium fluorosulfate. J Chem Soc Perkin Trans I:421–425

Djerassi C, Scholz CR (1948) 5-Methyldoisynolic acid and 1-methylestrone. J Org Chem 13: 697–706

Djerassi C, Rosenkraz G, Romo J, Pataki J, Kaufmann S (1950) The dienone-phenol rearrangement in the steroid series. Synthesis of a new class of estrogens. J Am Chem Soc 72:4540–4544

Doi T, Shimizu K, Takahashi T, Tsuji J, Yamamoto K (1990) Enantioselective approach to the steroid skeleton by conjugate addition of alkenylcopper-phosphine complexes. Total syntheses of estrone methyl ether and its 7-alkylated derivative. Tetrahedron Lett 31:3313–3316

Dryden HL (1971) Reduction of steroids by metal-ammonia solutions in Organic Reactions. In: Fried J, Edwards A (eds) Steroid Chemistry, Van Nostrand Reinhold Company, New York, pp 1–60

Eakins MN, Palmer AJ, Waters SL (1979) Studies in the rat with ^{18}F-4-fluoro-oestradiol and ^{18}F-4-fluoro-oestrone as potential prostate scanning agents: comparison with ^{125}I-2-iodo-oestradiol and ^{125}I-2,4-di-iodo-oestradiol. Int J Appl Radiat Isot 30:695–700

Eder U, Sauer G, Wiechert R (1971) Neuartige asymmetrische Cyclisierung zu optisch aktiven Steroid-CD-Teilstücken. Angew Chem 83:492–493

Eder U, Gibian H, Haffer G, Neef G, Sauer G, Wiechert R (1976) Synthese von Östradiol. Chem Ber 109:2948–2953

Eder U, Haffer G, Neef G, Prezewowsky K, Sauer G, Wiechert R (1978) Darstellung von 1,3,17ß-Triacetoxy-8α-östra-1,3,5(10)-trien, einem dissozierten Östrogen. Chem Ber 111:939–943

Elger W, Schwarz S, Hedden A, Reddersen G, Schneider B (1995) Sulfamates of various estrogens are prodrugs with increased systemic and reduced hepatic estrogenicity at oral application. J Steroid Biochem Molec Biol 55:395–403

Ferrer JC, Calzada V, Bonet JJ (1990) Microbiologic oxidation of estratrienes and estratetraenes by *Streptomyces roseochromogenes* ATCC 13400. Steroids 55:390–394

Fevig TL, Katzenellenbogen JA (1987) A short, stereoselective route to 16α-(substituted-alkyl)estradiol derivatives. J Org Chem 52:247–251

Fex HJ, Högberg KB, Könyves I, Kneip PHOJ (1967) Certain steroid N-bis(haloethyl)-carbamates. Belg Patent 646 319, 31 July; Chem Abstr (1965) 63:11675f

Fieser LF, Fieser M (1959) Steroids, Reinhold Publishing Corporation, New York

French AN, Wilson SR, Welch MJ, Katzenellenbogen JA (1993) A synthesis of 7α-substituted estradiols: synthesis and biological evaluation of a 7α-pentyl-substituted BODIPY fluorescent conjugate and a fluorine-18-labeled 7α-pentylestradiol analog. Steroids 58:157–169

Funk RL, Vollhardt KPC (1980) Transition-metal-catalyzed Alkyne Cyclizations. A cobalt-mediated total synthesis of dl-estrone. J Am Chem Soc 102:5253–5261

Gill JC, Marples BA, Traynor JR (1987) Regioselective 2-hydroxylation of 3-methoxyestra-1,3,5(10)-trienes via chromium carbonyl complexes. Tetrahedron Lett 28:2643–2644

Gold AM, Schwenk E (1958) Synthesis and reactions of steroidal quinols. J Am Chem Soc 80: 5683–5687

Goto G, Yoshioka K, Hiraga K, Miki T (1977) A stereoselective synthesis and nuclear magnetic resonance spectral study of four epimeric 17-hydroxy-16-ethylestranes. Chem Pharm Bull 25:1295–1301

Greene GL, Gilna P, Waterfield M, Baker A, Hort Y, Shine J (1986) Sequence and expression of human estrogen receptor complementary DNA. Science 231:1150–1154

Groen MB, Zeelen FJ (1979) Stereoselective synthesis of 15α-methyl-19-norsteroids. Rec Trav Chim Pays-Bas 98:239–242

Guy A, Doussot J, Lemaire M (1991) Regio- and stereoselective azidation of 19-norsteroids. Synthesis:460–462

Guy A, Doussot J, Guette J-P, Garreau R, Lemaire M (1992) Regio- and stereoselective cyanation of aromatic steroids in the 9α- and 12α-position by DDQ-TMSCN. Synlett:821–822

Hajos ZG, Parrish DR (1974) Asymmetric synthesis of bicyclic intermediates of natural product chemistry. J Org Chem 39:1615–1621

Hanson RN, Napolitano E, Fiaschi R (1990) Synthesis and estrogen receptor binding of novel 11ß-substituted estra-1,3,5(10)-triene-3,17β-diols. J Med Chem 33:3155–3160

Harreus A, Kunz H (1986) Stereoselektive Glycosylierung von Steroidalkoholen mit 2,3,4,6-Tetra-O-pivaloyl-α-D-glucopyranosylbromid (Pivalobromglucose) und 2,3,4,6-Tetra-O-(o-toluoyl)-α-D-glucopyranosylbromid. Liebigs Ann Chem: 717–730

Hochberg RB (1979) Iodine-125-labeled estradiol: a gamma-emitting analog of estradiol that binds to the estrogen receptor. Science 205: 1138–1140

Horiuchi CA, Haga A, Satoh JY (1986) Novel regioselective iodination of estradiol 17β-acetate. Bull Chem Soc Jpn 59:2459–2462

Howarth NM, Cooper G, Purohit A, Duncan L, Reed MJ, Potter BVL (1993) Phosphonates and thiophosphonates as sulfate surrogates: synthesis of estrone-3-methylthiophosphonate, a potent inhibitor of estrone sulfatase. Bioorg Med Chem Lett 3:313–318

Howarth NM, Purohit A, Reed MJ, Potter BVL (1994) Estrone sulfamates: potent inhibitors of estrone sulfatase with therapeutic potential. J Med Chem 37:219–221

Hutchinson JH, Money T (1987) Enantiospecific synthesis of estrone. Can J Chem 65:1–6

Hylarides MD, Leon AA, Mettler FA (1984) Synthesis of 1-chloroestradiol. Steroids 43:219–224

Inhoffen HH, Logemann W, Hohlweg W, Serini A (1938) Untersuchungen in der Sexualhormon-Reihe. Ber dtsch chem Ges 71:1024–1032

Inhoffen HH, Zühlsdorff G (1941) Übergang von Sterinen in aromatische Verbindungen, VI. Mitteil.: Die Darstellung des Follikelhormons Oestradiol aus Cholesterin. Ber dtsch chem Ges 74:1911–1916

Jacobsohn MK, Byler DM, Jacobsohn GM (1991) Isolation of estradiol-2,3-quinone and its intermediary role in melanin formation. Biochim Biophys Acta 1073:1–10

Jaouen G, Top S, Laconi A, Couturier D, Brocard J (1984) Regiospecific and stereospecific functionalization of benzylic sites by tricarbonylchromium arene complexation. J Am Chem Soc 106:2207–2208

Kakis FJ (1963) The Birch reduction and the partial synthesis of 19-nor-steroids. In: Djerassi C (ed) Steroid Reactions. Holden-Day Inc., San Francisco, pp 267–298

Kametani T (1979) Total Syntheses of natural products by thermolysis. Pure & Appl Chem 51:747–768

Kieslich K (1981) Steroid Conversions. In: Rose RH (ed) Microbiological Bioconversions. Academic Press, New York, pp 425

Kosmol H, Kieslich K, Vössing R, Koch H-J, Petzoldt K, Gibian H (1967) Totalsynthese optisch aktiver Steroide I: Mikrobiologische stereospezifische Reduktion von 3-Methoxy-8,14-seco-1,3,5(10)-östratetraen-14,17-dion, Liebigs Ann Chem 701:199–205

Künzer H, Thiel M (1988a) Regioselective functionalization of 3-oxygenated estra-1,3,5(10)-trienes at C-1 via (η^6-arene)Cr(CO)$_3$ complexes. Tetrahedron Lett 29:3223–3226

Künzer H, Thiel M (1988b) A novel route to 3-alkylated estra-1,3,5(10)-trienes. Tetrahedron Lett 29:1135–1136

Künzer H, Sauer G, Wiechert R (1991) Stereocontrolled derivatization of 3-methoxy-1,3,5(10),n-tetraenes via Lewis acids promoted Prins reactions, [n=7; 8(9)]. Tetrahedron Lett 32:1135–1136

Künzer H, Thiel M, Sauer G, Wiechert R (1994) A new, stereoselective approach to C(7)-alkylated estra-1,3,5(10)-triene derivatives. Tetrahedron Lett 35:1691–1694

Künzer H, Thiel M (1995) A highly stereocontrolled approach to 3-methoxy-estra-1,3,5(10),14-tetraen-17α-ol, an important intermediate for the synthesis of C(14)-substituted steroid derivatives. Tetrahedron Lett 36:1237–1238

Künzer H, Thiel M, Peschke B (1996) A vinyl sulfone/vinyl sulfoxide based route to C(6)-C(7) methylene-bridged derivatives of estradiol. Tetrahedron Lett 37:1771–1772

Kurosawa T, Tohma M, Oikawa Y, Yonemitsu O (1981) Synthesis of 12α- and 12β-carboxyestradiol derivatives via the thermal elimination of β-ketosulfoxide. Chem Pharm Bull 29:2101–2103

Laing SB, Sykes PJ (1968) Synthetic steroids. Part IX. A new route to 19-nor-steroids. J Chem Soc [C]:2915–2918

Lewis DFV, Parker MG, King RJB (1995) Molecular modelling of the human estrogen receptor and ligand interactions based on site-directed mutagenesis and amino acid sequence homology. J Steroid Biochem Molec Biol 52:55–65

Li PK, Pillai R, Dibbelt L (1995) Estrone sulfate analogs as estrone sulfatase inhibitors. Steroids 60:299–306

Lovely CJ, Gilbert NE, Liberto MM, Sharp DW, Lin YC, Brueggemeier RW (1996) 2-(Hydroxyalkyl)estradiols: Synthesis and biological evaluation. J Med Chem 39:1917–1923

Maggs JL, Morgan P, Hussain F, Page PCB, Parks BK (1990) The metabolism of 2,4-dibromo-17α-ethynyl[6,7-^{3}H]oestradiol in the rat. Xenobiotica 20:45–54

McGuire AH, Dehdashti F, Siegel BA, Lyss AP, Brodack JW, Mathias CJ, Mintun MA, Katzenellenbogen JA, Welch MJ (1991) Positron tomographic assessment of 16α-[^{18}F]fluoro-17β-estradiol uptake in metastatic breast carcinoma. J Nucl Med 32:1526–1531

Meert C, Wang J, De Clercq PJ (1997) Estramicins: a novel cyclic diyl precursor derived from estradiol. Tetrahedron Lett 38:2179–2182

Mikhail G, Demuth M (1989) Selective C-2 acetylation of steroidal aromatic rings. Synlett: 54–55

Montforts F-P, Meier A, Scheurich G, Haake G, Bats JW (1992) Maßgeschneiderte Chlorine für die photodynamische Tumortherapie and als Modellsysteme für die Photosynthese. Angew Chem 104:1650–1652

Nique F, Van de Velde P, Bremaud J, Hardy M, Philibert D, Teutsch G (1994) 11β-Amidoalkoxyphenyl estradiols, a new series of pure antiestrogens. J Steroid Biochem Molec Biol 50:21–29

Numazawa M, Kimura K (1983) Novel and regiospecific synthesis of 2-amino estrogens via Zincke nitration. Steroids 41:675–682

Numazawa M, Yoshimura A (1996) Synthesis and GC-MS of 6-alkyl estradiols, possible aromatase reaction products of 6-alkylandrostenediones. Chem Pharm Bull 44:1530–1534

Ohkubo T, Wakasawa T, Nambara T (1990) Synthesis of 2-hydroxyestriol monoglucuronides and monosulfates. Steroids 55:128–132

Oppolzer W (1978) Intramolecular cycloaddition reactions of ortho-quinodimethanes in organic synthesis. Synthesis:793–802

Patrick T, Mortezania R (1988) Synthesis of 16α- and 16β-fluoro-17ß-estradiol by fluorination of estrone enols. J Org Chem 53:5153–5155

Pennington WT, Resnati G, DesMarteau DD (1992) Para fluorination by N-Fluorobis[(trifluoromethyl)sulfonyl]imide: Synthesis of 10β-fluoro-3-oxo-1,4-estradiene steroids. J Org Chem 57:1536–1539

Pert DJ, Ridley DD (1987) An alternative route to 2-bromo and 2-iodo-estradiols from estradiol. Aust J Chem 40:303–309

Peters RH, Crowe DF, Avery MA, Chong WKM, Tanabe M (1989) 11β-Nitrate estrane analogues: potent estrogens. J Med Chem 32:2306–2310

Petzoldt K, Neef G, Eder U (1984) Verfahren zur Herstellung von 3,11α-Dihydroxy-1,3,5(10)-östratrien-Derivaten. Ger Offen DE 3 248 434, 28 Jun; Chem Abstr (1984) 101:149782 t

Poirier J-M, Vottero C (1989) Mononitration de phenols par des nitrates metalliques. Tetrahedron 45:1415–1422

Poirier D, Merand Y, Labrie F (1991) Synthesis of 17β-estradiol derivatives with n-butyl, n-methyl alkylamide side chain at position 15. Tetrahedron 47:7751–7766

Pomper MG, Van Brocklin H, Thieme AM, Thomas RD, Kiesewetter DO, Carlson KE, Mathias CJ, Welch MJ, Katzenellenbogen JA (1990) 11β-Methoxy-, 11β-ethyl- and 17α-ethynyl-substituted 16α-fluoroestradiols: receptor-based imaging agents with enhanced uptake efficiency and selectivity. J Med Chem 33:3143–3155

Prousa R, Schönecker B, Tresselt D, Ponsold K (1986) Synthese, Reaktivität und ¹H-NMR-Daten von 14,15-Methylenderivaten der Androstan- und Östratrienreihe. J prakt Chem 328:55–70

Quinkert G, Weber W-D, Schwartz U, Stark H, Baier H, Dürner G (1981) Hochselektive Totalsynthese von 19-Nor-Steroiden mit photochemischer Schlüsselreaktion: Racemische Zielverbindungen. Liebigs Ann Chem:2335–2371

Quinkert G, Stark H (1983) Stereoselektive Synthese enantiomerenreiner Naturstoffe – Beispiel Östron. Angew Chem 95:651–669

Ranjith H, Dharmaratne W, Kilgore JL, Roitman E, Shackleton C, Caspi E (1993) Biosynthesis of estrogens. Estr-5(10)-ene-3,17-dione: isolation, metabolism and mechanistic implications. J Chem Soc Perkin Trans I:1529–1535

Rao PN, Cessac JW, Kim HK (1994) Preparative chemical methods for aromatization of 19-nor-Δ^4-3-oxosteroids. Steroids 59:621–627

Rao PN, Wang Z (1997) New synthesis of Δ^6-estrogens. Steroids 62:487–490

Rathore R, Saxena N, Chandrasekaran S (1986) A convenient method of benzylic oxidation with pyridinium chlorochromate. Synth Commun 16:1493–1498

Rufer C, Schröder E, Gibian H (1967) Totalsynthese optisch aktiver Steroide II: Total-synthese von natürlichem Östradiolmethyläther. Liebigs Ann Chem 701:206–216

Ruhland T, Thiel M, Künzer H (1995) Total synthesis and reactivity of C(8)-C(14) methylene bridged derivatives in the 8α-estra-1,3,5(10)-triene series. Tetrahedron Lett 36:7651–7652

Santaniello E, Ravasi M, Ferraboschi P (1983) A-Ring nitration of estrone. J Org Chem 48:739–740

Sauer G, Eder U, Haffer G, Neef G, Wiechert R, Rosenberg D (1982) Darstellung eines 7α-Methylöstratriens durch stereoselektive Methylierung von Östratrien-6-on. Liebigs Ann

Schönecker VB, Tresselt D, Ponsold K (1975) ^{1}H-NMR-Untersuchungen. Konfigura-tionszuordnung 16,17-Disubstituierter Steroide. Tetrahedron 31:2845–2852

Schwarz S, Thieme I, Richter M, Undeutsch B, Henkel H, Elger W (1996) Synthesis of estrogen sulfamates: compounds with a novel endocrinological profile. Steroids 61:710–717

Selcer KW, Li P-K (1995) Estrogenicity, antiestrogenicity and estrone sulfatase inhibi-tion of estrone-3-amine and estrone-3-thiol. J Steroid Biochem Molec Biol 52:281–286

Shimizu I, Ishii H (1989) Short effective synthesis of α-fluoroketones by palladium-catalyzed decarboxylation reactions of allyl α-fluoro-β-keto carboxylates. Chem-istry Lett:577–580

Siemann H-J, Droescher P, Undeutsch B, Schwarz S (1995) A novel synthesis of 14α,15α-methylene estradiol (J 824). Steroids 60:308–315

Sih CJ, Bennett RE (1962) Steroid 1-Dehydrogenase of Norcardia Restrictus. Biochim Biophys Acta 56:584–592

Solo AJ, Singh B, Shefter E, Cooper A (1968) Ring-D-bridged analogs. VI. Proof of stereochemistry and further reactions of 14α,17α-etheno-16α-carbomethoxy-pregn-5-ene-3β-ol-20-one acetate. Steroids 11:637–648

Spyriounis M, Rekka E, Demopoulos VJ (1990) Phase transfer catalyzed aromatic nucleophilic substitution of triflate esters of 2- and 4-nitro-estrone. Synth Commun 20:2417–2421

Suzuki E, Saegusa K, Anjo T, Matsuki Y, Nambara T (1993) Enzymatic O-methylation of catechol estrogens in red blood cells: differences in animal species and strains. Steroids 58:540–546

Sykes PJ, Rutherford FJ (1971) Oxidation of ring A-aromatic steroids to 9,11β-diol 11-nitrates with ceric ammonium nitrate. Tetrahedron Lett:3393–3396

Takano S, Moriya M, Ogasawara K (1992) A concise stereocontrolled total synthesis of (+)-estrone. Tetrahedron Lett 33:1909–1910

Umemoto T, Tomita K, Kawada K (1990) N-Fluoropyridinium triflate: an electrophilic fluorinating agent. Org Synth 69:129–143

Utne T, Jobson RB, Babson RD (1968) The synthesis of 2- and 4-fluoroestradiol. J Org Chem 33:2469–2473

Van Brocklin HF, Liu A, Welch MJ, O'Neil JP, Katzenellenbogen JA (1994) The syn-thesis of 7α-methyl-substituted estrogens labeled with fluorine-18: potential breast tumor imaging agents. Steroids 59: 34–45

Weber A, Müller R, Kennecke M, Eder U, Wiechert R (1975) Verfahren zur Herstel-lung von 4-Androsten-3,17-dion-Derivaten. Ger Offen 2 558 090,19 Dec; Chem Abstr (1977) 87:165993 k

CHAPTER 3
Non-steroidal Estrogens

J.A. Dodge and C.D. Jones

A. Introduction

Endogenous estrogens, such as 17β-estradiol (**1**, Scheme 1), have long been recognized as the primary hormones involved in the development and maintenance of the female sex organs, mammary glands and other sexual characteristics. More recently, their involvement in the growth and/or function of a number of other tissues, such as the skeleton, the cardiovascular system and the central nervous system has been recognized. These natural steroids are all derived from a common structural platform represented by the 18-carbon estrogen ring system (**1**, Scheme 1). Critical structural features within this framework include: (i) a phenol at the C-3 position of the aromatic A-ring, (ii) a relatively flat and rigid hydrocarbon core and (iii) a ketone or alcohol function at the C-17 position. Of these features, a detailed pharmacophore model postulates the important contribution of the two hydroxy groups of 17β-estradiol (**1**) to receptor binding, with the C-3 hydroxy acting as the major contributor to the binding free energy (Anstead et al. 1997). This model is supported by recent X-ray crystallographic data showing 17β-estradiol bound to the ligand-binding domain of the estrogen receptor (ER) (Brzozowski et al. 1997).

   The discovery that compounds that deviate from the traditional estrogen structure (**1**) have the ability to mimic the biological effects of the natural steroid has proven of great significance in the field. Pioneering observations in the 1930s that subcutaneous administration of select non-steroidal compounds causes the onset of estrus in ovariectomized rats has, over the past 60 years, resulted in the identification of a diverse array of non-steroidal estrogens. In turn, the structural knowledge derived from this class of agents has resulted, both directly and indirectly, in the subsequent development of anti-estrogens, partial agonists/antagonists and selective estrogen receptor modulators (SERMs). In this context, we wish to review non-steroidal compounds that act as estrogen agonists, with particular focus on underlying structural themes among this unique class of chemical entities.

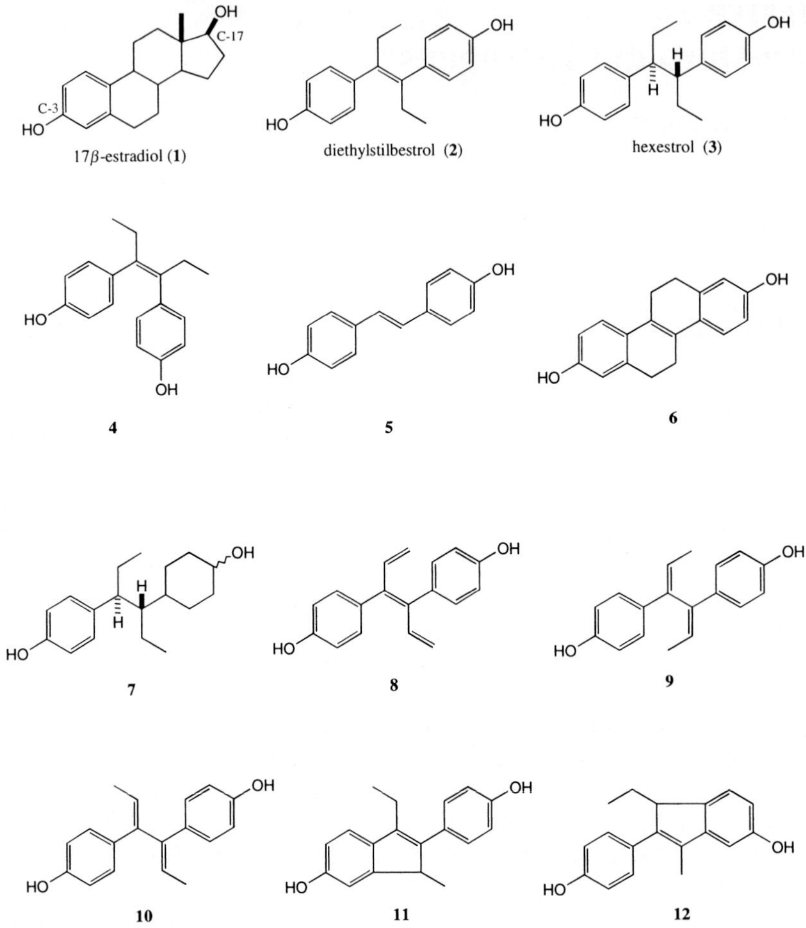

Scheme 1. Representative diaryl and diethylene non-steroidal estrogens

B. Structural Classifications of Non-Steroidal Estrogens

I. 1,2-Diarylethanes and Ethylenes

1. Diethylstilbestrol (DES), hexestrol (HES) and Analogs

Since their discovery in the late 1930s, DES (**2**) and HES (**3**) have attracted considerable scientific attention. These 1,2-diarylethylenes (DES) and -ethanes (HES) represent a key milestone in the identification of orally active non-steroidal agents with extremely potent estrogenic activity (Dodds et al. 1938a; Dodds et al. 1938b; Dodds et al. 1939). Contemporary accounts of this work have been published periodically (Solmssen 1945; Grundy 1957; Page 1991; Ruenitz 1997). The chemical properties, stability, crystal structure, synthesis, analytical methodology and medical uses, including maintenance of pregnancy, estrogen replacement therapy, suppression of lactation, post-coital contraception, cancer treatment and treatment of acne have also been reviewed (CRAWLEY 1980; GRUNDY 1957).

The early synthesis, purification and structural assignments for DES were complicated due to the recognized possibility of *cis/trans* isomerization between the DES geometrical isomers 2 and 4 (WINKLER et al. 1971; LEA et al. 1979; KATZENELLENBOGEN et al. 1985). However, the *trans* configuration 2, initially preferred on the basis of its similarity in overlay with the natural steroid molecule, was correctly assigned from the start to the more active isomer. More recent studies have demonstrated that the ER interacts exclusively with the *trans* isomer (KATZENELLENBOGEN et al. 1985), even in the presence of *cis* or *Z* DES (4). A wide variety of structures of both conformationally restricted ethylenes and conformationally energetically preferred ethanes, e.g., meso-hexesterol, exhibit binding and biological-response data that are consistent with this assignment.

It has been more difficult to ascertain the estrogenic potency of *Z* DES in solution, because of the facile isomerization which likely proceeds via a bimolecular reversible isomerization process involving tautomerization of the carbon–carbon double bond (C=C) and the hydroxyl (OH) group to form a quinone intermediate (WINKLER et al. 1971) Nevertheless, it seems clear that the estrogenicity of the *trans* compound far exceeds that of the *cis* isomer (WALTON and BROWNLEE 1943; WINKLER et al. 1971).

X-ray crystallographic determinations, in addition to confirming the isomeric structure beyond question, have provided additional information concerning the structural origins responsible for the potent estrogen-like pharmacology for this class of compounds. Although necessarily *trans* to each other, the phenolic rings of DES are not co-planar, but rather parallel with an approximate 62° rotation out of plane from the ethylene double bond. (SMILEY and ROSSMAN 1969; RUBAN and LUGER 1975; DUAX et al. 1985). This rotational alteration reduces the steric interference between the ethyl groups and the ortho-H of the phenolic rings, and appears optimal for high estrogenic potency, especially in view of the 10,000-fold and 10-fold diminished activity of compounds (5) and (6), respectively (EMMENS 1941; SOLMSSEN 1945).

2. Structure–Activity Relationships

A wide range of compounds based on the 1,2-diarylethylene and -diarylethane structural motifs of DES and HES, respectively, have been synthesized and evaluated with regard to estrogenic activity. These structure–activity studies indicate that both hydrogen-bonding groups of DES or HES (i.e., hydroxy) are required for optimal estrogenicity, and replacement of such groups by hydrogen greatly diminishes activity. Nevertheless, acylation of the aromatic-ring hydroxyl groups (SOLMSSEN 1945; GRUNDY 1957) provides ester derivatives with extended duration of action which parallels their ease of hydrolysis. Ether derivatives of DES and HES, which are more slowly cleaved in vivo, also exhibit protracted activity.

Positional isomers of the OH groups (KRANZFELDER et al. 1982; SCHNEIDER et al. 1980) result in greatly reduced estrogenic activity and, in some cases, give rise to antagonist properties. In addition to removal or repositioning of the essential OH groups, other extensive modifications of the aromatic substitution pattern have been investigated. Of the many functionalities examined (including F, Cl, Br, I, Me-, Et-, Allyl-, NO_2, NH_2, SMe, COOH, $COCH_3$ and CH_2OAc), introduction of a pair of methyl groups (at 3, 3′) and even a second pair (giving 2,2,5,5′-tetramethyl substitution) does not greatly diminish estrogenic activity (HARTMANN et al. 1981; GRUNDY 1957). Reduction of the aromatic rings is exemplified in the case of reducing HES to a mixture of cyclohexanol isomers (**7**), each isomer exhibiting diminished estrogenic activity relative to HES (GRUNDY 1957).

While additional functional groups (e.g., ethyl- or methyl-) on the carbon–carbon ethylene bridge of HES lower estrogenicity and binding affinity for the estrogen receptor (HARTMANN et al. 1980), the introduction of additional degrees of carbon–carbon double bond unsaturation into the DES structure conserves activity. The maximally unsaturated trienestrol (**8**), following subcutaneous administration to rats, exhibited estrogenicity equivalent to that of DES (GRUNDY et al. 1957).

The diethyl-substitution feature of DES is essentially optimal for highly potent estrogenic activity. Replacement of ethyl groups by hydrogen, methyl or larger alkyl moieties generally decreases the activity. An exception is the combination of (isopropyl + methyl) substitution, which is nearly as active as DES (GRUNDY 1957; HARTMANN 1981; MARGARIAN et al. 1994).

The difficulties encountered with double-bond isomerization were recognized to be avoidable through the use of cyclic structures. This feature was found to be quite well accommodated by the estrogen receptor and includes structural templates such as indenes (GRUNDY 1957; ANSTEAD et al. 1989), indoles (ROBINSON et al. 1988; VON ANGERER et al. 1984) and benzothiophenes, (VON ANGERER et al. 1992).

3. DES Metabolites

As expected for substances of such potent and profound biological activity, the metabolism of DES and, to a much lesser extent, HES have been subjects of intensive investigation and review (BOLT 1979; KORACH 1982; METZLER 1984; METZLER 1981). Metabolism of DES, which has been demonstrated in human and non-human primates, as well as other species, arise by several chemical pathways and intermediates (METZLER 1975; METZLER and MCLACHLAN 1978; BOLT 1979; ABUL-HAJJ et al. 1995).

The Z,Z- isomer of dienesterol (**9**) is the major urinary metabolite of DES in all species reported (METZLER 1981). However, it is E,E-dienesterol (**10**) which is, by far, the more estrogenic, leading some authors to describe Z,Z-dienesterol as being non-estrogenic (DEGEN et al. 1985). Nevertheless, 9 exhibits significant binding affinity, approximately 1/40 that of DES (HOSPITAL et al. 1975). *E,E*-dienesterol shows much greater binding affinity and estrogenicity, approximately one-third that of DES in the immature mouse, although it is clear that oxidation of DES to either dienesterol isomer decreases estrogenic activity (KORACH et al. 1978; KORACH 1982).

Oxidative metabolic transformation of DES also gives rise to other metabolites that incorporate the indene or indane ring system. Indenestrol A (**11**) and Indenestrol B (**12**) bind well to an estrogen receptor, but have only about 1/15 the uterotropic activity of DES in immature mice (DUAX et al. 1985). The *S* enantiomer of **11** binds to the mouse uterine cytosolic ER with four times the affinity of its enantiomeric counterpart (R-11) (KORACH et al. 1989; CHAE et al. 1991). Additional binding and hormonal assays for the isomers have been reported (KORACH et al. 1985; KORACH et al. 1987; KOHNO et al. 1996; CURTIS et al. 1997). Other DES estrogenic metabolites that have been reported include: (i) *E*-, and *Z* pseudo-DES isomers (DUAX et al. 1985; KORACH 1985; KORACH et al. 1987), (ii) a monomethyl ether derivative (SOLMSSEN 1945; MADYL et al. 1983) and (iii) an oxirane analog (E-DES-3,4-oxide) (KORACH et al. 1978; KORACH 1982).

II. Flavones and Isoflavones

A particularly diverse array of estrogenic activity has been demonstrated by plant-derived estrogens or phytoestrogens. The most studied of these compounds include genistein and coumestrol. Genistein (**13**, Scheme 2) competes with 17β-estradiol in receptor binding assays (SHUTT and COX 1972) and has been shown to have estrogenic activity on the uterus, mammary and hypothalamic axis in rats (SANTELL et al. 1997). Coumestrol (**14**) binds effectively to the ER, stimulates MCF-7 breast cancer cell proliferation and has demonstrated a number of estrogen agonist activities, including effects on uterine weight gain and bone (DODGE et al. 1996). Structural studies on coumestrol have revealed the importance of the phenolic hydroxy groups in dictating uterotrophic activity for this class of compounds, i. e., blocking of the phenolic group results in a decrease in uterine weight gain (BICKOFF et al. 1960). Other phytoestrogens include daidsein, prunetin, biochanin A and formention, as well as ring-opened forms, such as phloridzin, all of which have been postulated to have some degree of estrogen agonist activity. Recently, a systematic study of the structural requirements for estrogen flavinoids and isoflavinoids indicated that the molecular features most important for estrogenic activity include the diaryl ring and a minimum of one hydroxy group on each of these aromatic rings. The optimal substitution pattern of hydroxylation appears to be at the 4' and 7-positions of the flavone or isoflavone nucleus (MIKSICEK 1995).

III. Macrolactones

The estrogenic potential of a series of 14-membered rings was originally identified as a result of observed hyperestrogencity in swine following the

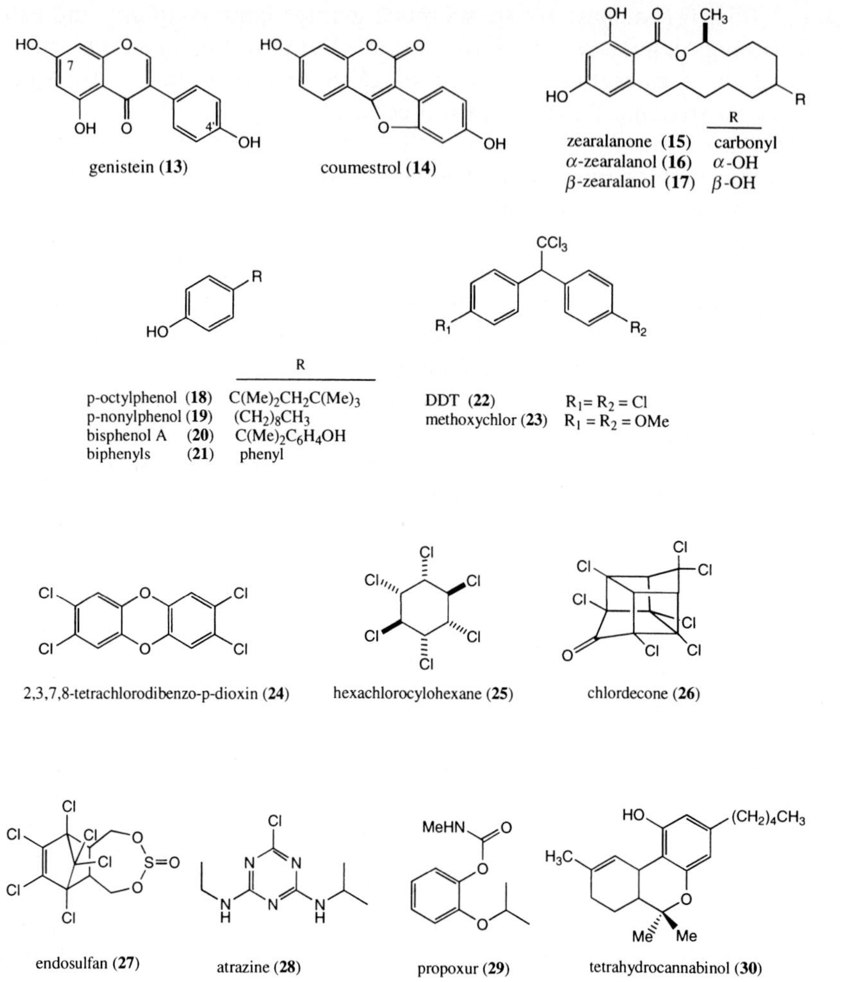

genistein (13)

coumestrol (14)

	R
zearalanone (15)	carbonyl
α-zearalanol (16)	α-OH
β-zearalanol (17)	β-OH

		R
p-octylphenol	(18)	C(Me)₂CH₂C(Me)₃
p-nonylphenol	(19)	(CH₂)₈CH₃
bisphenol A	(20)	C(Me)₂C₆H₄OH
biphenyls	(21)	phenyl

DDT (22)	R₁= R₂ = Cl
methoxychlor (23)	R₁ = R₂ = OMe

2,3,7,8-tetrachlorodibenzo-p-dioxin (24)   hexachlorocylohexane (25)   chlordecone (26)

endosulfan (27)   atrazine (28)   propoxur (29)   tetrahydrocannabinol (30)

Scheme 2. Representative non-steroidal estrogens

ingestion of mold-infected corn. These mycotoxins, exemplified by zearalenone (**15**), have been extensively studied for their estrogen agonist
behavior (KATZENELLENBOGEN et al. 1979). Competitive receptor-binding
experiments against estradiol have shown that zearalenone and analogs bind
to the ER with a relative binding order of zearalenone > α-zearalenol (**16**) >
β-zearalenol (**17**). In vivo studies indicate that α-zearalenol induces the most
uterotrophic response in rats. This compound has also been used to promote
growth in cattle (McMARTIN et al. 1973) and to relieve post-menopausal stress
in humans (UTIAN 1973). Despite its structural dissimilarity with estrogens,
these lactones possess a phenolic hydroxy group which is implicated as a

potential site for interaction with the ER in a manner similar to the 3-OH group of estrogens.

IV. Alkylphenols and Arylphenols

Soon after the identification of DES as a potent non-steroidal estrogen, other aryl compounds possessing a single phenolic moiety were also found to have estrogenic activity (DODDS et al. 1938). Subsequent work has shown that this simple structural template, a phenolic hydroxy group with an alkyl substituent at the *para* position, is a common theme among estrogenic agents. The most studied of these agents include (*para*-octyl phenol (**18**) and *para*-nonyl phenol (**19**) which have shown uterotrophic activity in vitro (MUELLER and KIM 1978; WHITE et al. 1994) and in vivo (LEE and LEE 1996; ODUM et al. 1997). A very thorough examination of the structural requirements for alkyl phenols has recently been published, indicating that both the position (*para* > *meta* > *ortho*) and the branching (tertiary > secondary) of the alkyl group dramatically affects estrogenicity (ROUTLEDGE and SUMPTER 1997). Optimal estrogenic agonism is dictated by molecules incorporating a single tertiary branched alkyl group with six to eight carbons at the 4-position relative to a phenolic moiety. The estrogenic activity of bicyclic structures having a phenolic substituent, such as 1- and 2-naphthol has also been described (MUELLER and KIM 1978).

In addition to alkyl substitution at the *para* position of a halogenated or hydroxy-substituted aromatic ring, aryl substituents also elicit an estrogenic response. Compounds in this class include: (i) bisphenol A (**20**); (ii) biphenyls (**21** and polychlorinated biphenyls); (iii) diphenylmethanes such as DDT (**22**) and methoxychlor (**23**); and (iv) tricyclic aromatic hydrocarbons, most commonly represented by dioxin (**24**). Quantitative structure–activity studies (QSAR) on hydroxylated biphenyl analogs indicate that the electron-donating properties of the hydroxy moiety and the aromatic component of estradiol correlate highly with ER binding affinity (BRADBURY et al. 1996). Structural studies have also identified geometric similarities between DDT analogs and estradiol or DES in which the internuclear distances of the relevant hydroxy groups are overlapping (BITMAN and CECIL 1970).

V. Non-Aromatic Estrogens

Although the majority of non-steroidal estrogens possess an aromatic ring that has often been speculated to mimic the aryl ring of the parent steroid nucleus, a number of compounds have been identified that lack this important structural feature. The majority of these agents are xenoestrogens derived from commercial sources, such as pesticides, and include halogenated carbo-cycles such as hexachlorocyclohexane (**25**), chlordecone (kepone, 26), toxaphene, dieldren, endosulfan (**27**) and chlordan. Studies with hexachloro-cyclohexane (**25**) indicate that it has estrogen agonist activity in an estrogen-responsive human breast cancer cell line, although it does not bind to the estrogen receptor (STEINMETZ et al. 1996). Comparative molecular field analysis of chlordecone and estradiol indicate that the optimal structural alignment is one in which the bulk of the chlordecone nucleus is positioned over the B-, C- and D-ring of estradiol, with the carbonyl moiety pointing toward the 3-hydroxy position of the steroid (WALLER et al. 1996). Although the origin(s) of the estrogen agonist behavior exhibited by this class of compounds is not entirely clear, the mode of action appears to occur via non-classical pathways (STEINMETZ et al. 1996).

VI. Miscellaneous Non-Steroidal Templates

A variety of additional structural frameworks have been identified as being estrogenic in some context. These include (i) atrazine (**28**) (COOPER et al. 1996); (ii) weakly estrogenic insecticidal carbamate insecticides (KLOTZ et al. 1997), such as aldicarb, propoxur (**29**), bendocarb, carbaryl, methamomyl and oxamyl; (iii) delta-9-tetrahydrocannabinol (**30**) (SOLOMON and SCHOENFELD 1980); (iv) phenol red and analogs (KYM et al. 1996); (v) azoresorcinol and other azobenzenes (KATZENELLENBOGEN et al. 1980); (vi) phthalates (JOBLING et al. 1995); (vii) *cis*-2,6-diphenylhexamethylcyclotetrasiloxane (KATZENELLENBOGEN et al. 1980); and (viii) secosteroids (GRESE and DODGE 1996).

C. Conclusions

Since the initial observations in the late 1930s that non-steroidal compounds can elicit similar biological responses to that of 17β-estradiol, a diverse array of such structural entities have been identified. These include 1,2-diarylethanes and ethylenes, flavones/isoflavones, macrolactones, aryl/alkyl phenols and halogenated carbocycles, as well as others. A pharmacophore model that has emerged from this vast body of structural information defines several molecular determinants of estrogen agonism for non-steroidal estrogens. These include, but are not limited to, a phenolic moiety, thought to be critical for high affinity interactions with the ER, and a hydrophobic alkyl or aryl group at the *para* position relative to the phenol. Significantly more diverse agents have also been identified that lack an aromatic component altogether. The recent discovery of a second ER (Erb) adds a further layer of complexity in understanding the structural requirements for estrogenicity, particularly since little is known about the interactions of non-steroidal estrogens with this ER isoform.

References

Abul-Hajj Y, Tabakovic K, Ojala W, Gleason WB (1995) Crystalline diethylstilbestrol (DES) quinone: synthesis, x-ray analysis, and stability. J Am Chem Soc 117:5600–5601

Anstead GM, Wilson SR, Katzenellenbogen JA (1989) 2-Arylindenes and 2-arylinde-nones: molecular structures and considerations in the binding orientation of unsymmetrical nonsteroidal ligands to the estrogen receptor. J Med Chem 32:2163–2171

Anstead GM, Carlson KE, Katzenellenbogen JA (1997) The estradiol pharmacophore: ligand structure-estrogen receptor binding affinity relationships and a model for the receptor binding site. Steroids 62:268–303

Bickoff EM, Livingston AL, Booth AN (1960) Estrogenic activity of coumestrol and related compounds. Arch Biochem Biophys 88:262–266

Bitman J, Cecil HC (1970) Estrogenic activity of DDT analogs and polychlorinated biphenyls. J Agr Food Chem 18:1108–1112

Bolt HM (1979) Metabolism of estrogens – natural and synthetic, pharmacology and therapeutics. 4:155–181

Bradbury SP, Mekenyan OG, Ankley GT (1996) Quantitative structure-activity rela-tionships for polychlorinated hydroxybiphenyl estrogen receptor binding affinity. An assessment of conformer flexibility. Environ Toxicol Chem 15:1945–1954

Brzozowski AM, Pike ACW, Dauter Z, Hubbard RE, Bonn T, Engstrom O, Ohman L, Greene GL, Gustafsson J-A, Carlquist M (1997) Nature 389:753–757

Chae K, Johnston SH, Korach KS (1991) Multiple estrogen binding sites in the uterus: stereochemistry of receptor and non-receptor binding of diethylstilbestrol and its metabolites. J Steroid Biochem Mol Biol 38:35–42

Cooper RL, Stoker TE, Goldman JM, Parrish MB, Tyrey (1996) Effect of atrazine on ovarian function in the rat. Reprod Toxicol 10:257–264

Crawley G (1980) Nonsteroidal estrogens. In: Grayson M (ed) Encyclopedia of chemical technology, Third Edition, Vol. 12, John Wiley & Sons, New York, NY, p 658

Curtis SW, Shi H, Teng C, Korach KS (1997) Promoter and species specific differential estrogen-mediated gene transcription in the uterus and cultured cells using struc-turally altered agonists. J Mol Endocrinol 18:203–211

Degen GH, McLachlan J (1985) Peroxidase-mediated in vitro metabolism of diethyl-stilbestrol and structural analogs with different biological activities. Chem-Biol Interact 54:363–365

Dodds EC, Goldberg L, Lawson W, Robinson R (1938a) Estrogenic activity of certain synthetic compounds. Nature (London) 141247–248

Dodds EC, Lawson W, Noble RL (1938b) Biological effects of the synthetic estrogenic substance 4,4′-dihydroxy-α,β-diethylstilbene. Lancet 234:1389–1391

Dodds EC, Golberg L, Lawson W, Robinson R (1939) Synthetic estrogenic compounds related to stilbene and diphenylethane. Part I Proc Roy Soc B 127:140–142

Dodge JA, Glasebrook AL, Magee DE, Phillips DL, Sato M, Short LL, Bryant HU (1996) Environmental estrogen: effects on cholesterol lowering and bone in the ovariectomized rat. J Steroid Biochem 59:155–161

Duax WL, Griffin JF, Weeks CM, Korach KS (1985) Molecular conformation, receptor binding, and hormone action of natural and synthetic estrogens and antiestrogens. Environ Health Perspect 62:11–21

Emmens C. (1941) Precursors of estrogens. J Endocrinol 2:444–449

Grese TA, Dodge JA (1996) Estrogen receptor modulators: effects in non-traditional target tissues. Ann Rep Med Chem 32:181–190

Grundy J (1957) Artificial estrogens. Chem Rev 57:281–416

Hartmann RW, Buchborn H., Kranzefelder G, Schonenberger H, Bogden AE (1981) Potential antiestrogens. synthesis and evaluation of mammary tumor inhibiting activity of 1,2-dialkyl-1,2-bis(3′-hydroxyphenyl)ethanes. J Med Chem 24:1192

Hartmann RW, Kranzefelder G, von Angerer E, Schonenberger HJ (1980) Antiestro-gens. Synthesis and evaluation of mammary tumor inhibiting activity of 1,1,2,2-tetraalkyl-1,2-diphenylethanes. J Med Chem 23:841–845

Hospital M, Busetta B, Courseille C, Precigoux G (1975) X-ray conformation of some estrogens and their binding to uiterine receptors. J Steroid Biochem 6:221–226

Jobling S, Reynolds T, White R, Parker MG, Sumpter JP (1995) A variety of environ-
    mentally persistent chemicals, including some phthalate plastisizers, are weakly
    estrogenic. Environ Health Perspec 103:582–587
Katzenellenbogen BS, Katzenellenbogen JA, Mordecai D (1979) Zearalenones: char-
    acterization of the estrogenic potencies and receptor interactions of a series of
    fungal β-resorcylic acid lactons. Endocrinology 105:33–40
Katzenellenbogen JA, Katzenellenbogen BS, Tatee T, Robertson DW, Landvatter SW
    (1980) The chemistry of estrogens and antiestrogens: relationship between struc-
    ture, receptor binding, and biological activity. In: McLachlan (ed) Estrogens in the
    environment. Elsevier, North Holland, p 33
Katzenellenbogen JA, Carlson KE, and Katzenellenbogen BS (1985) Facile geometric
    isomerization of phenolic nonsteroidal estrogens and antiestrogens: limitations to
    the interpretation of experiments characterizing the activity of individual isomers.
    J Steroid Biochem 22:589–597
Klotz DM, Arnold SF, McLachlan JA (1997) Inhibition of 17 beta-estradiol and prog-
    esterone activity in human breast cancer and endometrial cancer cells by carba-
    mate insecticides. Life Sci 60:1467–1457
Kohno H, Bocchinfuso WP, Gandini O, Curtis SW, Korach KS (1996) Mutational analy-
    sis of the estrogen receptor ligand-binding domain: influence of ligand structure
    and stereochemistry on transactivation. J Mol Endocrinol 16:77–285
Kym PR, Hummert KL, Nilsson AG, Lubin M, Katzenellenbogen JA (1996) J Med
    Chem 39:4897–4904
Korach KS, Metzler M, McLachlan JA (1978) Estrogenic activity in vivo and in vitro
    of some diethylstilbestrol metabolites and analogs. Proc Natl Acad Sc 75:468–471
Korach KS (1982), Biochemical and estrogenic activity of some diethylstilbestrol
    metabolites and analogs in the mouse uterus. Adv Exp Med Biol 138:39–62
Korach KS, Fox-Davies C, Quarmby VE, Swaisgood MH (1985) Diethylstilbestrol
    metabolites and analogs. Biochemical probes for differential stimulation of uterine
    estrogen responses. J Biol Chem 260:15420–15426
Korach KS, Levy LA, Sarver PJ (1987) Estrogen receptor stereochemistry: receptor
    binding and hormonal responses. J Steroid Biochem 27:281–290
Korach KS, Chae K, Levy LA, Duax WL, Sarver P (1989) Diethylstilbestrol metabo-
    lites and analogs. stereochemical probes for the estrogen receptor binding site.
    J Biol Chem 264:5642–5647
Kranzfelder G, Hartmann RW, von Angerer E, Schonenberger H, Bogden AE (1982)
    3,4-Bis(3'-hydroxyphenyl)hexane – a new mammary tumor-inhibiting compound.
    J Cancer Res Clin Oncol 103:165–180
Lea AR, Kayaba WJ, Hailey DM (1979) Analysis of diethylstilbestrol and its impuri-
    ties in tablets using reversed-phase high-performance liquid chromatography.
    J Chromatogr 177:61–66
Lee, P-C, Lee W (1996) In vivo estrogenic action of nonylphenol in immature female
    rats. Bull Environ Contam Toxicol 57:341–348
Magarian RA, Overacre LB, Singh S, Meyer L (1994) The medicinal chemistry of non-
    steroidal antiestrogens: a review. Current Medicinal Chemistry 1:61–104
Maydl R, Newbold, RR, Metzler M, McLachlan, JA (1983) Diethylstilbestrol metabo-
    lism by the fetal genital tract. Endocrinology (Baltimore) 13:146–151
McMartin KE, Kennedy KA, Greenspan P, Alam SN, Greiner P, Yam J (1973) Diethyl-
    stilbestrol: a review of its toxicity and use as a growth promotant in food-
    producing animals. J Environ Pathol Toxicol 1:279–313
Metzler M (1975) Metabolic activation of diethylstilbestrol: indirect evidence for the
    formation of a stilbene oxide intermediate in hamster and rat. Biochem Pharma-
    col 24:1449–1453
Metzler M, McLachlan JA (1978) Oxidative metabolites of diethylstilbestrol in the
    neonatal and adult mouse. Biochem Pharmacol 27:1087–1093
Metzler M (1981) The metabolism of diethylstilbestrol. CRC Crit Rev Biochem
    10:171–212

Metzler M (1987) Metabolic activation of xenobiotic stilbene estrogens. Fed Proc 46:1855–1857

Miksicek RJ (1995) Estrogenic flavinoids: structural requirements for biological activity. Proc Soc Exp Biol Med 208:44–50

Mueller GC, Kim U-H (1978) Displacement of estradiol from estrogen receptors by simple alky phenols. Endocrinology 102:1429–1435

Odum J, Lefevre PA, Tittensor S, Paton D, Routledge EJ, Beresford NA, Sumpter JP, Ashby J (1997) The rodent uterotrophic assay: critical protocol features, studies with nonyl phenols, and comparison with a yeast estrogencity assay. Reg Toxicol Pharm 25:176–188

Page SW (1991) Diethylstilboestrol – historical background and current regulator status. Australian Veterinary Journal 68:224–225

Robinson S, Koch R, Jordan VC (1988) In vitro estrogenic actions in rat and human cells of hydroxylated derivatives of D16726 (zindoxifene), an agent with known antimammary cancer activity in vivo. Cancer Res 48:784–788

Routledge EJ, Sumpter JP (1997) Structural features of alkylphenolic chemicals associated with estrogenic activity. J Biol Chem 272:3280–3288

Ruban G, Luger P (1975) Die Struktur des 4,4'-Dimethoxy-α,β-diäthylstilbens. Acta Crystallogr Sect. B 31:2658–2660

Ruenitz PC (1997) Female sex hormones and analogs. In: Wolff ME (ed) Burger's Medicinal Chemistry and Drug Discovery, Vol. 4, John Wiley & Sons, Inc. New York, p 553

Santell RC, Chang YC, Nair MG, Helferich WG (1977) Dietary genistein exerts estrogenic effects upon the uterus, mammary gland and the hypothalamic pituitary axis in rats. Am Soc Nutr Sci 22:263–269

Schneider M, von Angerer E, Kranzfelder G, Schonenberger H (1980) Mammatumorhemmende Antiöstrogene vom Typ des 3,3'-Dihydroxy-α,β-dialkylstilbens. Arch Pharm 313:919–925

Shutt DA, Cox RI (1972) Steroid and phytoestrogen binding to sheep uterine receptors in vitro. J Endocrinol 85:317–325

Smiley IE, Rossmann MG (1969) The crystal structure of α,α'-diethylstilbene-4,4'-diol. Chem Commun 198–199

Solmssen UV (1945) Synthetic estrogens and the relations between their structure and their activity, Chem Rev 37:481–498

Solomon J, Schoenfeld D (1980) The uterotrophic effect of THC in ovariectomized rats and mice. Res Comm Chem Path Pharmacol 29:603–608

Steinmetz R, Young PCM, Caperell-Grant A, Gize E, Madhykar V, Ben-Jonathan N, Bigsby RM (1996) Novel estrogenic action of the pesticide residue β-hexachlorocyclohexane in human breast cancer cells. Cancer Res 56:5403–5409

Utian WH (1973) Comparative trial of P-1496, a new non-steroidal estrogen analogue. Br Med J 1:579–581

von Angerer E, Prekajac J, Strohmeier J (1984) 2-Phenylindoles. Relationship between structure, estrogen receptor affinity, and mammary tumor inhibiting activity in the rat. J Med Chem 27:1439–1443

von Angerer E, Erber S (1992) 3-Alkyl-2-phenylbenzo[b]thiophenes: nonsteroidal estrogen antagonists with mammary tumor inhibiting activity. J Steroid Biochem Molec Biol 41:557–573

Waller CL, Oprea TI, Chae K, Park H-Y, Korach K, Laws SC, Weise TE, Kelce WR, Gray LE (1996) Ligand-based identification of environmental estrogens. Chem Res Toxicol 9:1240–1248

Walton E, Brownlee G (1943) Isomers of stilboestrol and its esters. Nature (London) 151:305–306

White R, Jobling S, Hoare SA, Sumpter JP, Parker MG (1994) Environmentally persistent alkylphenolic compounds are estrogenic. Endocrinology 135:175–182

Winkler VW, Nyman MA, Egan RS (1971) Diethylstilbestrol cis-trans isomerization and estrogen activity of diethylstilbestrol isomers. Steroids 17:197–201

CHAPTER 4
Antiestrogens and Partial Agonists

E. von Angerer

A. Introduction

For more than a decade, the antiestrogen tamoxifen dominated the field of estrogen antagonists despite its inherent estrogenicity. For a long time, efforts to replace tamoxifen or improve its pharmacological profile by structural modifications have failed. In the last few years, however, the situation has changed: antiestrogens that are devoid of any residual estrogenic activities were discovered, and mixed agonists/antagonists with a certain degree of tissue specificity were identified. Both types of agents have strongly stimulated the interest in novel antiestrogens. This review is confined to the chemistry of compounds that exert a varying degree of estrogen antagonism. Routes of synthesis leading to antiestrogens of clinical importance will be presented, but only a few biological aspects will be discussed here, because most of the antihormonal activities have been reviewed recently (von Angerer 1995).

The spectrum of agents covered in this chapter ranges from weak antagonists with substantial agonist activity to those that are completely devoid of residual estrogenic activity and are termed "pure antiestrogens". All of the compounds described in this section bind to the estrogen receptor, though their binding affinities are usually lower than that of the natural ligand 17β-estradiol. All of the binding data reported until now refer to the estrogen receptor α, and only a small number of compunds have been evaluated for binding affinity for the recently discovered estrogen receptor β (Kuiper et al. 1997; Kuiper and Gustafsson 1997). After binding the synthetic ligand, the receptor is understood to exert a lower or no hormonal activity. Thus, the displacement of the endogenous estrogen from its receptor by a partial agonist leads to antiestrogenic effects in classical endocrine assays, such as the uterine-weight test, vaginal-cornification test or in transfection assays which have become available recently.

The compounds with antiestrogenic activity fall into several different chemical categories, depending on their basic structure. Most of them are characterized by non-steroidal structures, but some of the most potent agents are based on estradiol. Since the chemistry of many of the new non-steroidal estrogen antagonists has been recently reviewed by Magarian et al. (1994), only the syntheses of antiestrogens in clinical use or with clinical potential will be discussed. The structure–activity relationships which provide the basis for further development in this field is the subject of the subsequent chapter.

B. Triphenylethylene Derivatives

I. Tamoxifen

For the rational design of non-steroidal antiestrogens, essentially two basic structures were applied and systematically altered: triphenylethylene from which tamoxifen is derived and stilbene which provided the skeleton for diethylstilbestrol (DES). The first non-steroidal antiestrogen discovered was ethamoxytriphetol MER 25 (Fig. 1) (LERNER and JORDAN 1990). It already contains a basic side chain that characterizes many of the triphenylethylene antiestrogens, although it is not essential (SCHNEIDER et al. 1982b). Its potency as an antiestrogen was rather low and serious central nervous system (CNS) effects have been reported (LERNER 1981). The hydrophilic hydroxy group in the center of this molecule probably interferes with the receptor binding, and its elimination should improve the activity.

Clomiphene (Fig. 1), the second drug that was developed, lacks the central hydroxy group and possesses a double bond with a chloro substituent which improves lipophilicity. Clomiphene, a mixture of *cis* (zuclomiphene) and *trans* isomers (enclomiphene), is still used as a gonad-stimulating agent in subfertile women (HUPPERT 1979). When nafoxidene, the third drug in this series (Fig. 1), was synthesized and tested, it became evident that a certain steric arrange-

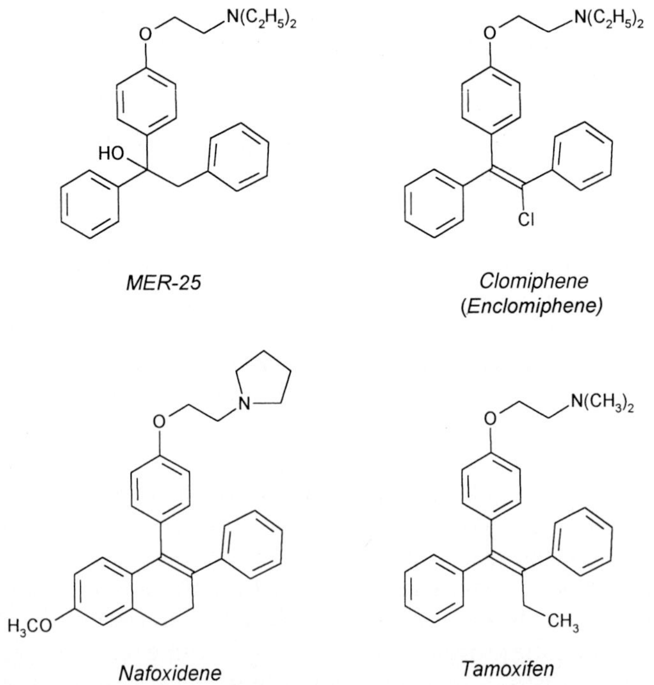

MER-25

Clomiphene
(Enclomiphene)

Nafoxidene

Tamoxifen

Fig. 1. Chemical structures of MER-25 and triphenylethylene antiestrogens

ment at the double bond favors the antiestrogenic activity. The stereoisomers of triphenylethylenes, in which the unsubstituted phenyl rings adopt the *trans* orientation, are partial agonists with antiestrogenic activity in the rat, whereas the corresponding *cis* isomers show no antagonism (HARPER and WALPOLE 1966; JORDAN et al. 1981).

The most prominent drug amongst these compounds is tamoxifen (Z-1-[4-(2-dimethylaminoethoxy)phenyl]-1,2-diphenylbut-1-ene) (PLOWMAN 1993) (Fig. 1) which was developed by ICI two decades ago and marketed as citrate salt under the name Nolvadex® (RICHARDSON 1988). In many clinical studies, it has proved its activity in postmenopausal patients suffering from hormone-dependent breast cancer and has become the treatment of choice for this malignancy (POWLES 1997). It is used both in advanced disease and in the adjuvant setting following the surgical removal of the primary tumor.

Tamoxifen, as the prototype of a triphenylethylene antiestrogen, can be synthesized using various methods. An important aspect in the synthesis is the fact that only the Z isomer is a potent antiestrogen and the E isomer is understood to be devoid of antagonistic activity and acts as weak estrogen. Therefore, a synthetic procedure that leads to production of the Z isomer would be desirable. Coupling of 4-(2-chloroethoxy) benzophenone with acetophenone in the presence of low-valent titanium (McMurry reaction) leads, preferentially, to the Z isomer (7:1) which can be purified by crystallization and converted to tamoxifen (Fig. 2) (COE and SCRIVEN 1986). Using a different approach, a bromo stilbene with the correct stereochemistry was used as a key intermediate, followed by the stereospecific displacement of the halogen atom by the third aryl group, using a zinc organic reagent in the presence of tetrakis (triphenylphosphine) palladium(0) (Fig. 3) (POTTER and McCAGUE 1990; MILLER and AL-HASSAN 1985). The amino function is usually introduced in one of the last steps of synthesis. Recently, the parallel synthesis of tamoxifen and its derivatives on a solid support has been reported (BROWN and ARMSTRONG 1997).

II. Triphenylethylene Derivatives Related to Tamoxifen

Since tamoxifen became a successful drug for the treatment of breast cancer, several groups modified its chemical structure. The main objectives for these modifications were: (i) alteration of the metabolic pathways by introduction of additional substituents; (ii) alteration of the pharmacodynamic profile by variation of the side chains; (iii) prevention of *cis/trans* isomerization by fixed-ring analogs. The chemical structure of tamoxifen offers various possibilities for alterations, such as the introduction of substituents into one or both unsubstituted phenyl rings, replacement of the ethyl group by other groups and variation of the functional side chain attached to the *para* position of the phenyl ring. The knowledge that 4-hydroxytamoxifen is more potent in vitro than the parent drug, but is rapidly cleared from the body, provided the basis for the synthesis of the 3-hydroxy derivative which was developed as droloxifene for

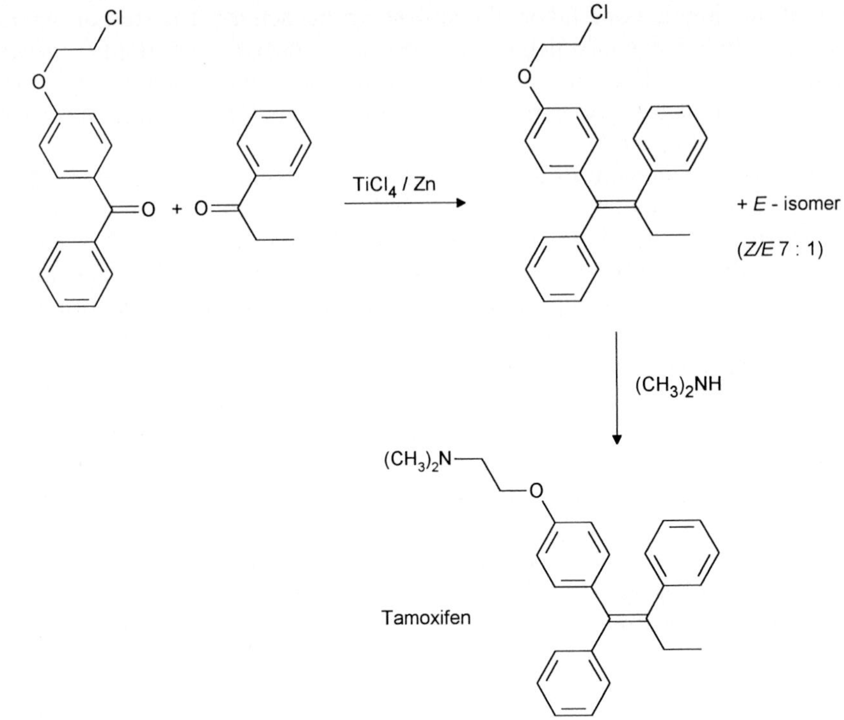

Fig. 2. Synthesis of tamoxifen by the McMurry reaction (COE and SCRIVEN 1986)

clinical use (Fig. 4) (DIETEL et al. 1989). It was synthesized in a conventional way by means of the Grignard reaction of α-ethyl-3-hydroxy-desoxybenzoin, followed by separation of the Z isomer from the inactive isomer by crystallization (RUENITZ et al. 1982). 4-Hydroxytamoxifen can be obtained in a similar fashion, but proved to be unstable with respect to *cis/trans* isomerization.

McCAGUE et al. (1989) blocked the 4-position of the geminal phenyl group with various substituents to prevent metabolic hydroxylation and rapid elimination. The iodo derivative idoxifene, with a pyrrolidino group as the amine function (Fig. 4), exerted a stronger growth inhibitory effect on MCF-7 breast cancer cells than tamoxifen. This effect could be explained by the directional interaction of this heavy halogen atom with the nucleophilic receptor site that interacts with the hydroxy function of 4-hydroxytamoxifen. Idoxifene is presently undergoing clinical trials (JOHNSTON et al. 1997). The search for an appropriate pro-drug for 4-hydroxytamoxifen, which is considered one of the important active metabolites, has guided the development of a phosporic-acid ester of 4-hydroxytamoxifen, with an additional 4-isopropyl group in the β-phenyl ring (TAT-59, Fig. 4) (IINO et al. 1994).

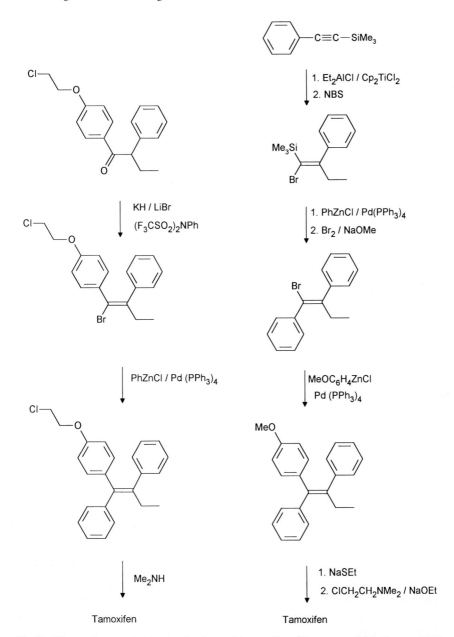

Fig. 3. Alternative routes to production of tamoxifen (POTTER and McCAGUE 1990; MILLER and AL-HASSAN 1985). *cp* cyclopentadienyl, *Et* ethyl, *Me* methyl, *NBS* N-bromosuccinimide, *Ph* phenyl

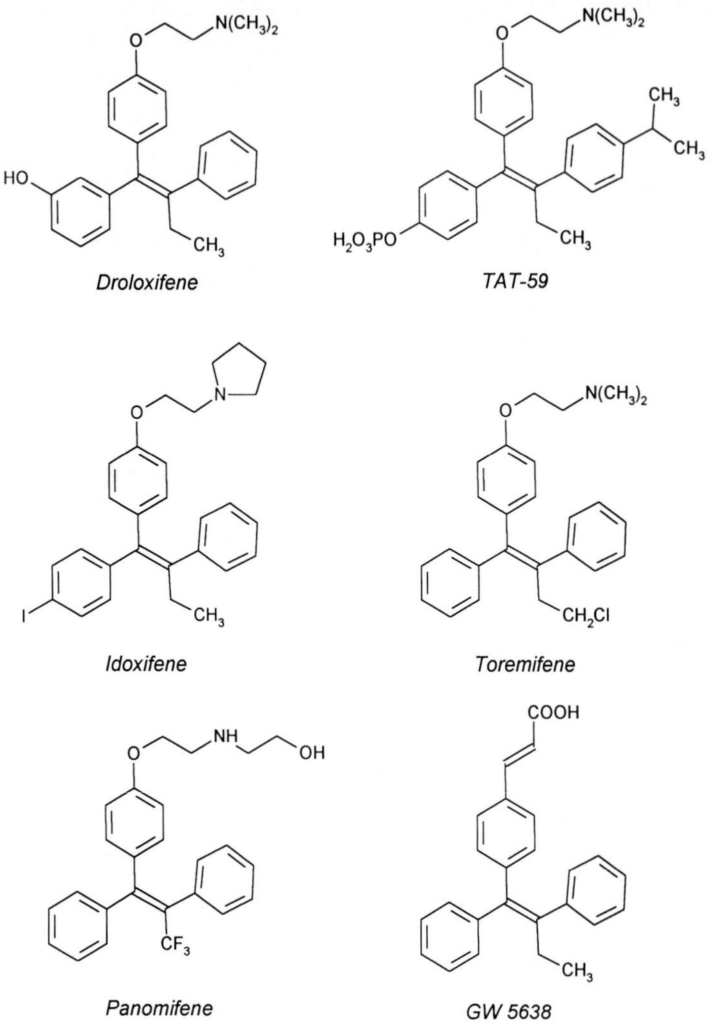

Fig. 4. Triphenylbutene antiestrogens in early clinical trials

Despite the fact that the basic side chain in tamoxifen is one of the important structural elements, only a very limited number of variations of this part of the molecule has been performed, mostly in combination with other alterations, such as fixation of the central double bond (Acton et al. 1983; McCague et al. 1986). When the 2-(dimethylamino) ethoxy fragment was replaced by acrylic acid, the agonist activity was lost in the uterus, but was retained in the bone and the cardiovascular system (Willson et al. 1994, 1997).

Concerning the non-aromatic substituent at the ethylene bond (an ethyl group in tamoxifen), only minor modifications were performed; except for those leading to cyclic structures. The closest tamoxifen analogue that has undergone clinical trials is toremifene (WISEMAN and GOA 1997) (Fig. 4) which differs from tamoxifen only by a chlorine atom in the terminal position of the ethyl group. As expected, its pharmacological profile is more or less identical with that of the parent drug (ROBINSON and JORDAN 1989; KAUFMANN et al. 1989). An interesting observation was that strong electron-withdrawing groups, such as nitro (nitromifene; RUENITZ et al. 1989), cyano (PONS et al. 1984) or trifluoromethyl (panomifene; BORVENDÉG 1985) functions can be introduced without changing the pharmacological profile much.

III. Fixed-Ring Analogues of Tamoxifen

As already pointed out, the two geometric isomers of tamoxifen display different pharmacological profiles, with the *E* isomer being a pure agonist (ROBERTSON et al. 1982). An interconversion between the *E* and *Z* form under in vivo conditions is not likely. However, the situation changes after the enzymatic introduction of an oxygen function into the *para* position of the geminal phenyl ring. The rotation around the central double bond in 4-hydroxytamoxifen via a tautomeric quinoid intermediate is quite facile (Fig. 5) and is

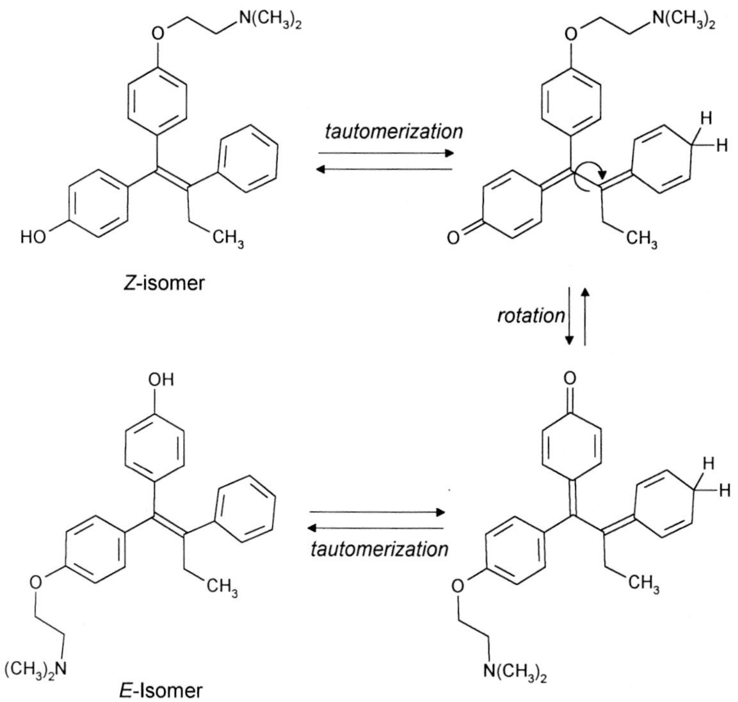

Fig. 5. Proposed mechanism of *cis/trans*-isomerization of 4-hydroxytamoxifen

known to occur even under cell-culture conditions to provide a mixture of *E* and *Z* isomers (KATZENELLENBOGEN et al. 1985). This ability to isomerize makes the biological evaluation of the active metabolite difficult. Therefore, it was of interest to synthesize non-isomerizable analogues and to determine their endocrine properties.

A common type of non-isomerizable derivative incorporates the double bond into a cyclic structure (McCAGUE et al. 1988). Nafoxidene (Fig. 1), which possesses a 3,4-dihydronaphthalene skeleton, was the first member of this class of compounds, but its development was not guided by these stereochemical considerations. ACTON et al. (1983) synthesized a number of triphenylethylene derivatives in which the two *cis*-orientated phenyl rings were linked by hetero atoms or an ethylene bridge (Fig. 6). McCAGUE et al. (1986) showed, by means of X-ray crystallography, that utilization of a seven-membered ring, which includes the adjacent bond of the phenyl ring, provides a conformation similar to that of the parent compound tamoxifen.

In the late 1960s, several cyclic systems, such as indenes (LEDNICER et al. 1965a), benzofurans, benzothiophenes and benzothiopyrans (CRENSHAW et al. 1971), were investigated (Fig. 6). The primary goal of these studies on non-steroidal antiestrogens was the discovery of antifertility agents. At that time,

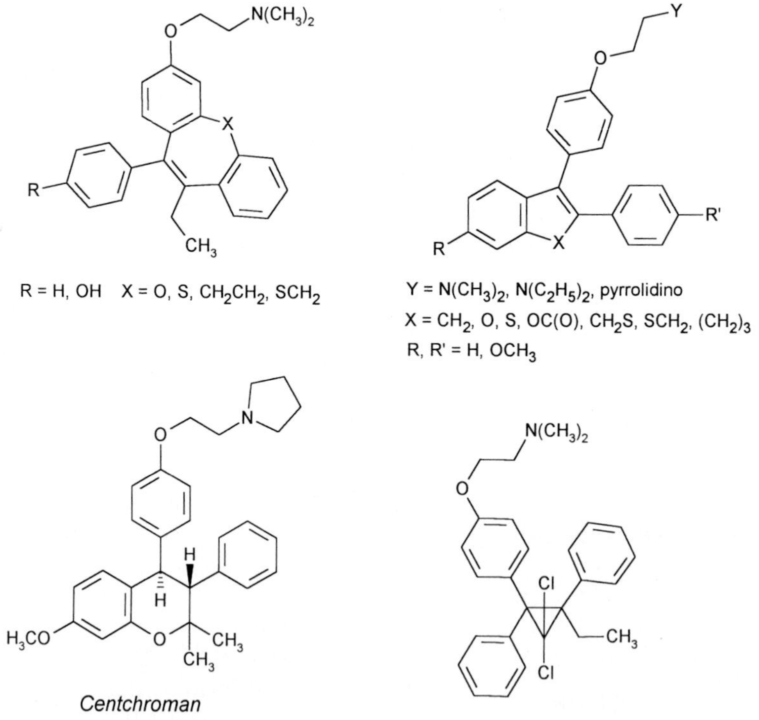

R = H, OH   X = O, S, CH₂CH₂, SCH₂

Y = N(CH₃)₂, N(C₂H₅)₂, pyrrolidino
X = CH₂, O, S, OC(O), CH₂S, SCH₂, (CH₂)₃
R, R' = H, OCH₃

Centchroman

Fig. 6. Fixed-ring analogues of tamoxifen

numerous modifications of the nafoxidine structure (Fig. 1) were performed, such as ring contraction (LEDNICER et al. 1965) and replacement of the cyclic ethane fragment by heteroatoms such as oxygen (CRENSHAW et al. 1971; DURANI et al. 1989) and sulfur (CRENSHAW et al. 1971) (Fig. 6).

It was shown that the double bond that links the three phenyl rings is not essential for receptor binding and antiestrogenic activity and that it can be replaced by a single bond (McCAGUE and LECLERCQ 1987). Thus, it was not surprising that centchroman, with a dihydropyran structure (Fig. 6), shares many endocrine properties with tamoxifen (SALMAN et al. 1986). Centchroman has recently been introduced into the market as the first non-steroidal oral contraceptive (KAMBOJ et al. 1992).

A very interesting alternative for preventing isomerization of tamoxifen is the fixation of the three phenyl rings in a 1,1-dichlorocyclopropane structure, as demonstrated by MAGARIAN and his group (Fig. 6) (DAY et al. 1991; JAIN et al. 1993). This structural modification did not change the binding affinity for the rat uterine estrogen receptor, the antiestrogenic activity or the antiproliferative effect in MCF-7 breast cancer cells very much.

IV. 1,2,3-Triarylpropenone-Derived Antiestrogens

Several studies have revealed that a geminal arrangement of two of the phenyl rings is not essential and can be replaced by structures in which all of the three phenyl groups are located at different carbon atoms. Examples of this modification are the triarylfurans (DURANI et al. 1989), 1,2,3-triarylpropenones (MITTAL et al. 1985), 1,2,3-triarylbutenones (DURANI et al. 1989) and cyclic structures that contain these fragments, such as 3-benzoyl-2-phenylbenzo[b]furan (Fig. 7) (DURANI et al. 1989). The pyrrolidinoethoxy group that was used as a basic side chain was preferably located in the *para* position of the benzoyl ring.

The starting point for these investigations was the compound trioxifene (LY 133,314; Fig. 7) which was synthesized by JONES et al. (1979) at Eli Lilly, in the 1970s. This antiestrogen displays several structural features similar to those of nafoxidine (Fig. 1). The search for a more potent drug led eventually to some very interesting compounds that comprise a benzothiophene substructure with the *para*-substituted benzoyl group in position 3. The first derivative that underwent a detailed study with respect to its endocrine properties and mammary tumor-inhibiting activity was LY 117,018 (JORDAN and GOSDEN 1983) which carried a pyrrolidino group as its amino function. This basic function was later replaced by the piperidine ring to give the drug keoxifene (LY 156,758) which is now known as raloxifene (LY 139,481 HCl; Fig. 10) (JONES et al. 1984; BLACK et al. 1994). An important structural element is the presence of two free hydroxy groups in the benzothiophene and the phenyl ring which give rise to higher binding affinities for the estrogen receptor than compounds that lack these polar functions, unless they had undergone metabolic hydroxylation. From the comparison of the chemical structure of raloxifene with

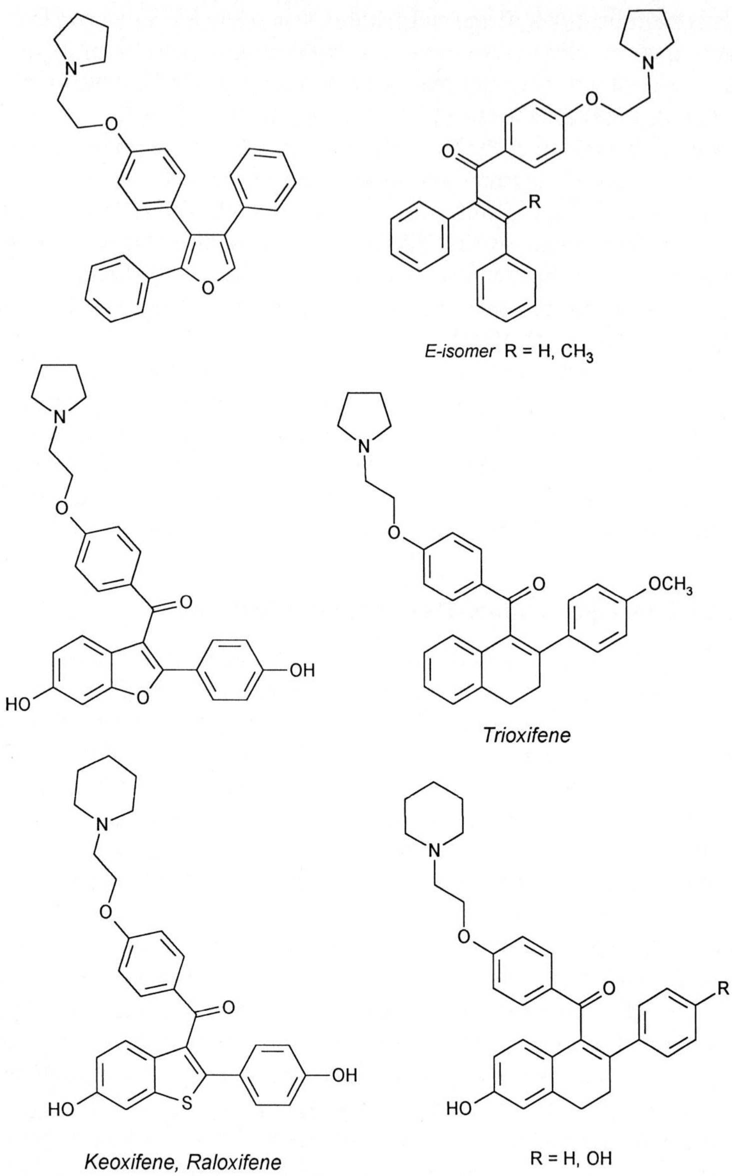

E-isomer R = H, CH₃

Trioxifene

Keoxifene, Raloxifene

R = H, OH

Fig. 7. 1,2,3-Triphenylpropenone-based antiestrogens

those of the antiestrogen tamoxifen and the non-steroidal estrogen DES, it becomes evident that raloxifene is a derivative of DES rather than a tamoxifen analogue. JONES et al. (1984) elaborated an elegant synthesis for this drug which involved a one-pot acylation of the methoxy-substituted 2-phenylbenzothiophene and selective cleavage of two of the three ether functions (Fig. 8).

In order to improve on the pharmacological properties of raloxifene with respect to tissue selectivity, its structure was systematically altered. The modification concerned both the oxygen functions (MARTIN et al. 1997) and the 2-phenyl ring (GRESE et al. 1997). An enhanced efficacy, however, could not been proven for these agents. An important step forward was the replacement of the carbonyl group by an oxygen atom (PALKOWITZ et al. 1997). Preliminary data showed that antiestrogenic activity and tissue specificity are higher than those of raloxifene. For the synthesis of this and similar analogues of raloxifene, the electron density in the thiophene substructure had to be decreased to make the system susceptible to a nucleophilic substitution

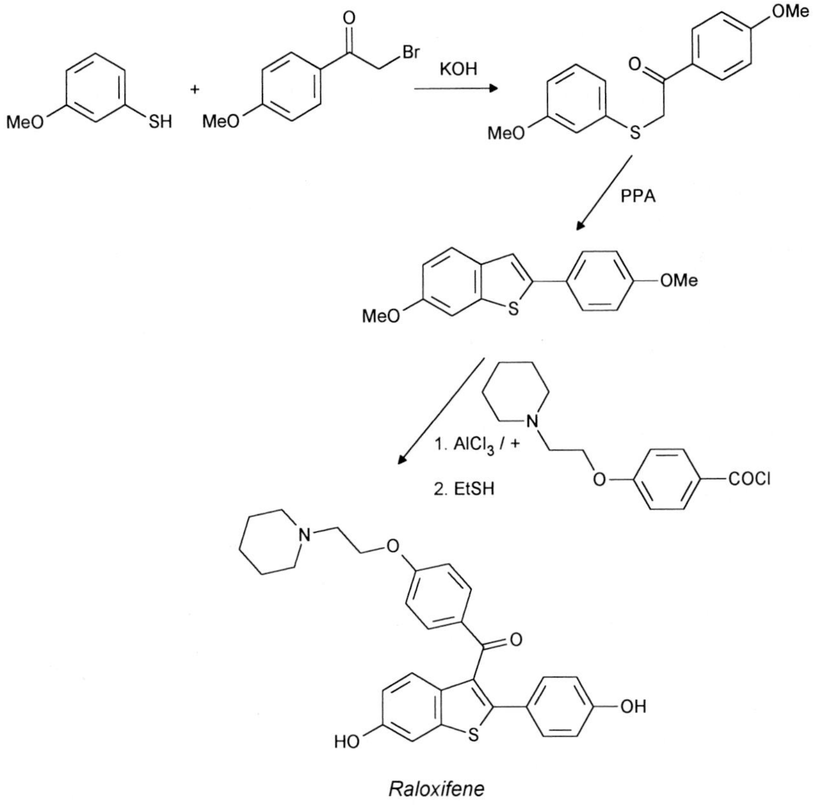

Fig. 8. Synthesis of raloxifene (JONES et al. 1984)

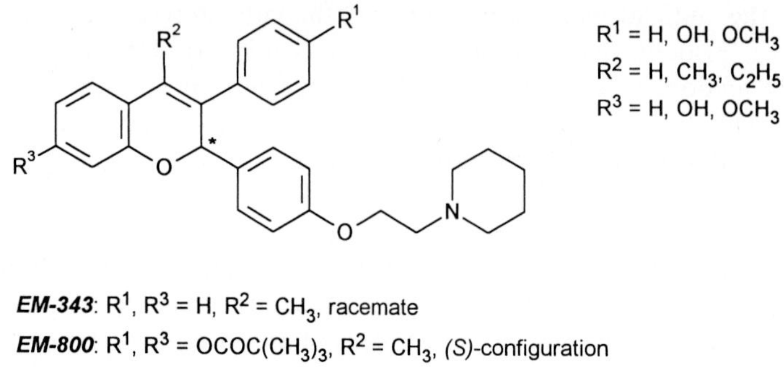

EM-343: R^1, R^3 = H, R^2 = CH_3, racemate
EM-800: R^1, R^3 = OCOC(CH_3)$_3$, R^2 = CH_3, (S)-configuration

Fig.9. General structure of 2,3-diphenyl-1–2H-benzopyran-derived antiestrogens

reaction. This was achieved by converting the 3-bromo-2-phenylbenzothio-phene to the corresponding sulfoxide, so that the bromine could be substituted by various nucleophiles, including the monosubstituted hydroquinone.

The successful application of the 1,2,3-triphenylpropene structure, as a part of a five-membered ring in the synthesis of compounds with binding affinity for the estrogen receptor and with antiestrogenic activity, suggested an extension of these studies to six-membered heterocyclic systems. Durani and coworkers (Saeed et al. 1990; Sharma et al. 1990a, b; Hajela and Kapil 1979) carried out a detailed study of structure–activity relationships in the class of 2,3-diaryl-2H-1-benzopyran (Fig. 9) which will be reviewed together with the work of other groups (Grese et al. 1996; Gauthier et al. 1997) in Chap. 5. C.VI. The combination of all of the favorable structural elements and the separation of the more-active stereoisomer from the less-potent one led to a non-steroidal antiestrogen with high potency (Gauthier et al. 1997). Esterification of the phenolic hydroxy groups with 2,2-dimethylpropanoic acid gave rise to the orally active drug EM-800 (Fig. 9).

C. Diphenylethylene Derivatives

I. 1,1-Diphenylethylene-Derived Agents

The triphenylethylene structure is composed of three aromatic rings linked by a double bond. A large number of investigators have shown that one of the two *cis*-orientated phenyl rings can be omitted without loss of the inherent hormonal activity. However, the endocrine profile changes with this structural modification. An example of a 1,1-diphenylethylene derivative is cyclofenyl (*bis* (4-acetoxyphenyl) cyclohexylidenemethane); (Fig. 10) (Devleeschouwer et al. 1978). Its relative binding affinity (RBA) is close to that of tamoxifen. Under in vivo conditions, the ester functions of cyclofenyl are likely to undergo hydrolytic cleavage by esterases to afford the free hydroxy derivative. The binding affinity of this metabolite for the rat estrogen receptor is about 7% of

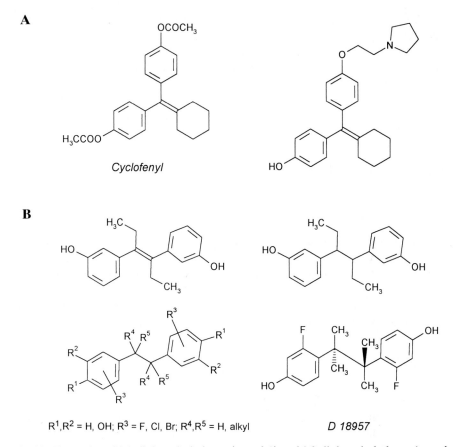

Fig. 10. Examples of 1,1-diphenylethylene- (*panel A*) and 1,2-diphenylethylene- (*panel B*) based antiestrogens

that of estradiol (DEVLEESCHOUWER et al. 1978). The magnitude of this value can only be rationalized by a contribution of both hydroxy groups to the receptor binding. The antagonistic activity of cyclofenyl increased considerably after one of the hydroxy groups had been converted to the pyrrolidinoethyl ether (Fig. 10) (GARG et al. 1983), a side chain that is found in many antiestrogens.

II. 1,2-Diphenylethylene Derivatives

It has already been pointed out that some of the triphenylethylene structures also contain the hydroxylated stilbene as pharmacophore. SCHÖNENBERGER and collaborators (KRANZFELDER et al. 1980, 1982; SCHNEIDER et al. 1980) were the first to exploit the latter structure for the development of antiestrogens with mammary tumor-inhibiting activity. When the hydroxy groups of DES or hexestrol were shifted from the *para* to the *meta* positions (Fig. 10), the estrogenic activity decreased considerably and a partial estrogen antagonism became apparent.

These findings prompted the same group to study the class of hydroxy-lated stilbenes and the corresponding dihydro derivatives, extensively. The structural modifications comprised the positions of the oxygen functions, the aliphatic backbone (HARTMANN et al. 1981, 1985, 1988a, 1988b) and the intro-duction of additional substituents into the aromatic rings (SCHNEIDER et al. 1982a; HARTMANN et al. 1983, 1984) (Fig. 10).

III. 2-Phenylindoles and Related Heterocycles

For the rational design of antiestrogens a versatile basic structure is required which allows both the independent modification of the oxygen functions and the introduction of various side chains, with or without functional groups. The 2-phenylindole system fulfills these requirements and was used as a starting structure for a variety of ligands for the estrogen receptor. It was shown that the aromatic ring mimics the A ring of the steroid. From extensive studies on structure–activity relationships, zindoxifene (5-acetoxy-2-(4-acetoxyphenyl)-1-ethyl-3-methylindole) (Fig. 11) emerged (VON ANGERER et al. 1984). Its endocrine profile is that of an antiestrogen with partial agonist activity. Its

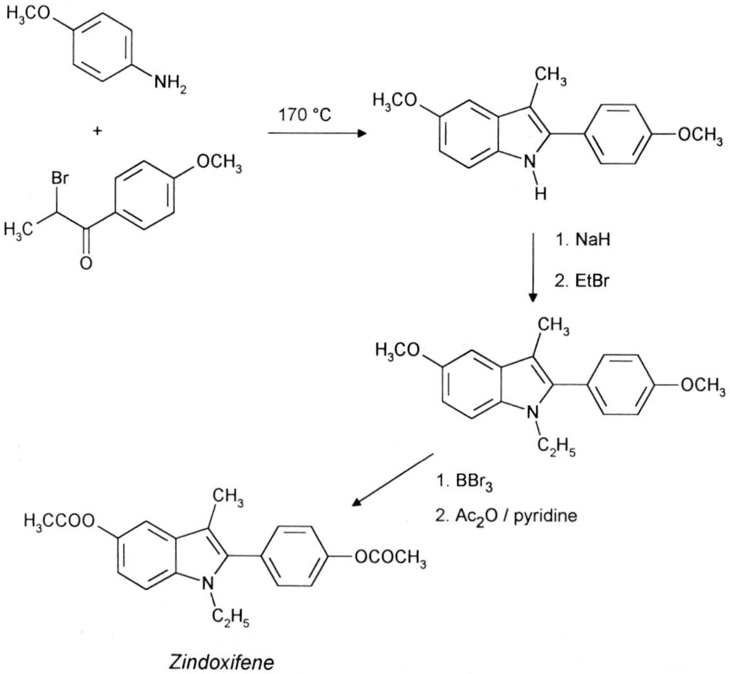

Fig. 11. Synthesis of the 2-phenylindole derivative zindoxifene (VON ANGERER et al. 1984)

potency in pre-clinical studies (VON ANGERER et al. 1984, 1985; SCHNEIDER et al. 1991) provided the basis for preliminary clinical trials (SINDERMANN et al. 1989; STEIN et al. 1990). In the first step of synthesis, the heterocycle with appropriate location of the oxygen functions was formed. The side chain was introduced by the nucleophilic substitution of the halide by the indole anion, followed by the ether cleavage with boron tribromide. For reasons of stability, the phenolic functions were converted into the acetates.

All data accumulated for zindoxifene and related structures showed that the 2-phenylindole system has to be considered as an interesting new lead structure for the design of drugs acting via the estrogen receptor. Early modifications comprised the introduction of chlorine atoms (VON ANGERER and PREKAJAC 1983), the N-benzylation (VON ANGERER and STROHMEIER 1987), and the formation of benzocarbazoles by linking the C-3 of the indole with the 2-phenyl ring (VON ANGERER and PREKAJAC 1986). Replacement of the substituted nitrogen atom by oxygen and sulfur, and concomitant transfer of the side chain to carbon 3 led to the isosteric 2-phenylbenzo[b]furans (ERBER et al. 1991) and 2-phenylbenzo[b]thiophenes (VON ANGERER and ERBER 1992) (Fig. 12). Both types of compounds share many pharmacological properties with the corresponding indoles.

The main disadvantage of these heterocycles was the partial agonist activity which became dominant in the benzothiophene series. In order to eliminate the estrogenic component, side chains with functional groups were

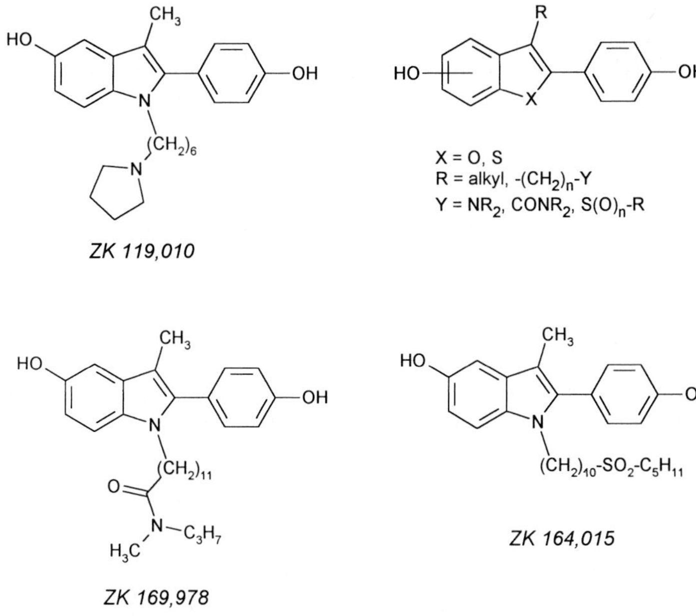

ZK 119,010

X = O, S
R = alkyl, -(CH$_2$)$_n$-Y
Y = NR$_2$, CONR$_2$, S(O)$_n$-R

ZK 169,978

ZK 164,015

Fig. 12. Analogues of zindoxifene

introduced in all of the heterocyclic systems (von Angerer et al. 1995) (Fig. 12). In the 2-phenylindole series, an extensive study on structure–activity relationships was performed (von Angerer et al. 1994; Biberger and von Angerer 1996, 1998) which is the subject of Chap. 5.C.II.

D. Steroidal Compounds

I. 7α-Substituted Estradiol Derivatives

For a long time, steroidal structures were regarded as less attractive for the development of antiestrogens than non-steroidal systems, such as the triphenylethylenes, despite the fact that anordiol (2α,17α-diethynyl-A-nor-5α-androstane-2β,17β-diol), an A-ring-modified steroid that is clinically used for contraception, acts as an antiestrogen and possesses only weak agonist activity (Chatterton Jr et al. 1989; Peters et al. 1995). One of the contributing reasons has been the difficulty of modifying the basic structure and introducing new substituents in appropriate positions. Another aspect that was thought to be a problem is the elimination of the inherent agonist activity of the natural estrogens. Therefore, it was not surprising that the first access to potent steroidal estrogen antagonists was discovered by chance, rather than by theoretical considerations.

Raynaud et al. (1974) demonstrated that 7α-estradiol derivatives with a long unbranched alkyl chain retained the high affinity for the estrogen receptor. Based on this knowledge, Baulieu's group prepared specific adsorbents for the purification of estrogen receptors (Bucourt et al. 1978) using estradiol linked to a polyamide or agarose matrix by long spacer chains (14 atoms). These investigations prompted Wakeling and Bowler (1988a, b; Bowler et al. 1989), at ICI, to use 7-substituted estradiol as a lead for novel antiestrogens). Three structural parameters were modified: the orientation at C-7 of the steroid, the functional group in the side chain and the length of the spacer group that linked both elements. ICI 164,384 (N-(n-butyl)-N-methyl-11-(3,17β-dihydroxyestra-1,3,5(10)-trien-7α-yl) undecanamide) (Fig. 13) with all the characteristics of a pure antiestrogen emerged from these studies (Wakeling and Bowler 1988a).

The original route to achieve this estradiol derivative is outlined in Fig. 14. The starting compound was 17β-acetoxy-estra-(4,6)-dien-3-one, which allowed the 1,6-addition of a copper(I)-activated Grignard reagent across the dienone system (Bowler et al. 1989). The resulting stereoisomeres were separated by chromatography. Aromatization of the A ring was achieved by the treatment with cupric bromide. The modification of the functional group in the side chain was performed according to standard procedures. This route of synthesis was used for a number of different compounds, the structure–activity relationships of which will be discussed in the Chapt. 5. The interesting biological profile of ICI 164,384 prompted other investigators to search for a stereoselective access to 7α-substituted estradiol derivatives. An appropriate

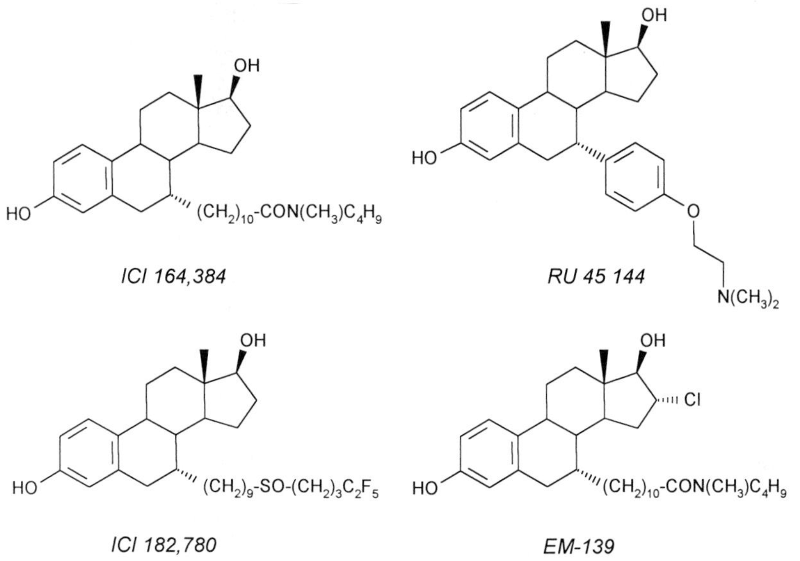

Fig. 13. 7α-substituted estradiol derivatives with antiestrogenic activity

approach has been the introduction of functional groups, such as an oxo function (TEDESCO et al. 1997) or a phenylsulfonyl group (KUENZER et al. 1994), into position 6 prior to the substitution reaction at carbon 7.

In one of the first studies, the side chain of tamoxifen was introduced into the 7α-position of estradiol leading to RU 45 144 (Fig. 13) which showed antiestrogenic activity (BOUHOUTE and LECLERCQ 1994). The discovery of ICI 164,384 as a new lead structure for the development of pure estrogen antagonists prompted LEVESQUE et al. (1991) to modify this structure slightly, by introducing halogen atoms into the 16α-position of the steroid. The most interesting compound synthesized was the chloro derivative EM-139 (Fig. 13).

The importance of the functional group in the side chain has already been stressed. Therefore, it had been a reasonable approach to increase the in vivo potency of ICI 164,384 by structural variation of this fragment. The most active compound that emerged from these studies was ICI 182,780 (7α-[9-(4,4,5,5,5-pentafluoropentylsulfinyl) nonyl] estra-1,3.5(10)-triene-3,17β-diol) (Fig. 13), in which the amide function was replaced by a sulfoxide group. This modification and the introduction of fluoro substituents at the terminal carbon atoms increased the in vivo potency considerably (WAKELING et al. 1991). The characteristics of a pure antiestrogen of the parent compound ICI 164,384 was retained. The sulfoxide ICI 182,780 is a mixture of two isomeric forms due to asymmetric sulfur function. No attempts, however, have been made to separate the diastereomers and study their activities separately.

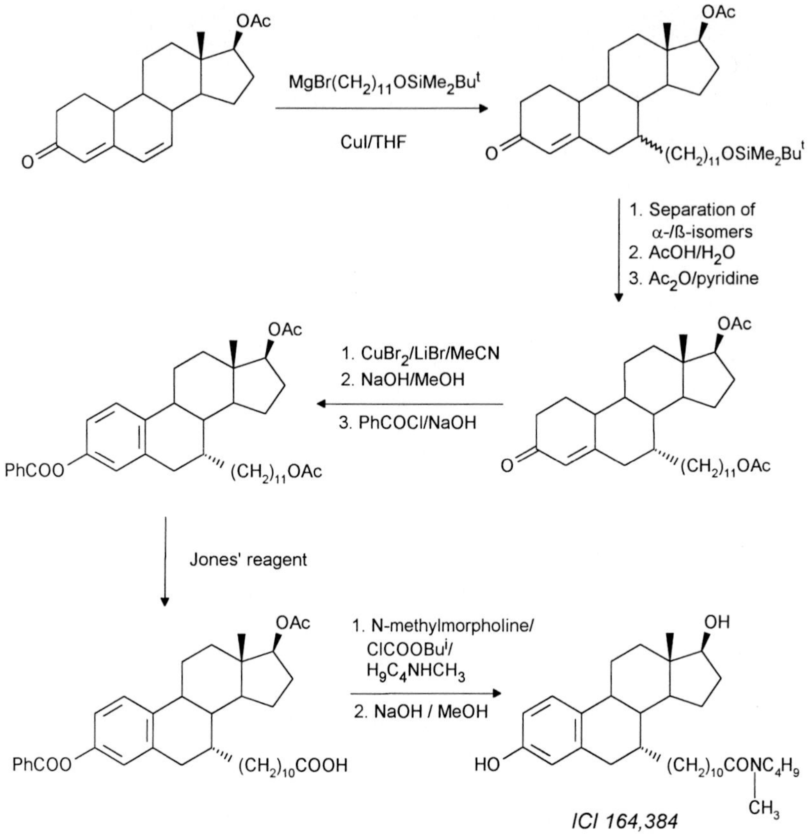

Fig. 14. Synthesis of the steroid ICI 164,384 (Bowler et al. 1989)

II. 11β-Substituted Estradiol Derivatives

The investigations of Wakeling and colleagues (1988a,b, 1991) and the experience in the development of antiprogestins with an aromatic substituent in the 11β-position of the steroid (Teutsch et al. 1988) prompted Van de Velde and collaborators, at Roussel UCLAF, to develop antiestrogens by attaching suitable side chains to carbon-11 of estradiol. The first 11β-substituted estradiol derivative studied (Claussner et al. 1992) carried the side chain of ICI 164,384 (Fig. 15).

Since the β-orientation of the side chain is essential for receptor binding and biological activity, a stereoselective synthetic route to these derivatives was developed. The key intermediate is the 5,10-epoxide of the norandrostenone system with an α-orientation. The oxygen function in position 17 can be protected as an acetoxy group or, in the case of a carbonyl function, as a trimethylsilyl cyanohydrine. The general procedure for the introduction of a side chain by a copper(I)-assisted Grignard reaction is outlined in Fig. 16. An

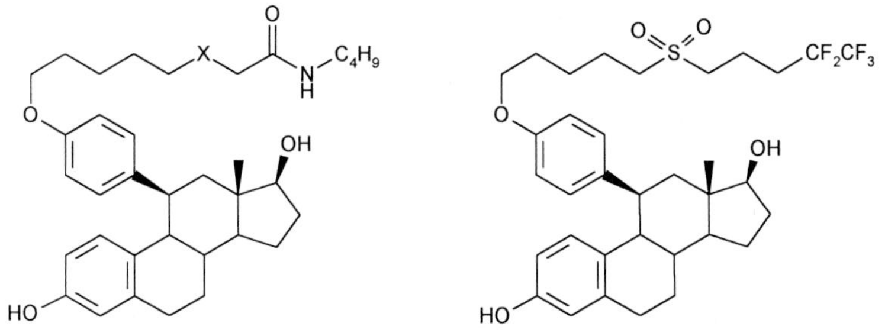

X = CH_2, O, S, SO, SO_2 *RU 58 668*

Fig. 15. 11β-substituted estradiol derivatives with antiestrogenic activity

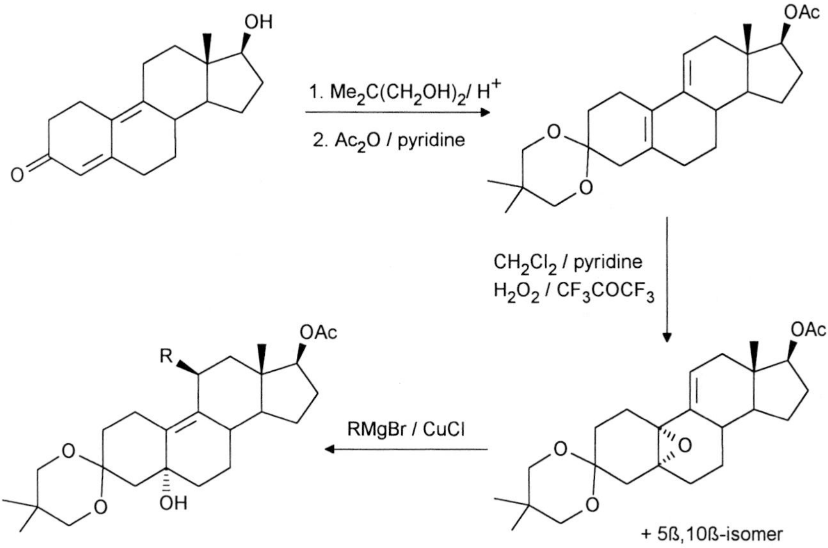

Fig. 16. General route to production of 11β-substituted estradiol derivatives

appropriate functional group in the organo-magnesium reagent allows further modification of the side chain.

The progress that was made in the series of 7α-estradiol derivatives, by the replacement of the amide function with a sulfoxide group, inspired the investigators at Roussel UCLAF to perform a similar modification in the 11β-series (VAN DE VELDE et al. 1994). The derivative that was selected for detailed investigations was RU 58 668 (11β-[4-[5-[(4,4,5,5,5-pentafluoropentyl) sulfonylpentyloxy] phenyl]-estra-1,3,5(10)-triene-3,17β-diol) (VAN DE VELDE et al. 1994, 1996). Starting from 11β-(4-hydroxyphenyl) estrone, the side chain was built up in several steps as outlined in Fig. 17.

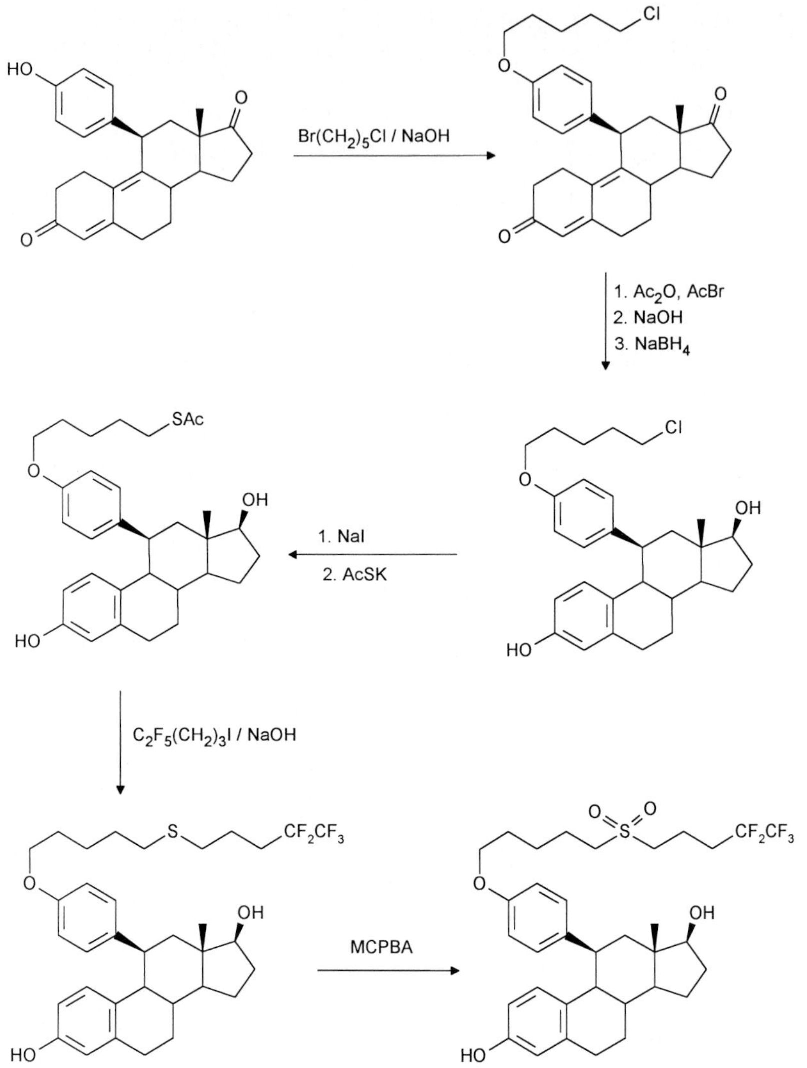

Fig. 17. Synthesis of the steroid RU 58 668 (van de Velde et al. 1994). *MCPBA* m-chloroperbenzoic acid

E. Conclusion

In concluding this chapter on antiestrogens and partial agonists, it can be stated that the goal of many, often futile, investigations over the last two decades to discover pure antiestrogens for the treatment of estrogen-related disorders has been reached. Both steroids, derived from 17β-estradiol and non-steroidal structures, were identified as pure estrogen antagonists in conventional endocrine assays, such as the uterine-weight test. Recently, more advanced techniques have been introduced that allow one to quantify transcriptional activity of estrogen-dependent genes in cellular systems. The complete lack of agonistic activity at the molecular level is in agreement with a complete suppression of estrogen action in animals (VON ANGERER et al. 1994). The clinical efficacy of pure antiestrogens, though expected from the preclinical data, has not yet been proven and their role in the management of breast cancer has to be defined in clinical studies.

References

Acton D, Hill G, Tait BS (1983) Tricyclic triarylethylene antiestrogens: Dibenz[b,f]oxepines, dibenzo[b,f]thiepins, dibenzo[a,e]cyclooctenes and dibenzo[b,f]thiocins. J Med Chem 26:1131–1137

Biberger C, von Angerer E (1996) 2-Phenylindoles with sulfur containing side chains. Estrogen receptor affinity, antiestrogenic potency, and antitumor activity. J Steroid Biochem Molec Biol 58:31–43

Biberger C, von Angerer E (1998) 1-Benzyl-2-phenylindole- and 1,2-diphenylindole-based antiestrogens. Estimation of agonist and antagonist activities in transfection assays. J Steroid Biochem Mol Biol 64:277–285

Black LJ, Sato M, Rowley ER, Magee DE, Bekele A, Williams DC, Cullinan GJ, Bendele R, Kauffman RF, Bensch WR, Frolik CA, Termine JD, Bryant HU (1994) Raloxifene (LY139481 HCl) prevents bone loss and reduces serum cholesterol without causing uterine hypertrophy in ovariectomized rats. J Clin Invest 93:63–69

Borvendég J (1985) GYKI-13504. Drugs Fut 10:395–396

Bouhoute A, Leclercq G (1994) Estradiol derivatives bearing the side-chain of tamoxifen antagonize the association between the estrogen receptor and calmodulin. Biochem Pharmacol 47:748–751

Bowler J, Lilley TJ, Pittam JD, Wakeling AE (1989) Novel steroidal pure antiestrogens. Steroids 54:71–99

Brown SD, Armstrong RW (1997) Parallel synthesis of tamoxifen and derivatives on solid support via resin capture. J Org Chem 62:7076–7077

Bucourt R, Vignau M, Torelli V, Richard-Foy H, Geynet C, Secco-Millet C, Redeuilh G, Baulieu EE (1978) New biospecific adsorbents for the purification of estradiol receptor. J Biol Chem 253:8221–8228

Chatterton RT, Jr., Berman C, Walters NN (1989) Anti-uterotrophic and folliculostatic activities of anordiol ($2\alpha,17\alpha$-diethynyl-A-nor-5α-androstane-$2\beta,17\beta$-diol). Contraception 39:291–297

Claussner A, Nédélec L, Nique F, Philibert D, Teutsch G, Van de Velde P (1992) 11β-Amidoalkyl estradiols, a new series of pure antiestrogens. J Steroid Biochem Molec Biol 41:609–614

Coe PL, Scriven CE (1986) Crossed coupling of functionalised ketones by low valent titanium (the McMurry reaction): a new stereoselective synthesis of tamoxifen. J C S Perkin Trans 1:475–477

Crenshaw RR, Jeffries AT, Luke GM, Cheney LC, Bialy G (1971) Potential antifertility agents. I. Substituted diaryl derivatives of benzo[b]thiophenes, benzofurans, 1-H-2-benzo-thiopyrans, 2-H-1-benzothiopyrans. J Med Chem 14:1185–1189

Day BW, Magarian RA, Jain PT, Pento JT, Mousissian GK, Meyer KL (1991) Synthesis and biological evaluation of a series of 1,1-dichloro-2,2,3-triarylcyclopropanes as pure antiestrogens. J Med Chem 34:842–851

Devleeschouwer N, Leclercq G, Danguy A, Heuson JC (1978) Antitumour effect of cyclofenil (F6066) on DMBA-induced rat mammary tumors. Europ J Cancer 14:721–723

Dietel M, Löser R, Röhlke P, Jonat W, Niendorf A, Gerding D, Kohr A, Hölzel F, Arps
    H (1989) Effect of continuous vs intermittent application of 3-OH-tamoxifen or
    tamoxifen on the proliferation of the human breast cancer cell line MCF-7 M1. J
    Cancer Res Clin Oncol 115:36–40
Durani N, Jain R, Saeed A, Dikshit DK, Durani S, Kapil RS (1989) Structure-activity
    relationship of antiestrogens: A study using triarylbutenone, benzofuran, and
    triarylfuran analogues as models for triarylethylenes and triarylpropenones. J Med
    Chem 32:1700–1707
Erber S, Ringshandl R, von Angerer E (1991) 2-Phenylbenzo[b]furans: Relationship
    between structure, estrogen receptor affinity and cytostatic activity against
    mammary tumor cells. Anti-Cancer Drug Des 6:417–426
Garg S, Bindal RD, Durani S, Kapil RS (1983) Structure-activity relationship of
    estrogens: A study involving cyclofenyl as model compound. J Steroid Biochem
    18:89–95
Gauthier S, Caron B, Cloutier J, Dory YL, Favre A, Larouche D, Mailhot J, Ouellet C,
    Schwerdtfeger A, Leblanc G, Martel C, Simard J, Mérand Y, Bélanger A, Labrie
    C, Labrie F (1997) (S)-(+)-[4-[7-(2,2-dimethyl-1-oxopropoxy)-4-methyl-2-[4-[2-(1-
    piperidinyl)ethoxy]phenyl-2H-1-benzopyran-3-yl]phenyl]-2,2-dimethyl-
    propanoate (EM-800): A highly potent, specific and orally active non-steroidal
    antiestrogen. J Med Chem 40:2117–2122
Grese TA, Cole HW, Magee DE, Phillips DL, Shetler PK, Short LL, Glasebrook AL,
    Bryant HU (1996) Conversion of the phytoestrogen coumestrol into a selective
    estrogen receptor modulator (SERM) by attachment of an amine-containing
    sidechain. Bioorg Med Chem Lett 6:2683–2686
Grese TA, Cho S, Finley DR, Godfrey AG, Jones CD, Lugar CW, Martin MJ, Matsumoto
    K, Pennington LD, Winter MA, Adrian MD, Cole HW, Magee DE, Phillips DL,
    Rowley ER, Short LL, Glasebrook AL, Bryant HU (1997) Structure-activity rela-
    tionships of selective estrogen receptor modulators: modifications to the 2-
    arylbenzothiophene core of raloxifene. J Med Chem 40:146–167
Hajela K, Kapil RS (1997) Synthesis and post-coital contraceptive activity of a new
    series of substituted 2,3-diaryl-2H-1-benzopyrans. Eur J Med Chem 32:135–142
Harper MJK, Walpole AL (1966) Contrasting endocrine activities of cis and trans
    isomers in a series of substituted triphenylethylenes. Nature (London) 212:87
Hartmann RW, Buchborn H, Kranzfelder G, Schönenberger H (1981) Potential antie-
    strogens. Synthesis and evaluation of mammary tumor inhibiting activity of 1,2-
    dialkyl-1,2-bis(3′-hydroxyphenyl)ethanes. J Med Chem 24:1192–1197
Hartmann RW, Schwarz W, Schönenberger H (1983) Ring-substituted 1,2-dialkylated
    1,2-bis(hydroxyphenyl)ethanes. 1. Synthesis and estrogen receptor binding affinity
    of 2,2′- and 3,3′-disubstituted hexestrols. J Med Chem 26:1137–1144
Hartmann RW, Heindl A, Schwarz W, Schönenberger H (1984) Ring-substituted 1,2-
    dialkylated 1,2-bis(hydroxyphenyl)ethanes. 2. Synthesis, estrogen receptor binding
    affinity, and evaluation of antiestrogenic and mammary tumor inhibiting activity
    of 2,2′-disubstituted butestrol and 6,6′-disubstituted metabutestrols. J Med Chem
    27:819–824
Hartmann RW, Schwarz W, Heindl A, Schönenberger H (1985) Ring-substituted
    1,1,2,2-tetraalkylated 1,2-bis(hydroxyphenyl)ethanes. 4. Synthesis, estrogen
    receptor binding affinity, and evaluation of antiestrogenic and mammary tumor
    inhibiting activity of symmetrically disubstituted 1,1,2,2-tetramethyl-1,2-
    bis(hydroxyphenyl)ethanes. J Med Chem 28:1295–1301
Hartmann RW, Schwarz W, Schneider MR, Engel J, Schönenberger H (1988a) D 18954.
    Drug Fut 13:720–721
Hartmann RW, vom Orde H-D, Heindl A, Schönenberger H (1988b) N-(4-Hydroxy-
    phenyl)-N-(1,1,1-trifluor-2-propyl)-4-hydroxybenzamid: Synthese und pharmakol-
    ogische Bewertung eines neuen Antiestrogens. Arch Pharm (Weinheim)
    321:497–501
Huppert LC (1979) Induction of ovulation with clomiphene citrate. Fert Steril 31:1–8

Iino Y, Takai Y, Ando T, Ohwada S, Yokoe T, Sugamata N, Takei H, Horiguchi J, Iijima K, Morishita Y (1994) A new triphenylethylene derivative, TAT-59; hormone receptors; insulin-like growth factor 1; and growth suppression of hormone-dependent MCF-7 tumors in athymic mice. Cancer Chemother Pharmacol 34: 372–376

Jain PT, Pento JT, Magarian RA (1993) The influence of a novel cyclopropyl antiestrogen (compound 7a) on human breast cancer cells in culture. Breast Cancer Res Treat 25:225–233

Johnston SR, Riddler S, Haynes BP, A'Hern R, Smith IE, Jarman M, Dowsett M (1997) The novel anti-estrogen idoxifene inhibits the growth of human MCF-7 breast cancer xenografts and reduces the frequency of acquired anti-estrogen resistance. Br J Cancer 75:804–809

Jones CD, Suarez T, Massey EH, Black LJ, Tinsley FC (1979) Synthesis and antiestrogenic activity of [3,4-dihydro-2-(4-methoxyphenyl)-1-naphthalenyl][4-(2-(1-pyrrolidinyl)ethoxy)-4-phenyl]-methanone, methanesulfonic acid salt. J Med Chem 22:962–966

Jones CD, Jevnikar MG, Pike AJ, Peters MK, Black LJ, Thompson AR, Falcone JF, Clemens JA (1984) Antiestrogens. 2. Structure-activity studies in a series of 3-aroyl-2-arylbenzo[b]thiophene derivatives leading to [6-hydroxy-2-(4-hydroxyphenyl)benzo[b]thien-3-yl][4-[2-(1-piperidinyl)ethoxy]phenyl]methanone hydrochloride (LY156758), a remarkably effective estrogen antagonist with only minimal intrinsic estrogenicity. J Med Chem 27:1057–1066

Jordan VC, Haldeman B, Allan KE (1981) Geometric isomers of substituted triphenylethylenes and antiestrogenic action. Endocrinology (Baltimore) 108:1353–1361

Jordan VC, Gosden B (1983) Inhibition of the uterotrophic activity of estrogens and antiestrogens by the short acting antiestrogen LY 117018. Endocrinology 113:463–468

Kamboj VP, Ray S, Dhawan BN (1992) Centchroman. Drugs Today 28:227–232

Katzenellenbogen JA, Carlson KE, Katzenellenbogen BS (1985) Facile geometric isomerization of phenolic non-steroidal estrogens and antiestrogens: limitations to the interpretation of experiments characterizing the activity of individual isomers. J Steroid Biochem 22:589–596

Kaufmann M, Possinger K, Illiger HJ, Schmid H, Hietanen T, Johansson R, Pyrhönen S, Valavaara R, Sindermann H (1989) Toremifene: clinical phase II/III trials of a new antiestrogen in patients with advanced breast cancer. Contrib Oncol 37:50–57

Kranzfelder G, Schneider M, von Angerer E, Schönenberger H (1980) Entwicklung neuer Antiöstrogene vom Typ des 3,3'-Dihydroxy-α,β-diäthylstilbens und ihre Prüfung am DMBA-induzierten, hormonabhängigen Mammacarcinom der SD-Ratte. J Cancer Res Clin Oncol 97:167–186

Kranzfelder G, Hartmann RW, von Angerer E, Schönenberger H, Bogden AE (1982) 3,4-Bis(3'-hydroxyphenyl)hexane- a new mammary tumor-inhibiting compound. J Cancer Res Clin Oncol 103:165–180

Kuiper GG, Carlsson B, Grandien K, Enmark E, Haggblad J, Nilsson S, Gustafsson JA (1997) Comparison of the ligand binding specificity and transcript tissue distribution of estrogen receptors alpha and beta. Endocrinology 138:863–870

Kuiper GG, Gustafsson JA (1997) The novel estrogen receptor-beta subtype: potential role in the cell- and promoter-specific actions of estrogens and anti-estrogens. FEBS Lett 410:87–90

Kuenzer H, Thiel M, Sauer G, Wiechert R (1994) A new stereoselective approach to C(7)-alkylated estra-1,3,5(10)-triene derivatives. Tetrahedron Lett 35:1691–1694

Lednicer D, Babcock JC, Marlati PE, Lyster SC, Duncan GW (1965) Mammalian antifertility agents. I. Derivatives of 2,3-diphenylindenes. J Med Chem 8:52–57

Lerner LJ (1981) The first non-steroidal antiestrogen – Mer-25. In: Sutherland RL, Jordan VC (eds) Non-steroidal antiestrogens: Molecular pharmacology and antitumor activity. Academic Press, Sidney, pp 1–16

Lerner LJ, Jordan VC (1990) Development of antiestrogens and their use in breast cancer: Eighth Cain memorial award lecture. Cancer Res 50:4177–4189

Levesque C, Merand Y, Dufour J-M, Labrie C, Labrie F (1991) Synthesis and biological activity of new halo-steroidal antiestrogens. J Med Chem 34:1624–1630

Magarian RA, Overacre LB, Singh S, Meyer KL (1994) The medicinal chemistry of nonsteroidal antiestrogens: a review. Curr Med Chem 1:61–104

Martin MJ, Grese TA, Glasebrook AL, Matsumoto K, Pennington LD, Philips DL, Short LL (1997) Versatile raloxifene triflates. Bioorg Med Chem Lett 7:887–892

McCague R, Kuroda R, Leclercq G, Stoessel S (1986) Synthesis and estrogen receptor binding of 6,7-dihydro-8-phenyl-9-[4-{2-(dimethylamino)ethoxy}phenyl]-5H-benzocycloheptene, a non-isomerizable analogue of tamoxifen. X-ray crystallographic studies. J Med Chem 29:2053–2059

McCague R, Leclercq G, Jordan VC (1988) Nonisomerizable analogues of (Z)- and (E)-4-hydroxytamoxifen. Synthesis and endocrinological properties of substituted diphenylbenzocycloheptenes. Med Chem 31:1285–1290

McCague R, Leclercq G, Legros N, Goodman J, Blackburn GM, Jarmann M, Foster AB (1989) Derivatives of tamoxifen. Dependence of antiestrogenicity on the 4-substituent. J Med Chem 32:2527–2533

McCague R, Leclercq G (1987) Synthesis, conformational considerations, and estrogen receptor binding of diastereoisomers and enantiomers of 1-[4-[2-(dimethylamino)ethoxy]phenyl]-1,2-diphenylbutane (dihydrotamoxifen). J Med Chem 30:1761–1767

Miller R, Al-Hassan MI (1985) Stereospecific synthesis of (Z)-tamoxifen via carbometalation of alkynylsilanes. J Org Chem 50:2121–2123

Mittal S, Durani S, Kapil RS (1985) Structure-activity relationship of estrogens: receptor affinity and estrogen antagonist activity of certain (E)- and (Z)-1,2,3-triaryl-2-propen-1-ones. J Med Chem 28:492–497

Palkowitz AD, Glasebrook AL, Thrasher KJ, Hauser KL, Short LL, Phillips DL, Muchl BS, Sato M, Shetler PK, Cullinan GJ, Pell TR, Bryant HU (1997) Discovery and synthesis of [6-hydroxy-3-[4-[2-(1-piperidinyl)ethoxy]phenoxy]-2-(hydroxyphenyl)]benzo[b]thiophene: a novel, highly potent, selective estrogen receptor modulator. J Med Chem 40:1407–1416

Peters AJ, Wentz AC, Kazer RR, Jeyendran RS, Chatterton RTJ (1995) Estrogenic and antiestrogenic activities of anordiol: a comparison of uterine and vaginal responses with those of clomiphene citrate. Contraception 52:195–202

Plowman PN (1993) Tamoxifen as adjuvant therapy in breast cancer. Current status. Drugs 46:819–833

Pons M, Michel F, de Paulet C, Gilbert J, Miquel JF, Précigoux G, Hospital M, Ojasoo T, Raynaud JP (1984) Influence of new hydroxylated triphenylethylene (TPE) derivatives on estradiol binding to uterine cytosol. J Steroid Biochem Molec Biol 20:137–145

Potter GA, McCague R (1990) Highly stereoselective access to an (E)-vinyl bromide from an aryl ketone leads to short syntheses of (Z)-tamoxifen and important substituted derivatives. J Org Chem 55:6184–6187

Powles TJ (1997) Efficacy of tamoxifen as treatment of breast cancer. Semin Oncol 24:S1–48

Raynaud J-P, Azadian-Boulanger G, Bucourt R (1974) Anticorps specifiques de l'estradiol. J Pharmacol (Paris) 5:27–40

Richardson DN (1988) The history of nolvadex. Drug Des Deliv 3:1–14

Robertson DW, Katzenellenbogen JA, Long DJ, Rorke EA, Katzenellenbogen BS (1982) Tamoxifen antiestrogens. A comparison of the activity, pharmacokinetics, and metabolic activation of the cis and trans isomers of tamoxifen. J Steroid Biochem 16:1–13

Robinson SP, Jordan VC (1989) Antiestrogenic action of toremifene on hormone-dependent, -independent, and heterogeneous breast tumor growth in athymic mice. Cancer Res 49:1758–1762

Ruenitz PC, Bagley JR, Mokler CM (1982) Estrogenic and antiestrogenic activity of monophenolic analogues of tamoxifen, (Z)-2-[p-(1,2-diphenyl-1-butenyl) phenoxy]-N,N-dimethylethylamine. J Med Chem 25:1056–1060

Ruenitz PC, Thompson CB, Srivatsan V (1989) Characterization of MCF 7 breast cancer cell growth inhibition by the antiestrogen nitromifene (CI 628) and selected metabolites. J Steroid Biochem Molec Biol 33:365–369

Saeed A, Sharma AP, Durani N, Jain R, Durani S, Kapil RS (1990) Structure-activity relationship of antiestrogens. Studies on 2,3-diaryl-1-benzopyrans. J Med Chem 33:3210–3216

Salman M, Ray S, Anand N, Agarwal AK, Singh MM, Selty BS, Kamboj VP (1986) Studies in antifertility agents. 50. Stereoselective binding of d- and l-centchromans to estrogen receptors and their antifertility activity. J Med Chem 19:1801–1803

Schneider M, von Angerer E, Kranzfelder G, Schönenberger H (1980) Mammatumorhemmende Antiöstrogene vom Typ des 3,3'-Dihydroxy-α,β-dialkylstilbens. Arch Pharm (Weinheim) 313:919–925

Schneider MR, Schönenberger H, Michel RT, Fortmeyer HP (1982a) Synthesis and evaluation of catechol analogs of diethylstilbestrol on a hormone-dependent human mammary carcinoma implanted in nude mice. J Cancer Res Clin Oncol 104:219–227

Schneider MR, von Angerer E, Schönenberger H, Michel RT, Fortmeyer HP (1982b) 1,1,2-Triphenylbut-1-enes: Relationship between structure, estradiol receptor affinity, and mammary tumor inhibiting properties. J Med Chem 25:1070–1077

Schneider MR, Schiller CD, Humm A, von Angerer E (1991) Effect of zindoxifene on experimental prostatic tumours of the rat. J Cancer Res Clin Oncol 117: 33–36

Sharma AP, Saeed A, Durani S, Kapil RS (1990a) Structure-activity relationship of antiestrogens. Effect of the side chain and its position on the activity of 2,3-diaryl-2H-1-benzopyrans. J Med Chem 33:3216–3222

Sharma AP, Saeed A, Durani S, Kapil RS (1990b) Structure-activity relationship of antiestrogens. Phenolic analogues of 2,3-diaryl-2H-1-benzopyrans. J Med Chem 33:3222–3229

Sindermann H, Coombes RC, Paridaens R, Peukert M, von Angerer E (1989) Zindoxifene: Clinical phase I studies in patients with advanced breast cancer. Contr Oncol 317:45–49

Stein RC, Dowsett M, Cunningham DC, Davenport J, Ford HT, Gazet J-C, von Angerer E, Coombes RC (1990) Phase I/II study of the antiestrogen zindoxifene (D 16726) in the treatment of advanced breast cancer. A Cancer Research Campaign Phase I/II Clinical Trials Commitee study. Br J Cancer 61:451–453

Tedesco R, Katzenellenbogen JA, Napolitano E (1997) An expeditious route to 7α-substituted estradiol derivatives. Tetrahedron Lett 38:7997–8000

Teutsch G, Ojasoo T, Raynaud JP (1988) 11-substituted steroids, an original pathway to antihormones. Steroid Biochem Molec Biol 31:549–565

Van de Velde P, Nique F, Bouchoux F, Brémaud J, Hameau M-C, Lucas D, Moratille C, Viet S, Philibert D, Teutsch G (1994) RU 58 668, a new pure antiestrogen inducing a regression of human mammary carcinoma implanted in nude mice. J Steroid Biochem Molec Biol 48:187–196

Van de Velde P, Nique F, Planchon P, Prevost G, Bremaud J, Hameau MC, Magnien V, Philibert D, Teutsch G (1996) RU 58 668: further in vitro and in vivo pharmacological data related to its antitumoral activity. J Steroid Biochem Mol Biol 59:449–457

von Angerer E, Prekajac J, Strohmeier J (1984) 2-Phenylindoles. Relationship between structure, estrogen receptor affinity, and mammary tumor inhibiting activity in the rat. J Med Chem 27:1439–1447

von Angerer E, Prekajac J, Berger MR (1985) The inhibitory effect of 5-acetoxy-2-(4-acetoxyphenyl)-1-ethyl-3-methylindole (D 16726) on estrogen-dependent mammary tumors. Eur J Cancer Clin Oncol 21:531–537 von Angerer E, Biberger C, Holler E, Koop R, Leichtl S (1994) 1-Carbamoylalkyl-2-phenylindoles: Relationship between side chain structure and estrogen antagonism. J Steroid Biochem Molec Biol 49:51–62

von Angerer E (1995) The estrogen receptor as a target for rational drug design. R.G. Landes Company, Austin, Texas

von Angerer E, Biberger C, Leichtl S (1995) Studies on heterocycle-based pure estrogen antagonists. Ann N Y Acad Sci 761:176–191

von Angerer E, Erber S (1992) 3-Alkyl-2-phenylbenzo[b]thiophenes: Nonsteroidal estrogen antagonists with mammary tumor inhibiting activity. J Steroid Biochem Molec Biol 41:557–562

von Angerer E, Prekajac J (1983) 2-(Hydroxyphenyl)indoles: a new class of mammary tumor inhibiting compounds. J Med Chem 26:113–116

von Angerer E, Prekajac J (1986) Benzo[a]carbazole derivatives. Synthesis, estrogen receptor binding affinities, and mammary tumor inhibiting activity. Med Chem 29:380–386

von Angerer E, Strohmeier J (1987) 2-Phenylindoles. Effect of N-benzylation on estrogen receptor affinity, estrogenic properties, and mammary tumor inhibiting activity. J Med Chem 30:131–136

Wakeling AE, Dukes M, Bowler J (1991) A potent specific pure antiestrogen with clinical potential. Cancer Res 51:3867–3873

Wakeling AE, Bowler J (1988a) Novel antiestrogens without partial agonist activity. J Steroid Biochem 31:645–653

Wakeling AE, Bowler J (1988b) Biology and mode of action of pure antiestrogens. J Steroid Biochem Molec Biol 30:141–147

Willson TM, Henke BR, Momtahen TM, Charifson PS, Batchelor KW, Lubahn DB, Moore LB, Oliver BB, Sauls HR, Triantafillou JA, Wolfe SG, Baer PG (1994) 3-[4-(1,2-diphenylbut-1-enyl)phenyl]acrylic acid: A non-steroidal estrogen with functional selectivity for bone over uterus in rats. J Med Chem 37:1550–1552

Willson TM, Norris JD, Wagner BL, Asplin I, Baer P, Brown HR, Jones SA, Henke B, Sauls H, Wolfe S, Morris DC, McDonnell DP (1997) Dissection of the molecular mechanism of action of GW5638, a novel estrogen receptor ligand, provides insights into the role of estrogen receptor in bone. Endocrinology 138:3901–3911

Wiseman LR, Goa KL (1997) Toremifene. A review of its pharmacological properties and clinical efficacy in the management of advanced breast cancer. Drugs 54:141–160

CHAPTER 5

Structure–Activity Relationships

E. von Angerer

A. Introduction

Estrogens and antiestrogens are understood to exert their hormonal action after binding to the estrogen receptor (ER), a member of the nuclear receptor superfamily. Until recently, the knowledge of the shape of the full-length ER and the structure of the hormone-binding domain has been rather vague, due to the lack of cristallographic data. Information on the binding mode has been obtained from the three-dimensional structure of various ligands (Wiese and Brooks 1994). In October 1997, Brzozowski et al. (1997) published the crystal structures of the ligand-binding domain of the human ER (residues Ser 301 to Thr 553) in complex with the endogenous ligand 17β-estradiol and raloxifene as representative for a non-steroidal antagonist. Agonist and antagonist bind at the same site within the core of the ligand-binding domain, but demonstrate different binding modes, which induce distinct conformations in the transactivating domain 2. These structural data provide an ideal basis for the elucidation of the mode of action of antiestrogens and the rational design of new ligands.

Most of the binding data have been generated in competitive binding assays with radioisotopically labeled estradiol. The affinity of the ligand to the ER is generally expressed as relative binding affinity (RBA) in relation to the natural hormone estradiol. Sources of the ER are usually uteri or breast cancer tissues of various species. The hormonal or antihormonal effects following the activation of the receptor can be assessed in various in vitro and in vivo systems.

The following review on structure–activity relationships (SARs) will mainly deal with binding affinities of representative series of steroidal and nonsteroidal ligands of the ER and their agonist and antagonist activities as the result of gene activation or repression. The pharmacological profile based on these characteristics provides the basis for the clinical use of these agents. Indications are control of fertility, treatment of estrogen-dependent diseases, such as carcinoma of the breast, and amelioration of postmenopausal symptoms such as osteoporosis. The activities in experimental models relevant for these specific applications will be discussed elsewhere.

A survey of the literature reveals a great variability of the binding data for a particular compound, mainly due to the use of tissues from different

species and varying experimental protocols. A typical example is the antiestrogen 4-hydroxytamoxifen, with RBA values ranging from 280% (rat
uterus) (Robertson et al. 1982) to 6.8% (calf uterus) (von Angerer et al.
1994). Therefore, it is sometimes difficult to establish structure–binding relationships if the data are derived from different laboratories. Reliable results
are obtained when all of the compounds are included in the same assay.

Recently, an additional ER (ERβ), with a tissue distribution different from
that of the α-form, was discovered (Kuiper and Gustafsson 1997). Since then,
a limited number of steroidal and nonsteroidal ligands have also been studied
for binding affinity for the ERβ (Kuiper et al. 1997). No significant differences
were observed for the endogenous ligands, but some environmental estrogens
show different preferences between the α- and β-forms. These data were used
to elaborate QSAR models for both receptors based on comparative molecular field analysis (CoMFA) (Tong et al. 1997). All of the binding data presented in this chapter were determined prior to the discovery of the ERβ.

A drug that binds to the ER can act in vivo either as agonist or as antagonist, or can display mixed agonist/antagonist activities. For many years, the
hormonal properties have been determined in classical endocrine assays, such
as the uterine-weight test in mice or rats. (Rubin et al. 1951; Black and Goode
1980; Odum et al. 1997). The increase of uterine weight in relation to body
weight is used as parameter for estrogenicity. Consequently, the inhibition of
estrogen-stimulated uterine weight gain by the simultaneous administration of
the test compound allows one to estimate the antiestrogenic potency of a compound. A pure antiestrogen is defined as a drug that shows no significant estrogenic activity, when given alone, and is capable of suppressing completely the
estrogen-stimulated uterine growth. Over the last 15 years, in vitro techniques
based on the proliferation of ER-positive breast cancer cells (MCF-7, T47-D,
ZR-75) have been developed. Cell lines transiently (Reese and Katzenellenbogen 1991; von Angerer et al. 1994) or stably (Demirpence et al. 1993;
Hafner et al. 1996) transfected with various reporter plasmids (Miksicek
1994, Vanderkuur et al. 1993a, Meyer et al. 1994; Newton et al. 1994) have
provided interesting alternatives. These methods are reviewed in Chap. 19.

B. Estrogens and Antiestrogens with a Modified Steroid Structure

Since the natural estrogens contain an aromatic A ring, their structure is
unique among the steroidal hormones and is responsible for their binding
specificity which is higher than that of other classes of steroidal hormones.
Several groups have studied this particular pharmacophore, both with respect
to the mode of binding (Anstead et al. 1997; von Angerer 1995) and the
expression of certain genes (Beato 1989). Most of the binding data obtained
with 17β-estradiol and analogues have recently been reviewed in great detail
by Anstead et al. (1997) and, therefore, only the most important aspects will

be discussed briefly in this section. For the binding interaction of the estra-1,3,5(10)-triene system with the ER, at least one hydroxy group, either in position 3 or 17β is essential (Table 1). Most favorable conditions are provided by the presence of both oxygen functions in the positions of the natural ligand estra-1,3,5(10)triene-3,17β-diol. Reduction of the aromatic ring to the tetrahydro (androstenediol) or hexahydro (androstanediol) derivative diminishes the binding affinity to 0.7% and 0.5% of estradiol if the 3-hydroxy group is in β-position or abolishes it (α-isomer) (VANDERKUUR et al. 1993b). A number of estradiol analogues with modified A- and D-ring structures have been synthesized and studied with respect to receptor binding and transactivation of estrogen-regulated genes, such as those encoding the progesterone receptor or pS2. In general, there was a good correlation between binding affinity and biological response, though exceptions were noted (VANDERKUUR et al.

Table 1. Relative binding affinities (RBA) of A-ring and D-ring hydroxylated estra-1,3,5(10)-triene derivatives for the estrogen receptor

R^1	R^2	R^3	R^4	R^5	R^6	R^7	RBA (MCF-7)[a]	RBA (calf)[b]
H	H	H	H	H	H	H	<0.05[c]	ND
H	H	OH	H	H	H	H	80[c]	2.48[d]
H	H	H	H	H	OH	H	11[c]	1.51[d]
OH	H	H	H	H	OH	H	0.5[c]	0.22[d]
H	OH	H	H	H	OH	H	71[c]	8.74[d]
H	H	OH	H	H	OH	H	100	100
H	H	H	OH	H	OH	H	0.7[c]	1.32[d]
H	H	OH	H	H	H	OH	22[c]	ND
H	H	OH	H	OH	H	H	80[c]	ND
H	OH	OH	H	H	OH	H	25[e]	ND
H	H	OH	OH	H	OH	H	48[e]	ND
H	H	OH	H	OH	OH	H	17[c]	16[f]

[a]RBA for the nuclear estrogen receptor from MCF-7 human breast cancer cells (17β-estradiol: RBA=100)
[b]RBA for the calf uterine estrogen receptor
[c]Data from VANDERKUUR et al. (1993b)
[d]Data from SCHWARTZ and SKAFAR (1993)
[e]Data from SCHÜTZE et al. (1994)
[f]Data from LEE et al.(1977)

1993a, b; Pilat et al. 1993). The structural elements responsible for receptor binding affinity and growth stimulation of human MCF-7 breast cancer cells were elucidated by a three-dimensional QSAR method that employed CoMFA (Wiese et al. 1997)

I. Modifications of the A Ring

Several authors have stressed the importance of the phenolic hydroxy group both as hydrogen-bridge donor and acceptor. Removal or shifting of the hydroxy function to other positions in the A-ring strongly reduced the binding of the steroid to the receptor, except for the 2-hydroxy isomer of 17β-estradiol. The data from various studies suggest an in-plane orientation of the phenolic hydrogen between positions 2 and 3 (Vanderkuur et al. 1993b; Anstead et al. 1997). Additional hydroxy groups next to the 3-hydroxy function had no major effect. Other substitutents, however, have been shown to interfere strongly with the binding (Vollmer et al. 1991; Brooks et al. 1987; Lovely et al. 1996; Quian and Abul-Hajj 1990a; Palomino et al. 1997; Cushman et al. 1995).

II. Modifications of the D Ring

The D ring carries the second hydroxy function which is believed to be somewhat less important for the binding interaction with the receptor protein than the 3-hydroxy group. Replacement by an amino group strongly decreased the binding affinity (Kaspar and Witzel 1985), whereas the introduction of an ethynyl group into the 17α-position had no negative effect. Due to the improved pharmacokinetic properties, this modified estradiol derivative is widely used in contraceptives. Systematic studies on the influence of substituents at C-17 on binding affinity for the ER and estrogenic activity have been performed by several groups and have revealed some interesting aspects. Depending on the nature of the 17α-substituent, the binding affinity decreases or increases. Methyl, ethynyl and vinyl in position 17α had no marked effect, whereas the ethyl group (Kaspar and Witzel 1985) and larger substituents (Anstead et al. 1997) dramatically reduced the binding. An important exception is the 2-iodovinyl group with a Z-configuration, which gives rise to a twofold (4°C) and eightfold (25°C) increase in binding to the lamb uterine ER (Napolitano et al. 1991). The high binding affinity and the halogen atom provided the basis for the development of new radio-ligands for the ER, especially after additional substituents had been introduced into position 11β (Symes et al. 1992; Napolitano et al. 1991; Ali et al. 1991; Souttou et al. 1993; Zeicher et al. 1996; Quivy et al. 1996). The binding data suggest that the ER tolerates small substituents that are linked directly to carbon-17 or larger ones, such as 2-iodovinyl, 2-phenylthiovinyl (Napolitano et al. 1996) or 2-styryl (Hanson et al. 1996), provided they possess Z-configuration. Poirier et al. (1990, 1991) showed that certain 17α-alkynyl derivatives of estradiol, with functional groups such as hydroxy and carbamoyl, can exert antiestrogenic effects, although their binding affinities are below 1% of that of estradiol.

Another easily accessible position for structural modifications is the 16α-position. The choice of substituents in this position is mainly limited by their size. FEVIG et al. (1988) synthesized a number of estradiol derivatives with halogen, halomethyl, and saturated and unsaturated carbon chains of up to three atoms at C16. They found that the binding affinity for the ER was retained. Bulky side chains in this position, however, diminished the binding to the receptor. An interesting feature of this type of compound is their ability to inhibit the 17β-hydroxysteroid dehydrogenase type 1 and exert antiestrogenic activity (PELLETIER et al. 1994; PELLETIER and POIRIER 1996).

Removal of the D ring or of both D- and C rings with concomitant aromatization of the B ring did not completely abolish endocrine activity, although the binding affinity for the ER decreased markedly, as demonstrated for (-)-Z-bisdehydrodoisynolic, (-)-allenolic acid and their esters. Despite the fact that these compounds are poor ligands for the ER, they exert a significant estrogenic action in animals (MEYERS et al. 1988) and act as agonists for both gene activation and gene repression of model reporter genes (MEYERS et al. 1997).

III. Modifications of the B Ring

Besides the two oxygen functions involved in the formation of hydrogen bonds to the carboxylate of Glu 353, the guanidinium group of Arg 394 (O–3), and an imidazol nitrogen of His 524 (O–17) (BRZOZOWSKI et al. 1997) the hydrophobic spacer that links both hydroxy groups contributes to the binding energy. Polar functions in the central part of a potential ligand strongly interfere with the binding. Thus, it is not surprising that steroidal structures related to estradiol benefit from additional lipophilic groups in the two central rings. One of the interesting positions is the carbon 7 in the B ring. The introduction of side chains into this position leads preferentially to the α-stereoisomer. The ER tolerates long side chains with different functionalities in the 7α-position of the steroid. Except for the methyl group, a substantial decrease of binding affinity is usually observed. The separation of the polar function from the steroid by spacer groups of sufficient length is important for receptor binding. The intriguing feature of these side chains is their ability to convert the agonist into potent antagonists (Table 2). In fact, the first antiestrogens discovered without any intrinsic estrogenic activity were 7α-substituted estradiol derivatives as exemplified by the amide ICI 164,384 (WAKELING and BOWLER 1988a).

Studies with pure 7α- and 7β-isomers of the N-butyl undecanamide showed that biological activity is confined to the 7α-isomer because its binding affinity for the rat uterine ER is almost 100-fold higher than that of the epimer (RBA: 13 vs 0.18). (WAKELING and BOWLER 1988b; WAKELING 1991). In derivatives with an alkyl chain containing ten methylene groups, the antiestrogenic effect of the N-butylamide exceeded that of the corresponding diethylaminomethyl derivative (BOWLER et al. 1989). Variation of the chain length in

Table 2. Estrogenic and antiestrogenic activities of 7α-substituted estradiol derivatives

ICI 182,780

n	R^1	R^2	Agonism[a] (%)	Antagonism[a] (%)	Dose (mg/kg) s.c.
2	H	$C_{12}H_{25}$	10	14	25
4	H	$C_{10}H_{21}$	−7	34	10
4	H	C_4H_9	18	104	10
7	H	C_7H_{15}	1	92	10
7	H	C_4H_9	28	60	10
10	H	C_4H_9	−3	100	10
10[b]	H	C_4H_9	0	0	25
10	H	$CH_2C(CH_3)_3$	−13	109	10
11	H	C_3H_{11}	26	59	10
10[c]	CH_3	C_4H_9	−7	105	10
10	C_2H_5	C_4H_9	10	93	10
10	C_4H_9	C_4H_9	−1	46	10
10	$CH(CH_3)_2$	C_4H_9	42	39	10
10	CH_3	$CH_2C_3F_7$	−7 (−9; p.o.)	81 (104; p.o.)	10
ICI 182,780[d]			−8 (−8; p.o.) −11 (p.o.)	104 (50; p.o.) 106 (p.o)	0.5 10 (p.o)

[a]Estrogenic and antiestrogenic activities were determined in the rat uterine-weight test and are expressed as a percentage of maximum increase (agonism) or % inhibition of estrogen-stimulated uterine growth. Data from Bowler et al. (1989)
[b]7β-isomer
[c]ICI 164,384
[d]Data from Wakeling et al. (1991)

the amide series revealed two distinct maxima for spacer groups of four and ten carbon atoms with respect to estrogen antagonism (Table 2). There was also a marked influence of the substituents at the nitrogen on the degree of antagonism that could be reached. Three compounds of this series were identified as pure antiestrogens: the estradiol-linked N-(n-butyl)-undecanamide, N-(neopentyl)-undecanamide and N-(n-butyl)-N-methyl-undecanamide (ICI 164,384) (Bowler et al. 1989). When the size of the second N-alkyl group in the N-butylamide function increased, a significant estrogenic effect appeared (Table 2). From a chemical point of view, it seems to be a pre-requisite for complete antagonism, that the functional group is capable of acting as a hydrogen-bridge acceptor (Katzenellenbogen et al. 1993) without being a nucleophile, such as an amine or a carboxylate. A precise spacial orientation of the hydrogen bond between ligand and receptor protein is necessary for the conformational alteration that causes the receptor to loose its transcriptional activity completely. These assumptions are supported by investigations with other polar functional groups, such as sulfoxides and sulfones. From these studies ICI 182,780 (7α-[9-(4,4,5,5,5-pentafluoropentyl-sulfinyl)nonyl]estra-1,3,5(10)-triene-3,17β-diol, Table 2) emerged as the first pure antiestrogen to undergo clinical trials (WAKELING et al. 1991; HOWELL et al. 1995).

Only a few attempts have been made to utilize position 6 for the introduction of side chains, except for the preparation of haptenes useful for generating estradiol-specific antibodies. Estradiol with a side chain similar to that of ICI 164,384, linked to position 6 via a sulfur atom, was shown to be a pure estrogen in ZR-75-1 breast cancer cells (AUGER et al. 1995).

IV. Modifications of the C Ring

Because of synthetic reasons, position 11 in the C ring was mainly used for the introduction of side chains. The attachment of small β-oriented alkyl substituents to C-11 led to derivatives with binding affinities similar to that of the parent drug, as exemplified by 11β-(chloromethyl)- (REINER et al. 1984; SASSON and KATZENELLENBOGEN 1989), 11β-ethyl-, and 11β-vinyl-17β-estradiol (HANSON et al. 1990). When the incubation temperature was increased from 4°C to 25°C, the affinity for the rat uterine ER sometimes increased dramatically and reached RBA values which exceeded that of the natural hormone by one order of magnitude, as shown for 11β-chloromethylestradiol (Table 3). Since the favorable effect of these substituents is due to the enhanced lipophilicity, polar functions at C-11, such as hydroxy or carbonyl groups, have a detrimental effect on receptor binding (PALOMINO et al. 1994). From the pharmocological point of view, the alkyl derivatives can be considered agonists.

The hormonal profile changes after functional side chains, such as that of tamoxifen, had been introduced (QUIAN and ABUL-HAJJ 1990b; GOTTARDIS et al. 1989) (Table 3). Knowledge regarding the influence of the functional group and its separation from the steroidal residue by spacer arms, as well as the experience with 7α-modified estradiol derivatives, led to the discovery of a new series of potent antiestrogens, some of which could be considered pure antagonists (NIQUE et al. 1994; VAN DE VELDE et al. 1994 1996). The most favorable characteristics were displayed by RU 58 668 (Table 3), which combines structural elements of the antiprogestins with those of the antiestrogen ICI 182,780.

V. Conclusion

The modifications of the chemical structure of the natural hormone 17β-estradiol were aimed in several different directions. Some of them, which have

Table 3. 11β-Substituted estradiol derivatives: binding affinities for the estrogen receptor and (anti)estrogenic activities

R group	RBA[a]	Agonism[b] % [dose (mg/kg); species]	Antagonism[b] % [dose (mg/kg); species]
—OCH$_3$	9.7 (0°C) 86 (25°C) (rat)[c]	Agonist[d]	
—CH$_2$Cl	110 (0°C) 1780 (25°C) (lamb)[e]	Agonist[d]	
O-CH$_2$CH$_2$-N(CH$_3$)$_2$	1.6 (4°C) (rat)[f]	60 (2.5; rat)[f]	15 (2.5; rat)[f]
O-(phenyl)-O-CH$_2$CH$_2$-N(CH$_3$)$_2$ RU 39,411	15 (4°C) (MCF-7)[g]	Partial agonist[d]	
-(CH$_2$)$_5$...C(=O)-N(CH$_3$)-C$_4$H$_9$	27 (25°C) (mouse)[h] 125 (25°C) (MCF-7)[h]	0 (30; mouse)[h]	55 (3; mouse)[h]
O-(phenyl)-O-(CH$_2$)$_5$...CH$_2$-C(=O)-NHC$_4$H$_9$	24 (25°C) (mouse)[i]	0 (30; mouse)[i]	120 (3; mouse)[i]

Table 3. (*Continued*)

R group	RBA[a]	Agonism[b] % [dose (mg/kg); species]	Antagonism[b] % [dose (mg/kg); species]
RU 58,668	56 (25°C) (mouse)[j] 3 (4°C) (MCF-7)[g]	–27 (30; mouse)[j]	120 (3; mouse)[j]

[a]Relative binding affinity for the estrogen receptor (estradiol: RBA=100)
[b]Estrogenic and antiestrogenic activities were determined in the uterine-weight test and are expressed as a percentage of maximum increase (agonism) or percentage inhibition of estrogen-stimulated uterine growth
[c]Data from POMPER et al. 1990
[d]Based on in vitro data
[e]Data from BINDAL et al. (1987)
[f]Data from QUIAN and ABUL-HAJJ (1990b)
[g]Data from JIN et al. (1995)
[h]Data from CLAUSSNER et al. (1992)
[i]Data from NIQUE et al. (1994)
[j]Data from VAN DE VELDE et al. (1994)

not been discussed in this chapter, comprised the improvement of pharmacokinetic parameters and the bioavailability. Others concerned the theoretical investigations on receptor binding and gene activation. Another important aspect was the development of pure antiestrogens based on estradiol which might be useful for the therapy of breast cancer and other estrogen-related disorders. It was interesting to see that this goal was reached by modification of two different sites of the steroid. The equivalence of the positions 7α and 11β with respect to antiestrogenic activity can be rationalized by two different binding models. The first implies two different residues in the receptor protein which can interact either with polar function of the side chain of the 7α- or the 11β-derivatives. If only one appropriate residue is present, however, one has to assume that the steroid rotates around its longitudinal axis to give the polar function access to the additional binding site. Since the rotation changes the spatial orientation of the 17β-hydroxy group and the adjacent methyl group, it should influence the binding affinity. Unfortunately, no data from experiments in which both types of estradiol derivatives with identical side chains were tested under the same conditions have been reported, except for 17β-estradiol with short alkyl groups in positions 7α and/or 11β. The RBA value of the 11β-ethyl derivative for the rat uterus ER exceeded that of the 7α-substituted analogue by one order of magnitude, and the value for the disubstituted derivative was in between these two values (TEDESCO et al. 1997).

C. Non-Steroidal Estrogens and Antiestrogens

In this and the preceding section, only estrogens and antiestrogens that are of therapeutic interest are discussed, but the inherent estrogenicity of various other compounds has become a major concern to the public. Several studies on SARs have been initiated after the (anti)estrogenic properties of a variety of environmental chemicals, such as polychlorinated biphenyls, bisphenol-A and nonylphenol, were discovered and related to reproductive abnormalities in animals (CONNOR et al. 1997; KRAMER et al. 1997; WALLER et al. 1996; ODUM et al. 1997; ROUTLEDGE and SUMPTER 1997). Other studies concerned the potential beneficial effects of phytoestrogens, such as coumestrol, genistein and β-zeraneol, which are found in various vegetables and fruits (LIEN and LIEN 1996; DEES et al. 1997).

I. Derivatives of Diethylstilbestrol

The prototype of a synthetic estrogen is diethylstilbestrol (DES) (MARSELOS and TOMATIS 1992). Its estrogenic potency is not affected if a second double bond has been introduced into the aliphatic backbone to give dienestrol. The saturated analogue hexestrol with a meso-configuration binds to the ER with a relative affinity of about 50% of that of DES, and acts as a very potent estrogen in vivo. Structural modifications of these agents, such as shifting of the phenolic hydroxy groups into the meta-positions, introduction of additional substituents into the phenyl rings and variation of the alkyl groups has led generally to a decrease of binding affinity, estrogenic potency and the appearance of estrogen antagonism as the result of reduced intrinsic activity (KRANZFELDER et al. 1980, 1982; SCHNEIDER et al. 1980; HARTMANN et al. 1981, 1983, 1984) (For chemical structures see Chap. 4.C.II. Fig. 10). When the central double bond in the stilbene system had been converted to a three-membered ring by cycloaddition of methylene, dichloromethylene or oxygen, the molecular shape remained more or less unaffected. Therefore, it was not unexpected that the transformation of DES and its meta-isomer into cyclopropane derivatives did not alter their biological properties much (SCHNEIDER et al. 1981, 1982a; PENTO et al. 1981).

II. 2-Phenylindole Derivatives

Most of the studies with analogues of DES were confined to symmetrical structures, whereas the natural hormones lack the symmetry of the molecule. Therefore, it was of great interest to investigate nonsteroidal ligands of the ER with an asymmetric skeleton. An appropriate system for these studies is the 2-phenylindole with two phenolic hydroxy groups (VON ANGERER et al. 1984) that proved to be more versatile than its carbocyclic counterpart indenestrol (SCHNEIDER et al. 1982b; SCHNEIDER and BALL 1986). The heterocyclic system allowed the independent variation of both of the positions of the hydroxy groups and the kind of alkyl substituents in positions 1 and 3. For a rational design, it was necessary to define the aromatic ring that corresponds to the A ring of the steroid. Studies in the mono-hydroxylated 2-phenylindole series clearly demonstrated that the phenyl ring that carries a para-hydroxy group mimics the A ring (Table 4). These findings are in agreement with other studies which showed that nonsteroidal ligands for the ER with a phenol ring bind in a fashion where the nonannelated phenol corresponds to the aromatic A ring in estradiol (ANSTEAD et al. 1989; DUAX and GRIFFIN 1987; DUAX et al. 1984; KYM et al. 1993). The crystal structure of the ligand-binding domain with raloxifene bound, however, revealed a different orientation, in which the phenyl ring of the 2-phenyl-benzothiophene moiety mimics the D ring. Possibly, this

Table 4. Relative binding affinities (RBAs) of mono- and dihydroxylated 2-phenylindole derivatives for the calf uterine estrogen receptor

X^1	X^2	R^1	R^2	Y	RBA[a]
OH	H	H	CH_3	OH	0.06
OH	H	CH_3	H	OH	3.8
OH	H	C_2H_5	H	OH	16
OH	H	C_2H_5	CH_3	OH	33
H	OH	H	CH_3	OH	0.06
H	OH	CH_3	H	OH	0.8
H	OH	C_2H_5	H	OH	5.8
H	OH	C_2H_5	CH_3	OH	9.5
H	OH	C_3H_7	CH_3	OH	16
H	OH	C_5H_{11}	CH_3	OH	2.3
OH	H	C_2H_5	CH_3	H	0.08
H	H	C_2H_5	CH_3	OH	1.25

[a]RBA determined at 4°C (estradiol: RBA = 100). Data from VON ANGERER et al. (1984) and VON ANGERER (1995)

orientation is necessary to enable the amino function in the side chain to form a hydrogen bond with the carboxylate of Asp 351 (BRZOZOWSKI et al. 1997) and might be different in ligands that lack this functional element.

The best binding conditions for the calf uterine ER were provided by two hydroxy groups, one in the *para*-position of the 2-phenyl ring and the other at C6 or C5 of the indole, and short alkyl groups in positions 1 and 3 to guarantee sufficient lipophilicity in the central part of the molecule (VON ANGERER et al. 1984). Derivatives with hydrogen at C3 exhibited very low affinity and 1-H-2-phenylindoles were devoid of binding affinity (Table 4). By computer-aided molecular modeling, it can be shown that the energetically minimized structures of 17β-estradiol and the 2-phenylindole derivatives, with high binding affinities, can be superimposed in a way in which both hydroxy groups of the indole adopt nearly the same spatial positions as those of the steroid. Also, the shapes of the lipophilic elements are rather similar in both structures.

The estrogenic activity of the 2-phenylindole derivatives was influenced by both the position of the oxygen function in the indole part and the size of the alkyl group. In general, the 6-hydroxy derivatives acted as estrogens in the mouse uterus, whereas the corresponding 5-hydroxy compounds were characterized by a considerably reduced estrogenicity and a significant antiestrogenic activity (VON ANGERER et al. 1984). The analysis of the SAR in this series showed that the agonist activity of this heterocyclic stilbene analog can only be minimized to a certain extent by shifting the hydroxy functions and varying the length of the alkyl groups.

A substantial increase in antiestrogenic activity could be achieved by the introduction of side chains with functional groups into position 1 of the indole. From these studies, the 6-pyrrolidinohexyl derivative ZK 119,010 emerged as a potent antiestrogen (VON ANGERER et al. 1990). It possesses high binding affinity for the calf uterine ER (RBA: 21% of estradiol) and was shown to suppress the uterine growth in immature mice completely (Table 5). In rats, a weak estrogenic effect was observed when high doses were applied (NISHINO et al. 1991). An important extension of the structural variations in this class of compounds comprised the introduction of an amide function into the side chain. When ZK 119,010 had been converted to the corresponding amide, the antiestrogenic potency was reduced. It could be restored by the elongation of the alkyl spacer group by up to 10–11 methylene groups.

The most potent derivative of this series was the 2-phenylindole with a *N*-methyl-*N*-propyldodecanamide side chain (Table 5). As in the steroidal series, it was possible to increase the potency further by replacing the amide functions by sulfoxide or sulfone groups (Table 5). The experimental data showed that the oxidation state of the sulfur has only a minor influence on the in vivo activity (BIBERGER and VON ANGERER 1996). The number of 9 or 10 methylene groups in the spacer arm appeared to be most favorable with respect to antiestrogenic potency and antitumor activity in MCF-7 cells. The fluoro substitu-

Table 5. 5-Hydroxy-2-(4-hydroxyphenyl)-3-methylindoles with functional side chains. Relative binding affinities (RBAs) for the estrogen receptor and (anti)estrogenic activities

R group	RBA[a]	Agonism[b] (%)	Antagonism[b] (%)	Dose (mg/kg, sc)
$-C_2H_5$	9.5^c	43	46	6
$-(CH_2)_6-NH_2$	25^d	2	99	6
$-(CH_2)_6$-pyrrolidino (ZK 119,010)	21^d	-3	117	6
$-(CH_2)_6$-morpholino	10^d	-4	102	6
$-(CH_2)_{11}-N(CH_3)C_4H_9$	15^e	36	18	6
$-(CH_2)_5-CO$-pyrrolidino	7.0^e	47	44	30
$-(CH_2)_7-CON(CH_3)C_7H_{15}$	3.5^e	10	57	30
$-(CH_2)_{11}-CON(CH_3)C_3H_7$	2.1^e	1	82	30
$-(CH_2)_{10}-S-C_5H_{11}$	0.3^f	-14	106	30
$-(CH_2)_{10}- SO_2-C_5H_{11}$ (ZK 164,015)	4.7^f	-13	108	30
$-(CH_2)_9-SO-(CH_2)_3C_2F_5$	2.4^f	0	88	6

[a]RBA for the calf uterus estrogen receptor determined at 4°C (estradiol: RBA=100)
[b]Estrogenic and antiestrogenic activities were determined in the mouse uterine-weight test and are expressed as a percentage of maximum increase (agonism) or inhibition of estrogen-stimulated uterine growth
[c]Data from VON ANGERER et al. (1984)
[d]Data from VON ANGERER et al. (1990)
[e]Data from VON ANGERER et al. (1994)
[f]Data from BIBERGER and VON ANGERER (1996)

tion in the terminal position found in both ICI 182,780 and RU 58 668 is not a prerequisite for the in vitro activity, but might influence the metabolic stability of these compounds under in vivo conditions. The most potent derivative in this series of 2-phenylindoles was ZK 164,015 (5-hydroxy-2-(4-hydroxyphenyl)-3-methyl-1-[10-(pentylsulfonyl)decyl]indole) (BIBERGER and VON ANGERER 1996). It was devoid of any estrogenic activity in the mouse uterine-weight test and suppressed the stimulatory effect of estrone completely (Table 5).

From these data and the comparison with the estradiol-based agents, it became evident that the structure of the side chain determines the endocrine profile of the compound, and the carrier of the side chain is mainly responsible for binding affinity or, speaking in chemical terms, to provide the correct spatial orientation of the side chain to allow the interaction with a distinct amino acid in the receptor protein.

III. 2-Phenylbenzo[b]thiophene Derivatives

Another versatile structure for the synthesis of ligands for ER with a varying degree of estrogen antagonism is 2-phenylbenzo[b]thiophene. This system was approached from two sides. First, in the early eighties, Jones et al. (1984) investigated cyclic derivatives of 1,2,3-triarylpropenone, including 3-benzoyl-2-phenylbenzothiophene, as potential drugs for the treatment of breast cancer. In the second approach, von Angerer and Erber (1992) considered the 2-phenylbenzothiophene as the isosteric analogue of the 2-phenylindole structure. Both groups modified the basic structure and the side chain in position 3. The high binding affinities for the ER can be rationalized by the 4,4'-dihydroxystilbene structure and lipophilicity of the sulfur atom (Table 6). The analogy to DES is also reflected by the agonist activity of the 3-ethyl derivative. The estrogenicity can be considerably reduced by the introduction of functional side chains, as exemplified by raloxifene, a drug that exhibits tissue specificity to a certain degree and is presently undergoing phase-III clinical studies for the treatment of osteoporosis (Black et al. 1994). Very recently, Palkowitz et al. (1997) at Eli Lilly, showed that the pharmacological profile of raloxifene (substantial agonist activity on the bone and strong antagonism in the uterus) can be improved by replacing the carbonyl function by an oxygen atom (Table 6). The most potent antiestrogen in the benzothiophene series with an aliphatic side chain was the one with a sulfone group (6-hydroxy-2-(4-hydroxyphenyl)-3-[10-(pentylsulfonyl)decyl]benzo[b]thiophene (von Angerer et al. 1995).

IV. Triphenylethylene Derivatives

For a long time, the triphenylethylene structure with an amino function in the side chain was the only chemical structure associated with antiestrogenic activity. The most prominent example is tamoxifen, the only antiestrogen that is used world wide for the treatment of breast cancer (Fig. 1). Therefore, it was

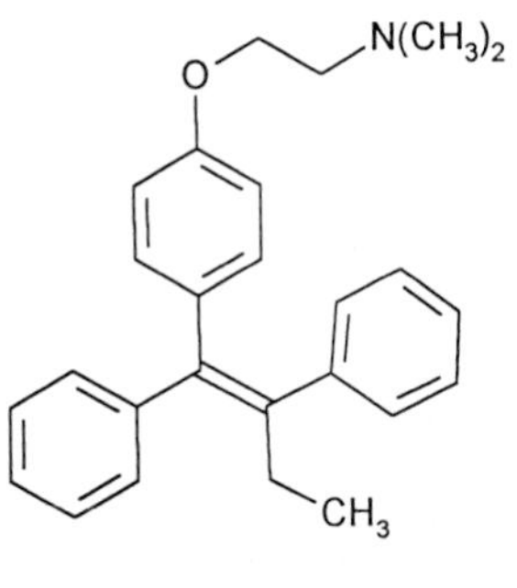

Tamoxifen

Fig. 1. Chemical structure of tamoxifen

Table 6. Effect of side chain variation on the hormonal activity of 6-hydroxy-2-(4-hydroxyphenyl)benzo[b]thiophenes

R group	RBA[a]	Agonism[b] (%)	Antagonism[b] (%)	Dose[c] (mg/kg)/species
CH$_3$	28[d]	74	31	6 /mouse, s.c.
Raloxifene	7.1[e]	52	61	6 /mouse, s.c.
	34[f]	30	87	2.5 /rat, s.c.
	~20(4°C) ~100[g] (37°C)	4	98	1.0 /rat, p.o.

Table 6. (*Continued*)

R group	RBA[a]	Agonism[b] (%)	Antagonism[b] (%)	Dose[c] (mg/kg)/species
—(CH₂)₉—N(CH₃)—CH₂CH₂CH₃	5.5[h]	21	54	6 /mouse, s.c.
—(CH₂)₉—N(C=O)(CH₃)—CH₂CH₂CH₃	3.2[h]	10	57	30 /mouse, s.c.
—(CH₂)₉—S(=O)(=O)—CH₂CH₂CH₃	1.8[h]	12	87	30 /mouse,s.c.

[a]Relative binding affinity for the calf uterine estrogen receptor except for raloxifene and analogue for which MCF-7 human breast cancer cells were used. 17β-Estradiol: RBA = 100
[b]Estrogenic and antiestrogenic activities were determined in the uterine-weight test after administration of the drug alone or in combination with a standard dose of an estrogen
[c]Daily dose for three consecutive days
[d]Data from von Angerer and Erber (1992)
[e]Data from Erber (1989)
[f]Data from Grese et al. (1997) and Jones et al. (1984)
[g]Data from Palkowitz et al. (1997)
[h]Data from von Angerer et al. (1995)

not surprising that the first studies on the SARs with antiestrogens were carried out in the triphenylethylene series. A variety of structural modifications were performed; some of them have already been addressed in the preceding chapter. From a chemical point of view, there were several questions to be answered. What is the influence of *cis/trans* isomerization of the central double bond? Can the amino residue be replaced by other functional groups? Does the hormonal profile change following the introduction of additional substituents into the unsubstituted aromatic rings?

Many studies have demonstrated a marked difference between the geometric isomers of tamoxifen, both with respect to receptor binding and endocrine activity. The binding affinity of the *E*-isomer is lower by one order of magnitude than that of the *Z*-form, which is used clinically. The difference observed for the 4-hydroxy metabolites was even more pronounced (ROBERTSON et al. 1982; DORÉ et al. 1992) despite the fact that they are easily interconverted (KATZENELLENBOGEN et al. 1985). The endocrine activity of the *E*-isomer of tamoxifen is much lower than that of the *Z*-form, presumably due to the lower binding affinity for the receptor. The antiestrogenic activity of tamoxifen is mainly confined to the Z-isomer. These findings were supported by data obtained with non-isomerizable derivatives of tamoxifen (McCAGUE et al. 1988). Obviously, the partial agonist activity does not arise from isomerization or conversion to estrogenic metabolites.

Introduction of additional substituents into the aromatic rings influences binding affinity, but does not change the typical endocrine profile of an antiestrogen with substantial agonist activity, as demonstrated by McCAGUE et al. (1989), who introduced halogen atoms (Cl, Br, I) and various functional groups, such as MeS-, MeSO-, $MeSO_2$-, -SH, -CHO, -CH_2OH and -CH(O) into the *para*-position of the geminal phenyl ring. Generally, the binding affinities for the calf uterine ER were higher than that of tamoxifen (Table 7). The results from the MCF-7 proliferation assay showed that the antiestrogenic activity of tamoxifen was retained in most of these derivatives except for the sulfoxide and the hydroxymethyl derivatives which appeared to be pure estrogens. When the stereochemistry was changed from *trans* to *cis*, the binding affinities decreased by one order of magnitude (Table 7) (McCAGUE et al. 1989).

The ratio of agonist/antagonist activity does not only depend on the chemical structure, but also on the biological system used for evaluation: for example, tamoxifen loses most of the antiestrogenic properties observed in rats when tested in mice. It appears that the estrogenic activity is an inherent feature of the triphenylethylene structure with a basic side chain, and is modulated by an additional antiestrogenic effect which reflects the sensitivity of a given tissue towards estrogen antagonists. This assumption is supported by studies on a molecular level which showed that 4-hydroxytamoxifen can only block TAF-2 activity of the ER, but not TAF-1 (VON ANGERER et al. 1994). For some purposes, tissue selectivity would be desirable. When the (dimethylamino)ethoxy group of tamoxifen had been replaced by the acrylic acid

Table 7. Relative binding affinities (RBAs) of various 4-substituted tamoxifen derivatives and their E-isomers for the estrogen receptor

R group	RBA[a] (Z-isomer[b])	RBA[a] (E-isomer[b])
H	1	0.1
I	5	0.2
Br	5	0.2
Cl	1	ND
SCH_3	3	0.1
$SOCH_3$	3	0.3
SO_2CH_3	6	0.6
SH	1	ND
CH=O	20	ND
CH_2OH	80	ND

[a]RBAs for the calf uterine estrogen receptor measured in a competitive binding assay after incubation for 30 min at 18°C (17β-estradiol: RBA=100); Data from McCague et al. (1989)
[b]Refers to the parent structure of the unsubstituted tamoxifen

fragment, significant estrogenic effects were only observed in the bone of rats, but not in the classical target organ, the uterus (Willson et al. 1997) (for chemical formulae of triphenylethylene derivatives see Chap. 4. A.II. and III.).

V. 2,3-Diphenyl-2H-1-Benzopyrans and Related Structures

Several studies on nonsteroidal antiestrogens have revealed that a geminal arrangement of two of the phenyl rings is not essential and can be replaced by structures in which all of the three phenyl groups are located on different carbon atoms. Examples of this modification are the triarylfurans (Durani et al. 1989), 1,2,3-triarylpropenones (Mittal et al. 1985), 1,2,3-triarylbutenones (Durani et al. 1989) and cyclic structures that contain these fragments, such as 3-benzoyl-2-phenylbenzo[b]furan (Durani et al. 1989) and 3-benzoyl-2-phenylbenzo[b]thiophene which has already been discussed in Sect. 5. C.IV. The pyrrolidinoethoxy group which was used as the basic side chain was usually located in the *para*-position of the benzoyl ring. Both the binding affinities for the ER and the endocrine activities in the rat uterus were very low for this kind of tamoxifen analogue. When the basic side chain, however, had been replaced by a hydroxy function, a considerable increase in estrogenic activity was observed due to the enhanced binding affinities (DURANI et al. 1989).

The successful application of the 1,2,3-triphenylpropenone structure, as a part of a five-membered ring in the synthesis of compounds with binding affinity for the ER and with antiestrogenic activity, suggested an extension of these studies onto six-membered heterocyclic systems. DURANI and COLLEAGUES (SAEED et al. 1990; SHARMA et al. 1990a, b) performed a detailed study of SARs in the class of 2,3-diaryl-2H-1-benzopyran (Table 8), which provided the basis for the investigations of other groups (GAUTHIER et al. 1997; GRESE et al. 1996). The analysis of the SARs in this series of hormonal active compounds clearly revealed the structural elements important for receptor binding and estrogen antagonism. High binding affinity can be achieved by the 2-(piperidino)ethoxy group in the *para*-position of the 2-phenyl ring and additional hydroxy groups in the two other aromatic rings. The high binding affinities of the dihydroxy derivatives can be rationalized by the 4,4'-dihydroxystilbene structure that is part of the molecule (Fig. 2). The antiestrogenic activity arises from the amino function in the side chain, which is rather similar to that of tamoxifen, although its antagonistic effect is much stronger than that of tamoxifen or its active metabolite 4-hydroxytamoxifen. All compounds except one were tested as racemates and no effort was made to separate the optical isomers. GAUTHIER et al. (1997) were the only group who studied both enantiomers after separation and found a marked difference in both binding affinity and antiestrogenic potency (Table 8).

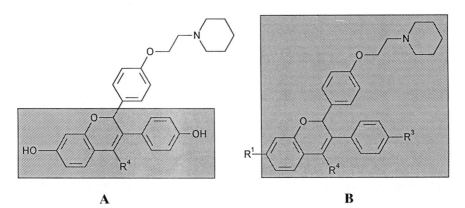

A   **B**

Fig. 2A,B. Structural analogies of 1-benzopyran-based antiestrogens to diethylstilbestrol (**A**) and tamoxifen (**B**), as indicated by shaded rectangles

Table 8. Relative binding affinities (RBAs) and (anti)estrogenic activities of substituted 2,3-diphenyl-2H-1-benzopyrans

R^1	R^2	R^3	R^4	RBA[a]	Agonism[b] %	Antagonism[b] %	Dose/ species
H	OH	H	H	0.05[c]	3	0	10μg/rat (sc)
H	PyEtO[d]	H	Me	9.9[c]	47	56	10μg/rat (sc)
H	PiEtO[e]	H	H	0.3[c]	10	75	10μg/rat (sc)
H	PiEtO	OH	H	4.2[f]	−3	94	10μg/rat (sc)
OH	PiEtO	H	H	9.5[f]	11	65	10μg/rat (sc)
OMe	PiEtO	H	H	0.02[f]	33	6	10μg/rat (sc)
OH	PiEtO	OH	H	49[f]	−3	98	10μg/rat (sc)
OH	PiEtO	OH	Me	284[g]	9	75	47μg/mouse (po)
OH	PiEtO	OH	Me	6.3[h]	4	30	47μg/mouse (po)
OH	PiEtO	OH	O[i]	7.2[j]	−9	77	10mg/kgrat (po)

[a]RBA for the rat estrogen receptor determined at 4°C, except were noted (estradiol: RBA=100)
[b]Estrogenic and antiestrogenic activities were determined in the uterine weight test and are expressed as a percentage of maximum increase (agonism) or inhibition of estrogen-stimulated uterine growth
[c]Data from Sharma et al. (1990b)
[d]2-(pyrrolidino)ethoxy
[e]2-(piperidino)ethoxy
[f]Data from Saeed et al. (1990)
[g]Data for the S-isomer (Gauthier et al. 1997): RBA determined at 37°C with human uterus cytosol; in vivo evaluation was performed with the pivaloyl derivatives
[h]Data for the R-isomer (Gauthier et al. 1997): RBA determined at 37°C with human uterus cytosol; in vivo evaluation was performed with the pivaloyl derivatives
[i]Linked to the ortho-position of the 3-phenyl ring
[j]Data from Grese et al. (1996): RBA determined with MCF-7 human breast cancer cells

VI. Conclusion

Most of the nonsteroidal antiestrogens feature two important structural elements responsible for binding and activity: a 1,2-diphenylethylene fragment or its saturated analogue and a functional group capable of forming a hydrogen bridge with an appropriate residue in the receptor protein. Although both elements contribute to the binding, high affinity can only be achieved when hydroxy functions are been introduced into one or both phenyl rings, preferably into the *para*-position. Compounds that lack the functional side chain act either as pure agonist, as DES does, or as partial antiestrogen after the arrangement of the hydroxy function and the alkyl substitution have been altered in an appropriate way. The antagonism of the latter type of compound can be explained by their lower intrinsic estrogenic activity due to an incomplete activation of the ER.

A substantial increase in antiestrogenic activity can be reached by functional groups separated from the diphenylethylene structure by a lipophilic spacer element. It can be assumed that lipophilicity in the vicinity of the basic structure that mimics the steroid is necessary, but the extension of the lipophilic area has not been defined. In the early studies, the spacer arm contained a phenyl ring, as exemplified by tamoxifen and congeners. In the more recent studies of antiestrogens, aliphatic side chains were successfully applied. The use of this kind of spacer group or its combination with aromatic rings allows the precise spatial orientation of the functional group required for the interaction with a distinct residue in the receptor protein. Compounds with amide or sulfur functions separated from the basic structure by 9–12 methylene groups or equivalent were shown to be completely devoid of agonist activity and can be considered pure antiestrogens. Any location of the functional group that does not meet these steric requirements led to an incomplete antagonism and appearance of estrogenic side effects. This holds also true for suboptimal functional groups, such as the tertiary amino function present in tamoxifen. In compounds of the latter type, the degree of antagonism that can be reached is partly determined by the basic structure. Tamoxifen and its close analogues still possess substantial estrogenic activity, which is considerably reduced in the 2,3-diphenyl-1-benzopyran series. Complete antagonism was achieved in several classes of steroidal and nonsteroidal compounds, including 7α- and 11β-substituted estradiol derivatives, 2-phenylindoles and 2-benzo[b]thiophenes after the side chains had been optimized.

The knowledge accumulated over the last decade on SARs of ligands of ERs has opened the path to investigations on tissue- and gene-specific agonists and antagonists. In vitro studies with A ring-modified steroids have already shown that differential activation of genes is possible. Tissue specificity can arise from differences in cell sensitivity, drug distribution and gene activation. The development of tailor-made (anti)estrogens for important clinical indications, such as breast cancer or osteoporosis, is now conceivable. After the successful crystallization of the hormone-binding domain of the human ER in complex with both the endogenous agonist 17β-estradiol and the antagonist raloxifene, some of the known prerequisites for binding could be rationalized and influences of the ligand on the structure of the transactivating domain have become apparent.

References

Ali H, Rousseau J, Ghaffari MA, Van Lier JE (1991) Synthesis, receptor binding, and tissue distribution of 7α- and 11β-substituted $(17\alpha,20E)$- and $(17\alpha,20Z)$-21-[^{125}I]iodo-19 norpregna-1,3,5(10),20-tetraene-3,17-diols. J Med Chem 34:854–860

Anstead GM, Wilson SR, Katzenellenbogen JA (1989) 2-Arylindenes and 2-arylinde-
nones: Molecular structures and considerations in the binding orientation of
unsymmetrical nonsteroidal ligands to the estrogen receptor. J Med Chem
32:2163–2171

Anstead GM, Carlson KE, Katzenellenbogen JA (1997) The estradiol pharmacophore:
ligand structure-estrogen receptor binding affinity relationships and a model for
the receptor binding site. Steroids 62:268–303

Auger S, Merand Y, Pelletier JD, Poirier D, Labrie F (1995) Synthesis and biological
activities of thioether derivatives related to the antiestrogens tamoxifen and ICI
164384. J Steroid Biochem Mol Biol 52:547–565

Beato M (1989) Gene regulation by steroid hormones. Cell 56:335–344

Biberger C, von Angerer E (1996) 2-Phenylindoles with sulfur containing side chains.
Estrogen receptor affinity, antiestrogenic potency, and antitumor activity. J Steroid
Biochem Molec Biol 58:31–43

Bindal RD, Carlson KE, Reiner GCA, Katzenellenbogen JA (1987) 11β-Chloromethyl-
[^{3}H]estradiol-17β: A very high-affinity, reversible ligand for the estrogen receptor.
J Steroid Biochem 28:361–370

Black LJ, Goode RL (1980) Uterine bioassay of tamoxifen, trioxifene and a new estrogen antagonist (LY 117018) in rats and mice. Life Sci 26:1453–1458

Black LJ, Sato M, Rowley ER, Magee DE, Bekele A, Williams DC, Cullinan GJ,
Bendele R, Kauffman RF, Bensch WR, Frolik CA, Termine JD, Bryant HU (1994)
Raloxifene (LY139481 HCl) prevents bone loss and reduces serum cholesterol
without causing uterine hypertrophy in ovariectomized rats. J Clin Invest 93:63–69

Bowler J, Lilley TJ, Pittam JD, Wakeling AE (1989) Novel steroidal pure antiestrogens.
Steroids 54:71–99

Brooks SC, Wappler NL, Corombos JD, et al. (1987) Estrogen structure-receptor function relationships. In: Moudgil VK (ed) Recent Advances in Steroid Hormone
Action. Moudgil VK., Berlin-New York, pp 443–466

Brzozowski AM, Pike AC, Dauter Z, Hubbard RE, Bonn T, Engstrom O, Ohman L,
Greene GL, Gustafsson JA, Carlquist M (1997) Molecular basis of agonism and
antagonism in the estrogen receptor. Nature 389:753–758

Claussner A, Nédélec L, Nique F, Philibert D, Teutsch G, Van de Velde P (1992) 11β-
Amidoalkyl estradiols, a new series of pure antiestrogens. J Steroid Biochem
Molec Biol 41:609–614

Connor K, Ramamoorthy K, Moore M, Mustain M, Chen I, Safe S, Zacharewski T,
Gillesby B, Joyeux A, Balaguer P (1997) Hydroxylated polychlorinated biphenyls
(PCBs) as estrogens and antiestrogens: structure–activity relationships. Toxicol
Appl Pharmacol 145:111–123

Cushman M, He HM, Katzenellenbogen JA, Lin CM, Hamel E (1995) Synthesis, anti-
tubulin and antimitotic activity, and cytotoxicity of analogs of 2-methoxyestradiol,
an endogenous mammalian metabolite of estradiol that inhibits tubulin polymer-
ization by binding to the colchicine binding site. J Med Chem 38:2041–2049

Dees C, Foster JS, Ahamed S, Wimalasena J (1997) Dietary estrogens stimulate human
breast cells to enter the cell cycle. Environ Health Perspect 105[Suppl 3]:633–636

Demirpence E, Duchesne M-J, Badia E, Gagne D, Pons M (1993) MVLN cells: A bio-
luminescent MCF-7-derived cell line to study the modulation of estrogenic activity. J Steroid Biochem Molec Biol 46:355–364

Doré J-C, Gilbert J, Bignon E, Crastes de Paulet A, Ojasoo T, Pons M, Raynaud J-P,
Miquel J-F (1992) Multivariate analysis by the minimum spanning tree method of
the structural determinants of diphenylethylenes and triphenylacrylonitriles implicated in estrogen receptor binding, protein kinase C activity, and MCF7 cell proliferation. J Med Chem 35:573–583

Duax WL, Swenson DC, Strong PD, Korach KS, McLachlan J, Metzler M (1984)
Molecular structures of metabolites and analogues of diethylstilbestrol and
their relationship to receptor binding and biological activity. Mol Pharmacol
26:520–525

Duax WL, Griffin JF (1987) Structural features which distinguish estrogen agonist and antagonist. J Steroid Biochem Molec Biol 27:271–280

Durani N, Jain R, Saeed A, Dikshit DK, Durani S, Kapil RS (1989) Structure-activity relationship of antiestrogens: A study using triarylbutenone, benzofuran, and triarylfuran analogues as models for triarylethylenes and triarylpropenones. J Med Chem 32:1700–1707

Erber S (1989) Synthese und Testung mammatumorhemmender Derivate des 2-Phenylindols, -benzo[b]furans und benzo[b]thiophens. Ph D Theses, University of Regensburg

Fevig TL, Mao MK, Katzenellenbogen JA (1988) Estrogen receptor binding tolerance of 16α-substituted estradiol derivatives. Steroids 51:471–497

Gauthier S, Caron B, Cloutier J, Dory YL, Favre A, Larouche D, Mailhot J, Ouellet C, Schwerdtfeger A, Leblanc G, Martel C, Simard J, Mérand Y, Bélanger A, Labrie C, Labrie F (1997) (S)-(+)-[4-[7-(2,2-dimethyl-1-oxopropoxy)-4-methyl-2-[4-[2-(1-piperidinyl)ethoxy]phenyl-2H-1-benzopyran-3-yl]phenyl]-2,2-dimethyl-propanoate (EM-800): A highly potent, specific and orally active non-steroidal antiestrogen. J Med Chem 40:2117–2122

Gottardis MM, Jiang SY, Jeng MH, Jordan VC (1989) Inhibition of tamoxifen-stimulated growth of an MCF-7 tumor variant in athymic mice by novel steroidal antiestrogens. Cancer Res 49:4090–4093

Grese TA, Cole HW, Magee DE, Phillips DL, Shetler PK, Short LL, Glasebrook AL, Bryant HU (1996) Conversion of the phytoestrogen coumestrol into a selective estrogen receptor modulator (SERM) by attachment of an amine-containing sidechain. Bioorg Med Chem Lett 6:2683–2686

Grese TA, Cho S, Finley DR, Godfrey AG, Jones CD, Lugar CW, Martin MJ, Matsumoto K, Pennington LD, Winter MA, Adrian MD, Cole HW, Magee DE, Phillips DL, Rowley ER, Short LL, Glasebrook AL, Bryant HU (1997) Structure-activity relationships of selective estrogen receptor modulators: modifications to the 2-arylbenzothiophene core of raloxifene. J Med Chem 40:146–167

Hafner F, Holler E, von Angerer E (1996) Effect of growth factors on estrogen receptor mediated gene expression. J Steroid Biochem Molec Biol 58:385–393

Hanson RN, Napolitano E, Fiaschi R, Onan KD (1990) Synthesis and estrogen receptor binding of novel 11β-substituted estra-1,3,5(10)-triene-3,17β-diols. J Med Chem 33:3155–3160

Hanson RN, Herman LW, Fiaschi R, Napolitano E (1996) Stereochemical probes for the estrogen receptor: synthesis and receptor binding of (17α,20E/Z)-21-phenyl-19-norpregna-1,3,5(10),20-tetraene-3,17β-diols. Steroids 61:718–722

Hartmann RW, Buchborn H, Kranzfelder G, Schönenberger H (1981) Potential antiestrogens. Synthesis and evaluation of mammary tumor inhibiting activity of 1,2-dialkyl-1,2-bis(3′-hydroxyphenyl)ethanes. J Med Chem 24:1192–1197

Hartmann RW, Schwarz W, Schönenberger H (1983) Ring-substituted 1,2-dialkylated 1,2-bis(hydroxyphenyl)ethanes. 1. Synthesis and estrogen receptor binding affinity of 2,2′- and 3,3′-disubstituted hexestrols. J Med Chem 26:1137–1144

Hartmann RW, Heindl A, Schwarz W, Schönenberger H (1984) Ring-substituted 1,2-dialkylated 1,2-bis(hydroxyphenyl)ethanes. Synthesis, estrogen receptor binding affinity, and evaluation of antiestrogenic and mammary tumor inhibiting activity of 2,2′-disubstituted butestrol and 6,6'-disubstituted metabutestrols. J Med Chem 27:819–824

Howell A, DeFriend D, Robertson J, Blamey R, Walton P (1995) Response to a specific antiestrogen (ICI 182780) in tamoxifen-resistant breast cancer. Lancet 345:29–30

Jin L, Borras M, Lacroix M, Legros N, Leclercq G (1995) Antiestrogenic activity of two 11β-estradiol derivatives on MCF-7 breast cancer cells. Steroids 60:512–518

Jones CD, Jevnikar MG, Pike AJ, Peters MK, Black LJ, Thompson AR, Falcone JF, Clemens JA (1984) Antiestrogens. 2. Structure-activity studies in a series of 3-aroyl-2-arylbenzo[b]thiophene derivatives leading to [6-hydroxy-2-(4-hydroxy-

phenyl)benzo[b]thien-3-yl][4-[2-(1-piperidinyl)ethoxy]phenyl]methanone
hydrochloride (LY156758), a remarkably effective estrogen antagonist with only
minimal intrinsic estrogenicity. J Med Chem 27:1057–1066

Kaspar P, Witzel H (1985) Steroid binding to the cytosolic estrogen receptor from rat
uterus. Influence of the orientation of substituents in the 17-position of the 8β- and
8α-series. J Steroid Biochem 23:259–265

Katzenellenbogen JA, Carlson KE, Katzenellenbogen BS (1985) Facile geometric isomerization of phenolic non-steroidal estrogens and antiestrogens: limitations to
the interpretation of experiments characterizing the activity of individual isomers.
J Steroid Biochem 22:589–596

Katzenellenbogen BS, Fang H, Ince BA, Pakdel F, Reese JC, Wooge CH, Wrenn CK
(1993) William L. McGuire Memorial Symposium. Estrogen receptors: ligand dis-
crimination and antiestrogen action. Breast Cancer Res Treat 27:17–26

Kramer VJ, Halferich WG, Bergman A, Klasson Wheler E, Giesy JP (1997) Hydroxylated polychlorinated biphenyl metabolites are antiestrogenic in a stably transfected human breast adenocarcinoma (MCF7) cell line. Toxicol Appl Pharmacol
144:363–376

Kranzfelder G, Schneider M, von Angerer E, Schönenberger H (1980) Entwicklung
neuer Antiöstrogene vom Typ des 3,3'-Dihydroxy-α,β-diäthylstilbens und ihre
Prüfung am DMBA-induzierten, hormonabhängigen Mammacarcinom der SD-
Ratte. J Cancer Res Clin Oncol 97:167–186

Kranzfelder G, Hartmann RW, von Angerer E, Schönenberger H, Bogden AE (1982)
3,4-Bis(3'-hydroxyphenyl)hexane- a new mammary tumor-inhibiting compound. J
Cancer Res Clin Oncol 103:165–180

Kuiper GG, Gustafsson JA (1997) The novel estrogen receptor-beta subtype: potential
role in the cell- and promoter-specific actions of estrogens and anti-estrogens.
FEBS Lett 410:87–90

Kuiper GG, Carlsson B, Grandien K, Enmark E, Haggblad J, Nilsson S, Gustafsson JA
(1997) Comparison of the ligand binding specificity and transcript tissue distribution of estrogen receptors alpha and beta. Endocrinology 138:863–870

Kym PR, Anstead GM, Pinney KG, Wilson SR, Katzenellenbogen JA (1993) Molecular structures, conformational analysis, and preferential modes of binding of 3-
aroyl-2-arylbenzo[b]thiophene estrogen receptor ligands: LY117018 and aryl azide
photoaffinity labeling analogs. J Med Chem 36:3910–3922

Lee YJ, Notides AC, Tsay Y-G, Kende AS (1977) Coumestrol, NBD-norhexestrol, and
dansy-norhexestrol, fluorescent probes for estrogen-binding proteins. Biochemistry 16:2896–2901

Lien LL, Lien EJ (1996) Hormone therapy and phytoestrogens. J Clin Pharm Ther 21:
101–111

Lovely CJ, Gilbert NE, Liberto MM, Sharp DW, Lin YC, Brueggemeier RW (1996) 2-
(Hydroxyalkyl)estradiols: Synthesis and Biological Evaluation. J Med Chem
39:1917–1923

Marselos M, Tomatis L (1992) Diethylstilboestrol: I, pharmacology, toxicology and car-
cinogenicity in humans. Eur J Cancer 28 A:1182–1189

McCague R, Leclercq G, Jordan VC (1988) Nonisomerizable analogues of (Z)- and
(E)-4-hydroxytamoxifen. Synthesis and endocrinological properties of substituted
diphenylbenzocycloheptenes. J Med Chem 31:1285–1290

McCague R, Leclercq G, Legros N, Goodman J, Blackburn GM, Jarmann M, Foster
AB (1989) Derivatives of tamoxifen. Dependence of antiestrogenicity on the 4-
substituent. J Med Chem 32:2527–2533

Meyer T, Koop R, von Angerer E, Holler E (1994) A rapid luciferase transfection assay
for transcription activation effects and stability control of estrogenic drugs in cell
culture. J Cancer Res Clin Oncol 120:359–364

Meyers CY, Lufti HG, Adler S (1997) Transcriptional regulation of estrogen-
responsive genes by non-steroidal estrogens: doisynolic and allenolic acids. J
Steroid Biochem Molec Biol 62:477–489

Meyers CY, Kolb VM, Gass GH, Rao BR, Roos CF, Dandliker WB (1988) Doisynolic-type acids – uterotropically potent estrogens which compete poorly with estradiol for cytosolic estradiol receptors. J Steroid Biochem 31:393–404

Miksicek RJ (1994) Interaction of naturally occurring nonsteroidal estrogens with expressed recombinant human estrogen receptor. J Steroid Biochem Molec Biol 49:153–160

Mittal S, Durani S, Kapil RS (1985) Structure-activity relationship of estrogens: receptor affinity and estrogen antagonist activity of certain (E)- and (Z)-1,2,3-triaryl-2-propen-1-ones. J Med Chem 28:492–497

Napolitano E, Fiaschi R, Hanson RN (1991) Structure-activity relationships of estrogenic ligands: synthesis and evaluation of $(17\alpha,20E)$- and $(17\alpha,20Z)$-21-halo-19-norpregna-1,3,5(10),20-tetraene-3,17β-diols. J Med Chem 34:2754–2759

Napolitano E, Fiaschi R, Herman LW, Hanson RN (1996) Synthesis and estrogen receptor binding of $(17\alpha,20E)$- and $(17\alpha,20Z)$-21-phenylthio- and 21-phenyl-19-nor-pregna-1,3,5(10),20-tetraene-3,17β-diols. Steroids 61:384–389

Newton CJ, Buric R, Trapp T, Brockmeier S, Pagotto U, Stalla GK (1994) The unliganded estrogen receptor (ER) transduces growth factor signals. J Steroid Biochem Molec Biol 48:481–486

Nique F, Van de Velde P, Brémaud J, Hardy M, Philibert D, Teutsch G (1994) 11β-amidoalkoxyphenyl estradiols, a new series of pure antiestrogens. J Steroid Biochem Molec Biol 50:21–29

Nishino Y, Schneider MR, Michna H, von Angerer E (1991) Pharmacological characterisation of a novel estrogen antagonist, ZK 119010, in rats and mice. J Endocrinol 130:409–414

Odum J, Lefevre PA, Tittensor S, Paton D, Routledge EJ, Beresford NA, Sumpter JP, Ashby J (1997) The rodent uterotrophic assay: critical protocol features, studies with nonyl phenols, and comparison with a yeast estrogenicity assay. Regul Toxicol Pharmacol 25:176–188

Palkowitz AD, Glasebrook AL, Thrasher KJ, Hauser KL, Short LL, Phillips DL, Muchl BS, Sato M, Shetler PK, Cullinan GJ, Pell TR, Bryant HU (1997) Discovery and synthesis of [6-hydroxy-3-[4-[2-(1-piperidinyl)ethoxy]phenoxy]-2-(hydroxy-phenyl)]benzo[b]thiophene: a novel, highly potent, selective estrogen receptor modulator. J Med Chem 40:1407–1416

Palomino E, Heeg MJ, Horwitz JP, Polin L, Brooks SC (1994) Skeletal conformations and receptor binding of some 9,11-modified estradiols. J Steroid Biochem Molec Biol 50:75–84

Palomino E, Heeg MJ, Pilat MJ, Hafner M, Polin L, Brooks SC (1997) Crystal structure, receptor binding, and gene regulation of 2- and 4-nitroestradiols. Steroids 61:670–676

Pelletier JD, Labrie F, Poirier D (1994) N-butyl, N-methyl, 11-[3′,17′β-(dihydroxy)-1′,3′,5′(10′)-estratrien-16′α-yl]-9(R/S)-bromo undecanamide: Synthesis and 17β-HSD inhibiting, estrogenic and antiestrogenic activities. Steroids 59:536–547

Pelletier JD, Poirier D (1996) Synthesis and evaluation of estradiol derivatives with 16α-(bromoalkylamide), 16α-(bromoalkyl or 16α-(bromoalkynyl) side chain as inhibitors of 17-hydroxysteroid dehydrogenase type 1 without estrogenic activity. Biorg Med Chem 4:1617–1628

Pento JT, Magarian RA, Wright RJ, King MM, Benjamin EJ (1981) Nonsteroidal estrogens antiestrogens: Biological activity of cyclopropyl analogs of stilbene and stilbenediol. J Pharm Sci 70:399–403

Pilat MJ, Hafner MS, Kral LG, Brooks SC (1993) Differential induction of pS2 and cathepsin D mRNAs by structurally altered estrogens. Biochemistry 32:7009–7015

Poirier D, Labrie C, Mérand Y, Labrie F (1990) Derivatives of ethynylestradiol with oxygenated 17α-alkyl side chain: synthesis and biological activity. J Steroid Biochem Molec Biol 36:133–142

Poirier D, Labrie C, Merand Y, Labrie F (1991) Synthesis and biological activity of 17α-alkynylamide derivatives of estradiol. J Steroid Biochem Molec Biol 38:759–774

Pomper MG, VanBrocklin H, Thieme AM, Thomas RD, Kiesewetter DO, Carlson KE, Mathias CJ, Welch MJ, Katzenellenbogen JA (1990) 11β-Methoxy-, 11β-ethyl- and 17α-ethynyl-substituted 16α-fluoroestradiols: Receptor-based imaging agents with enhanced uptake efficiency and selectivity. J Med Chem 33:3143–3155

Quian X, Abul-Hajj YJ (1990a) Synthesis and biological activity of 4-methylestradiol. J Steroid Biochem 35:745–747

Quian X, Abul-Hajj YJ (1990b) Synthesis and biologic activities of 11β-substituted estradiol as potential antiestrogens. Steroids 55:238–241

Quivy J, Leclerq G, Deblaton M, Henrot P, Velings N, Norberg B, Ervard G, Zeicher M (1996) Synthesis, structure and biological properties of Z-17α-(2-iodovinyl)-11-β=chloromethyl estradiol-17β (Z-CMIV), a high affinity ligand for the characterization of estrogen receptor-positive tumors. J Steroid Biochem Mol Biol 59:103–117

Reese JC, Katzenellenbogen BS (1991) Differential DNA-binding abilities of estrogen receptor occupied with two classes of antiestrogens: studies using human estrogen receptor overexpressed in mammalian cells. Nucleic Acids Res 19:6595–6602

Reiner GCA, Katzenellenbogen BS, Bindal RD, Katzenellenbogen JA (1984) Biological activity and receptor binding of a strongly interacting estrogen in human breast cancer cells. Cancer Res 44:2302–2308

Robertson DW, Katzenellenbogen JA, Long DJ, Rorke EA, Katzenellenbogen BS (1982) Tamoxifen antiestrogens. A comparison of the activity, pharmacokinetics, and metabolic activation of the cis and trans isomers of tamoxifen. J Steroid Biochem 16:1–13

Routledge EJ, Sumpter JP (1997) Structural features of alkylphenolic chemicals associated with estrogenic activity. J Biol Chem 272:3280–3288

Rubin BL, Dorfman AS, Black L, Dorfman RI (1951) Bioassay of estrogens using the mouse uterine response. Endocrinology 49:429–439

Saeed A, Sharma AP, Durani N, Jain R, Durani S, Kapil RS (1990) Structure-activity relationship of antiestrogens. Studies on 2,3-diaryl-1-benzopyrans. J Med Chem 33:3210–3216

Sasson S, Katzenellenbogen JA (1989) Reversible, positive cooperative interaction of 11β-chloromethyl-[^{3}H]estradiol-17β with the calf uterine estrogen receptor. J Steroid Biochem 33:859–865

Schneider M, von Angerer E, Kranzfelder G, Schönenberger H (1980) Mammatumorhemmende Antiöstrogene vom Typ des 3,3'-Dihydroxy-α,β-dialkylstilbens. Arch Pharm (Weinheim) 313:919–925

Schneider MR, Kranzfelder G, von Angerer E, Schönenberger H, Metzler M, Michel RT, Fortmeyer HP, Ruckdeschel G (1981) The tumor-inhibiting effect of diethylstilbestrol-3,4-oxide. J Cancer Res Clin Oncol 100:247–254

Schneider MR, Schönenberger H, Michel RT (1982a) Mammary tumor inhibiting [(1,2-diethyl-1,2-cyclopropanediyl)bis(phenyl)] diacetates. Eur J Med Chem – Chim Ther 17:491–495

Schneider MR, von Angerer E, Schönenberger H (1982b) 5-Acetoxy-2-(3'-acetoxyphenyl)-3-ethyl-1-methyl-1H-indene: a new mammary tumor inhibiting compound. Eur J Med Chem -Chim Ther 17:245–248

Schneider MR, Ball H (1986) 2-Phenylindenes: Development of a new mammary tumor inhibiting antiestrogen by combination of estrogenic side effect lowering structural elements. J Med Chem 29:75–79

Schütze N, Vollmer G, Wünsche W, Grote A, Feit B, Knuppen R (1994) Binding of 2-hydroxyestradiol and 4-hydroxyestradiol to the estrogen receptor of MCF-7 cells in cytosolic extracts and in nuclei of intact cells. Exp Clin Endocrinol 102:399–408

Schwartz JA, Skafar DF (1993) Ligand-mediated modulation of estrogen receptor conformation by estradiol analogs. Biochemistry 32:10109–10115

Sharma AP, Saeed A, Durani S, Kapil RS (1990a) Structure-activity relationship of antiestrogens. Effect of the side chain and its position on the activity of 2,3-diaryl-2H-1-benzopyrans. J Med Chem 33:3216–3222

Sharma AP, Saeed A, Durani S, Kapil RS (1990b) Structure-activity relationship of antiestrogens. Phenolic analogues of 2,3-diaryl-2H-1-benzopyrans. J Med Chem 33:3222–3229

Souttou B, Moretti J-L, Gros J, Guilloteau D, Crepin M (1993) Receptor binding and biological effects of three ^{125}I-iodinated estrogen derivatives in human breast cancer cells (MCF-7). J Steroid Biochem Molec Biol 44:105–112

Symes EK, Bishop PB, Coulson WF, Davies AG (1992) 17α-Z-[125I]iodovinyloestradiol and its 3-acetate: chemical synthesis in vivo distribution studies in the rat. Comparison of tissue accumulation and metabolic stability with 17α-E-[125I]iodovinyl and 16α-[125I]iodo oestradiols. Biochem Pharmacol 44:741–746

Tedesco R, Katzenellenbogen JA, Napolitano E (1997) 7α,11β-Disubstituted estrogens: probes for the shape of the ligand binding pocket in the estrogen receptor. Bioorg Med Chem Lett 7:2919–2924

Tong W, Perkins R, Xing L, Welsh WJ, Sheehan DM (1997) QSAR models for binding of estrogenic compounds to estrogen receptor alpha and beta subtypes. Endocrinology 138:4022–4025

Van de Velde P, Nique F, Bouchoux F, Brémaud J, Hameau M-C, Lucas D, Moratille C, Viet S, Philibert D, Teutsch G (1994) RU 58 668, a new pure antiestrogen inducing a regression of human mammary carcinoma implanted in nude mice. J Steroid Biochem Molec Biol 48:187–196

Van de Velde P, Nique F, Planchon P, Prevost G, Bremaud J, Hameau MC, Magnien V, Philibert D, Teutsch G (1996) RU 58668: further in vitro and in vivo pharmacological data related to its antitumoral activity. J Steroid Biochem Molec Biol 59:449–457

VanderKuur JA, Hafner MS, Christman JK, Brooks SC (1993a) Effects of estradiol-17β analogues on activation of estrogen response element regulated chloramphenicol acetyltransferase expression. Biochemistry 32:7016–7021

VanderKuur JA, Wiese T, Brooks SC (1993b) Influence of estrogen structure on nuclear binding and progesterone receptor induction by the receptor complex. Biochemistry 32:7002–7008

Vollmer G, Wünsche W, Schütze N, Feit B, Knuppen R (1991) Methyl and bromo derivatives of estradiol are agonistic ligands for the estrogen receptor of MCF-7 breast cancer cells. J Steroid Biochem Molec Biol 39:359–366

von Angerer E, Prekajac J, Strohmeier J (1984) 2-Phenylindoles. Relationship between structure, estrogen receptor affinity, and mammary tumor inhibiting activity in the rat. J Med Chem 27:1439–1447

von Angerer E, Knebel N, Kager M, Ganß B (1990) 1-Aminoalkyl-2-phenylindoles as novel pure estrogen antagonists. J Med Chem 33:2635–2640

von Angerer E, Erber S (1992) 3-Alkyl-2-phenylbenzo[b]thiophenes: Nonsteroidal estrogen antagonists with mammary tumor inhibiting activity. J Steroid Biochem Molec Biol 41:557–562

von Angerer E, Biberger C, Holler E, Koop R, Leichtl S (1994) 1-Carbamoylalkyl-2-phenylindoles: Relationship between side chain structure and estrogen antagonism. J Steroid Biochem Molec Biol 49:51–62

von Angerer E (1995) The estrogen receptor as a target for rational drug design. R.G. Landes Company, Austin, Texas

von Angerer E, Biberger C, Leichtl S (1995) Studies on heterocycle-based pure estrogen antagonists. Ann N Y Acad Sci 761:176–191

Wakeling AE, Bowler J (1988a) Novel antiestrogens without partial agonist activity. J Steroid Biochem 31:645–653

Wakeling AE, Bowler J (1988b) Biology and mode of action of pure antiestrogens. J Steroid Biochem Molec Biol 30:141–147

Wakeling AE (1991) Steroidal pure antiestrogens. In: Lippman M, Dickson R (eds) Regulatory mechanisms in breast cancer. Kluwer Academic Publishers, Boston, pp 239–257

Wakeling AE, Dukes M, Bowler J (1991) A potent specific pure antiestrogen with clinical potential. Cancer Res 51:3867–3873

Waller CL, Oprea TI, Chae K, Park HK, Korach KS, Laws SC, Wiese TE, Kelce WR,
    Gray LEJ (1996) Ligand-based identification of environmental estrogens. Chem
    Res Toxicol 9:1240–1248
Wiese TE, Brooks SC (1994) Molecular modeling of steroidal estrogens: novel con-
    formations and their role in biological activity. J Steroid Biochem Molec Biol
    50:61–73
Wiese TE, Polin LA, Palomino E, Brooks SC (1997) Induction of the estrogen specific
    mitogenic response of MCF-7 cells by selected analogues of estradiol-17β: a 3D
    QSAR study. J Med Chem 40:3659–3669
Willson TM, Norris JD, Wagner BL, Asplin I, Baer P, Brown HR, Jones SA, Henke B,
    Sauls H, Wolfe S, Morris DC, McDonnell DP (1997) Dissection of the molecular
    mechanism of action of GW5638, a novel estrogen receptor ligand, provides
    insights into the role of estrogen receptor in bone. Endocrinology 138:3901–3911
Zeicher M, Delcorde A, Quivy J, Dupuis Y, Vervist A, Fruhling J (1996) Radioimaging
    of human breast carcinoma xenografts in mice by [^{123}I]-labeled Z-17α-iodovinyl-
    11β-chloromethyl-estradiol. Nucl Med Biol 23:69–73

Part 2
Molecular Biology of Estrogenic Action

CHAPTER 6
Structure and Function of the Estrogen Receptor

A.K. Hihi and W. Wahli

A. Introduction

The mechanisms which allow steroid hormones to exert their actions within organisms are multiple and complex. These hormones – including estrogens – participate in many events, such as cell proliferation and differentiation. They also have subtle influences on organogenesis. Disturbance of steroid hormone signaling pathways can lead to adverse consequences as severe as cancer.

To understand how steroid hormones are able to deliver their messages, investigators in the 1960s hypothesized that there was a mediator – a cellular hormone receptor – capable of binding the hormone and subsequently interacting with the cellular machinery (JENSEN et al. 1968). The identification and molecular cloning of intracellular receptors for steroid hormones in the last 15 years opened a new field of research devoted to members of the nuclear hormone receptor superfamily as transcription factors (MANGELSDORF et al. 1995). Although estrogen and other steroid hormones can elicit changes in cellular behavior via transcription-independent mechanisms, we will focus here on the intracellular estrogen receptor (ER), for which two isotypes are known, ERα and ERβ. A look at ER structure and amino acid organization in structural domains will permit us to delineate the functional regions of the ER. Its characteristics as a transcription factor, such as DNA binding, ligand binding, transactivation properties, interaction with co-factors and cross-talk with other signaling routes, will be discussed in relation to its structure. Finally, a view of ER functions in the organism will be proposed which takes advantage of a knock-out mouse model in which the ER gene has been invalidated.

B. Structure of the Estrogen Receptor

The ER belongs to the steroid/thyroid hormone-receptor family of transcription factors. It was one of the first nuclear receptors to be cloned in humans (GREEN et al. 1986) and has been identified in many species, such as mouse (WHITE et al. 1987), rat (KOIKE et al. 1987) and Xenopus (WEILER et al. 1987). A new isotype of ER, termed ERβ, was recently discovered in human (MOSSELMAN et al. 1996), mouse (TREMBLAY et al. 1997) and rat (KUIPER et al. 1996). Both isotypes, α (for the "classic" ER) and β, are encoded by a single-copy gene. These genes are localized on different chromosomes. ERα is on chromosome 6 in human and on chomosome 10 in mouse (Sʟᴜʏsᴇʀ et al. 1988), while ERβ is on chromosome 14 in human (Eɴᴍᴀʀᴋ et al. 1997) and on chromosome 12 in mouse (Tʀᴇᴍʙʟᴀʏ et al. 1997).

Like other members of the nuclear receptor family, the ER structure was found to be modular, as first shown for the glucocorticoid receptor (GR) (Eᴠᴀɴs 1988). Limited proteolysis of the GR first demonstrated that the DNA-binding activity and the hormone-binding activity can be separated, which is reflected at the gene-structure level. The organization of ER protein, like that of most of the other nuclear receptors, comprises six regions A–F (Mᴀɴɢᴇʟs-ᴅᴏʀꜰ et al. 1995) (Fig. 1). The N-terminal A/B domain is weakly conserved between receptors and has a ligand-independent transactivation function (AF-1), the highly conserved C domain is the DNA-binding domain and the C-terminal ligand-binding domain (E/F region) comprises the ligand-dependent transcriptional activation function (AF-2) and allows multiple interactions with cellular proteins by providing interaction surfaces. This latter function, however, is not limited to the ligand-binding domain.

I. The A/B Domain

This region is variable in length and in amino acid composition among different species (Gʀᴇᴇɴ and Cʜᴀᴍʙᴏɴ 1991). At first sight, no conserved structure could be deduced from this domain. However, there are regions of similarity, in particular some phosphorylation sites implicated in ligand-independent transactivation function (Kᴀᴛᴏ et al. 1995; Bᴜɴᴏɴᴇ et al. 1996) and DNA binding (our unpublished results). Since this region is implicated in cross-talk with growth factors and in transcriptional activity of ERs, investigating how cellular or intramolecular constraints modulate its three-dimensional structural properties will help us to better understand the structure–function relationship of the whole ER molecule.

II. The DNA-Binding Domain

The DNA-binding domain (DBD) is the most conserved domain between ERs and other nuclear receptors. It has a core region of 66 amino acids, which folds into two "zinc fingers", a DNA-binding motif also found in other DNA-binding proteins (for a schematic representation, see Fig. 1). Each zinc finger coordinates a zinc atom via two pairs of cysteine residues (Fig. 1) (Gʀᴇᴇɴ and Cʜᴀᴍʙᴏɴ 1991). The three-dimensional structure of this domain has been resolved (Fig. 2A) by nuclear magnetic resonance (NMR) (Sᴄʜᴡᴀʙᴇ et al. 1990) and X-ray crystallography (Sᴄʜᴡᴀʙᴇ et al. 1993). The overall structure is globular and can be subdivided into two parts, each containing one zinc finger. The first part begins with a short antiparallel β sheet, which orients the residues contacting the DNA phosphate backbone, and is followed by a helical structure. ER binds DNA as a dimer (Kᴜᴍᴀʀ and Cʜᴀᴍʙᴏɴ 1988), and the helical region at the C-terminus of the first zinc finger of each partner in the

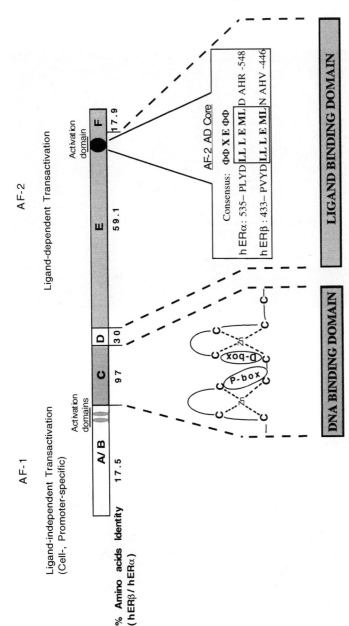

Fig. 1. Structure of estrogen receptor (ER). Structural and functional domains are depicted, with the zinc-finger motif of the DNA-binding domain drawn schematically. Activation functions 1 and 2 (AF-1 and AF-2, respectively) are located in the N-terminal A/B domain (AF-1) and in the C-terminal E/F domain (AF-2). The relative amino acid identity of the human ERα and ERβ domains is presented. The amino acid sequence of ER AF-2 core autonomous domain (AF-2 AD core) is compared between human ERα and ERβ. Φ corresponds to hydrophobic amino acids in this region

A

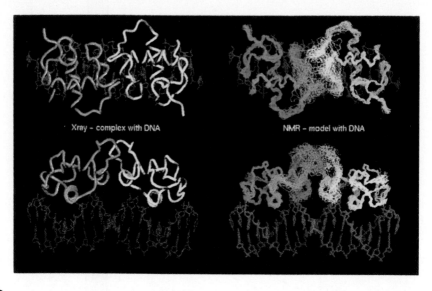

B

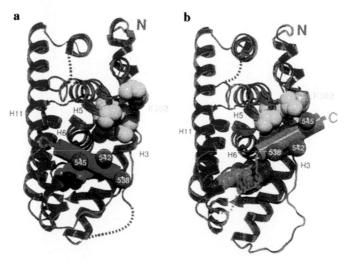

Fig. 2A,B. Three-dimensional structure of the estrogen receptor (ER) DNA-binding domain and ligand-binding domain. **A** Orthogonal views (*top* and *bottom*) of the X-ray structure of the ER DNA-binding domain (DBD)–DNA complex (*left*) and nuclear magnetic resonance structures modeled as a dimer with DNA (*right*). The two monomers are shown in *blue* and *yellow* (reproduced from Schwabe et al. 1993). **B** Crystal structure of liganded ER ligand-binding domain with estradiol (*a*) and raloxifene (*b*). The position of helix 12 (identified by amino acids 538, 542 and 545) is favorable for transcriptional activity with estradiol, while its different position induced by raloxifene blocks its transcriptional activity (reproduced from Brzozowski et al. 1997)

dimer specifically contacts nucleotides in the DNA major groove. This region comprises the P-box and determines the DNA-binding properties of the receptor in terms of target specificity. For ERs, the P-box is composed of the amino acids CEGCKA and mutations in this region alter ER binding to DNA. The second zinc finger of the DBD encompasses a region called the D-box between the first two cysteines of this zinc finger. The D-box is important for the dimerization of the DBD moieties of ER. This protein–protein interaction is important in maintaining the proper spacing of each ER molecule on DNA (GREEN and CHAMBON 1991).

NMR studies of the DBD bound to DNA in solution revealed that DNA binding induces an ordering of the molecule. Indeed, internal residues with no apparent structure in solution contribute to the cooperative interaction of DBD monomers (Fig. 2A).

III. The Ligand-Binding Domain

This so-called E/F domain is the region of the receptor which binds the ligand and provides some interacting surfaces on the receptor to allow interconnections with other cellular proteins. The ligand-binding domain (LBD) is a hydrophobic region of about 250 amino acids whose integrity is necessary for hormone binding. The LBD can be obtained as a trypsin-resistant fragment only when it is loaded with the hormone, indicating a ligand-dependent change in ER structure. Upon this transition, new interfaces are likely to be provided for interaction with co-factors (GREEN and CHAMBON 1991). The crystal structure of the liganded ERα has been resolved (BRZOZOWSKI et al. 1997). The LBD consists of an antiparallel α-helical sandwich containing 12 α-helices. This corresponds to the canonical structure for the LBD of nuclear receptors (WURTZ et al. 1996). The LBD contains a ligand-binding pocket which is kept away from the external environment. This hormone cavity corresponds to an important part of the LBD hydrophobic core and its volume is twice that of estradiol, which may somehow explain the ER's ability to bind different steroidal and non-steroidal ligands.

Ligand-dependent activity of nuclear receptors relies on a correct positioning of helix 12. In the estradiol–ER complex, helix 12 "locks" the ligand-binding cavity and is positioned close to helices 3, 5/6 and 11. This folds the LBD into a transcriptionally competent AF-2 configuration, which is able to interact with co-factors and activate gene expression. In contrast, the ER antagonist raloxifene induces a different positioning of helix 12, which prevents a proper alignment with the ligand cavity and leads to a transcriptionally incompetent AF-2 (Fig. 2B). The F domain of the ER was not included in the X-ray studies and shows no real sequence similarities when ER from different species are compared. Its length, however, is conserved within each ER isotype, α or β, from different species. This might reflect some important structural constraints for receptor activity (NICHOLS et al. 1998).

C. ER Functions and Transcription

The ER structure reveals a modular organization which is found throughout the nuclear receptor family. This allows us to delineate functions conveyed by each domain. While activities such as DNA binding and ligand binding are separable, it is reasonable to think that each domain will influence the activity of other parts of the receptor. It is possible to distinguish three kinds of functions and activities corresponding to the ER domains, as defined previously: a DNA-binding activity, a ligand-independent activity controlled by growth factors, and a ligand-dependent activity modulated by co-factor interactions.

I. ER DNA-Binding Properties

As a transcription factor, ER modulates the transcriptional activity of target genes in a ligand-dependent manner or under the influence of growth factors (see Part C, Sect. II). To do so, ER has to bind the target promoter by recognizing a specific DNA called hormone response element (HRE). ER, which binds DNA as a dimer, recognizes a palindromic DNA element of 13 bp called estrogen response element (ERE). The consensus ERE contains a repeat of the core AGGTCA motif organized as a palindrome with a three-nucleotide spacing between the half-sites (AGGTCANNNTGACCT, MARTINEZ and WAHLI 1991). EREs are found in ER-regulated genes such as vitellogenin, ovalbumin, prolactin and c-fos (WALKER et al. 1984; MARTINEZ and WAHLI 1991). However, EREs have not been found in all estrogen-regulated genes. This might reflect ER- or ERE-independent effects of estrogen, which await further investigation.

To act through EREs via an efficient DNA binding, ER needs its DBD with its specific structural determinants (P-box and D-box) (GREEN and CHAMBON 1991). Bacterially expressed DBD is capable of binding target sequences independently of the rest of the molecule. In the context of the whole molecule, however, different constraints require the intervention of other cellular factors not yet cloned (LANDEL et al. 1994) as well as a correct phosphorylation status (DENTON et al. 1992). The phosphorylation site important for DNA binding lies close to but outside of the DBD (our unpublished results) and illustrates intramolecular ER-domain cross-talk. An intermolecular cross-talk can also take place on EREs with other transcription factors able to bind these DNA sequences. Within the nuclear receptor family, peroxisome proliferator-activated receptor (PPAR), a nuclear receptor for fatty acid and fatty-acid derivatives heterodimerizes with retinoid-X receptor (RXR), the receptor for 9-cis-retinoic acid (LEMBERGER et al. 1996). The PPAR–RXR heterodimer is able to bind an ERE, but less efficiently than an ER, which can lead to a situation of competition for the response element. The PPAR–RXR complex does not stimulate transcription in the context of the vitellogenin promoter, which contains an ERE, but inhibits ER-mediated transactivation in transfection experiments (KELLER et al. 1995). Genes influenced by estrogens and fatty acids could then be co-regulated by PPAR and ER, most likely in a promoter-dependent manner.

II. ER Ligand-Independent Activity

While ligand binding triggers major conformational changes in ERs, receptor activity is possible in the absence of any hormone bound to the LBD (CHO and KATZENELLENBOGEN 1993). It is important to distinguish the latter situation from antagonist binding, which elicits an active inhibition of the ER transactivation capacity. Ligand-independent activity of ERs is supported by its *N*-terminal A/B domain. This domain contains a separable transcriptional function (AF-1) able to activate transcription, exhibiting cell-type and promoter-context specificity (GRONEMEYER 1991). Studies using chimeric molecules composed of the A/B region of human ERs and the heterologous Gal4 DNA-binding domain allowed researchers to define this AF1 function. Dissection of this region indicated that independent activation regions were located in the A/B domain. These regions were shown to synergize with the hormone-dependent activity of ERs, illustrating the intramolecular domain cross-talk mentioned above (METZGER et al. 1995).

Besides this intramolecular cross-talk, a ligand-independent activity of ER is triggered by growth factors such as epidermal growth factor (EGF). A first hint of EGF–ER convergence was suggested by the inhibition of EGF-stimulated uterine DNA synthesis, when ER transcriptional potency was totally blocked by antiestrogens. This was reinforced by the observation that EGF treatment was able to modulate ER phosphorylation in ovariectomized mice, ruling out that estrogen was necessary to allow EGF to signal through ER (IGNAR-TROWBRIDGE et al. 1992). A decisive step in the understanding of the molecular mechanism of EGF's effect on ER was accomplished with the unraveling of the pathway linking EGF, its membrane receptor activation, the mobilization of the mitogen-activated protein kinase (MAPK) cascade and direct ER phosphorylation by MAPK. The phosphorylation site is located in the A/B domain (serine 118 in human ERs) and is conserved among species (KATO et al. 1995; BUNONE et al. 1996). Cross-coupling of ER ligand-independent function with other growth factors can also occur using the same mechanism of phosphorylation (KATO et al. 1995), which is prone to regulation because of the rapidity and reversibility of the phosphorylation process.

III. Ligand-Dependent Transcriptional Activity

In order to activate transcription, the DNA-bound ER needs to be in a conformation that allows favorable interactions with the transcriptional machinery. Agonist ligands such as estrogen are capable of inducing such structural changes, which unmask a ligand-dependent activation function (AF-2) (BRZOZOWSKI et al. 1997). A careful analysis of AF-2 led to the identification of a conserved region responsible for this activity lying in the C-terminal part of the LBD. In this region, mutations impairing hormone binding have a negative impact on ER transcriptional activity. This helped to define a core autonomous domain (AF-2 AD core, see Fig. 1) able to mediate gene transcription (DANIELAN et al. 1992). Another set of mutations, which had an effect only on transactivation, suggested that there might be some proteins interacting with the ER that are necessary for transcriptional activation. Such co-factors were discovered using the yeast two-hybrid system (ONATE et al. 1995) or protein–protein interaction assays (CAVAILLÈS et al. 1994; HALACHMI et al. 1994). These investigations revealed two kinds of interacting molecules. The first category consists of molecules that interact with ER and other nuclear receptors in a ligand-dependent way, but do not directly help ER-driven transcription. An example is transcription intermediary factor 1 (TIF1), which is thought to mediate ER interactions with those proteins implicated in chromatin remodeling that indirectly act on transcription (GLASS et al. 1997).

The second category corresponds to real co-activators, and we will focus on three of them: steroid receptor coactivator 1/nuclear-receptor co-activator 1 (SRC1/NCoA1) (ONATE et al. 1995; KALKHOVEN et al. 1998), transcription intermediary factor 2/glucocorticoid-receptor interacting protein 1 (TIF2/GRIP1) (VOEGEL et al. 1996) and cyclic AMP response element-binding protein and its p300 homologue (CBP/p300) (CHAKRAVARTI et al. 1996; KAMEI et al. 1996).

SRC1/NCoA1 and TIF2/GRIP1 are proteins of 120 kDa/156 kDa and 160 kDa, respectively. They have the same general organization (Fig. 3) and they enhance, in a ligand-dependent way, transcription mediated by various steroid receptors (estrogen, glucocorticoid and progesterone receptor), thyroid hormone receptor and RXR. The domain organization of these co-activators (Fig. 3) shows the presence in the N-terminal region of a basic helix loop helix region (bHLH) and a PER gene–aryl hydrocarbon nuclear translocator–SIM gene (PAS) region capable of functioning as a dimerization motif, a central serine/threonine-rich region which interacts with ER, and a C-terminal region which can interact with CBP/p300 (GLASS et al. 1997). How these co-activators are regulated spatio-temporally and how they are shared among the nuclear receptor family members remains to be discovered.

CBP/p300 are large proteins (300 kDa) which are conserved from *Caenorhabditis elegans* to mammals. They are thought to act as an integrator complex, interpreting many signals through their several domains. Indeed, they contact proteins such as activator protein 1 transcription factor complex (AP-1), cyclic AMP response element-binding protein (CREB), myogenic transcription factor D (MyoD) and nuclear receptors. Nuclear-receptor interaction occurs via the N-terminal region of CBP/p300, while the SRC-1 interaction domain is found in the C-terminus. Since SRC-1 and CBP/p300 can also interact together (GLASS et al. 1997), an additional level of complexity for

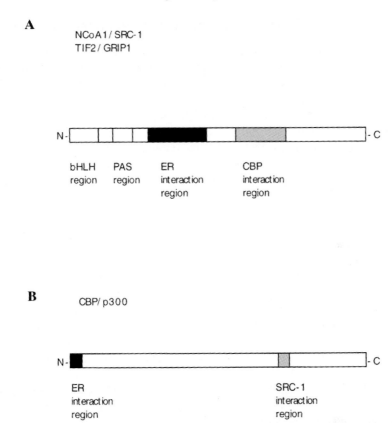

A

NCoA1/SRC-1
TIF2/GRIP1

N -[ ] - C

bHLH   PAS   ER   CBP
region   region   interaction   interaction
                       region   region

B

CBP/p300

N - [ ] - C

ER   SRC-1
interaction   interaction
region   region

Fig. 3A,B. Schematic structure of estrogen receptor (ER) co-activators. The localization of ER-interacting regions on nuclear-receptor co-activator 1 (*NCoA1*)/steroid receptor coactivator 1 (*SRC1*), transcription intermediary factor 2 (*TIF2*)/glucocorticoid receptor interacting protein 1 (*GRIP1*) (**A**) as well as on cyclic AMP response element-binding protein and its homologue (CBP/p300) (**B**) are schematically represented

ER-transcription activation is provided. The molecular mechanism of co-activation is not yet clarified, but could involve the histone acetyl transferase activity of CBP and SRC-1.

Ligand-dependent transcriptional activity of ERs can also be stimulated by general transcription factors. For instance, CCAAT-sequence transcription factor/nuclear factor 1 synergistically interacts with ERs, supposedly through a soluble bridging factor (MARTINEZ et al. 1991), while in the context of the vitellogenin A1 io (upstream) promoter, Sp1 synergistically acts in concert with ERs to activate estrogen-dependent transcription (BATISTUZZO DE MEDEIROS et al. 1997). How the ER interacts with general transcription factors in the presence of co-factors is not known at present.

D. ER Isotype Diversity Generates Specificity

Convergence of different transcription factors with ER signaling adds complexity to the study of specific ER effects. This complexity is enhanced by the occurrence of the two ER isotypes. Indeed, ERα and ERβ are capable of interacting directly, forming heterodimers in addition to homodimers (Cowley et al. 1997; Pettersson et al. 1997). Their tissue expression sometimes overlaps but their ligand-binding properties are not exactly the same (Kuiper et al. 1997). These observations suggest that ERα and ERβ are capable of modulating gene transcription depending on a specific local situation. Below, we will examine some features of ERα and ERβ, giving insights into ER signaling.

I. ERα and ERβ Heterodimerization

Co-expression of ERα and ERβ might find a special meaning with the discovery that the two isotypes form heterodimeric complexes. The heterodimer has the same specificity as ER homodimers for the palindromic ERE. Moreover, the co-expression in cellular systems of both isotypes showed no inhibitory effects on ERα-driven transcription, indicating that the heterodimers are likely transcriptional effectors in vivo (Cowley et al. 1997; Pettersson et al. 1997).

Investigations of whether potential alternative signaling pathways exist in cells expressing both subtypes and whether novel EREs, in addition to the "classic" ERE, are responsive to the heterodimers, remain to be carried out.

II. Tissue Distribution and Ligand Binding

The control of different estrogen-responsive promoters must depend on the selective effects of the two ER isotypes. In rat, ERα and ERβ transcripts were found to be differentially expressed to a certain extent. While they are both well expressed in ovary and uterus, their relative amounts vary in other tissues. For instance, ERα is relatively more abundant in testis and epididymis, while ERβ is expressed at high levels in prostate, bladder and lung. Furthermore, there is only one isotype detected in some organs, such as liver, kidney and adrenal gland for ERα, and brain for ERβ. Such a differential pattern is not fully understood with respect to functionality, but is probably linked to specific target-gene expression (Kuiper et al. 1997).

An additional level of diversity is based on the ligand-binding capacity of the two isotypes. In vitro ligand-binding studies were performed for the rat isotypes. ERα and ERβ displayed the same rank of affinity for at least two antiestrogens, with ERβ binding more strongly than ERα to hydroxytamoxifen ($2.5\leftrightarrow$) and to the compound N-n-butyl-N-methyl-11-(3,17beta-dihydroxyoestra-1,3,5(10)-trien-7-α-yl) undecamide (ICI-164384) ($2.3\leftrightarrow$). Interestingly, a reversed order of affinity was observed for estrogens (Kuiper et al. 1997). An alternative strategy was used for the human isotypes to sort

Table 1. Effective concentration (EC_{50}) transcription activation values of different estrogens with human estrogen receptors ERα and ERβ. The EC_{50} values were deduced from dose–response analyses in a cell-based system using transfection procedures (adapted from BARKHEM et al. 1998)

	Estradiol (E2)	Diethylstilbestrol	17α-E2	17α-Ethynyl, 17β-E2	$16\beta, 17\alpha$- Epiestrol	Genistein
Human ERα	50 pM	0.2 nM	12 nM	0.06 nM	72 nM	38 nM
Human ERβ	0.2 nM	0.4 nM	25 nM	2.1 nM	11 nM	9 nM

selective ligands. A reporter gene under the control of EREs was stably introduced into a human embryonic kidney cell line (HEK 293 cell line). Human ERα and ERβ were then transiently transfected. The ligand dose–response curves obtained allowed us to establish preferences of agonist ligands for ERα or ERβ (Table 1). Furthermore, in this same system, it was demonstrated that 4-hydroxy-tamoxifen and ICI-164384 have a partial agonist effect on ERα, but behave as pure antagonists with ERβ (BARKHEM et al. 1998).

The design of selective ligands for ERα and ERβ will allow more refinement in analyzing isotype-specific functions, especially in tissues where they are both expressed.

III. ERα and ERβ Differential Activity

While ERα and ERβ exhibit almost the same type of transcriptional profile on a classical ERE, it appears that a differential activity can be detected in the context of estrogen-responsive promoters containing other types of binding sites. Indeed, in the case of the insulin-like growth factor gene, the effects of estrogens are mediated through an AP-1 binding site (UMAYAHARA et al. 1994). The ER is thought to have protein–protein interactions with the DNA-bound AP-1 complex without necessarily having direct contacts with DNA itself. In this model, tamoxifen was found to have agonistic effects in uterine cell lines but not in breast cell lines, which could explain some estrogen-like effects of tamoxifen in uterus (WEBB et al. 1995). The behavior of ERβ was also examined and has revealed some interesting features. When considering an AP-1 dependent response element, ERα and ERβ were both capable of mediating a transcriptional activation in response to tamoxifen. Surprisingly, however, the ligand-dependent activity profile of ERβ was reversed compared with that of ERα when using another antiestrogen, raloxifene, which had antagonistic effects on ERα and agonistic effects on ERβ. Estradiol, however, was able to activate ERα but had no effect on ERβ-mediated transcription (PAECH et al. 1997). This suggests that estrogens could be interpreted differently by the cell, depending on which of the two ER isotypes is active at complex promoters, such as those showing an AP-1-dependent activity. The consequences of this model of ER action could be dramatic in terms of antiestrogen treatments and could open perspectives to search for selective drugs and differential effects of endogenous estrogen derivatives.

Finally, the existence of two ER isotypes brings back the question of the physiological role of each of them. The use of cellular and animal models for the purpose of solving this question will be very informative.

E. ERα Knock-Out Mouse: A Functional Model

The effects of estrogens on gene transcription are supposed to occur mainly through the ER pathway. To investigate this pathway, a knock-out mouse model was developed (ERKO), where the ERα gene was disrupted (LUBAHN et al. 1993). ERKO mice are viable, but are sterile and show important defects in the reproductive system in both sexes. Female ERKO have hypoplastic uteri and regressed ovaries, while spermatogenesis is severely affected in males (EDDY et al. 1996). These defects point out that ERα plays a central role in reproductive organ formation. However, the growth of small follicles was not blocked, which raised the possibility of an alternative intraovarian pathway of estrogen action.

Another surprising finding came from observation of the mouse cardio-vascular system, where estrogen exerts protective effects by inhibiting the development of atherosclerotic and injury-induced vascular lesions. Estrogen is still capable of inhibiting vascular injury in the ERKO mice to the same extent as in wild-type animals. Since ERβ is expressed in aortic tissues, the current hypothesis proposes the existence of a novel estrogen-responsive pathway with ERβ as a likely mediator (IAFRATI et al. 1997).

Finally, a still different situation is observed in ERKO mice for a particular estrogen-responsive gene, lactoferrin, expressed in uterus. In ERKO, lactoferrin is induced by catecholestrogens, which are ER ligands and are implicated in embryo implantation. The ER antagonist 7-α-[9-(4,4,5,5,5-pentafluoro-pentylsulfinyl)nonyl]estra-1,3,5(10)-triene-3,17-β-diol (ICI-182780) is not capable of repressing this catecholestrogen-dependent induction. This suggests that the regulation of an estrogen-responsive gene can occur through pathways distinct from those of both ERα and ERβ (DAS et al. 1997).

Altogether, the knowledge gained from the ERKO mouse is useful at different levels. This model allowed LUBAHN and co-workers (1993) to postulate the existence of a second ER and to identify a few ERα-dependent mechanisms. The major surprise is the relatively mild ERKO phenotype, which might reflect a functional redundancy due to ERβ, along with putative ER-independent effects of estrogen. Further use of this animal model will probably open new perspectives in understanding ER functions in the whole organism.

F. Future Directions

The careful study of ER structures, including regions and amino acids important for receptor functions, has already led to many breakthroughs since the discovery of ERs. The generation of tissue-specific ER modulators should be of importance in selectively targeting the desired organs. A first glimpse was given by raloxifene, which has therapeutic effects on heart and bone disorders in women without raising the likelihood of uterine cancer (GUSTAFSSON 1998). Future efforts will make use of the three-dimensional structure of the LBD to design structure-based ligands with desired characteristics.

Knowledge of selective functions of the two isotypes will involve the generation of ERβ knock-out mice, which will continue to address the question of functional redundancy versus isotype-specific roles.

Modern methods for studying differential gene expression will help to trace the molecular action of ERs in complex diseases, in which estrogen is implicated. Indeed, in pathologies such as osteoporosis, atherosclerosis or Alzheimer's disease, preventative effects are attributed to estrogen but the mode of action of the hormone in these processes is not clear as yet.

Finally, ER-function elucidation will also depend on a global analysis of the combined effects of ligands and growth factors on co-activators. Since these latter molecules are in close contact with ERs, it will be important to address how their transcriptional capacity is modulated – for instance, via phosphorylation mechanisms – and how it affects ER functions.

Acknowledgements. The authors wish to thank Daniel Robyr, Gwendal Lazennec and Béatrice Desvergne for careful reading of the manuscript. Work done in the authors' laboratory was supported by the Swiss National Science Foundation and the Etat de Vaud.

References

Barkhem T, Carlsson, B, Nilsson Y, Enmark E, Gustafsson JA, Nilsson S (1998) Differential response of estrogen receptor α and estrogen receptor β to partial agonists/antagonists. Mol Pharmacol 54:105–112

Batistuzzo de Medeiros SR, Krey G, Hihi AK, Wahli W (1997) Functional interactions between the estrogen receptor and the transcription activator Sp1 regulate the estrogen-dependent transcriptional activity of the vitellogenin A1 io promoter. J Biol Chem 272:18250–18260

Brzozowski AM, Pike ACW, Dauter Z, Hubbard RE, Bonn T, Engström O, Öhman L, Greene G, Gustafsson JA, Carlquist M (1997) Molecular basis of agonism in the estrogen receptor. Nature 389:753–758

Bunone G, Briand PA, Miksicek RJ, Picard D (1996) Activation of the unliganded estrogen receptor by EGF involves the MAP kinase pathway and direct phosphorylation. EMBO J 15:2174–2183

Cavaillès V, Dauvois S, Daniellan PS, Parker MG (1994) Interaction of proteins with transcriptionally active estrogen receptors. Proc Natl Acad Sci USA 91: 10009–10013

Chakravarti D, Lamorte VJ, Nelson MC, Nakajima T, Schulman IG, Juguilon H, Montminy M, Evans RM (1996) Role of CBP/p300 in nuclear receptor signalling. Nature 383:99–103

Cho H, Katzenellenbogen B (1993) Synergistic activation of estrogen receptor mediated transcription by estradiol and protein kinase activators. Mol Endocrinol 7:441–452

Cowley SM, Hoare S, Mosselman S, Parker MG (1997) Estrogen receptors alpha and beta form heterodimers on DNA. J Biol Chem 272:19858–19862

Danielan PS, White R, Lees JA, Parker MG (1992) Identification of a conserved region required for hormone-dependent transcriptional activation by steroid hormone receptors. EMBO J 11:1025–1033

Das SK, Taylor, JA, Korach KS, Paria BC, Dey SK, Lubahn DB (1997) Estrogenic responses in estrogen receptor-α deficient mice reveal a distinct estrogen signaling pathway. Proc Natl Acad Sci USA 94:12786–12791

Denton RR, Koszewski NJ, Notides AC (1992) Estrogen receptor phosphorylation. Hormonal dependence and consequence on specific DNA binding. J Biol Chem 267:7263–7268

Eddy EM, Washburn TF, Bunch DO, Goulding EH, Gladen BC, Lubahn DB, Korach KS (1996) Targeted disruption of the estrogen receptor gene in male mice causes alteration of spermatogenesis and infertility. Endocrinology 137:4796–4805

Enmark E, Peltohuikko M, Grandien K, Lagercrantz S, Lagercrantz J, Fried G, Nordenskjold M, Gustafsson JA (1997) Human estrogen receptor beta-gene structure, chromosomal localisation and expression pattern. J Clin Endocrinol Metab 82:4258–4265

Evans RM (1988) The steroid and thyroid hormone receptor superfamily. Science 240:889–895

Glass CK, Rose DW, Rosenfeld MG (1997) Nuclear receptor coactivators. Curr Opin Cell Biol 9:222–232

Green S, Chambon P (1991) The estrogen receptor: from perception to mechanism. In: Parker MG (ed) Nuclear hormone receptors. Academic, London, pp 15–38

Green S, Walter P, Kumar V, Krust A, Bornert JM, Argos P, Chambon P (1986) Human estrogen receptor cDNA: sequence, expression and homology to v-erb-A. Nature 320:134–139

Gronemeyer H (1991) Transcription activation by estrogen and progesterone receptors. Annu Rev Genet 25:89–123

Gustafsson JA (1998) Raloxifene: magic bullet for heart and bone? Nat Med 4:152–153

Halachmi S, Marden E, Martin G, Mac Kay HCA, Brown M (1994) Estrogen receptor-associated proteins: possible mediators of hormone-induced transcription. Science 264:1455–1458

Iafrati MD, Karas RH, Aronovitz M, Kim S, Sullivan TR, Lubahn DB, O'Donnell TF, Korach KS, Mendelsohn ME (1997) Estrogen inhibits the vascular injury response in estrogen receptor α-deficient mouse. Nat Med 3:545–548

Ignar-Trowbridge DM, Nelson KG, Bidwell MC, Curtis SW, Washburn,TF, McLachlan JA, Korach KS (1992) Coupling of dual signaling pathways: epidermal growth factor action involves the estrogen receptor. Proc Natl Acad Sci USA 89: 4658–4662

Jensen EV, Suzuki T, Kawashima T, Stumpf WE, Jungblut PW, de Sombre ER (1968) A two-step mechanism for the interaction of estradiol with rat uterus. Proc Natl Acad Sci USA 59:632–638

Kalkhoven E, Valentine JE, Heery DM, Parker MG (1998) Isoforms of steroid receptor co-activator 1 differ in their ability to potentiate transcription by the estrogen receptor. EMBO J 17:232–243

Kamei Y, Xu L, Heinzel T, Torchia J, Kurokawa R, Gloss B, Lin SC, Heyman RA, Rose DW, Glass CK, Rosenfeld MG (1996) A CBP integrator complex mediates transcriptional activation and AP-1 inhibition by nuclear receptors. Cell 85:403–414

Kato S, Endoh H, Masuhiro Y, Kitamoto T, Uchiyama S, Sasaki H, Masushige S, Gotoh Y, Nishida E, Kawashima H, et al. (1995) Activation of the estrogen receptor through phosphorylation by mitogen-activated protein kinase. Science 270:1491–1494

Keller H, Givel F, Perroud M, Wahli W (1995) Signaling cross-talk between peroxisome-proliferator activated receptor retinoid x receptor and estrogen receptor through estrogen response elements. Mol Endocrinol 9:794–804

Koike S, Sakai M, Maramatsu M (1987) Molecular cloning and characterization of rat estrogen receptor cDNA. Nucleic Acids Res 15:2499–2513

Kuiper GG, Enmark E, Pelto-Huikko M, Nilsson S, Gustafsson JA (1996) Cloning of a novel receptor expressed in rat prostate and ovary. Proc Natl Acad Sci USA 93:5925–5930

Kuiper GG, Carlsson B, Grandien K, Enmark E, Haggblad J, Nilsson S, Gustafsson JA (1997) Comparison of the ligand binding specificity and transcript tissue distribution of estrogen receptors alpha and beta. Endocrinology 138:863–870

Kumar V, Chambon P (1988) The estrogen receptor binds tightly to its responsive element as a ligand-induced homodimer. Cell 55:145–156

Landel CC, Kushner PJ, Greene GL (1994) The interaction of human estrogen receptor with DNA is modulated by receptor-associated proteins. Mol Endocrinol 8:1407–1419

Lemberger T, Desvergne B, Wahli W (1996) Peroxisome proliferator-activated receptors: a nuclear receptor signaling pathway in lipid physiology. Ann Rev Cell Dev Biol 12:335–363

Lubahn DB, Moyer JS, Golding TS, Couse JF, Korach KS, Smithies O (1993) Alteration of reproductive function but not prenatal sexual development after insertional disruption of the mouse estrogen receptor gene Proc Natl Acad Sci USA 90: 11162–11166

Mangelsdorf DJ, Thummel C., Beato M, Herrlich P, Schütz G, Umesono K, Blumberg BP, Kastner P, Mark M, Chambon P, Evans RM (1995) The nuclear receptor superfamily: the second decade. Cell 83:835–839

Martinez E, Wahli W (1991) Characterization of hormone response element. In: Parker, M G (ed) Nuclear hormone receptors. Academic, London, pp 125–153

Martinez E, Dusserre Y, Wahli W, Mermod N (1991) Synergistic transcriptional activation by CTF/NF-I and the estrogen receptor involves stabilized interactions with a limiting target factor. Mol Cell Biol 11:2937–2945

Metzger D, Ali S, Bornert J, Chambon P (1995) Characterization of the amino-terminal transcriptional activation function of the human estrogen receptor in animal and yeast cells. J Biol Chem 270:9535–9542

Mosselman S, Polman J, Dijkema R (1996) ERβ: identification and characterization of a novel human estrogen receptor. FEBS Lett 392:49–53

Nichols M, Rientjes JMJ, Stewart FA (1998) Different positioning of the ligand-binding domain helix 12 and the F domain of the estrogen receptor accounts for functional differences between agonists and antagonists. EMBO J 17:765–773

Onate SA, Tsai SY, Tsai M, O'Malley BW (1995) Sequence and characterization of a coactivator for the steroid hormone receptor superfamily. Science 270:1354–1357

Paech K, Webb P, Kuiper GG, Nilsson S, Gustafsson J, Kushner PJ, Scanlan TS (1997) Differential ligand activation of estrogen receptors ERalpha and ERbeta at AP1 sites. Science 277:1508–1510

Pettersson K, Grandien K, Kuiper GG, Gustafsson JA (1997) Mouse estrogen receptor beta forms estrogen response element-binding heterodimers with estrogen receptor alpha. Mol Endocrinol 11:1486–1496

Schwabe JWR, Neuhaus D, Rhodes D (1990) Solution structure of the DNA-binding domain of the estrogen receptor. Nature 348:458–461

Schwabe JW, Chapman L, Finch JT, Rhodes D (1993) The crystal structure of the estrogen receptor DNA-binding domain bound to DNA: how receptors discriminate between their response elements. Cell 75:567–578

Sluyser M, Rijkers AW, de Goeij CC, Parker M, Hilkens J (1988) Assignment of estradiol receptor gene to mouse chromosome 10. J Steroid Biochem 31:757–761

Tremblay GB, Tremblay A, Copeland NG, Gilbert DJ, Jenkins NA, Labrie F, Giguere V (1997) Cloning, chromosomal localization and functional analysis of the murine estrogen receptor β. Mol Endocrinol 11:353–365

Umayahara Y, Kawamori R, Watada H, Imano E, Iwama N, Morishima T, Yamasaki Y, Kajimoto Y, Kamada T (1994) Estrogen regulation of the insulin-like growth factor I gene transcription involves an AP-1 enhancer. J Biol Chem 269:16433–16442

Voegel JJ, Heine MJS, Zechel C, Chambon P, Gronemeyer H (1996) TIF2, a 160 kDa transcriptional mediator for the ligand-dependent activation function AF-2 of nuclear receptors. EMBO J 15:3667–3675

Walker P, Germond JE, Brown-Luedi M, Givel F, Wahli W (1984) Sequence homologies in the region preceding the transcription initiation site of the liver estrogen-responsive vitellogenin and apo-VLDLII genes. Nucleic Acids Res 12:8611–8626

Webb P, Lopez GN, Uht RM, Kushner PJ (1995) Tamoxifen activation of the estrogen receptor/AP-1 pathway: potential origin for the cell-specific estrogen-like effects of antiestrogens. Mol Endocrinol 9:443–456

Weiler IJ, Lew D, Shapiro DJ (1987) The Xenopus laevis estrogen receptor: sequence homology with human and avian receptors and identification of multiple estrogen receptor messenger ribonucleic acids. Mol Endocrinol 1:355–362

White R, Lees JA, Needham M, Ham J, Parker M (1987) Structural organization and expression of the mouse estrogen receptor. Mol Endocrinol 1:735–744

CHAPTER 7

Estrogen-Regulated Genes

A. WEISZ

A. Introduction

Estrogens are small molecules of relatively simple structure, endowed however with the ability to trigger a vast array of physiological responses of significant biological complexity. These responses range from the expression of cell type-specific differentiated cellular responses to induction or inhibition of the expression and function of specific gene products, to cell proliferation, to a direct control of the cell cycle's regulatory machinery, and to modifications of metabolic and endocrine pathways involving multiple cellular effectors. Since the early investigations of the molecular basis of estrogen's actions, it has become evident that responsive or target cells are endowed with specific ERs – molecular adapters capable of binding with high affinity to these hormones and, thereby, activated to transduce the hormonal signal to multiple cellular components. The first identification and characterization of such molecules in rodent uterus was made possible in the late 1950's by the availability of tritium-labeled estrogens of sufficiently high specific activity (JENSEN 1960; GLASCOCK and HOEKSTRA 1959). This led to the proposal by JENSEN and COLLEAGUES (1968) and GORSKI et al. (1968) of the original, and still current, model of estrogens and other steroid hormones' actions. According to this model, the hormone diffuses through the plasma membrane of all cells, to be retained only by target cells as stable complexes with specific intracellular receptor proteins. Each receptor binds its cognate steroid with high affinity, with equilibrium dissociation constants (Kds) in the nanomolar range, compatible with the physiological concentrations of free hormones. The stable but reversible interaction between steroids and their receptors results in a functional activation of these proteins that appears to involve their release from ternary and quaternary complexes with heat shock and other proteins (recently reviewed by PRATT and TOFT 1997). Conformational and post-translational changes enable them to accumulate in an active form in the nucleus, where they interact with the genome and, thereby, modulate the rate of gene transcription (YAMAMOTO and ALBERTS 1976). The products of the steroid-responsive genes, in turn, are responsible for the phenotypic changes which are the expression of the specific cellular responses to each hormone (MUELLER et al. 1961; O'MALLEY et al. 1969). According to this model, selectivity of steroid-hormone action in a given cell is essentially ligand- and receptor-based. Therefore, pharmacological parameters, such as potency, agonist/antagonist activities or cell specificity, are dictated by the concentration of a given receptor in each cell type and its relative affinity for a ligand. This in turn affects the stability of the steroid–protein interactions and, thus, the degree and kinetics of receptor activation.

This simplistic picture of estrogen's action is essentially correct and has been quite useful in the past to identify, characterize and purify ERs from different sources, and eventually to clone the corresponding cDNAs and genes. However, it hides the fact that most cellular and genomic responses to these hormones can be explained only by taking into consideration a higher degree of complexity of these regulatory pathways (DEAN and SANDERS 1996; KATZENELLENBOGEN et al. 1996). Moreover, the effects of estrogens, as those of any other hormone, can be influenced by the metabolic and cell-cycle stage of the cell, which implies that estrogen-mediated signal transduction is also highly dependent upon a variety of other cellular parameters.

It is difficult to explain the cellular specificity of the hormonal effect on a given gene, the variety of gene response kinetics which include both rapid and readily reversible changes in gene activity as well as delayed or sustained gene regulation events, and the differences in the sensitivity of estrogen target tissues, cells and genes to the inhibitory actions of different classes of estrogen antagonists or to modulation by agonists. These latter aspects, which are of particular relevance for the establishment of efficient antagonist-based therapeutic protocols for treatment of hormone-responsive neoplasms, or for the optimization of tissue-selective hormone replacement therapies (MCDONNELL et al. 1995a; GRAINGER and METCALFE 1996), have been extensively studied both in vivo and in vitro, and led to the definition of variegated patterns of pharmacological responses to antiestrogens (DAUVOIS and PARKER 1993) that further defined the complexity of the gene responses to estrogens. For example, in estrogen-responsive human breast cancer cells in culture, the non-steroidal triphenylethylene derivative tamoxifen is an estrogen antagonist toward induction of plasminogen-activator activity (KATZENELLENBOGEN et al. 1984) and transforming growth factor-β synthesis (KNABBE et al. 1987), but a weak agonist for induction of pS2 and progesterone receptor genes (WEAVER et al. 1988; ECKERT and KATZENELLENBOGEN 1982).

In rodents, tamoxifen and the dihydronaphtylene derivative nafoxidine, another synthetic antiestrogen, act as partial agonists/antagonists in the modulation of pituitary prolactin production and dopamine turnover in the medial basal hypothalamus; on progesterone-receptor gene expression; stimulation of plasminogen activator and peroxidase activities; or on weight gain in the uterus (TONEY and KATZENELLENBOGEN 1986; ROBERTSON et al. 1982; KNEIFEL et al. 1982); however, they act as full agonists in inducing plasma renin substrate secretion by the liver (KNEIFEL and KATZENELLENBOGEN 1981). In pituitary tumor cells, tamoxifen acts both as full agonist on some gene responses and as estrogen antagonist on others (SHULL et al. 1992; RAMKUMAR and ADLER 1995). In analogy, raloxifene, a benzothiophene compound analogue of tamoxifen with a substituent at the 4-position of the α-phenyl ring, shows markedly different effects in vivo, with strong agonist activities on bone density and blood lipid profiles, but no proliferative actions in the uterus (BLACK et al. 1994; SATO et al. 1994; FUCHS-YOUNG et al. 1995). Finally, in human MG63 osteosarcoma cells, this same compound shows an even more striking gene promoter-specific action when associated with ERs: it acts as a full agonist in inducing transforming growth factor-β3 promoter activation (YANG et al. 1996a; YANG et al. 1996c), but antagonizes stimulation by estrogens of a structurally different ER-responsive recombinant gene (YANG et al. 1996a). Although part of these differences might be due to the interaction of anti-estrogen molecules with cellular effectors other than the ERs, they clearly reflect the intricacies of gene regulation by ER ligands (McDONNELL et al. 1995b) and the multi-layered mechanisms of gene response to estrogens.

For all of the reasons stated above, it is evident that one of the major issues in the field of estrogens and other steroid hormones' actions concerns the mechanisms by which distinct gene responses are elicited in different cells. With the identification of an ever-growing number of genes regulated by estrogens in various tissues and organs, the genetic bases of the wide range of biological activities of these hormones start to be elucidated. At the same time, a deeper understanding of ERs and their target gene's structure and function is leading to novel insights into the mechanisms underlying specific, context-dependent gene responses to these hormones.

B. Transcriptional Control of Gene Activity by Estrogens

Since the early observations showing that labeled estrogens concentrate on the nucleus of cells in target organs and the finding that ERs are activated by their ligands to a DNA-binding conformational status (JENSEN et al. 1968; YAMAMOTO and ALBERTS 1972), the attention of researchers in this field has been focused on both the identification of genes regulated by these hormones in various organs and cell types and the mechanisms mediating this regulation. This has led not only to the description of a vast array of gene responses to estrogens, but also to the notion that the basis for their specificity is achieved through the integration of at least three distinct levels of control: the ERs themselves, the structure and organization of each target gene and the status of target cells that greatly affects the former two components (TSAI and O'MALLEY 1994; KATZENELLENBOGEN et al. 1996).

The role of the receptors is determined first by the variable levels of expression of each of their known subtypes in different cell types. Moreover, it is specified by intrinsic structural features of the receptors that allow the physical and functional transduction of the hormonal stimulus to the target genes and their variegated regulatory components (GRONEMEYER and LAUDET 1995). Responsive genes, for their part, include specific determinants that dictate not only the targeting of the estrogen-receptor complexes to well defined chromatin sites, but also the nature (activation or repression), the extent and kinetics of reaction (rapid or delayed, transient or sustained, etc.) and the specificity of the response (Gronemeyer 1991; Glass 1994). The cellular setting or background, for its part, is dependent upon the genetic "programming" of the cell through differentiation which conditions the gene networks accessible and responsive to hormonal signaling. This includes the presence and activation status of specific components of the basal transcription machinery, of other transactivators and of transcriptional co-regulators (Anzick et al. 1997; Beato and Sanchez-Pacheco 1996; Gelman et al. 1997; Glass et al. 1997; Horwitz et al. 1996; Milgrom et al. 1997). In this latter concept epigenetic factors such as the cell-cycle phase (Weisz and Bresciani 1993) or the activity of other signal-transduction pathways have been found to condition ERs' functions directly (O'Malley et al. 1995; Smith et al. 1995; Katzenellenbogen 1996); for example, through their phosphorylation (Auricchio 1989; Orti et al. 1992; Weigel et al. 1996) or by affecting other transcription factors co-operating or interfering with the receptors' actions (Truss and Beato 1993).

I. Molecular and Cellular Determinants of Gene Responses to Estrogens

1. The Estrogen Receptors

Estrogens affect nearly all vertebrate tissues to some degree (Baysal and Losordo 1996; Cicatiello et al. 1995; Ciocca and Vargas-Roig 1995; Cutolo et al. 1995; Demay et al. 1996; Farnsworth 1996; Greco et al. 1993; Hess et al. 1997; Abry-White et al. 1995; McEwen et al. 1997; Pelzer et al. 1996; Romer et al. 1997; Simerly et al. 1990; Turner et al. 1994; Weisz and Bresciani, 1993; Wickelgren 1997), and the classical concept of estrogen target organs, as first applied to those of the female reproductive tract, is now considered outdated. The biological responses to these steroids are mediated by intracellular receptors of the nuclear receptor's superfamily (Gronemeyer and Laudet 1995; Mangelsdorf et al. 1995; Kuiper and Gustasfsson 1997), although it has been proposed that some estrogenic effects might be mediated by signalling pathways independent of these proteins (Matsuda et al. 1993; Das et al. 1997; Wehling 1997).

The first cloning of a human estrogen-receptor complementary DNA (cDNA) (Walter et al. 1985) and comparative analysis of the encoded protein sequence with that of its chicken homologue (Krust et al. 1986) led to identification, in these proteins, of multiple regions, based on the degree of evolutionary conservation, referred to as A through F (Kumar et al. 1986; Kumar et al. 1987; Kumar and Chambon 1988; Tora et al. 1989). These domains are endowed with different functional specificity. Region C, located at the center of the molecule, encodes the DNA-binding domain (DBD) of the receptor, which targets it to specific DNA sequences present in most responsive genes and known as estrogen response elements (EREs). The DBD displays the highest degree of conservation among receptors from different species and receptor subtypes, and includes two "zinc-fingers" structures. In each of these, two pairs of cysteines chelate one zinc ion, contributing to the establishment of the DNA-binding module characteristic of these proteins (SCHWABE et al. 1993; ZILLIACUS et al. 1995).

The second most conserved region (E) is located in the C-terminal half of the molecule and harbors the ligand-binding domain (LBD); the hydrophobic "pocket" that accommodates the hormone and antihormones; a strong dimerization function that allows homo- and heterodimerization of the receptors or their interaction with other transacting factors; a ligand-dependent transcription activating function (AF-2) that mediates cell- and promoter-specific effects of the receptor on gene transcription through interactions with specific transcriptional co-regulators and components of the basal transcription machinery; and two determinants of the intracellular trafficking of receptors, i.e., the ligand-dependent nuclear localization function and a binding surface for molecular chaperones. The less conserved A/B region encodes an additional transactivation domain of the receptor (AF-1) that appears also to function independently of the hormone and of AF-2, and a recently identified third autonomous activation domain (AF-2a) functional both in yeast and mammalian cells. The role of regions D and F are less defined, but region D, which is considered as a flexible joint between regions C and E, also includes structural determinants involved in DNA binding. (GREEN and CHAMBON 1988; GRONEMEYER and LAUDET 1995; GUIOCHON-MANTEL et al. 1996; MANGELSDORF et al. 1995; NORRIS et al. 1997; PARKER 1995).

Since cloning of the first ER cDNA, there has been general acceptance that, unlike most other members of the steroid/nuclear receptor superfamily that are present in multiple subtypes, only one ER exists. Recently, however, a second estrogen-receptor subtype (ER-β) was identified (KUIPER et al. 1996; MOSSELMAN et al. 1996). The structural organization of ER-α (the first receptor subtype to be cloned and the best characterized so far) and ER-β are quite similar. However, significantly different sequence features allow them to predict differences in their functional interaction with estrogen target genes. Their DBDs are virtually identical, differing by only one amino acid, a feature indicating that they are likely to interact with similar DNA response elements. However, the sequences of their N- and C-terminal transactivation domains differ quite significantly, suggesting that their effects on transcription of different estrogen responsive genes may show distinct patterns.

TREMBLEY et al. (1997), PAECH et al. (1997) and WATANABE et al. (1997) recently described different transcriptional activities of ER-α and ER-β bound to estrogen agonists or antagonists. The relative binding potency of the two receptors for a number of estrogenic substances is comparable (KUIPER et al. 1997), as is the structure of their LBDs bound to 17β-estradiol (WITKOWSKA et al. 1997). Thus, different gene- and cell-specific responses to different classes of estrogen agonists and antagonists cannot be explained only by their selective interaction with one of the two receptor subtypes. A clue for explaining differences in gene and cell response to estrogen is given, perhaps, by the findings that the tissue distribution and relative levels of ER-α and ER-β are quite different in several organs (BRANDENBERGER et al. 1997; BYERS 1997; CAMPBELL-THOMPSON 1997; COUSE et al. 1997; ENMARK et al. 1997; KUIPER et al. 1997; ONOE et al. 1997) and that the two receptors can heterodimerize with each other and the complex can bind efficiently to EREs (COWLEY et al. 1997; PACE et al. 1997; PETTERSSON et al. 1997). Variable combinations of different transcription activation functions through homo- and heterodimerization of the two receptor subtypes are likely to increase the range of possible gene responses, as shown by the case of retinoid and thyroid hormone receptors (CHAMBON 1996; MINUCCI and OZATO 1996).

2. Estrogen-Responsive Gene Elements

a) Estrogen Response Elements and Other Estrogen-Responsive DNA Elements

Regulation of gene expression by steroid hormones can be a direct, or primary genetic response to ligand-activated receptors interacting with key molecular components of the gene regulatory machinery or an indirect or secondary response, resulting from the activities of hormone-induced gene products (YAMAMOTO and ALBERTS 1976). The ability to respond to the regulatory signals specified by ligand-activated ERs is an intrinsic property of primary genes, as demonstrated by the fact that it is maintained despite molecular cloning and re-introduction in widely different ER-expressing cell types.

The original hypothesis concerning steroid hormone regulation of gene expression foresaw the existence in steroid target genes of elements physically and functionally interacting with ligand-activated receptors which convey the hormonal signal to gene promoters (YAMAMOTO and ALBERTS 1976). Based on the above working hypothesis, several laboratories focused on the identification of DNA elements within or near estrogen target genes by binding studies in cell-free systems of ERs to DNA and investigation of hormonal regulation of the test gene promoters in transfected cells. Systematic analysis of these DNA elements from a number of primary estrogen target genes revealed that they indeed exist in multiple variants and can be broadly classified as two separate groups: (1) the elements that bind directly to the ER through protein-DNA interactions (EREs); and (2) binding sites for other DNA-binding factors that may target the receptor to specific gene promoters via protein–protein interactions and in the absence of detectable receptor binding to DNA (ER-responsive elements).

Following the first identification of a common palindromic sequence in the 5'-flanking regions of estrogen-responsive frog and chicken liver-specific genes by comparative DNA sequence analysis (WALKER et al. 1984), experimental evidences indicating specific binding of the ER to defined DNA regions of target genes started accumulating (JOST et al. 1984; MAURER 1985; WEISZ et al.

1986). The first experimental evidence of an estrogen-responsive enhancer in the 5'-flanking DNA region of the hormone-responsive *Xenopus* vitellogenin A2 gene led the way to the identification of the first ERE (KLEIN-HITPASS et al. 1986; KLEIN-HITPASS et al. 1988) and to the finding that ligand-induced ER homodimers were able to bind tightly to this DNA element in vitro (KUMAR and CHAMBON 1988; KLEIN-HITPASS et al. 1989). This ERE, now referred to as "canonical" or perfectly palindromic, consists of the inverted repeat (Pu)GGTCANNNTGACC(Py), where the two arms of the palindrome are separated by 3 bp [inverted repeat (IR)3].

Crystallographic analysis of the ER-α DBD bound to the above sequence (SCHWABE et al. 1993) confirmed that the protein binds as a symmetrical dimer to the palindromic binding site. However, it is worth mentioning that the ER can also bind as a monomer to one arm of the palindrome (MEDICI et al. 1991). As more EREs were identified in ERs' target genes, it became clear that in nature these elements are most often imperfect palindromes, where the nucleotide sequence of one "arm" of the element diverges in one or more positions from that reported above (for reviews see ANOLIK et al. 1993; TRUSS and BEATO 1993; ZILLIACUS et al. 1993), or even only half palindromes (KATO et al. 1992). These imperfect palindromes bind the ER with definitely lower affinity in vitro. The finding that natural EREs are usually imperfect palindromes might seem a paradox, since one would expect that high-affinity DNA binding sites should have been subjected to highly selective positive evolutionary pressure, and that their presence in estrogen target genes would have been favored. However, the existence of multiple ERE variants could be explained by the necessity of achieving a wider range of possibilities for gene regulation by estrogens. First of all, one may envisage that different response elements induce different, element-specific, conformational changes in DNA-bound receptors; the DNA binding site would serve, in this case, as a determinant of receptor conformation and, as a consequence, of its function.

According to this view, the EREs would not simply specify estrogen-responsive genes by targeting the receptor to certain gene promoters, but would be an integral part of the transcriptional control exerted by receptors. A specific receptor conformation, dependent on the sequence of the response element, would affect, for example, its interaction with specific transcriptional co-activators, co-repressors and/or basal transcription factors and with other transcription enhancer factors. This hypothesis was proposed by YAMAMOTO and COLLEAGUES (1997; STARR et al. 1996) to explain the divergent activities of wild-type and point-mutated glucocorticoid receptors on the same glucocorticoid response element (GRE), but still awaits confirmation in the case of the ERs.

Alternatively, non-palindromic response elements could function as integrators between the action of receptors and that of other cellular components. For example, many cell types express proteins that appear to act as dimerization partners of ERs or even modulators of their binding to DNA

(CHURCH-LANDEL et al. 1995; CRAWFORD and CHAPMAN 1990; FEAVERS et al. 1987; HAMADA 1989; JOHNSTON et al. 1997; MUKHERJEE and CHAMBON 1990; YANG et al. 1996b). Such partners include other transcription-enhancer factors whose response elements are often found in proximity to EREs and that co-operate with, and in some cases antagonize, the transactivating functions of nuclear receptors. It is possible, in fact, that apparent subtle sequence differences between EREs might be responsible for the binding of the receptors only in the presence, or absence, of specific molecular partners. In addition, since multiple members of the steroid/nuclear-receptor gene superfamily can recognize the core (Pu)GGTCA sequence (GRONEMEYER and LAUDET 1995), cell type-specific promiscuous transcriptional regulation by estrogen and other nuclear receptors and transcriptional "cross-talks" and interference may be more likely achieved through imperfectly or half-palindromic EREs (GRAUPNER et al. 1991; KATO et al. 1995A; KHARAT and SAATCIOGLU 1997; KLINGE et al. 1997; LIU et al. 1993; PFAFF et al. 1994; SCOTT et al. 1997; SEGARS et al. 1993). The existence of two receptor subtypes, both responding to the same hormonal signal and able to dimerize with each other, could also furnish a reason for the existence of multiple ERE sequences, should it be possible to demonstrate differences in DNA binding specificity between homo- and het-erodimers of receptor subtypes. Finally, the chromatin structure in and around EREs, which appears to play a role in receptor-ERE recognition, may inter-vene in modulating interaction of hormone-activated receptors with certain EREs (reviewed in BEATO et al. 1996).

Recently, gene elements other than classical EREs were found to mediate ER action on certain responsive genes. YANG et al. (1996c) identified a poly-purine element in the human transforming growth factor-β3 gene promoter that is activated by ER agonists and antagonists. Interaction of ERs with this element does not require their DBD and appears to occur through a cellular adapter protein. Also, NORRIS ET AL. (1995) suggested that a new class of Alu DNA repeats can function as ER-dependent transcription enhancers and ELGORT et al. (1996) identified a novel DNA sequence that mediates hormonal regulation of the human retinoic acid receptor-α1 promoter in Hep G2 cells. Furthermore, non-canonical, cell type-specific mechanisms for ER-mediated transcriptional activation of the brain creatine kinase gene (SUKOVICH et al. 1994) and transcriptional repression of the rat prolactin gene (ADLER et al. 1988) in the absence of EREs have been suggested.

In line with these findings, the best examples of regulation of gene tran-scription by the ERs in the absence of direct interactions with EREs involve their physical association with other transcription-enhancer factors, with the result being inhibition or, in a few documented cases, enhancement of tran-scription. In this case, the final effect on gene transcription is thought to depend on the formation of a complex between the two factors that affects their respective binding to DNA and/or their transactivating potential. The ever-growing number of transcription-enhancer factors that can associate with the ERs include the AP-1 complex (AMBROSINO et al. 1993; DOUCAS et al. 1991;

GAUB et al. 1990; PAECH et al. 1997; PHILIPS et al. 1993; SCHMITT et al. 1995; SHEMSHEDINI et al. 1991; TZUKERMAN et al. 1991; UHT et al. 1997; UMAYAHARA et al. 1994; WEISZ et al. 1990; WEISZ and ROSALES 1990; WEISZ et al. 1991); the chicken ovalbumin upstream-promoter transcription factor COUP-TF (KLINGE et al. 1997); C/EBPβ(STEIN and YANG 1995); the erythroid-specific GATA-1 factor (BLOBEL et al. 1995; BLOBEL and ORKIN 1996); the neurotrophic factor (NF)-κB complex (GALIEN and GARCIA 1997; KUREBAYASHI et al. 1997; RAY et al. 1997; STEIN and YANG 1995); NF-IL6 (RAY et al. 1997); Sp1 (BATIS-TUZZO DE MEDEIROS et al. 1997; DUAN et al. 1998; KRISHNAN et al. 1994; PORTER et al. 1996; PORTER et al. 1997); and cyclic adenosine monophosphate (cAMP)-responsive factors (MCCARTHY et al. 1997).

b) Target Gene Promoters

Transcription initiation from eukaryotic gene promoters depends on the assembly of the multi-component basal transcription apparatus that con-tributes to maintain a constitutive level of promoter activity and provides a conduit for particular signal transduction pathways that influence the tran-scription rate. It has been demonstrated that the interactions among protein factors at a given promoter play a key role in defining the activity of tran-scription regulators, including the nuclear receptors (BEATO and SANCHEZ-PACHECO 1996; BURATOWSKI 1994; PTASHNE and GANN 1997; SANGUEDOLCE et al. 1997; TJIAN and MANIATIS 1994), and different examples of promoter-specific transcriptional regulation by estrogens have been reported (BERRY et al. 1990; GRAHAM et al. 1995; METZGER et al. 1992; MILGROM et al. 1997; TORA et al. 1988; TZUKERMAN et al. 1994).

Several explanations have been put forward to account for the influence of the promoter context on gene responsiveness to estrogen and other nuclear receptors. They include positive and negative interactions between receptors and basal and enhancer transcription factors, which are often promoter- and/or cell type-specific, as well as the existence of different receptor co-activators and co-repressors. The composition of the basal transcription machinery can vary since, for example, different subsets of TFIID complexes have been reported and the same is likely to apply to any other of its components, and receptor activity could require specific combinations of these factors. More-over, the promoters of many known estrogen-regulated genes are complex and include binding sites for other transactivators. Particular combinations and arrangements of *cis*-acting elements near the transcription start site appear to be conditioning a promoter response to estrogens.

3. The Target Cell Environment

Estrogen's effects on target genes are cell type-specific. However, the basis of this specificity cannot be ascribed only to differences in expression of the two ER subtypes in different cells. In mammalian uterus, for example, expression of cell-cycle regulatory genes in response to 17β-estradiol is restricted to the few cell types that respond to the mitogenic action of this hormone (CICATIELLO et al. 1993; PAPA et al. 1991; WEISZ and BRESCIANI 1988; WEISZ and BRESCIANI 1993), despite the fact that functional ERs are present in all uterine compartments. Also, the chicken very low density apolipoprotein-II gene is induced by estrogens in liver cells through receptor binding to an ERE present in its promoter-near region. However, it is not induced in oviduct tubular glands that nevertheless contain functional ERs (WIJNHOLDS et al. 1988). The exact opposite occurs with the chicken transferrin gene, which is stimulated by estradiol in the oviduct, but not in the liver cells (LEE et al. 1988). Finally, stable transfection of expression vectors encoding ERs in different receptor-negative cell lines clearly indicates that the cellular background is a primary determinant for the specificity of gene responses to estrogens (KANEKO et al. 1993; LEVENSON and JORDAN 1994; LUNDHOLDT et al. 1996; SEILER-TUYNS et al. 1988; TOUITOU et al. 1991; WATSON and TORRES 1990; ZAJCHOWSKI and SAGER 1991).

There are several possible explanations for the cell specificity of the gene response to ER action that are not mutually exclusive. First of all, as mentioned in the previous section, transcriptional regulation involves a complexity of protein–protein and protein–DNA interactions, where the combinatorial aspects, also involving cell type-specific co-activators and co-repressors, are critical determinants of whether a particular transcription regulatory factor will elicit a gene response. For this reason, cell specificity of ERs' action on target genes may, in part, be explained by the existence of different sets of cell- and gene-specific factors that co-operate or interfere with each other on specific gene promoters. In addition, the chromatin structure, DNA topology and methylation status of target genes, which also depend on the cell type and its relative functional status, determine the accessibility of their regulatory DNA sequences to nuclear receptors and/or to other transacting factors that co-operate with them. In this way, they may greatly condition the ability of estrogen-activated receptors to interact with specific genomic sites (BEATO and SANCHEZ-PACHECO 1996; DARBRE and KING 1987; KIM et al. 1997; MARTIN et al. 1997; NGO et al. 1995; SMITH and HAGER 1997; SIEGFRIED and CEDAR 1997; YING and GORSKI 1994).

Several lines of evidence suggest that certain cellular components, which were not thought to be directly involved in regulation of gene transcription, may exert strong influences on ER functions in certain cell types. For example, calreticulin, a 46-kDa Ca^{2+}-binding protein, present mainly in the endoplasmic reticulum, was found to modulate steroid-sensitive gene expression, inhibiting nuclear hormone receptor activity by binding to their DBDs and, thereby, preventing their interaction with DNA (BURNS et al. 1994; DEDHAR et al. 1994; KRAUSE and MICHALAK 1997). Another example is provided by the cell cycle phase-specific cyclins D1 and A, which have recently been found to stimulate the transcriptional activity of ER-α by forming cyclin–receptor complexes with enhanced transactivating functions (NEUMAN et al. 1997; TROWBRIDGE et al. 1997; ZWIJSEN et al. 1997). Since these cyclins are expressed at significant levels in actively cycling cells during G1 (cyclin D1) or S-G2 (cyclin A) phases, and their levels are positively regulated by estrogens in growth-responsive cells (ALTUCCI et al. 1996; ALTUCCI et al. 1997; CICATIELLO et al. 1998; WEISZ et al. 1996), formation of these complexes might represent a determinant of cell cycle phase-specific gene regulation by estrogens, a central aspect of their mitogenic activity.

Finally, post-translational modifications of ERs, in particular protein phosphorylation, greatly affect both their hormone and DNA binding, as well as stability and transcriptional activities (reviewed by WEIGEL 1996). In this way, receptor functions are also regulated by the effectors of multiple signalling pathways, the activities of which can be cell-type specific and, more importantly, depend on the functional status of cells. These pathways have been found to include those stimulated by the neurotransmitter dopamine (POWER et al. 1991; SMITH et al. 1993); by growth factors such as epidermal growth factor, transforming growth factor-α, heregulin, insulin and insulin-like growth factor I (ARONICA and KATZENELLENBOGEN 1993; BUNONE et al. 1996; IGNAR-TROWBRIDGE et al. 1992; IGNAR-TROWBRIDGE et al. 1996; KATO et al. 1995b; NEWTON et al. 1994; PATRONE et al. 1996; PEITRAS et al. 1995;); by cyclic nucleotides (ARONICA and KATZENELLENBOGEN 1993); by tumor promoters (JOEL et al. 1995; MARTIN et al. 1995); and by estrogens themselves (AURICCHIO et al. 1987; ARONICA and KATZENELLENBOGEN 1993; JOEL et al. 1995; LAHOOTI et al. 1994; MIGLIACCIO et al. 1996).

C. Non-transcriptional Control of Gene Activity by Estrogens

A growing body of literature is documenting the effects of estrogens that appear to be triggered, at least in certain cases, through mechanism(s) independent of the "classical" nuclear pathway of steroid receptors action. These effects include post-transcriptional regulation of specific messenger RNA (mRNA) stability and translation efficiency, as well as rapid activation of signal transduction pathways through extra-genomic actions of estrogens. These two aspects of estrogen action must be considered separately, since the underlying mechanisms appear to be quite different.

Circumstantial evidence suggests that modulation of the half-life and activity of specific RNAs is a late event in estrogen-stimulated cells, affected by products of primary estrogen-responsive genes. However certain rapid responses of estrogen-stimulated cells, such as inhibition of voltage-gated Ca^{2+} channels (MERMELSTEIN et al. 1996), membrane hyperpolarization and increased potassium conductance (NABEKURA et al. 1986), stimulation of cyclic nucleotides accumulation (SZEGO and DAVIS 1967) or induction of mitogen-activated protein (MAP)-kinase's activation (MIGLIACCIO et al. 1996; WATTERS et al. 1997), appear to be mediated by extra-genomic actions of different hormone-activated effectors.

I. Post-transcriptional Effects of Estrogens

The notion that estrogens can affect individual mRNA turnover stemmed from the observation of BROCK and SHAPIRO (1983) that the exposure of primary cultures of *Xenopus* liver cells to estradiol induced a dramatic stabilization of vitellogenin RNA, while total poly(A)-RNA exhibited the same half-life either in the presence or absence of hormone. This observation was later extended to other mRNAs, including the chicken very low density apolipoprotein II (COCHRANE and DEELY 1988), apolipoprotein B and vitellogenin II (MARGOT and WILLIAMS 1996) gene transcripts, which are all stabilized by estrogens, while the half-lives of frog serum albumin (SCHOENBERG et al. 1989) and rat peptidylglycine α-amidating monooxygenase mRNAs were found to be reduced (MESKINI et al. 1997).

Relevant to these last findings is the fact that estrogen administration was found to induce cellular effectors involved in the rapid degradation of avian very low density apolipoprotein II and vitellogenin mRNAs (GORDON et al. 1988). Therefore, in addition to stabilizing specific mRNAs, estrogens are also capable of setting forth mechanisms for their rapid destabilization upon hormone withdrawal that are likely to mediate degradation of other RNA transcripts in hormone-stimulated cells. Although it has been proposed that these effects were induced only by estrogens at pharmacological concentrations (MCKENZIE and KNOWLAND 1990), this phenomenon has been analyzed in detail and compared with other known cases of regulation of cytoplasmic mRNA stability (NIELSON and SHAPIRO 1990). This allowed the identification of estrogen-induced proteins that bind to the 3'-untranslated region of certain target mRNAs and which are believed to be responsible for their increased stability in hormone-stimulated cells (DODSON and SHAPIRO 1994; MARGOT and WILLIAMS 1996). One of these proteins was purified and found to be vigilin – a KH domain protein the mechanism of action of which is still not fully defined (DODSON and SHAPIRO 1997).

An alternative or complementary mechanism to the one described above could involve 2',5'-oligoadenylates, potent ribonuclease activators involved in RNA destabilization by interferons in virus-infected cells (BAGLIONI 1979) the cellular concentration of which was found to be reduced by estrogens in different rat organs (SUZUKI et al. 1990; SUZUKI et al. 1992). In this case, a reduction of the cellular concentration of these oligonucleotides, resulting in inhibition of 2',5'-oligoadenylate-dependent RNases, could explain the reduced rate of RNA degradation in hormone-treated cells.

An additional post-transcriptional control is exerted by estrogens on the translational efficiency of certain mRNAs, through enhanced polysomal recruitment, as shown in the case of the ornitine-δ-aminotransferase gene transcript in retinoblastoma cells (FAGAN and ROZEN 1993) and apolipoprotein E mRNA in liver cells (SRIVASTAVA et al. 1997). In the latter case, the effect was found to be dependent on the genetic background of certain "responder" mouse strains and to occur through an ER-α-mediated pathway, because it was not observed in mice lacking this receptor subtype due to inactivation of the corresponding chromosomal gene by gene targeting (SRIVASTAVA et al. 1997).

II. Extra-Genomic Effects of Estrogens

Cellular responses to estrogens, showing a lag time so short to exclude the possibility that they can be mediated by a genomic effect of the ERs, have been known for a long time (MEBRY-WHITE et al. 1995; NEMERE et al. 1993; WEHLING 1997). They appear to involve multiple cytoplasmic or membrane-bound enzymes and a number of intracellular mediators. No definitive answer has yet been given as to whether they are mediated by still-undefined membrane ERs, by an extra-nuclear pool of the classical receptor subtypes or by other estrogen-binding molecules. The current knowledge concerning these aspects of estrogen's action is addressed in Chapt. 4.4 and, therefore, it will not be discussed further here. However, it is worth mentioning that the possible involvement of these extra-genomic pathways have to be considered when addressing the issue of gene regulation by estrogens and other sex-steroid hormones, since their known impact on gene activity could help explain certain less-known aspects of hormone action.

Acknowledgments. The author is thankful to Francesco Bresciani for helpful suggestions and for critically reading the manuscript, and to Paola de Fazio for secretarial help. Work was supported by: Associazione Italiana per la Ricerca sul Cancro (A.I.R.C.); Consiglio Nazionale delle Ricerche and Ministero per l'Università e la Ricerca Scientifica e Tecnologica of Italy; and the European Commission (BIOMED2 Program: Contract BMH4-CT98-3433).

References

Adler S, Waterman ML, He X, Rosenfeld MG (1988) Steroid receptor-mediated inhibition of rat prolactin gene expression does not require the receptor DNA-binding domain. Cell 52:685–695

Altucci L, Addeo R, Cicatiello L, Dauvois S, Malcolm GP, Truss M, Beato M, Sica V, Bresciani F, Weisz A (1996) 17β-Estradiol induces cyclin D1 gene transcription, p36D1-p34cdk4 complex activation and p105Rb phosphorylation during mitogenic stimulation of G1-arrested human breast cancer cells. Oncogene 12:2315–2324

Altucci L, Addeo R, Cicatiello L, Germano D, Pacilio C, Battista T, Cancemi M, Belsito Patrizzi V, Bresciani F, Weisz A (1997) Estrogen induces early and timed activation of cyclin-dependent kinases 4, 5 and 6 and increases cyclin messenger ribonucleic acid expression in rat uterus. Endocrinology 138:978–984

Ambrosino C, Cicatiello L, Cobellis G, Addeo R, Sica V, Bresciani F, Weisz A (1993) Functional interference between the estrogen receptor and Fos in the regulation of c-fos protooncogene transcription. Mol Endocrinol 7:1472–1483

Anolik JH, Klinge CM, Bambara RA, Hilf R (1993) Differential impact of flanking sequences on estradiol- vs. 4-hydroxytamoxifen-liganded estrogen receptor binding to estrogen responsive element DNA. J Steroid Biochem Mol Biol 46:713–730

Aronica SM, Katzenellenbogen BS (1993) Stimulation of estrogen receptor-mediated transcription and alteration of the phosphorylation state of the rat uterine estrogen receptor by estrogen, cyclic adenosine monophosphate, and insulin-like growth factor I. Mol Endocrinol 7:743–752

Auricchio F (1989) Phosphorylation of steroid receptors. J Steroid Biochem 32:613–622

Auricchio F, Migliaccio A, Di Domenico M, Nola E (1987) Oestradiol stimulates tyrosine phosphorylation and hormone binding of its own receptor in a cell-free system. EMBO J 6:2923–2929

Baglioni C (1979) Interferon-induced enzymatic activities and their role in the antiviral state. Cell 17:255–264

Batistuzzo de Medeiros SR, Krey G, Hihi AK, Wahli W (1997) Functional interactions between the estrogen receptor and the transcription activator Sp1 regulate the estrogen-dependent transcriptional activity of the vitellogenin A1 promoter. J Biol Chem 272:18250–18260

Baysal K, Losordo DW (1996) Estrogen receptors and cardiovascular disease. Clin Exp Pharmacol Physiol 23:537–548

Beato M, Candau R, Chàvez S, Möws C, Truss M (1996) Interaction of steroid hormone receptors with transcription factors involves chromatin remodelling. J Steroid Biochem Mol Biol 56:47–59

Beato M, Sànchez-Pacheco A (1996) Interaction of steroid hormone receptors with the transcription initiation complex. Endocr Rev 17:587–609

Berry M, Metzger D, Chambon P (1990) Role of the two activating domains of the estrogen receptor in the cell-type and promoter-context dependent agonistic activity of the antiestrogen 4-hydroxytamoxifen. EMBO J 9:2811–2818

Black LJ, Sato M, Rowley ER, Magee DE, Bekele A, Williams DC, Cullinan GJ, Bendele R, Kauffman RF, Bensch WR, Frolik CA, Termine JD, Bryant HU (1994) Raloxifene (LY1394881 HCl) prevents bone loss and reduces serum cholesterol without causing uterine hypertrophy in ovariectomized rats. J Clin Invest 93:63–69

Blobel GA, Sieff CA, Orkin SH (1995) Ligand-dependent repression of the erythroid transcription factor GATA-1 by the estrogen receptor. Mol Cell Biol 15:3147–3153

Blobel GA, Orkin SH (1996) Estrogen-induced apoptosis by inhibition of the erythroid transcription factor GATA-1. Mol Cell Biol 16:1687–1694

Brandenberger AW, Tee MK, Lee JY, Chao V, Jaffe RB (1997) Tissue distribution of estrogen receptor α(ER-α) and β(ER-β) mRNA in the midgestation human fetus. J Clin Endocrinol Metab 82:3509–3512

Brock ML, Shapiro DJ (1983) Estrogen stabilizes vitellogenin mRNA against cytoplasmic degradation. Cell 34:207–214

Bunone G, Briand P-A, Miksicek RJ, Picard D (1996) Activation of the unliganded estrogen receptor by EGF involves the MAP kinase pathway and direct phosphorylation. EMBO J 15:2174–2183

Buratowski S (1994) The basics of basal transcription by RNA polymerase II. Cell 77:1–3

Burns K, Duggan B, Atkinson EA, Famulski KS, Nemer M, Bleackley RC, Michalak M (1994). Modulation of gene expression by calreticulin binding to the glucocorticoid receptor. Nature 367:476–480

Byers M, Kuiper GGJM, Gustafsson J-A, Park-Sarge OK (1997) Estrogen receptor-β mRNA expression in rat ovary: down-regulation by gonadotropins. Mol Endocrinol 11:172–182

Campbell-Thompson ML (1997) Estrogen receptor α and β expression in upper gastrointestinal tract with regulation of trefoil factor family 2 mRNA levels in ovariectomized rats. Biochem Biophys Res Commun 240:478–483

Chambon P (1996) A decade of molecular biology of retinoic acid receptors. FASEB J 10:940–954

Cicatiello L, Sica V, Bresciani F, Weisz A (1993) Identification of a specific pattern of "immediate-early" gene activation induced by estrogen during mitogenic stimulation of rat uterine cells. Receptor 3:17–30

Cicatiello L, Cobellis G, Addeo R, Papa M, Altucci L, Sica V, Bresciani V, LeMeur M, Kumar VL, Chambon P, Weisz A (1995) In vivo functional analysis of the mouse estrogen receptor gene promoter: a transgenic mouse model to study tissue-specific and developmental regulation of estrogen receptor gene transcription. Mol Endocrinol 9:1077–1090

Cicatiello L, Addeo R, Altucci L, Belsito Petrizzi V, Germano D, Pacilio C, Salzano S, Bresciani F, Weisz A (1998) The antiestrogen ICI 182,780 inhibits proliferation of human breast cancer cells by interfering with multiple, sequential estrogen-regulated processes required for cell cycle progression. Cancer Res (in press)

Ciocca DR, Vargas Roig LM (1995) Estrogen receptors in human nontarget tissues: biological and clinical implications. Endocr Rev 16:35–62

Cochrane AW, Deely RG, Parker MG (1988) Estrogen-dependent activation of the avian very low density apolipoprotein II and vitellognine genes. Transient alterations in mRNA polyadenylation and stability during induction. J mol Biol 303:555–567

Couse JF, Lindzey J, Grandien K, Gustafsson J-A, Korach KS (1997) Tissue distribution and quantitative analysis of estrogen receptor-α(ERα) and estrogen receptorβ(ERβ) messenger ribonucleic acid in the wild-type and ERα-knockout mouse. Endocrinology 138:4613–4621

Cowley SM, Hoare S, Mosselman S, Parker MG (1997) Estrogen receptors $\alpha$and β form heterodimers on DNA. J Biol Chem 272:19858–19862

Church Landel C, Kushner PJ, Greene GL (1995) Estrogen receptor accessory proteins: effects on receptor-DNA interactions. Environ Health Perspect 103 (suppl 7):23–28

Cutolo M, Sulli A, Seriolo B, Accardo S, Masi AT (1995) Estrogens, the immune response and autoimmunity. Clin Exp Rheumatol 13:217–226

Crawford L, Chapman K (1990) Identification of a high molecular weight steroid response element binding protein. Mol Endocrinol 4:685–692

Darbre PD, King RJB (1987) Progression to steroid insensitivity can occur irrespective of the presence of functional steroid receptors. Cell 51: 521–528

Das SK, Taylor JA, Korach KS, Paria BC, Dey SK, Lubahn DB (1997) Estrogenic responses in estrogen receptor-α deficient mice reveal a distinct estrogen signaling pathway. Proc Natl Acad Sci USA 94:12786–12791

Dauvois S, Praker MG (1993) Mechanism of action of hormone antagonists. In: Parker MG (ed) Frontiers in molecular biology. Steroid hormone action. Oxford University Press, Oxford, pp 166–185

Dean DM, Sanders MM (1996) The years after: reclassification of steroid-sensitive genes. Mol Endocrinol 10:1489–1495

Dedhar S, Rennie PS, Shago M, Hagesteijn C-YL, Yang H, Filmus J, Hawley RG, Bruchowsky N, Cheng H, Matusik RJ, Giguère V (1994) Inhibition of nuclear hormone receptor activity by calreticulin. Nature 367:480–483

Demay F, Geffroy S, Tiffoche C, de Monti M, Thieulant ML (1996) Cell-specific mechanisms of estrogen receptor in the pituitary gland. Cell Biol Toxicol 12:317–324

Dodson RE, Shapiro DJ (1994) An estrogen-inducible protein binds specifically to a sequence in the 3′ untranslated region of estrogen-stabilized vitellogenin mRNA. Mol Cell Biol 15: 3130–3138

Dodson RE, Shapiro DJ (1997) Vigilin, a ubiquitous protein with 14K homology domains, is the estrogen-inducible vitellogenin mRNA 3′-untranslated region-binding protein. J Biol Chem 272:12249–12252

Doucas V, Spyrou G, Yaniv M (1991) Unregulated expression of c-jun or c-fos proteins but not jun D inhibits estrogen receptor activity in human breast cancer derived cells. EMBO Journal 10:2237–2245

Duan R, Porter W, Safe S (1998) Estrogen-induced c-Fos protooncogene expression in MCF-7 human breast cancer cells: role of estrogen receptor Sp1 complex formation. Endocrinology (in press)

Eckert RL, Katzenellenbogen BS (1982) Effects of estrogens and antiestrogens on estrogen receptor dynamics and the induction of progesterone receptor in MCF-7 human breast cancer cells. Cancer Res 42:139–144

El Meskini R, Boudouresque F, Ouafik L (1997) Estrogen regulation of peptidylglycine α-amidating monooxygenase messenger ribonucleic acid levels by a nuclear post-transcriptional event. Endocrinol 138:5256–5265

Elgort MG, Zou A, Marschke KB, Allegretto EA (1996) Estrogen and estrogen receptor antagonists stimulate transcription from the human retinoic acid receptor-α1 promoter via a novel sequence. Mol Endocrinol 10:477–487

Enmark E, Pelto-Huikko M, Grandien K, Legercrantz S, Lagercrantz J, Frien G, Nordenskjold M, Gustafsson J-A (1997) Human estrogen receptor β-gene structure, chromosomal localization, and expression pattern. J Clin Endocrinol Metab 82:4258–4265

Fagan RJ, Rozen R (1993) Translational control of ornithine-δ-aminotransferase (OAT) by estrogen. Mol Cell Endocrinol 90:171–177

Farnsworth WE (1996) Roles of estrogen and SHBG in prostate physiology. Prostate 28:17–23

Feavers M, Jiricny J, Moncharmont B, Saluz HP, Jost JP (1987) Interaction of two non-histone proteins with the estradiol response element of the avian vitellogenin gene modulates the binding of estradiol-receptor complex. Proc Natl Acad Sci USA 84:7453–7457

Fuchs-Young R, Glasebrook AL, Short LL, Draper MW, Rippy MK, Cole HW, Magee DE, Termine JD, Bryant HU (1995) Raloxifene is a tissue-selective agonist/antagonist that functions through the estrogen receptor. Ann NY Acad Sci 761:355–360

Galien R, Garcia T (1997) Estrogen receptor impairs interleukin-6 expression by preventing protein binding on the NF-κB site. Nucleic Acids Res 252424–2429

Gaub M-P, Bellard M, Scheuer I, Chambon P, Sassone-Corsi P (1990) Activation of the ovalbumin gene by the estrogen receptor involves the Fos-Jun complex. Cell 63:1267–1276

Gelman L, Staels B, Auwerx J (1997) Rôle des co-facteurs transcriptionnels dans la transduction des signaux hormonaux par les récepteurs nucléaires. Med Sci 13:961–969

Glascock RF, Hoekstra WG (1959) Selective accumulation of tritium-labeled hexoestrol by the reproductive organs of immature of immature female goat and sheep. Biochem J 72:673–682

Glass CK (1994) Differential recognition of target genes by nuclear receptor monomers, dimers and eterodimers. Endocr Rev 15:391–407

Glass CK, Rose DW, Rosenfeld MG (1997) Nuclear receptor coactivators. Curr Opin Cell Biol 9:222–232

Gordon DA, Shellness GS, Nicosia M, Williams DL (1988) Estrogen-induced destabilization of yolk precursor protein mRNA in avian liver. J Biol Chem 263: 2625–2631

Gorski J, Toft DO, Shyamala G, Smith D, Notides A (1968) Studies on the interaction of estrogen with the uterus. Rec Progr Horm Res 24:45–80

Graham JD, Roman SD, McGowan E, Sutherland RL, Clarke CL (1995) Preferential stimulation of human progesterone receptor B expression by estrogen in T-47D human breast cancer cells. J Biol Chem 270:30693–30700

Grainger DJ, Metcalfe JC (1996) Tamoxifen: Teaching an old drug new tricks? Nature Med 2:381–385

Graupner G, Zhang X-K, Tzukerman M, Wills K, Hermann T, Pfahl M (1991) Thyroid hormone receptors repress estrogen receptor activation of a TRE. Mol Endocrinol 5:365–372

Greco TL, Duello TM, Gorski J (1993) Estrogen receptor, estradiol, and diethylstilbestrol in early development: the mouse as a model for the study of estrogen receptors and estrogen sensitivity in embryonic development of male and female reproductive tracts. Endocr Rev 14:59–71

Green S, Chambon P (1988) Nuclear receptors enhance our understanding of transcription regulation. Trends Genet 4:309–314
Gronemeyer H, Laudet V (1995) Transcription factors 3: nuclear receptors. Protein
  profile, vol. 2. Academic Press, London, New York
Gronemeyer H (1991) Transcription activation by estrogen and progesterone receptors. Annu Rev Genet 25:89–123
Guiochon-Mantel A, Delabre K, Lescop P, Milgrom E (1996) Intracellular traffic of
  steroid hormone receptors. J Steroid Biochem Molec Biol 56:3–9
Hess RA, Bunik D, Lee K-H, Bahr J, Taylor JA, Korach KS, Lubahn DB (1997) A role
  for estrogens in the male reproductive system. Nature 390:509–512
Hamada K, Gleason SL, Levi B-Z, Hirschfeld S, Appella E, Ozato K (1989) H-2RIIBP,
  a member of the nuclear hormone receptor superfamily that binds to both the
  regulatory element of major histocompatibility class I genes and the estrogen
  response element. Proc Natl Acad Sci USA 86:8289–8293
Horwitz KB, Jackson TA, Bain DL, Richer JK, Takimoto GS, Tung L (1996) Nuclear
  receptor coactivators and corepressors. Mol Endocrinol 10:1167–1177
Ignar-Trowbridge DM, Nelson KG, Bidwell MC, Curtis SW, Washburn TF, McLachlan
  JA, Korach KS (1992) Coupling of dual signaling pathways: epidermal growth
  factor action involves the estrogen receptor. Proc Natl Acad Sci USA
  89:4658–4662
Ignar-Trowbridge DM, Pimentel M, Parker MG, McLachlan JA, Korach KS (1996)
  Peptide growth factor cross-talk with the estrogen receptor requires the A/B
  domain and occurs independently of protein kinase C or estradiol. Endocrinology
  137:1735–1744
Jensen EV (1959) Studies of growth phenomena using tritium-labeled steroids. Proc
  4th International Congress Biochem, Vienna. Vol 15, p 119
Jensen EV, Suzuki T, Kawashima T, Stumpf WE, Jungblut PW, DeSombre E (1968) A
  two step mechanism for the interaction of estradiol with rat uterus. Proc Natl Acad
  Sci USA 59:632–638
Joel PB, Traish AM, Lannigan DA (1995) Estradiol and phorbol ester cause phosphorylation of serine 118 in the human estrogen receptor. Mol Endocrinol 9:1041–1052
Johnston SD, Lui X, Zuo F, Eisenbraun TL, Wiley SR, Kraus RJ, Mertz JE (1997) Estrogen-related receptor α1 functionally binds as a monomer to extended half-site
  sequences including ones contained within estrogen-response elements. Mol
  Endocrinol 11:342–352
Jost J-P, Seldran M, Geiser M (1984) Preferential binding of estrogen-receptor complex
  to a region containing the estrogen-dependent hypomethylation site preceding the
  chicken vitellogenin II gene. Proc Natl Acad Sci USA 81:429–433
Kaneko KJ, Gélinas C, Gorski J (1993) Activation of the silent progesterone receptor
  gene by ectopic expression of estrogen receptors in a rat fibroblast cell line. Biochemistry 32:8348–8359
Kato S, Tora L, Yamauchi J, Masushige S, Bellard M, Chambon P (1992) A far upstream
  estrogen response element of the ovalbumin gene contains several half-
  palindromic 5'-TGACC-3' motifs acting synergistically. Cell 68:731–742
Kato S, Endoh H, Masuhiro Y, Kitamoto T, Uchiyama S, Sasaki H, Masushige S, Gotoh
  Y, Nishida E, Kawashima H, Metzger D, Chambon P (1995a) Activation of estro-
  gen receptor through phosphorylation by mitogen-activated protein kinase.
  Science 270:1491–1494
Kato S, Sasaki H, Suzawa M, Masushige S, Tora L, Chambon P, Gronemeyer H (1995b)
  Widely spaced, directly repeated PuGGTCA elements act as promiscuous
  enhancers for different classes of nuclear receptors. Mol Cell Biol 15:5858–
  5867
Katzenellenbogen BS (1996) Estrogen receptors: bioactivities and interactions with cell
  signaling pathways. Biol Reprod 54: 287–293
Katzenellenbogen BS, Norman MJ, Eckert RL, Peltz SW, Mangel SF (1984) Bioactiv-
  ities, estrogen receptor interactions and plasminogen activator inducing activities

of tamoxifen and hydroxytamoxifen isomers in MCF-7 human breast cancer cells. Cancer Res 44:112–119

Katzenellenbogen JA, O'Malley BW, Katzenellenbogen BS (1996) Tripartite steroid hormone receptor pharmacology: Interaction with multiple effector sites as a basis for the cell-and promoter-specific action of these hormones. Mol Endocrinol 10:119–131

Kharat I, Saatcioglu F (1996) Antiestrogenic effects of 2,3,7,8-tetrachlorodibenzo-p-dioxin are mediated by direct transcriptional interference with the liganded estrogen receptor. Cross-talk between aryl hydrocarbon- and estrogen-mediated signaling. J Biol Chem 271:10533–10537

Kim J, de Haan G, Nardulli AM, Shapiro DJ (1997) Prebending the estrogen response element destabilizes binding of the estrogen receptor DNA binding domain. Mol Cell Biol 17:3173–3180

Klein-Hitpass L, Schorpp M, Wagner U, Ryffel GU (1986) An estrogen-responsive element derived from the 5' flanking region of the *Xenopus* vitellogenin A2 gene functions in transfected human cells. Cell 46:1053–1061

Klein-Hitpass L, Ryffel GU, Heitlinger E, Cato A.C.B. (1988) A 13 bp palindrome is a functional estrogen responsive element and interacts specifically with estrogen receptor. Nucleic Acids Res 16:647–663

Klein-Hitpass L, Tsai SY, Greene GL, Clark JH, Tsai M-J, O'Malley BW (1989) Specific binding of estrogen receptor to the estrogen response element. Mol Cell Biol 9: 43–49

Klinge CM, Silver BF, Driscoll MD, Sathya G, Bambara RA, Hilf R (1997) Chicken ovalbumin upstream promoter-transcription factor interacts with estrogen receptor, binds to estrogen response elements and half-sites, and inhibits estrogen-induced gene expression. J Biol Chem 272:31465–31474

Kabbe C, Lippman ME, Wakefield LM, Flanders KC, Kasid A, Derynck R, Dickson RB (1987) Evidence that the transforming growth factor-β is a hormonally regulated negative growth factor in human breast cancer cells. Cell 48:417–428

Kneifel MA, Katzenellenbogen BS (1981) Comparative effects of estrogen and anti-estrogen on plasma renin substrate levels and hepatic estrogen receptor in the rat. Endocrinology 108:545–552

Kneifel MA, Leytus SP, Fletcher E, Weber T, Mangel WF, Katzenellenbogen BS (1982) Uterine plasminogen activator activity: modulation by steroid hormones. Endocrinology 111:493–499

Krause K-H, Michalak M (1997) Calreticulin. Cell 88:439–443

Krishnan V, Wang X, Safe S (1994) Estrogen receptor-Sp1 complexes mediate estrogen-induced cathepsin D gene expression in MCF-7 human breast cancer cells. J Biol Chem 269:15912–15917

Krust A, Green S, Argos P, Kumar V, Walter P, Bornert J-M, Chambon P (1986) The chicken estrogen receptor sequence: homology with v-erbA and the human estrogen and glucocorticoid receptors. EMBO J 5:891–897

Kuiper GGJM, Enmark E, Pelto-Huikko M, Nilsson S, Gustafsson J-A (1996) Cloning of a novel estrogen receptor expressed in rat prostate and ovary. Proc Natl Acad Sci USA 93:5925–5930

Kuiper GGJM, Carlsson B, Grandien K, Enmark E, Haggblad J, Nilsson S, Gustafsson J-A (1997) Comparison of the ligand binding specificity and transcript tissue distribution of estrogen receptors α and β. Endocrinology 138:863–870

Kuiper GGJM, Gustafsson J-A (1997) The novel estrogen receptor-β subtype: potential role in the cell-and promoter-specific actions of estrogens and anti-estrogens. FEBF Lett 410:87–90

Kumar V, Green S, Staub A, Chambon P (1986) Localisation of the oestradiol-binding and putative DNA-binding domains of the human estrogen receptor. EMBO J 5:2231–2236

Kumar V, Green S, Stack G, Berry M, Jin J-R, Chambon P (1987) Functional domains of the human estrogen receptor. Cell 51:941–951

Kumar V, Chambon P (1988) The estrogen receptor binds tightly to its responsive element as a ligand-induced homodimer. Cell 55:145–156

Kurebayashi S, Miyashita Y, Hirose T, Kasayama S, Akira S, Kishimoto T (1997) Characterization of mechanisms of interleukin-6 gene repression by estrogen receptor. J Steroid Biochem Mol Biol 60:11–17

Lahooti H, White R, Danielian PS, Parker MG (1994) Characterization of ligand-dependent phosphorylation of the estrogen receptor. Mol Endocrinol 8:182–188

Lee DC, McKnight GS, Palmiter RD (1988) The action of estrogen and progesterone on the expression of the transferrin gene. A comparison of the response in chick liver and oviduct. J Biol Chem 253:3494–3503

Levenson AS, Jordan VC (1994) Transfection of human estrogen receptor (ER) cDNA into ER-negative mammalian cell lines. J Steroid Biochem Mol Biol 51:229–239

Liu Y, Yang N, Teng CT (1993) COUP-TF acts as a competitive repressor for estrogen receptor-mediated activation of the mouse lactoferrin gene. Mol Cell Biol 13:1836–1846

Lundholt BK, Madsen MW, Lykkesfeldt AE, Petersen OW, Briand P (1996) Characterization of a nontumorigenic human breast epithelial cell line stably transfected with the human estrogen receptor (ER) cDNA. Mol Cell Endocrinol 119:47–59

Mabry White M, Zamudio S, Stevens T, Tyler R, Lindenfeld J, Leslie K, Moore LG (1995) Estrogen, progesterone, and vascular reactivity: potential cellular mechanisms. Endocr Rev 16:739–751

Mangelsdorf DJ, Thummel C, Beato M, Herrlich P, Schütz G, Umesono K, Blumberg B, Kastner P, Mark M, Chambon P, Evans RM (1995) The nuclear receptor superfamily: the second decade. Cell 83:835–839

Margot JB, Williams DL (1996) Estrogen induces the assembly of a multiprotein messenger ribonucleoprotein complex on the 3′-untranslated region of chicken apolipoprotein II mRNA. J Biol Chem 271:4452–4460

Martin MB, Garcia-Morales P, Stoica A, Solomon HB, Pierce M, Katz D, Zhang S, Danielsen M, Saceda M (1995) Effects of 12-O-tetradecanoylphorbol-13-acetate on estrogen receptor activity in MCF-7 cells. J Biol Chem 270:25244–25251

Martin V, Ribieras S, Song-Wang XG, Lasne Y, Frappart L, Rio MC, Dante R (1997) Involvement of DNA methylation in the control of the expression of an estrogen-induced breast-cancer-associated protein (pS2) in human breats cancer. J Cell Biochem 65:95–106

Matsuda S, Kadowaki Y, Ichino M, Akiyama T, Toyoshima K, Yamamoto T (1993) 17β-estradiol mimics ligand activity of the c-erbB2 protooncogene product. Proc Natl Acad Sci USA 90:10803–10807

Maurer R (1985) Selective binding of the estradiol receptor to a region at least one kilobase upstream from the rat prolactin gene. DNA 4:1–9

McCarthy T, Ji C, Shu H, Casinghino S, Crothers K, Rotwein P, Centrella M (1997) 17β-estradiol potently supresses cAMP-induced insulin-like growth factor-I gene activation in primary rat osteoblast cultures. J Biol Chem 272:18132–18139

McDonnell DP, Lieberman BA, Norris J (1995a) Development of tissue-selective estrogen receptor modulators. In: Baird DT, Schutz G, Krattenmacher R (eds) Organselective actions of steroid hormones. Springer-Verlag, Berlin Heidelberg New York, pp 1–28

McDonnell DP, Clemm DL, Hermann T, Goldman ME, Pike JW (1995b) Analysis of estrogen receptor function in vitro reveals three distinct classes of antiestrogens. Mol Endocrinol 9:659–669

McEwen BS, Alves SE, Bulloch K, Weiland NG (1997) Ovarian steroids and the brain: implications for cognition and aging. Neurology 48:S8-S15

McKenzie EA, Knowland J (1990) High concentrations of estrogen stabilize vitellogenin mRNA against cytoplasmic degradation but physiological concentrations do not. Mol Endocrinol 4:807–811

Medici N, Nigro V, Abbondanza C, Moncharmont B, Molinari AM, Puca GA (1991) In vitro binding of the purified hormone-binding subunit of the estrogen receptor to oligonucleotides containing the natural or modified sequences of an estrogen-responsive element. Mol Endocrinol 5:555–563

Mermelstein PG, Becker JB, Surmeier DJ (1996) Estradiol reduces calcium current in rat neostriatal neurons via a membrane receptor. J Neurosci 16:595–604

Metzger D, Losson R, Bornert J-M, Lemoine Y, Chambon P (1992) Promoter specificity of the two transcriptional activation functions of the human estrogen receptor in yeast. Nucleic Acids Res 20:2813–2817

Migliaccio A, Di Domenico M, Castoria G, de Falco A, Bontempo P, Nola E, Auricchio F (1996) Tyrosine kinase/p21ras/MAP-kinase pathway activation by estradiol-receptor complex in MCF-7 cells. EMBO J 15:1292–1300

Milgrom E, Savouret JF, Mantel A, Perrot-Applanat M, Delabre K, Lescop P (1997) Promoter- and cell-specific responses to sex steroids. Osteoporos Int 7 (suppl 1):S28–28

Minucci S, Ozato K (1996) Retinoid receptors in transcriptional regulation. Curr Opin Gen Dev 6:567–574

Mosselman S, Polman J, Dijkerma R (1996) ERβ: identification and characterization of a novel estrogen receptor. FEBS Lett 392:49–53

Mueller GC, Gorski J, Aizawa Y (1961) The role of protein synthesis in early estrogen action. Proc Natl Acad Sci USA 47:164–169

Mukherjee R, Chambon P (1990) A single-stranded DNA-binding protein promotes the binding of the purified estrogen receptor to its responsive element. Nucleic Acids Res 18:5713–5716

Nabekura J, Oomura Y, Minami T, Mizuno Y, Fukuda A (1986) Mechanism of the rapid effect of 17β-estradiol on medial amygdala neurons. Science 233:226–228

Nemere I, Zhou L-X, Norman AW (1993) Nontranscriptional effects of steroid hormones. Receptor 3:277–291

Neuman E, Ladha MH, Lin N, Upton TM, Miller SJ, DiRenzo J, Pestell RG, Hinds PW, Dowdy SF, Brown M, Ewen ME (1997) Cyclin D1 stimulation of estrogen receptor transcriptional activity independent of cdk4. Mol Cell Biol 17:5338–5347

Newton CJ, Buric R, Trapp T, Brockmeier S, Pagotto U, Stalla GK (1994) The unliganded estrogen receptor (ER) transduces growth factor signals. J Steroid Biochem Mol Biol 48:481–486

Ngo VM, Laverriere JN, Gourdji D (1995) CpG methylation represses the activity of the rat prolactin promoter in rat GH3 pituitary cell lines. Mol Cell Endocrinol 108:95–105

Nielsen DA, Shapiro DJ (1990) Insights into hormonal control of messenger RNA stability. Mol Endocrinol 4:953–957

Norris JD, Fan D, Aleman C, Marks JR, Futreal PA, Wiseman RW, Iglehart JD, Deininger PL, McDonnell DP (1995) Identification of a new subclass of Alu DNA repeats which can function as estrogen receptor-dependent transcriptional enhancers. J Biol Chem 270:22777–22782

Norris JD, Fan D, Kerner SA, McDonnell DP (1997) Identification of a third autonomus activation domain within the human estrogen receptor. Mol Endocrinol 11:747–754

O'Malley BW, McGuire WL, Kohler PO, Koreman SG (1969) Studies on the mechanism of steroid hormone regulation of synthesis of specific proteins. Recent Progr Horm Res 25:105–160

O'Malley BW, Schrader WT, Mani S, Smith C, Weigel NL, Conneely OM, Clark JH (1995) An alternative ligand-independent pathway for activation of steroid receptors. Rec Progr Horm Res 50:333–347

Onoe Y, Miyaura C, Ohta H, Nozawa S, Suda T (1997) Expression of estrogen receptor β in rat bone. Endocrinology 138:4509–4512

Orti E, Bodwell JE, Munck A (1992) Phosphorylation of steroid hormone receptors. Endocr Rev 13:105–128

Pace P, Taylor J, Suntharalingam S, Coombes RC, Ali S (1997) Human estrogen receptor β binds DNA in a manner similar to and dimerizes with estrogen receptor α. J Biol Chem 272:25832–25838

Paech K, Webb P, Kuiper GGJM, Nilsson S, Gustafsson J, Kushner PJ, Scanlan TS (1997) Differential ligand activation of estrogen receptors ERα and ERβ at AP1 sites. Science 277:1508–1510

Papa M, Mezzogiorno V, Bresciani F, Weisz A (1991) Estrogen induces c-fos expression specifically in the luminal and glandular epithelia of adult rat uterus. Biochem Biophys Res Commun 175:480–485

Parker MG (1995) Structure and function of estrogen receptors. Vitam Horm 51:267–287

Patrone C, Ma ZQ, Pollio G, Agrati P, Parker MG, Maggi A (1996) Cross-coupling between insulin and estrogen receptor in human neuroblastoma cells. Mol Endocrinol 10:499–507

Pelzer T, Shamim A, Neyses L (1996) Estrogen effects in the heart. Mol Cell Biochem 160:307–313

Pettersson K, Grandien K, Kuiper GGJM, Gustafsson J-A (1997) Mouse estrogen receptor β forms estrogen response element-binding heterodimers with estrogen receptor α. Mol Endocrinol 11:1486–1496

Pfaff DW, Freidin MM, Wu-Peng XS, Yin J, Zhu YS (1994) Competition for DNA steroid response elements as a possible mechanism for neuroendocrine integration. J Steroid Biochem Mol Biol 49:373–379

Philips A, Chalbos D, Rochefort H (1993) Estradiol increases and anti-estrogens antagonize the growth factor-induced activator protein-1 activity in MCF-7 breast cancer cells without affecting c-fos and c-jun synthesis. J Biol Chem 268:14103–14108

Pietras RJ, Arboleda J, Reese DM, Wongvipat N, Pegram MD, Ramos L, Gorman CM, Parker MG, Sliwkowski MX, Slamon DJ (1995) HER-2 tyrosine kinase pathway targets estrogen receptor and promotes hormone-independent growth in human breast cancer cells. Oncogene 10:2435–2446

Porter W, Wang F, Wang W, Duan R, Safe S (1996) Role of estrogen receptor/Sp1 complexes in estrogen-induced heat shock protein 27 gene expression. Mol Endocrinol 10:1371–1378

Porter W, Saville B, Hoivik D, Safe S (1997) Functional synergy between the transcription factor Sp1 and the estrogen receptor. Mol Endocrinol 11:1569–1580

Power RF, Mani SK, Codina J, Conneely OM, O'Malley BW (1991) Dopamine and ligand-independent activation of steroid hormone receptors. Science 254:1636–1639

Pratt WB, Toft DO (1997) Steroid receptor interactions with heat shock protein and immunophilin chaperones. Endocr Rev 18:306–360

Ptashne M, Gann A (1997) Transcriptional activation by recruitment. Nature 386:569–577

Ramkumar T, Adler S (1995) Different positive and negative transcriptional regulation by tamoxifen. Endocrinology 136:536–542

Ray P, Ghosh SK, Zhang DH, Ray A (1997) Repression of interleukin-6 gene expression by 17β-estradiol: inhibition of the DNA-binding activity of the transcription factors NF-IL6 and NF-κB by the estrogen receptor. FEBS Lett 409:79–85

R DW, Katzenellenbogen JA, Long DJ, Rorke EA, Katzenellenbogen BS (1982) Tamoxifen antiestrogens. A comparison of the activity, pharmacokinetics and metabolic activation of the cis and trans isomers of tamoxifen. J Steroid Biochem 16:1–13

Romer J, Lund LR, Dano K (1997) Healing hormones. Nature Med 3:1195–1196

Sanguedolce MV, Leblanc BP, Betz JL, Stunnenberg HG (1997) The promoter context is a decisive factor in establishing selective responsiveness to nuclear class II receptors. EMBO Journal 16:2861–2873

Sato M, Glasebrook AL, Bryant HU (1994) Raloxifene: a selective estrogen receptor modulator. J Bon Miner Met 12:S9–S20

Schmitt M, Bausero P, Simoni P, Queuche D, Geoffroy V, Marschal C, Kempf J, Quirin-Stricker C (1995) Positive and negative effects of nuclear receptors on transcription activation by AP-1 of the human choline acetyltransferase proximal promoter. J Neurosci Res 40:152–164

Schwabe JWR, Chapman L, Finch JT, Rhodes D (1993) The crystal structure of the estrogen receptor DNA-binding domain bound to DNA: how receptor discriminate between their response elements. Cell 75:567–578

Scott REM, Wu-Peng XS, Yen PM, Chin WW, Pfaff DW (1997) Interactions of estrogen- and thyroid hormone receptors on a progesterone receptor estrogen response element (ERE) sequence: a comparison with the vitellogenin A2 consensus ERE. Mol Endocrinol 11:1581–1592

Segars JH, Marks MS, Hirschfeld S, Driggers PH, Martinez E, Grippo JF, Wahli W, Ozato K (1993) Inhibition of estrogen-responsive gene activation by the retinoid X receptor β: evidence for multiple inhibitory pathways. Mol Cell Biol 13:2258–2268

Seiler-Tuyns A, Mérillat A-M, Nardelli Haefliger D, Wahli W (1988) The human estrogen receptor can regulate exogenous but not endogenous vitellogenin gene promoters in a Xenopus cell line. Nucleic Acids Res 16:8291–8305

Shemshedini L, Knauthe R, Sassone Corsi P, Pornon A, Gronemeyer H (1991) Cell-specific inhibitory and stimulatory effects of Fos and Jun on transcription activation by nuclear receptors. EMBO J 10:3839–3849

Shoenberg DR, Moskaitis JE, Smith LH, Pastori RL (1989) Extranuclear estrogen-regulated destabilization of Xenopus Laevis serum albumin. Mol Endocrinol 3:805–814

Shull JD, Beams FE, Baldwin TM, Gilchrist CA, Hrbek MJ (1992) The estrogenic and antiestrogenic properties of tamoxifen in GH4C1 pituitary tumor cells are gene specific. Mol Endocrinol 6:529–535

Siegfried Z, Cedar H (1997) DNA methylation: a molecuar lock. Curr Biol 7:R305–R307

Simerly RB, Chang C, Muramatsu M, Swanson LW (1990) Distribution of androgen and estrogen receptor mRNA-containing cells in the rat brain: an in situ hybridization study. J Comp Neurol294:76–95

Smith CL, Conneely OM, O'Malley BW (1993) Modulation of ligand-independent activation of the human estrogen receptor by hormone and antihormone. Proc Natl Acad Sci USA 90:6120–6124

Smith CL, Conneely O.M., O'Malley BW (1995) Estrogen receptor activation in the absence of ligand. Biochem Soc Trans 23:935–939

Smith CL, Hager GL (1997) Transcriptional regulation of mammalian genes <in vivo>. A tale of two templates. J Biol Chem 272:27493–27496

Srivastava RAK, Srivastava N, Averna M, Lin RC, Korach KS, Lubahn DB, Schonfeld G (1997) Estrogen up-regulates apolipoprotein E (Apo E) gene expression by increasing Apo E mRNA in the translating pool via the estrogen receptor α-mediated pathway. J Biol Chem 272:33360–33366

Starr DB, Matsui W, Thomas JR, Yamamoto KR (1996) Intracellular receptors use a common mechanism to interpret signaling information at response elements. Genes Dev 10:1271–1283

Stein B, Yang MX (1995) Repression of the interleukin-6 promoter by estrogen receptor is mediated by NF-κB and C/EBPβ. Mol Cell Biol 15:4971–4979

Sukovich DA, Mukherjee R, Benfield PA (1994) A novel, cell-type specific mechanism for estrogen receptor-mediated gene activation in the absence of an estrogen-responsive element. Mol Cell Biol 14:7134–7143

Suzuki H, Tornese Buonamassa D, Weisz A (1990) Inverse relationship between poly (ADP-ribose) polymerase activity and 2',5'-oligoadenylates core level in estrogen-treated immature rat. Mol Cell Biochem 99:33–39

Suzuki H, Tornese Buonamassa D, Cicatiello L, Weisz A (1992) Variation in poly (ADP-ribose) polymerase activity and 2',5'-oligoadenylates core concentration in estrogen-stimulated uterus and liver of immature and adult rats. In: Poirier G, Moireau P (eds). Springer-Verlag, New York, pp 133–136

Szego CM, Davis JS (1967) Adenosine 3',5'-monophosphate in rat uterus: acute elevation by estrogen. Proc Natl Acad Sci USA 58:1711–1718

Tjian R, Maniatis T (1994) Transcriptional activation: a complex puzzle with few easy pieces. Cell 77: 5–8

Toney TW, Katzenellenbogen BS (1986) Antiestrogen action in the medial basal hypothalamus and pituitary of immature female rats: insights concerning relationships among estrogen, dopamine and prolactin. Endocrinology 119:2661–2669

Tora L, Gaub M-P, Mader S, Dierich A, Bellard M, Chambon P (1988) Cell-specific activity of a GGTCA half-palindromic estrogen-responsive element in the chicken ovalbumin gene promoter. EMBO J 7:3771–3778

Tora L, White J, Brou C, Tasset D, Webster N, Scheer E, Chambon P (1989) The human estrogen receptor has two indipendent nonacidic transcriptional activation functions. Cell 59: 477–487

Touitou I, Vignon F, Cavailles V, Rochefort H (1991) Hormonal regulation of cathepsin D following transfection of the estrogen or progesterone receptor into three steroid hormone resistant cancer cell lines. J Steroid Biochem Mol Biol 40:231–237

Tremblay GB, Tremblay A, Copeland NG, Gilbert DJ, Jenkins NA, Labrie F, Giguère V (1997) Cloning, chromosomal localization, and functional analysis of the murine estrogen receptor. Mol Endocrinol 11:353–365

Trowbridge JM, Rogatsky I, Garabedian MJ (1997) Regulation of estrogen receptor transcriptional enhancement by the cyclin A/Cdk2 complex. Proc Natl Acad Sci USA 94:10132–10137

Truss M, Beato M (1993) Steroid hormone receptors: interaction with deoxyribonucleic acid and transcription factors. Endocr Rev 14:459–479

Tsai M-J, O'Malley BW (1994) Molecular mechanisms of action of steroid/thyroid receptor superfamily members. Annu Rev Biochem 63:451–486

Turner RT, Riggs BL, Spelsberg TC (1994) Skeletal effects of estrogen. Endocr Rev 15:275–299

Tzukerman M, Zhang X-K, Pfahl M (1991) Inhibition of estrogen receptor activity by the tumor promoter 12-O-tetradeconylphorbol-13-acetate: a molecular analysis. Mol Endocrinol 5:1983–1992

Tzukerman MT, Esty A, Santiso-Mere D, Danielian P, Parker MG, Stein RP, Pike JW, McDonnell DP (1994) Human estrogen receptor transactivational capacity is determined by both cellular and promoter context and mediated by two functionally distinct intramolecular regions. Mol Endocrinol 8:21–30

Uht RM, Anderson CM, Webb P, Kushner P (1997) Transcriptional activities of estrogen and glucocorticoid receptors are functionally integrated at the AP-1 response element. Endocrinology 138:2900–2908

Umayahara Y, Kawamori R, Watada H, Imano E, Iwama N, Morishima T, Yamasaki Y, Kajimoto Y, Kamada T (1994) Estrogen regulation of the insulin-like growth factor I gene transcription involves an AP-1 enhancer. J Biol Chem 269:16433–16442

Walker P, Germond J-E, Brown-Luedi M, Givel F, Wahli W (1984) Sequence homologies in the region preceding the transcriptional initiation site of the liver estrogen-responsive vitellogenin and apo-VLDL II genes. Nucleic Acids Res 12:8611–8626

Walter P, Green S, Greene GL, Krust A, Bornert JM, Jeltsch JM, Staub A, Jensen EV, Scrace G, Waterfield M, Chambon P (1985) Cloning of the human estrogen receptor cDNA. Proc Natl Acad Sci USA 82:7889–7893

Watanabe T, Inoue S, Ogawa S, Ishii Y, Hiroi H, Ikeda K, Orimo A, Muramatsu M (1997) Agonistic effect of tamoxifen is dependent on cell type, ERE-promoter context, and estrogen receptor subtype: functional difference between estrogen receptors α and β. Biochem Biophys Res Commun 236:140–145

Watson CS, Torres T (1990) Expression and translocation of cloned human estrogen receptor in the Xenopus oocyte does not induce expression of the endogenous oocyte vitellogenin genes. Mol Endocrinol 4:565–572

Watters JJ, Campbell JS, Cunningham WJ, Krebs EG, Dorsa DM (1997) Rapid membrane effects of steroids in neuroblastoma cells: effects of estrogen on mitogen activated protein kinase signalling cascade and c-fos immediate early gene transcription. Endocrinology 138:4030–4033

Weaver CA, Springer PA, Katzenellenbogen BS (1988) Regulation of pS2 gene expression by affinity labeling and reversibly binding estrogens and antiestrogens: comparison of effects on the native gene and on pS2-chloramphenicol acetyltransferase fusion genes transfected into MCF-7 human breast cancer cells. Mol Endocrinol 2:936–945

Wehling M (1997) Specific, nongenomic actions of steroid hormones. Annu Rev Physiol 59:365–393

Weigel NL (1996) Steroid hormone receptors and their regulation by phosphorylation. Biochem J 319:657–667

Weisz A, Coppola L, Bresciani F (1986) Specific binding of estrogen receptor to sites upstream and within the transcribed region of the chicken ovalbumin gene. Biochem Biophys Res Commun 139:396–402

Weisz A, Cicatiello L, Persico E, Scalona M, Bresciani F (1990) Estrogen stimulates transcription of c-jun protooncogene. Mol Endocrinol 4:1041–1050

Weisz A, Rosales R (1990) Identification of an estrogen response element upstream of the human c-fos gene that binds the estrogen receptor and the AP-1 transcription factor. Nucleic Acids Res 18:5097–5106

Weisz A, Cicatiello L, Rosales R, Bresciani F (1991) Estrogen control of cell proliferation: molecular mechanism of activation of c-fos proto-oncogene transription by the estrogen receptor. In: Maggi M, Geenen V (eds) Horizons in Endocrinology, vol. II. Raven Press, New York, pp 116–120

Weisz A, Bresciani F (1988) Estrogen induces expression of c-fos and c-myc protooncogenes in rat uterus. Mol Endocrinol 2:816–824

Weisz A, Bresciani F (1993) Estrogen regulation of proto-oncogenes coding for nuclear proteins. Crit Rev Oncogenesis 4:361–388

Weisz A, Addeo R, Altucci L, Battista T, Boccia V, Cancemi M, Cicatiello L, Germano D, Mancini A, Pacilio C, Bresciani F (1996) Molecular mechanisms for estrogen control of cell cycle progression during G1. In: Kuramoto H, Gurpide E (eds) In vitro Biology of Sex Steroid Hormone Action. Churchill Livingstone Japan, Tokyo, pp 1–16

Wickelgren I (1997) Estrogen stakes claim recognition. Science 276:675–678

Wijnholds J, Philipsen JNJ, Ab G(1988) Tissue-specific and steroid-dependent interaction of transcription factors with the estrogen-inducible apoVLDL II promoter in vivo. EMBO J 7:2757–2763

Witkowska HE, Carlquist M, Engstrom O, Carlsson B, Bonn T, Gustafsson J-A, Shackleton CH (1997) Characterization of bacterially expressed rat estrogen receptor β ligand binding domain by mass spectroscopy: structural comparison with estrogen receptor α. Steroids 62:621–631

Yamamoto KR (1997) Multilayered control of intracellular receptor function. Harvey Lect 91:1–19

Yamamoto KR, Alberts BM (1972) In vitro conversion of estradiol-receptor protein to its nuclear form: Dependence on hormone and DNA. Proc Natl Acad Sci USA 69:2105–2109

Yamamoto KR, Alberts BM (1976) Steroid receptors: elements for modulation of eukaryotic transcription. Annu Rev Biochem 45:721–746

Yang NN, Bryant HU, Hardikar S, Sato M, Galvin RJS, Glasebrook AL, Termine JD (1996a) Estrogen and Raloxifene stimulate transforming growth factor-$\beta 3$ gene expression in rat bone: Potential mechanism for estrogen- or Raloxifene-mediated bone maintenance. Endocrinology 137:2075–2084

Yang N, Shigeta H, Shi H, Teng CT (1996b) Estrogen-related receptor, hERR1, modulates estrogen receptor-mediated response of human lactoferrin gene promoter. J Biol Chem 271:5795–5804

Yang NN, Venugopalan M, Hardikar S, Glasebrook AL (1996c) Identification of an estrogen response element activated by metabolites of 17β-estradiol and Raloxifene. Science 273:1222–1225

Ying C, Gorski J (1994) DNA topology regulates rat prolactin gene transcription. Mol Cell Endocrinol 99:183–192

Zajchowski DA, Sager R (1991) Induction of estrogen-regulated genes differs in immortal and tumorigenic human mammary epithelial cells expressing a recombinant estrogen receptor. Mol Endocrinol 5:1613–1623

Zilliacus J, Wright APH, Carlstedt-Duke J, Gustafsson J-A (1995) Structural determinants of DNA-binding specificity by steroid receptors. Mol Endocrinol 9:389–400

Zwijsen RML, Wientjens E, Klopmaker R, van der Sman J, Bernards R, Michalides RJAM (1997) CDK-independent activation of estrogen receptor by cyclin D1. Cell 88:405–415

CHAPTER 8

Regulation of Constitutive and Inducible Nitric Oxide Synthase by Estrogen

K. KAUSER and G.M. RUBANYI

A. Introduction

Nitric oxide synthases (NOSs) catalyze the oxidative metabolism of L-arginine to nitric oxide (NO). NO is an important free radical with multiple functions, including endothelium-dependent relaxation, neurotransmission, immunoprotection and inflammation. In addition to the numerous regulatory functions of NO, impaired synthesis of NO seems to be a key element in several diseases, such as atherosclerosis, pulmonary hypertension, and pyloric stenosis. In contrast, excessive NO production has been associated with pathophysiological states, such as endotoxemia, stroke, multiple sclerosis. Understanding the role of NO in specific disease states could provide important therapeutic approaches in the future. Gender-dependent differences in NO synthesis suggested that NOS expression and/or activity is under the control of sex steroid hormones. Indeed, animal and clinical studies showed that estrogen treatment leads to modulation of NO production. Regulation of NO production by the different NOS isoenzymes may contribute to the observed benefits of estrogen replacement therapy (ERT) on the cardiovascular as well as on the central nervous system. This chapter will discuss the possible mechanisms of 17β-estradiol-induced regulation of NO production by the three different isoforms of NOS.

B. Isoenzymes of NOS

NOSs are a unique family of enzymes that form homodimers and contain a heme oxygenase domain and a cytochrome P-450 reductase domain. NOS-I (neuronal NOS) and NOS-III (endothelial NOS) are constitutively expressed, calcium–calmodulin-dependent enzymes. NOS-II (inducible NOS) is induced upon stimulation and does not require additional calcium for activity. NOSs represent unique cytochrome P-450s, as they require tetrahydrobiopterin (BH_4) for activity. Other cofactors include reduced nicotinamide adenine dinucleotide phosphate (NADPH) and flavins. The three isoenzymes for NOS-I, II and III are encoded by genes located on human chromosomes 12, 17 and 7, respectively (for review see FÖRSTERMANN et al. 1994; SESSA 1994). NO is synthesized in large quantities by NOS-II and NOS-I to mediate cytotoxic effects that are part of inflammatory reactions, or in smaller amounts by the endothelial isoform, NOS-III, to exert antioxidant, anti-inflammatory and anti-thrombotic effects and to mediate endothelium-dependent vasodilation of blood vessels (Rubanyi 1993).

I. Nitric Oxide Synthase-I

NOS-I is constitutively expressed in brain and was first purified from rat cerebellum (Bredt and Snyder 1990). Anatomical mapping demonstrated NOS-I messenger ribonucleic acid (mRNA) and protein localization to specific central and peripheral neuronal structures and to specific cells of the kidney and skeletal muscle (Bredt et al. 1991; Nakane et al. 1993).

NOS-I has been implicated in neurotoxicity. Studies with NOS-I knock-out mice support the involvement of NOS-I-derived NO in neuronal injury associated with stroke (Huang et al. 1994). NOS-I-derived NO can be released in response to the influx of calcium into neurones which occurs when postsynaptic glutamate receptors are stimulated by the excessive amount of glutamate that is released after ischemia. The calcium influx can also activate calcineurin, which may reactivate NOS-I by dephosphorylation, leading to further NO production (Dawson et al. 1993).

It has recently been shown that the post-synaptic density protein, PSD-95, NOS-I and N-methyl-D-aspartate (NMDA) are co-expressed in several neuronal populations, and that a NOS-I/PSD-95 complex exists in the cerebellum (Kornau et al. 1995; Brenman et al. 1996). The interaction of NOS-I with PSD-95 represents a unique, selective way of regulating NOS-I activity.

II. Nitric Oxide Synthase-II

Lipopolysaccharide (LPS) treatment of macrophages leads to significant nitrite and nitrate production (Stuehr and Marletta 1985). The induction of NOS-II by cytokines or LPS has been reported in various cell types, including vascular smooth muscle cells, hepatocytes, mesangial cells, myocytes, astrocytes and keratinocytes (for review see Sessa 1994). Treatment of these cells with actinomycin D and cycloheximide demonstrated that NOS-II synthesis is dependent on de novo mRNA transcription and protein synthesis (Xie et al. 1992; Geller et al. 1993). Analysis of the NOS-II gene promoter revealed the existence of several consensus sequence motifs of transcription factors that are involved in the induction of cytokine-responsive genes, such as neurotrophic factor (NF)-κB sites and the interferon (INF)γ response element (Xie et al. 1993).

Treatment with different growth factors, e.g., transforming growth factor-β (TGF-β) or platelet-derived growth factor AB and BB (PDGF-AB and -BB) reduces cytokine-induced NOS activity (Förstermann et al. 1992; Scott-Burden et al. 1992). Steroids, such as dexamethasone and progesterone also inhibit NOS-II induction (Dirosa et al. 1990; Schmidt et al. 1992). Retinoids have also been shown to downregulate NO production by cytokine-treated smooth muscle cells (HIROKAWA et al. 1994). The exact mechanism of these inhibitory effects on NOS-II is not entirely understood. Potential mechanisms include alteration of gene expression, mRNA stability or protein biosynthesis. Recent studies with the antioxidant pyrrolidine dithiocarbamate revealed the important role of NF-κB in the regulation of NOS-II expression (HECKER et al. 1996).

III. Nitric Oxide Synthase-III

Endothelial or constitutive NOS accounts for the synthesis of the endothelium-derived relaxing factor (EDRF) (FURCHGOTT and ZAWADZKI 1980). Results of studies by RUBANYI et al. (1986) and MILLER et al. (1986) revealed physiologically important rapid and long-term regulation of NOS-III by shear stress. The presence of AP-1, AP-2, NF-1, p53, sterol regulatory elements in the NOS-III promoter argues for the potential regulation of NOS-III expression by several different factors (for review see SESSA 1994). Downregulation of NOS-III has also been demonstrated by cytokine treatment and oxidized low-density lipoprotein (LDL) incubation, involving destabilization of mRNA (YOSHIZUMI et al. 1993).

Recent studies indicate the importance of post-translational modifications in the regulation of NOS-III activity (GARCIA-GARDENA et al. 1996). Cell membrane localization of NOS-III is mediated by post-translational myristoylation and palmitoylation, which target the enzyme to the plasmalemmal caveolae (SHAUL et al. 1996). This localization has been shown to significantly affect the activity of the enzyme, probably by interfering with the binding of calmodulin to the enzyme. Caveolin binding to the isolated enzyme inhibits its activity, which can be restored by the addition of calmodulin (MICHEL et al. 1997). Phosphorylation has also been shown to play an important role in shear-stress-induced activation of NOS-III, by changing the calcium dependence of the enzyme by an unknown mechanism (AYAJIKI et al. 1996).

C. Regulation of NOS Isoenzymes by 17β-Estradiol

All three isoforms of NOS have been shown to be regulated by 17β-estradiol. NOS-III has been studied the most extensively, but the NOS-I and NOS-II isoenzymes have also been shown to be affected by estrogen treatment in vivo (WEINER et al. 1994b; KAUSER et al. 1997). Inhibition of NOS-II has been studied in vitro in our laboratory using cytokine-treated isolated rat aortic rings (KAUSER et al. 1994, 1998).

I. Inhibition of NOS-II Activity by 17β-Estradiol

Excessive production of nitrite/nitrate by vascular smooth muscle cells mediates endothelium-independent suppression of blood-vessel contractility during septic shock, contributing to the severe hypotension (Thiemermann and Vane 1990). The source of excessive NO production is NOS-II, induced by endotoxin and cytokines, such as interleukin (IL)-1β and tumor necrosis factor (TNFα).

Although the effect of estrogen on spontaneous NO production in cultured hepatocytes (Pittner and Spitzer 1993) and rat peritoneal macrophages (Chao et al. 1994) has been investigated, the effect of the ovarian sex steroid on cytokine-induced NOS-II activity and expression has not yet been studied. Our laboratory was the first to study the effects of 17β-estradiol on IL-1β-induced nitrite/nitrate production and inhibition of contractile reactivity in isolated rat aortic rings. We found that, similar to the selective NOS-II inhibitor, aminoguanidine, co-incubation of the aortic rings with the cytokines and 17β-estradiol significantly reduced nitrite accumulation and increased contractile reactivity to phenylephrine. These results show that, in contrast to augmentation of NOS-III expression/activity in endothelial cells, 17β-estradiol inhibits IL-1β-induced excessive nitrite/nitrate generation and restores contractile responsiveness of vascular smooth muscle in isolated rat aortic rings (Fig. 1).

Nuclear estrogen receptors (ERs) are present in the rat aorta (Lin et al. 1986; Knauthe et al. 1996). It is possible that the effect of 17β-estradiol is mediated by the activation of ERs, given that the non-steroidal, partial ER agonist 4-OH-tamoxifen also inhibited nitrite production and attenuated the suppression of phenylephrine-induced contraction (Fig. 2).

We also found, for the first time, that physiological substitution doses of 17β-estradiol reduced nitrite/nitrate accumulation in the plasma of lipopolysaccharide (LPS)-treated ovariectomized rats (Kauser et al. 1997) (Fig. 3). Ovariectomized and sham-operated female rats were injected with LPS, and excessive NO generation was estimated by measuring plasma nitrite levels 12 h after the injection. An LPS-induced increase in circulating plasma nitrite levels was significantly higher in the ovariectomized animals compared with sham-operated females, suggesting that endogenous ovarian sex steroid hormones may also suppress excessive NO production by NOS-II in this model. Our finding suggests that estrogens may be beneficial in pathological conditions associated with excessive NO generation via the inducible NOS isoform (NOS-II).

II. Upregulation of Endothelial NOS-III Activity by 17β-Estradiol

Epidemiological, clinical, and experimental data showed that ERT in postmenopausal women reduces the morbidity and mortality associated with cardiovascular diseases by ~50% (Stampfer et al. 1991; Isles et al. 1992; Farhat et al. 1996; Kauser and Rubanyi 1997a,b). The exact mechanism of cardiovascular protection by estrogen is not known. Beneficial effects on the plasma lipoprotein profile is responsible for only ~25–40% of the protection. It is postulated that direct effects of estrogen on the vascular wall must play an

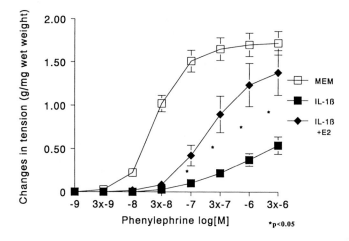

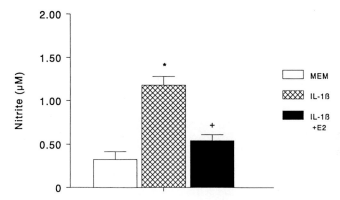

Fig. 1. Effect of 17β-estradiol on interleukin (IL)-1β-induced suppression of contractility and nitrite production by isolated rat thoracic aortic rings. The *upper panel* of the figure illustrates dose-response curves to phenylephrine in controls (*open square*), IL-1β (20 U/ml; *filled square*) and IL-1β(20 U/ml) + 17β-estradiol (1 μM; *filled diamond*) incubated de-endothelialized rat aortic rings. IL-1β significantly attenuated smooth muscle contraction. Treatment with 17β-estradiol significantly (*$P < 0.05$) reversed the inhibition of contractility by IL-1β. The *lower panel* demonstrates nitrite accumulation in the incubation media of the vessels. IL-1β induced significant (*$P < 0.05$) nitrite production (*cross-hatched bar*) by the aortic rings in the organ culture, which was significantly inhibited (+$P < 0.05$) by 17β-estradiol treatment of the cytokine incubated rings (*filled bar*)

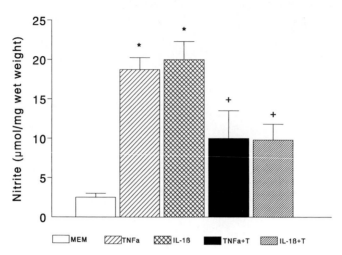

Fig. 2. Effect of 4-OH-tamoxifen on nitrite production in tumor necrosis factor (TNF)α- and interleukin (IL)-1β-treated rat aortic rings. TNFα (1 ng/ml; *shaded bars*) and IL-1β (20 U/ml; *cross-hatched bar*) treatment resulted in a significant (*$P < 0.05$) increase in nitrite accumulation in the incubation media of rat aortic rings when compared with that of control rings, minimal essential media (*open bars*). The partial estrogen receptor is agonist 4-OH-tamoxifen (1 μM) significantly inhibited nitrite accumulation by both cytokines. TNFα + 4-OH-tamoxifen (*T*) is illustrated by the *filled bar* and IL-1β + 4-OH-tamoxifen (*T*) is represented by the *double shaded bar*. Data represent mean ± SEM of $n = 6$ rings from different animals in each treatment group is

important role. NO, produced by NOS-III, seems to be an ideal mediator of the beneficial cardiovascular effects of 17β-estradiol, due to its multiple vasculoprotective actions.

1. Gender Difference and the Effect of 17β-Estradiol on Endothelial Function

The effect of sex steroid hormones on endothelial NO synthesis was suggested by studies demonstrating gender difference in endothelium-dependent modulation of vascular tone (Maddox et al. 1987; Stallone et al. 1991; Hayashi et al. 1992; Kauser and Rubanyi 1995a). These studies showed that the vascular endothelium of females is able to produce a large amount of NO. Our own studies, utilizing a superfusion bioassay system, showed that more NO is released from the perfused aortas of female than of male rats (Fig. 4) (Kauser and Rubanyi 1994b). The effect of ovariectomy and/or 17β-estradiol treatment on endothelium-dependent vascular responses (Gisclard et al. 1988; Hayashi et al. 1992; Kauser and Rubanyi 1995b) proved that gender differences are due (at least in part) to ovarian estrogens. Facilitation of endothelial NO-mediated vascular responses were found in rabbit femoral arteries (Gisclard et al. 1988) and in coronary arteries isolated from 17β-estradiol treated monkeys (Williams et al. 1990). Hayashi et al. (1992) demonstrated that basal

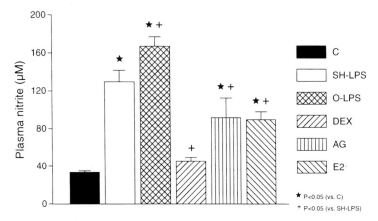

Fig. 3. Plasma nitrite level in sham operated and ovariectomized female rats. Animals were divided into six groups ($n = 8$ in each): *C*, control; *SH-LPS*, sham-operated female + lipopolysaccharide (LPS) (5 mg/kg) injection; *O-LPS*, ovariectomized female + LPS injection; *DEX*, ovariectomized female + LPS + dexamethasone (3 mg/kg, i.p.) injection 1 h prior to LPS; *AG*, ovariectomized female + LPS + aminoguanidine (200 µM/kg, i.p.) 1 h after LPS injection; *E2*, ovariectomized female + LPS + 17β-estradiol (3 µg/rat, s.c.) injection 48 h before LPS. LPS injection resulted in a four- to fivefold increase in plasma nitrite concentration compared with the control group. Nitrite levels of the plasma were determined using the Griess reaction. The nitrite level was significantly higher in the *O*-LPS than in the SH-LPS group. Dexamethasone completely prevented the nitrite increase ($+P < 0.05$). Relative to the *O*-LPS group, 17β-estradiol treatment (*E2*) inhibited nitrite formation to a similar degree ($+P < 0.05$) as aminoguanidine (*AG*). Reproduced with the permission of the American Physiological Society (from KAUSER et al. 1997c)

(unstimulated) release of NO from female rabbits was greater than that released from male and ovariectomized female rabbits.

These experimental results have been confirmed by clinical investigations demonstrating attenuation of abnormal vasomotor responses by administration of ethinyl-estradiol (REIS et al. 1994) or physiological doses of 17β-estradiol (GILLIGAN et al. 1994). A 2-year follow-up study reported increased circulating NO levels in postmenopausal women on hormone replacement therapy compared with women not taking estrogen (ROSSELLI et al. 1995). More recently, increased basal NO release (in forearm circulation) has been shown in response to 17β-estradiol treatment of perimenopausal women (SUDHIR et al. 1996).

2. Potential Mechanisms of Estrogen-Induced Increased Endothelial NO Production

a) NOS-III Gene Expression

One of the possible mechanisms by which 17β-estradiol can enhance endothelial NO synthesis is via activation of NOS-III gene expression, which leads to increased amounts of the NOS-III enzyme. Indeed, data obtained from several

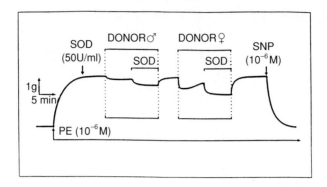

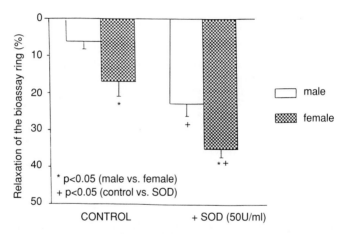

Fig. 4. Gender difference in bioassayable endothelium-derived nitric oxide (NO) generation in rat aorta. The *upper panel* illustrates an original tracing of a typical bioassay experiment. Perfusate from the donor segment isolated from female Wistar rats (*DONOR* is ♀) evoked greater relaxation of a denuded rat thoracic aortic ring than the perfusate from the male donor (*DONOR* is ♂). Superoxide dismutase (SOD; 50 U/ml) further enhanced the NO-mediated relaxation in both cases, but did not affect the gender difference. The *lower panel* demonstrates the mean values of eight experiments. *Open bars* represent responses (mean ± SEM) of the de-endothelialized bioassay rings to perfusate from male donor segments and the *cross-hatched bars* illustrate the responses evoked by the perfusate from the female donors. Relaxation of the bioassay ring is expressed as percentage inhibition of the phenylephrine-induced contraction. Asterisks indicate significantly (*$P < 0.05$) greater relaxation of the same assay ring in response to the perfusate from the female donor segments. Reproduced with the permission of the American Physiological Society (from KAUSER and RUBANYI 1994a)

species suggest that 17β-estradiol can facilitate NOS-III gene expression and increase the amount of NOS-III enzyme (GOETZ et al. 1994; WEINER et al. 1994b). Pregnancy has been associated with a fourfold increase in NOS-III activity in the guinea pig uterine artery (WEINER et al. 1994a). Similarly, using semi-quantitative polymerase chain reaction (PCR), increases of NOS-III mRNA have been demonstrated in the aortas of pregnant rats (GOETZ et al. 1994). Besides pregnancy, chronic treatment with 17β-estradiol also increased Ca^{2+}-dependent NO activity in female guinea pigs (WEINER et al. 1994b).

Regulation of NOS-III gene expression by 17β-estradiol has been studied extensively in cultured endothelial cells (HAYASHI et al. 1995; HISHIKAWA et al. 1995; MACRITCHIE et al. 1997). In early passages of human aortic and ovine fetal pulmonary endothelial cells, physiological concentrations of 17β-estradiol caused significant increases in NOS-III protein levels and, as a consequence, augmented NO production (HISHIKAWA et al. 1995; MACRITCHIE et al. 1997). Bovine aortic endothelial cells in culture contain functional ERs (BAYARD et al. 1995), which mediate 17β-estradiol-induced transactivation of a luciferase reporter gene; an estrogen response element (ERE) acts as a transcriptional enhancer (BAYARD et al. 1995). The existence of the two half palindromic sites of ERE in the promoter region of the human NOS-III gene supports a potential receptor-mediated effect of estrogen on NOS-III gene expression in endothelial cells (VENEMA et al. 1994).

b) NOS-III Enzyme Activity

Increased synthesis of NO can be a result of elevated enzyme activity without changes in gene expression or enzyme protein levels. Activity of NOS-III can be augmented by increased availability of the substrate L-arginine or cofactors, such as BH_4, calmodulin or intracellular free cytosolic calcium level. Studies have demonstrated that the administration of exogenous L-arginine, the physiological precursor of NO production, restored endothelium-dependent relaxation in hypercholesterolemic humans (CREAGER et al. 1992) and decreased aortic lesion formation in cholesterol-fed rabbits (COOKE et al. 1992). These findings indicate that, under certain pathological conditions, substrate availability of NOS-III may be limited and that supplementing L-arginine can provide vasculoprotective benefit. However, a recent report by HAYASHI et al. (1995) argues against the possibility that sex steroid hormones regulate the levels of circulating L-arginine.

BH_4, a cofactor of NO synthesis for all NOS enzymes, restores NO-mediated vasodilation in hypercholesterolemic patients (STROES et al. 1997). BH_4 synthesis is shown to be a rate-limiting step for NO production in cultured human endothelial cells (SCHMIDT et al. 1992; WERNER-FELMAYER et al. 1993; ROSENKRANZ-WEISS et al. 1994). It is possible that BH_4 synthesis can be regulated by 17β-estradiol via modulation of the level and/or activity of the guanosine triphosphate (GTP)-cyclohydrolase enzyme, which is responsible for the intracellular level of BH_4 (HATTORI and GROSS 1993). However, there is no data available describing the effect of 17β-estradiol on BH_4 synthesis.

It is also probable that 17β-estradiol facilitates the activity of NOS-III via calmodulin (BOUHOUTE and LECLERCQ 1995). The influence of 17β-estradiol on intracellular calmodulin synthesis has been described in rabbit myometrium (MATSUI et al. 1983), but it has not yet been investigated in vascular endothelial cells. In contrast, 17β-estradiol was reported to increase the levels of free cytosolic calcium in cultured endothelial cells (MOINI et al. 1997).

c) Bioactivity of NO

In addition to its rate of synthesis, the level of bioavailable NO can be augmented by inhibiting its inactivation/degradation. Increased amounts of guanylate cyclase have been measured in ethinyl-estradiol-treated bovine aortic endothelial cells, without any increases in NOS-III mRNA, proteins, or enzyme activity (ARNAL et al. 1996). Measurement of superoxide anion radical production from the same cells revealed that estrogen treatment dramatically reduced superoxide generation. The superoxide anion is an effective inactivator of NO (RUBANYI and VANHOUTTE 1986). Decreasing the level of superoxide anions could lead to increased amounts of bioavailable NO without changing its synthetic rate. Besides regulating superoxide production in endothelial cells, estrogens may stabilize NO by virtue of their well-documented antioxidant properties (SUBBIAH et al. 1993; RUIZ-LARREA et al. 1995). They have been shown to inhibit copper-induced and cell-mediated oxidative modification of LDL (MAZIERE et al. 1991), and they are also potent inhibitors of high-density lipoprotein (HDL) oxidation (BANKA 1996). It is conceivable that the antioxidant/radical scavenging properties of 17β-estradiol may contribute to the increased bioactivity of NO. However, unpublished data from our laboratory demonstrated that 17β-estradiol does not directly scavenge the superoxide anion and, in contrast with superoxide dismutase (SOD) (RUBANYI and VANHOUTTE 1986), does not prolong the half-life of endothelium-derived NO in superfusion bioassay experiments.

3. Role of the Estrogen Receptor in the Regulation of NOS-III

The presence of functional ERs has been demonstrated in endothelial cells (BAYARD et al. 1995; HAYASHI et al. 1995; KIM-SCHULZE et al. 1996; VENKOV et al. 1996; MACRITCHIE et al. 1997). The ER antagonists, ICI 182780 and tamoxifen inhibited 17β-estradiol-induced, dose-dependent increases in NOS-III mRNA and protein, as well as enzyme activity in human umbilical vein endothelial cells and bovine aortic endothelial cells in culture (BAYARD et al. 1995; HAYASHI et al. 1995; MACRITCHIE et al. 1997). These findings strongly suggest that the ER is involved in regulation of NOS-III by 17β-estradiol.

We have recently studied the role of ERs in modulating the release of endothelium-derived NO in a homozygous ERα mutant mouse (ERKO) (RUBANYI et al. 1997). In ERKO mice, the ERα gene was disrupted, resulting in the lack of functional ER expression (LUBAHN et al. 1993). We found that the lack of expression led to significant impairment in basal endothelial NO release (Fig. 5), without affecting the amount of NOS-III enzyme. This finding

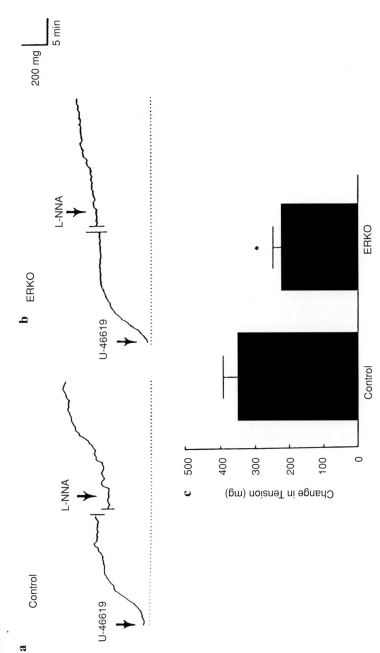

Fig. 5a–c. Decreased basal endothelial NO production in estrogen-receptor-deficient (ERKO) mouse aorta. Endothelium-dependent contraction of the thoracic aorta to N^G-nitro-L-arginine (L-NNA) was significantly ($*P < 0.05$) diminished in rings isolated from ERKO mice (**b**), compared with age-matched wild-type (WT) controls (**a**). The *upper two panels* (**a,b**) of the graph illustrate original tracing of the organ chamber experiments and the *lower panel* is a *bar graph* (**c**) of the average data (mean ± SEM) obtained from $n = 5$ animals. The *bar graphs* represent total contraction of the isolated mouse thoracic aortic rings to the thromboxane receptor agonist U-46619 and L-NNA. Reproduced with the permission of the American Society of Clinical Investigation (from RUBANYI et al. 1997)

demonstrated that ERα probably increases NOS-III enzyme activity by affecting the level of free cytosolic calcium and/or participating in post-translational modifications of NOS-III.

Since the introduction of the ERKO animal, a second ER, termed ERβ, has been identified (KUIPER et al. 1996; MOSSELMAN et al. 1996). Cultured human endothelial cells express both the ERα and ERβ genes (personal communication, Dr. ZAJCHOWSKI, Berlex Biosciences). However, the potential role of ERβ in the regulation of NOS-III remains to be elucidated.

4. Role of Increased Endothelial NO in the Anti-Atherosclerotic Effect of 17β-Estradiol

Endothelial cell injury or endothelial dysfunction is considered to play an important role in the development of atherosclerotic plaques (RUBANYI 1993; BUSSE and FLEMING 1996). Endothelial dysfunction is manifested mainly in impaired (reduced) NO production. The role of NO in mediating vasculoprotection has been demonstrated by accelerating atherosclerosis in hypercholesterolemic rabbits through chronic inhibition of NO formation (CAYATTE et al. 1994; NARUSE et al. 1994). Enhanced acetylcholine-induced, endothelium-dependent relaxation of coronary arteries following 17β-estradiol treatment of cholesterol-fed cynomolgus monkeys (WILLIAMS et al. 1990) suggested that the effect of estrogen may be mediated by increased NO production.

Recently, we investigated the role of NO in mediating the protective effect of 17β-estradiol on atherosclerotic lesion development in ovariectomized, hypercholesterolemic rabbits. Rabbits were treated with 17β-estradiol and/or N^w-nitro-L-arginine methylester (L-NAME) for 8 weeks. Diminished aortic plaque formation in the estrogen-treated animals was accompanied by enhanced endothelium-dependent vasorelaxation of isolated aortic rings, suggesting that the anti-atherosclerotic effect of 17β-estradiol could be mediated, at least in part, by increased NO formation in the aortic endothelium (Fig. 6). Chronic, systemic administration of L-NAME to 17β-estradiol-treated hypercholesterolemic rabbits significantly attenuated the vasculoprotective effects of estrogen, but did not completely prevent the reduction in plaque/surface ratio evoked by 17β-estradiol (Fig. 6). Our results demonstrated a strong correlation between elevated NO production and anti-atherosclerotic effects of estrogen in this model. These findings support the role of endothelial NO in estrogen-induced vasculoprotection. However, other factors/mechanisms may also be involved that are able to mediate estrogen-induced protection against atherosclerotic plaque development in the presence of an NOS inhibitor.

In good agreement with our study, it was reported recently that long-term inhibition of NO synthesis by L-NAME significantly reduced the anti-atherogenic effect of 17β-estradiol and levormeloxifene, a partial ER agonist (HOLM et al. 1997). In this study, plasma cholesterol was maintained at a fixed level, demonstrating that the beneficial effect of these estrogens in this model is dependent on endothelial NO production and independent of lipid lowering.

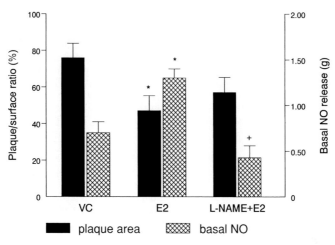

Fig. 6. Effect of chronic nitric oxide synthase (NOS) inhibition on the anti-atherosclerotic effect of 17β-estradiol in hypercholesterolemic rabbits. *Filled bars* illustrate plaque/surface area ratio (*left Y axis*) of thoracic aortae isolated from control, 17β-estradiol-treated and 17β-estradiol- plus N^w-nitro-L-arginine methylester (L-NAME)-treated cholesterol-fed rabbits. Six-week treatment of the rabbits with 17β-estradiol resulted in a significant (*$P < 0.05$) reduction in the plaque/surface area, which was diminished by chronic NOS inhibition. Endothelium-dependent contraction to L-NNA is represented by *cross-hatched bars* (*right Y axis*). 17β-Estradiol significantly (*$P < 0.05$) augmented the endothelium-mediated contraction to L-NNA (estimation of basal NO production), whereas L-NAME treatment was accompanied by significantly (+$P<0.05$) attenuated basal NO production. In each group $n = 8$ rabbits were studied. *Bars* represent mean ± SEM plaque/surface area (*left Y axis*) and NO mediated contraction (*right Y axis*). *VC*, vehicle control

D. Summary

The ovarian sex steroid hormone 17β-estradiol significantly augments endothelial NO production by NOS-III, and suppresses excessive NO generation by the inducible NOS isoenzyme (NOS-II). The mechanism(s) by which 17β-estradiol increases bioavailable endothelial NO is not entirely understood. The role of the ER has been demonstrated, but it is not clear whether the regulation of NOS-III takes place at the transcriptional, translational or post-translational level. Due to its anti-atherogenic properties, NO is an ideal mediator of the vasculoprotective effect of 17β-estradiol. Inhibition of excessive NO production has been shown to provide therapeutic benefits (NAVA et al. 1992); however, non-specific inhibition of NOSs in general can be detrimental (NAVA et al. 1991; HARBRECHT et al. 1992), emphasizing the need for isoenzyme-specific NOS inhibition. Since 17β-estradiol upregulates NOS-III in the rat aorta and inhibits NOS-II activity, this ovarian sex steroid hormone is an ideal candidate for NOS isoenzyme regulation in pathological conditions, in which the therapeutic goal is inhibition of NOS-II expression/activity without reduction of endothelial NO production.

References

Arnal JF, Clamens S, Pechet C, Negre-Salvayre A, Allera C, Girolami J-P et al. (1996) Ethinylestradiol does not enhance the expression of nitric oxide synthase in bovine endothelial cells but increases the release of bioactive nitric oxide by inhibiting superoxide anion production. Proc Natl Acad Sci USA 93:4108–4113

Ayajiki K, Kindermann M, Hecker M, Fleming I, Busse R (1996) Intracellular pH and tyrosine phosphorylation but not calcium determine shear stress-induced nitric oxide production in native endothelial cells. Circ Res 78:750–758

Banka CL (1996) High density lipoprotein and lipoprotein oxidation Curr Opin Lipidol 7:139–142

Bayard F, Clamens S, Delson G, Blaes N, Maret A, Faye J-C (1995) Estrogen synthesis, estrogen metabolism and functional estrogen receptors in bovine aortic endothelial cells. In: Non-reproductive actions of sex steroids 191. Ciba foundation symposium, John Wiley & Sons, Chichester, pp 122–138

Bouhoute A, Leclercq G (1995) Modulation of estradiol and DNA binding to estrogen receptor upon association with calmodulin. Biochem Biophys Res Commun 208:748–755

Bredt DS, Snyder SH (1990) Isolation of nitric oxide synthetase, a calmodulin-requiring enzyme. Proc Natl Acad Sci USA 87:682–685

Bredt DS, Glatt CE, Hwang PM, Fotuhi M, Dawson TM, Snyder SH (1991) Nitric oxide synthase protein and mRNA are discretely localized in neuronal populations of the mammalian CNS together with NADPH diaphorase. Neuron 7:615–624

Brenman JE, Chao DS, Gee SH, McGee AW, Craven SE, Santillano DR, Wu Z, Huang F, Xia H, Peters MF, Froehner SC, Bredt DS (1996) Interaction of nitric oxide synthase with the synaptic density protein PSD-95 and a-1 syntrophin mediated by PDZ domains. Cell 84:757–767

Busse R, Fleming I (1996) Endothelial dysfunction in atherosclerosis. J Vasc Res 33:181–194

Cayatte AJ, Palacino JJ, Horten K, Cohen RA (1994) Chronic inhibition of nitric oxide production accelerates neointima formation and impairs endothelial function in hypercholesterolemic rabbits. Arterioscler Thromb 14:753–759

Chao T-C, VanAlten PJ, Walter RJ (1994) Steroid sex hormones and macrophage function: modulation of reactive oxygen intermediates and nitrite release. Am J Reprod Immunol 32:43–52

Cooke JP, Singer AH, Tsao P, Zera P, Rowan RA, Billingham ME (1992) Antiatherogenic effects of L-arginine in the hypercholesterolemic rabbit. J Clin Invest 90:168–1172

Creager MA, Girerd XJ, Gallagher SH, Coleman S, Dzau VJ, Cooke JP (1992) L-arginine improves endothelium-dependent vasodilation in hypercholesterolemic humans. J Clin Invest 90:1248–1253

Dawson VL, Dawson TM, Bredt DS. Snyder SH (1993) Inhibitors of nitric oxide synthase and use thereof to prevent glutamate neurotoxicity. U.S. Patent No. 5,266,594

DiRosa M, Radomski M, Carnuccio R, Moncada S (1990) Glucocorticoids inhibit the induction of nitric oxide synthase in macrophages. Biochem Biophys Res Commun 172:1246–1252

Farhat MY, Lavigne MC, Ramwell PW (1996) The vascular protective effects of estrogen. FASEB J 10:615–624

Försterman U, Mugge A, Alheid U, Haverich A, Frolich JC (1988) Selective attenuation of endothelium-mediated vasodilation in atherosclerotic human coronary arteries. Circ Res 62:185–190

Förstermann U, Schmidt HHHW, Kohlman KL, Murad F (1992) Induced RAW 264.7 macrophages express soluble and particulate nitric oxide synthase: inhibition by transforming growth factor-b. Eur J Pharmacol 225:161–165

Förstermann U, Closs EI, Pollock JS, Nakane M, Schwarz P, Gath I, Kleinert H (1994) Nitric oxide synthase isozymes. Characterization, purification, molecular cloning, and functions. Hypertension 23:1121–1131

Furchgott RF, Zawadzki JV (1980) The obligatory role of endothelial cells in the relaxation of arterial smooth muscle by acetylcholine. Nature 288:373–376

Garcia-Cardena G, Oh P, Liu J, Schnitzer JE, Sessa WC (1996) Targeting of nitric oxide synthase to endothelial cell caveolae via palmitoylation: implications for nitric oxide signaling. Proc Natl Acad Sci USA 93:6448–6453

Geller DA, Nussler AK, Di Silvio M, Löwenstein CJ, Shapiro RA, Wang SC, Simmons RL, Billiar TR (1993) Cytokines, endotoxin and glucocorticoids regulate the expression of inducible nitric oxide synthase in hepatocytes. Proc Natl Acad Sci USA 90:522–526

Gilligan DM, Quyyumi AA, Cannon III RO, Johnson GB, Schenke WH (1994) Effects of physiological levels of estrogen on coronary vasomotor function in post-menopausal women. Circulation 89:2545–2551

Gisclard V, Miller VM, Vanhoutte PM (1988) Effect of 17β-estradiol on endothelium-dependent responses in the rabbit. J Pharmacol Exp Ther 244:19–22

Goetz RM, Morano I, Calvoni T, Studer R, Holtz J (1994) Increased expression of endothelial constitutive nitric oxide synthase in rat aorta during pregnancy. Biochem Biophys Res Commun 205:905–910

Harbrecht BG, Billiar TR, Stadler J (1992) Inhibition of nitric oxide synthesis during endotoxemia promotes intrahepatic thrombosis and an oxygen radical mediated hepatic injury. J Leukoc Biol 52:390–394

Hattori Y, Gross SS (1993) GTP-cyclohydrolase I mRNA is induced by LPS in vascular smooth muscle: characterization, sequence and relationship to nitric oxide synthase. Biochem Biophys Res Commun 195:435–441

Hayashi T, Fukuto JM, Ignarro LJ, Chaudhuri G (1992) Basal release of nitric oxide from aortic rings is greater in female rabbits than in male rabbits: implications for atherosclerosis. Proc Natl Acad Sci USA 89:11259–11263

Hayashi T, Fukuto JM, Ignarro LJ, Chaudhuri G (1995) Gender differences in atherosclerosis: possible role of nitric oxide. J Cardiovasc Pharmacol 26:792–802

Hayashi TK, Yamada K, Esaki T, Kuzuya M, Satake S, Ishikawa T, Hidaka H, Iguchi A (1995) Estrogen increases endothelial nitric oxide by a receptor-mediated system. Biochem Biophys Res Commun 214:847–855

Hecker M, Preiss C, Klemm P, Busse R (1996) Inhibition by antioxidants of nitric oxide synthase expression in murine macrophages: role of nuclear factor kappa B and interferon regulatory factor 1. Br J Pharmacol 118:2178-2184

Hirokawa K, O'Shaughnessy KM, Ramrakha P, Wilkins MR (1994) Inhibition of nitric oxide synthesis in vascular smooth muscle by retinoids. Brit J Pharmacol 113:1448–1454

Hishikawa K, Makaki T, Marumo T, Suzuki H, Kato R, Saruta T (1995) Up-regulation of nitric oxide synthase by estradiol in human aortic endothelial cells. FEBS Lett 36:291–293

Holm P, Korsgaard N, Shalmi M, Andersen HL, Hougaard P, Skouby SO (1997) Significant reduction of the antiatherogenic effect of estrogen by long-term inhibition of nitric oxide synthesis in cholesterol-clamped rabbits. J Clin Invest 100:821–828

Huang Z, Huang PL, Panahian N, Dalkara T, Fishman MC, Moskowitz MA (1994) Effects of cerebral ischemia in mice deficient in neuronal nitric oxide synthase. Science 265:1883–1885

Isles CG, Hole DJ, Hawthorne VM, Lever AF (1992) Relation between coronary risk and coronary mortality in women of the Renfrew and Paisley survey: comparison with men. Lancet 339:702–706

Kauser K, Rubanyi GM (1994a) Gender difference in bioassayable endothelium-derived nitric oxide release from isolated rat aortae. Am J Physiol 267:H2311–H2317

Kauser K, Rubanyi GM (1994b) 17β-estradiol and endothelial nitric oxide synthase. Endothelium 2:203–208

Kauser K, Rubanyi GM (1995a) Gender difference in endothelial dysfunction in the aorta of spontaneously hypertensive rats. Hypertension 25:517–523

Kauser K, Rubanyi GM (1995b) Effect of 17β-estradiol on endothelial dysfunction in the aorta of spontaneously hypertensive rats. Endothelium 25:517–523

Kauser K, Rubanyi GM (1997a) Vasculoprotection by estrogen contributes to gender difference in cardiovascular diseases; potential mechanism and role of endothelium. In: Rubanyi GM, Dzau VJ (eds) The endothelium in clinical practice. Marcel Dekker, New York pp 439–467

Kauser K, Rubanyi GM (1997b) Potential cellular signaling mechanisms mediating upregulation of endothelial nitric oxide production by estrogen. J Vasc Res 34: 229–236

Kauser K, Sonnenberg D, Henrichmann F, Rubanyi GM (1994) 17β-estradiol inhibits IL-1β-induced suppression of contractile reactivity in isolated rat thoracic aorta. Circulation 90:1–577

Kauser K, Sonnenberg D, Tse J, Rubanyi GM (1997) 17β-Estradiol attenuates endotoxin-induced excessive nitric oxide production in ovariectomized rats in vivo. Am J Physiol 273:H506–H509

Kauser K, Sonnenberg D, Diel P, Rubanyi GM (1998) Effect of 17beta-oestradiol on cytokine-induced nitric oxide production in rat isolated aorta. Br J Pharmacol 123:1089–1096

Kim-Schulze S, McGowan KA, Hubchak SC, Cid MC, Martin MB, Kleinman HK et al. (1996) Expression of an estrogen receptor by human coronary artery and umbilical vein endothelial cells. Circulation 94:1402–1407

Knauthe R, Diel P, Hegele-Hartung CH, Engelhaupt A, Fritzemeier K-H (1996) Sexual dimorphism of steroid hormone receptor messenger ribonucleic acid expression and hormonal regulation in rat vascular tissue. Endocrinology 137: 3220–3227

Kornau H-C, Schenker LT, Kennedy MB, Seeburg PH (1995) Domain interaction between NMDA receptor subunits and the postsynaptic density protein PSD-95. Science 269:1737–1740

Kuiper GGJ, Enmark E, Pelto-Huikko M, Nilsson S, Gustafsson J-A (1996) Closing of a novel estrogen receptor expressed in rat prostate and ovary. Proc Natl Acad Sci USA 9:5925–5030

Lin AL, Shain SA, Gonzales R (1986) Sexual dimorphism characterizes steroid hormone modulation of rat aortic steroid hormone receptors. Endocrinology 119:296–302

Lubahn DB, Moyer JS, Golding TS, Course JF, Korach KS, Smithies O (1993) Alteration of reproductive function but not prenatal sexual development after insertional disruption of the mouse estrogen receptor gene. Proc Natl Acad Sci USA 90:11162–11166

MacRitchie AN, Jun SS, Chen Z, German Z, Yuhanna IS, Sherman TS, Shaul PW (1997) Estrogen upregulates endothelial nitric oxide synthase gene expression in fetal pulmonary artery endothelium. Circ Res 81:355–362

Maddox YT, Falcon JG, Ridinger M, Cunard CM, Ramwell PW (1987) Endothelium-dependent gender differences in the response of the rat aorta. J Pharmacol Exper Ther 240:392–395

Matsui K, Higashi K, Fukunga K, Miyazaki K, Maeyama M, Miyamoto E (1983) Hormone treatments and pregnancy alter myosin light chain kinase and calmodulin levels in rabbit myometrium. J Endocrinol 97:11–16

Maziere C, Ronveaus MF, Salmon S, Santus R, Mazier JC (1991) Estrogens inhibit copper and cell-mediated modification of low density lipoprotein. Atherosclerosis 89:175–182

Michel JB, Feron O, Sase K, Michel T (1997) Reciprocal regulation of endothelial NO synthase by calmodulin and caveolin: structural determinants of enzyme inhibition. Circulation 96:1–47

Miller VM, Aarthus LL, Vanhoutte PM (1986) Modulation of endothelium-dependent responses by chronic alterations of blood flow. Am J Physiol 251:H520–H527

Moini H, Bilsel S, Bekdemir T, Emerk K (1997) 17β-Estradiol increases intracellular free calcium concentrations of human vascular endothelial cells and modulates its responses to acetylcholine. Endothelium 5:11–21

Mosselman S, Polman J, Dijkema R (1996) ER-β: identification and characterization of a novel human estrogen receptor. FEBS Lett 392:49–53

Nakane M, Schmidt HHHW, Pollock, JS, Forstermann U, Murad F (1993) Cloned human brain nitric oxide synthase is highly expressed in skeletal muscle. FEBS Lett 316:175–180

Naruse KM, Shimizu M, Muramatsu M, Toki Y, Miyazaki Y, Okurama K, Hashimoto H, Ito, T (1994) Long-term inhibition of NO synthesis promotes atherosclerosis in the hypercholesterolemic rabbit thoracic aorta. Arterioscler Thromb 14:746–752

Nava E, Palmer RMJ, Moncada S (1991) Inhibition of nitric oxide synthesis in septic shock: how much is beneficial? Lancet 338:1555–1557

Nava E, Palmer RMJ, Moncada S (1992) The role of nitric oxide in endotoxic shock: effects of N^G-methyl-L-arginine. J Cardiovasc Pharmacol 20[Suppl 12]:132–134

Pittner RA, Spitzer JA (1993) Steroid hormones inhibit induction of spontaneous nitric oxide production in cultured hepatocytes without changes in arginase activity or urea production. Proc Soc Exp Biol Med 202:499–504

Reis SE, Gloth ST, Blumenthal RS, Resar JR, Zacur HA, Gerstenblith G, Brinker JA (1994) Ethinyl estradiol acutely attenuates abnormal coronary vasomotor responses to acetylcholine in postmenopausal women. Circulation 89:52–60

Rosenkranz-Weiss P, Sessa WC, Milstien S, Kaufman S, Watson CA, Pober JS (1994) Regulation of nitric oxide synthesis by proinflammatory cytokines in human umbilical vein endothelial cells. Elevations in tetrahydrobiopterin levels enhance endothelial nitric oxide synthase specific activity. J Clin Invest 93:2236–2243

Rosselli M, Imthurn B, Keller PJ, Jackson EK, Dubey RK (1995) Circulating nitric oxide (nitrite/nitrate) levels in postmenopausal women substituted with 17β-estradiol and norethisterone acetate. Hypertension 25:848–853

Rubanyi GM (1993) The role of endothelium in cardiovascular homeostasis and diseases. J Cardiovasc Res 22[Suppl 4]:1–14

Rubanyi GM, Vanhoutte PM(1986) Oxygen-derived free radicals, endothelium, and responsiveness of vascular smooth muscle. Am J Physiol 250:H815–H821

Rubanyi GM, Romero JC, Vanhoutte PM (1986) Flow-induced release of endothelium-derived relaxing factor. Am J Physiol 250:H1145–H1149

Rubanyi GM, Freay AD, Kauser K, Sukovich D, Burton G, Lubahn DB, Couse JF, Curtis SW, Korach KS (1997) Vascular estrogen receptors and endothelium-derived nitric oxide production in the mouse aorta gender difference and effect of estrogen receptor gene disruption. J Clin Invest 99:2429–2437

Ruiz-Larrea B, Leal A, Martin C, Martinez R, Lacort M (1995) Effects of estrogens on the redox chemistry of iron: a possible mechanism of the antioxidant action of estrogens. Steroids 60:780–783

Schmidt HHHW, Warner TD, Nakane M, Forstermann U, Murad F (1992) Regulation and subcellular location of nitrogen oxide synthases in RAW264.7 macrophages. Mol Pharmacol 41:615–624

Schmidt K, Werner ER, Mayer B, Wachter H, Kukovetz WR (1992) Tetrahydrobiopterin-dependent formation of endothelium-derived relaxing factor (nitric oxide) in aortic endothelial cells. Biochem J 281:297–300

Scott-Burden T, Schini VB, Elizondo E, Junquero DC, Vanhoutte PM (1992) Platelet-derived growth factor suppresses and fibroblast growth factor enhances cytokine-induced production of nitric oxide by cultured smooth muscle cells. Effects on cell proliferation. Circ Res 71:1088-1100

Sessa WC (1994) The nitric oxide synthase family of proteins. J Vasc Res 31:131–143

Shaul PW, Smart EJ, Robinson LJ, German Z, Yuhanna IS, Ying Y, Anderson RG, Michel T (1996) Acylation targets emdothelial nitric-oxide synthase to plasmalemmal caveolae. J Biol Chem 271:6518–6522

Stallone JN, Crofton JT, Share L (1991) Sexual dimorphism in vasopressin-induced contraction of rat aorta. Am J Physiol 260:H453–H458

Stampfer MJ, Colditz GA, Willett WC, Manson JE, Rosner B, Speizer FE, Hennekens, CH (1991) Postmenopausal estrogen therapy and cardiovascular disease: ten-year follow-up from the Nurses' Health Study. N Engl J Med 325:756–762

Stroes E, Kastelein J, Cosentino F, Erkelens W, Wever R, Koomans H, Luscher T, Rabelink T (1997) Tetrahydrobiopterin restores endothelial function in hypercholesterolemia. J Clin Invest 99:41–46

Stuehr DJ, Marletta MA (1985) Mammalian nitrate biosynthesis: mouse macrophages produce nitrite and nitrate in response to Escherichia coli lipopolysaccharide. Proc Natl Acad Sci USA 82:7738–7742

Subbiah MTR, Kessel B, Agrawal M, Rajan R, Ablanalp W, Rymaszewski Z (1993) Antioxidant potential of specific estrogens on lipid peroxidation. J Clin Endocrinol Metab 77:1095–1097

Sudhir K, Jennings GL, Funder JW, Komesaroff PA (1996) Estrogen enhances basal nitric oxide release in the forearm vasculature in perimenopausal women. Hypertension 28:330–334

Thiemermann C, Vane JR (1990) Inhibition of nitric oxide synthesis reduces the hypotension induced by bacterial lipopolysaccharides in the rat in vivo. Eur J Pharmacol 182:591–595

Venema RC, Nishida K, Alexander RW, Harrison DG, Murphy TJ (1994) Organization of the bovine gene encoding the endothelial nitric oxide synthase. Biochem Biophys Acta 1218:413–420

Venkov CD, Rankin AB, Vaughan DE (1996) Identification of authentic estrogen receptor in cultured endothelial cells. A potential mechanism for steroid hormone regulation of endothelial function. Circulation 94:727–733

Weiner CP, Knowles RG, Moncada S (1994a) Induction of nitric oxide synthases early in pregnancy. Am J Obstet Gynecol 171:838–843

Weiner CP, Lizasoain I, Baylis SA, Knowles RG, Charles IG, Moncada S (1994b) Induction of calcium-dependent nitric oxide synthases by sex hormones. Proc Natl Acad Sci USA 91:5212–5216

Werner-Felmayer G, Werner ER, Fuchs D, Hausen A, Reibnegger G, Schmidt K, Weiss G, Wachter H (1993) Pteridine biosynthesis in human endothelial cells. Impact on nitric oxide-mediated formation of cyclic GMP. J Biol Chem 268:1842–1846

Williams JK, Adams MR, Klopfenstein HS (1990) Estrogen modulates responses of atherosclerotic coronary arteries. Circulation 81:1680–1687

Xie QW, Cho HJ, Calaycay J, Mumford RA, Swiderek KM, Lee TD, Ding A, Troso T, Nathan C (1992) Cloning and characterization of inducible nitric oxide synthase from mouse macrophages. Science 256:225–228

Xie Q, Whisnant R, Nathan C (1993) Promoter of the mouse gene encoding calcium-independent nitric oxide synthase confers inducibility by interferon-γ and bacterial lipopolysaccharide. J Exp Med 177:1779–1784

Yoshizumi M, Perrella MA, Burnett JC, Lee ME (1993) Tumor necrosis factor downregulates an endothelial nitric oxide synthase messenger RNA by shortening its half-life. Circ Res 73:205–209

CHAPTER 9
Non-Genomic Effects of Estrogens

V.D. Ramirez and J. Zheng

List of Abbreviations

AMPH	amphetamine
BNST	bed nucleus of the striae terminalis
cAMP	cyclic adenosine monophosphate
cGMP	cyclic guanyl monophosphate
CNS	central nervous system
CREB	cAMP-response-element binding protein
DA	dopamine
DAD2	dopamine D_2 receptor
DAG	diacylglycerol
DAMGO	Tyr-D-Ala-Gly-MePhe-Gly-col
DES	diethylstilbestrol
DOPAC	dihydroxyphenyl acetic acid
E-6-BSA	17β-estradiol-6 (0-carboxymethyl) oxime-bovine serum albumin
EPSP	excitatory post-synaptic potential
ER	estrogen receptor
ERK	extracellular signal-regulated kinase
FITC	florescein isothiocyanate
GABA	γ-amino-butyric acid
GTP	guanosine triphosphate
G3PD	glyceraldehyde-3-phosphate dehydrogenase
HDL	high-density lipoprotein
ICI-182780	pure estrogen antagonist
IP_3	inositol, 1,4,5-triphosphate
Kd	binding affinity constant
LDL	low-density lipoprotein
M.W.	molecular weight
MAPKK	mitogen activated protein kinase kinase
MAPK	mitogen activated protein kinase
MEK	mitogen extracellular kinase
mPOA	medial pre-optic area
NAD^+	nicotinamide dinucleotide oxidative form

NADH   nicotinamide dinucleotide reduced form
NE   norepinephrine
NMDA   N-methyl-D-aspartic acid
NSD-1015   3-hydroxybenzylhydrazine dihydrochloride
OSCP   oligomycin sensitivity conferring protein
ovx   ovariectomized
P2   crude synaptosomal fraction
PKA   protein kinase A
PLC   phospholipase
SDS-PAGE   Sodium Dodecyl Sulfate-polyacrylamide gel electrophoresis
SHBG   sex-hormone binding globulin
T-3-BSA   testosterone 3-(0-carboxymethyl) oxime-bovine serum albumin
TH   tyrosine hydroxylase

A. Introduction

The primary fast, direct action of steroids at the cellular membrane is considered the non-genomic action of the steroid. However, biological responses that are initiated at the cellular membrane can be either non-genomic or genomic in nature. This primary effect can lead to activation of a cascade of three major categories of second messengers: (1) trimeric G-protein-coupled receptors, (2) receptor tyrosine kinases and (3) cytokine receptor-activated kinases with secondary slower activation of genomic responses.

Estradiol, a natural and potent sexual steroid hormone, mainly secreted by the ovary, but also produced locally in different tissues by the aromatase estrogen synthase which converts testosterone to estrogen (ROSELLI et al. 1985), is a particular example of the continuous cellular action of steroids and will be the focus of this chapter. This lipid-soluble molecule can act at different levels in the cellular membrane: (1) by intercalating among the phospholipid components of the membrane, the fluidity can be altered and changes in permeability and excitability are produced (MAKRIYANNIS et al. 1990); (2) by binding to specific sites at particular ion channels, it can modify the permeability of specific ions such as Ca^{2+} (PIETRAS and SZEGO 1975; AUDY et al. 1996), Cl$^-$ (HARDY and VALVERDE 1994) and K$^+$ (LAGRANGE et al. 1997); (3) by binding to well-established receptors, such as G-protein-mediated receptors (CONKLIN and BOURNE 1993), it can induce changes in secondary messengers; and (4) by binding to the putative membrane estrogen receptor (mER) it can induce "cognate responses" in particular cells.

This chapter will present evidence for the fast estradiol-evoked biological responses, the existence of specific binding sites in cellular membranes, the evidence for diverse protein estradiol binders and the different membrane mechanisms affected by estradiol. Last, we will present an overview of the continuum theory of estradiol's action at the cellular level.

B. Evidence for Fast Estradiol-Evoked Biological Responses

I. Central Biological Responses

Historically, the CNS is the site for the rapid effects of hormones, in general, (BEYER and SAWYER 1969; JOELS 1997) and estradiol, in particular (Moss 1997). Here, we will discuss the rapid actions of estradiol on extra-hypothalamic tissues, since we have recently described effects on the hypothalamic-pituitary axis (RAMIREZ and ZHENG 1996).

1. Fast Estradiol Effects on the Catecholamine System

In the nigrostriatal DA system, a variety of behavioral and functional indices are dependent on the gonadal status of the animal. For instance, estradiol was reported to rapidly influence striatal DA release, DA receptor concentration and behavior mediated by the striatum (HRUSKA and SILBERGELD 1980; DI PAOLO 1994, DI PAOLO et al. 1985). BECKER (1990a, 1990b) showed that within 20 min, physiological concentrations ($0.22–3.7\,\mu M$) of estradiol or the synthetic estrogen DES increased the AMPH-stimulated striatal DA release from striatal tissue of ovx female rats as measured by microdialysis. This fast estradiol effect was correlated with fast effects of estradiol on rat-rotational behavior. 17α-estradiol had much less of an effect, indicating a selective effect of estradiol. The rapid effect of estradiol on DA release is sexually dimorphic, since the same effect was not observed in the striatal tissue of intact male rats (BECKER 1990a). A subsequent study also revealed that estrogen acutely inhibited the striatal DAD2 receptor binding within 30 min (BAZZETT and BECKER 1994), probably due to a rapid conversion of high-affinity D2 receptors to low-affinity D2 receptors, as shown by LEVESQUE and DI PAOLO (1988).

Recently, the effect of estradiol on the expression of TH in cultured mesencephalic dopaminergic neurons from 14-day-old rat embryos was examined (RAAB et al. 1995). Curiously, estradiol at 1 pM was shown to increase TH mRNA, in these cells from male rats, without expression of the classical ER. No effect was observed in cultured cells from female rats. Therefore, estradiol regulates TH mRNA in a gender-specific fashion by a mechanism other than the classical ER pathway. Further acute effects of low doses of estradiol on striatal DA activity, most likely mediated by a non-genomic mechanism, were reported by PASQUALINE et al. (1995, 1996).

Subcutaneous doses of estradiol but not 17α-estradiol were able to accumulate DOPA in the striatum after i.p. administration of NSD-1015, an inhibitor of dopa decarboxylase. TH increased within minutes of the estradiol injections. The effect was probably direct on the striatum, since in a push–pull experiment addition of 1 nM estradiol to the superfusing fluid (but not 17α-estradiol) immediately elicited a robust increase in ^{3}H-DA and ^{3}H-DOPAC in the extracellular compartment with total DA or DOPAC concentrations remaining constant. Apparently, the effect of estradiol in this experiment was to decrease TH susceptibility approximately twofold to end product inhibition by DA, presumably due to phosphorylation of the enzyme.

Using a different approach, we recently reported that E-6-BSA, a complex that acts at the extracellular side of the membranes, was capable of acutely releasing endogenous DA from striatal tissue superfused in vitro and derived from cycling female rats. The maximal release within 10 min post-infusion of the complex (over twofold) was determined in tissue from rats sacrificed in proestrus morning, with a lower release but significantly different from controls in estrus and with no response in diestrus. Interestingly, striatal tissue derived from male rats were unresponsive to a range of E-6-BSA doses (1–100 nM). Finally, striatal tissue derived from spontaneous, constant-estrus rats was highly responsive to the complex (Ramirez and Zheng 1999).

A non-genomic effect of estradiol on DA release in the nucleus accumbens, studied by in vivo microdialysis, was also observed, in addition to a genomic effect (Thompson and Moss 1994). 17β-estradiol hemisuccinate infusion resulted in an initial increase in K^+-stimulated DA release within 2 min, followed by a late increase about 1 h after infusion. This, according to the authors, was probably secondary to a genomic activation.

Overall, these data indicate that the DA system in the CNS is an important central site of estradiol action, with robust stimulatory effects on dopaminergic activity which is, however, dependent on a particular hormonal milieu. The data may be significant in our understanding of the non-genomic role of estradiol on brain activity in other species, including humans, and particularly on those functions related to motor skill and mood behavior.

2. Fast Estradiol Effects on the Hippocampus

Fast changes in neural activity, i.e., extracellular field potential, of the hippocampus were shown in vitro after bath application of estradiol to hippocampal neurons (Teyler et al. 1980; Foy and Teyler 1983). The action was stereospecific since 17α-estradiol did not have an effect. Using intracellular recording techniques, Wong and Moss (1991) reported that estradiol, but not 17α-estradiol, triggered a rapid depolarization in a small percentage of CA1 neurons. This depolarization was not blocked by the glutamate receptor antagonist, kynurenic acid.

Superfusion of 10 nM estradiol, but not 17α-estradiol, caused a rapid and reversible increase in the amplitude of the Schaffer collateral-activated EPSP (Wong and Moss 1992). The inhibitory post-synaptic potential was not affected by estradiol. This potentiation of EPSP by estradiol was blocked by a non-NMDA receptor antagonist, but not by an NMDA receptor antagonist. However, further analysis by patch-clamp studies showed that the effect of estradiol in the non-NMDA receptor was not a direct action, but was mediated by some other factor(s), since estradiol had no effect on the kainate-induced current using excised patch-clamp recordings from CA1 neurons

(WONG and Moss 1994). In a recent paper from the same laboratory, using whole-cell voltage-clamp recording, it was found that under the perforated patch configuration, estradiol between 10 nM and 10 μM potentiated kainate induced currents within minutes in 38% of the tested neurons. The effect was stereospecific since 17α-estradiol did not have any effect. The authors concluded that, under this condition, in which intracellular components remained intact, the rapid effect of estradiol involved activation of a G-protein-coupled cAMP-dependent phosphorylation of kainate receptors (GU and Moss 1996). Interestingly, the authors indicate that estradiol conjugated to BSA did not potentiate kainate-induced currents, indicating that the steroid would have to cross the membrane to exert its effect (Moss 1997). TAUBOLL et al. (1994) were unable to observe the effect of estradiol (0.1–100 nM) on synaptic activation and inhibition, as well as the response to iontophoretically applied GABA in hippocampal slices of intact or ovx female rats within 30 min. Only a small significant increase in population spike amplitude was observed with 1 nM estradiol in male rats, but not with lower or higher concentrations of the steroid.

A rapid effect of estradiol on the growth of hippocampal neurons in culture has also been reported (BRINTON 1993). Within 1–10 min, estradiol induced a significant increase in the number of filopodia decorating neuritic extensions, as well as an increase in the length of existing and newly formed filopodia. The minimal concentration required for this action was in the range 0.25–1 μM. Testosterone had a similar effect on the number of filopodia, without a significant effect on the length. Similar concentrations of 17α-estradiol, progesterone and corticosterone lacked a rapid effect. Although it is not known whether these rapid effects of estradiol are responsible for the observation of change in synapse density in the hippocampus during the estrus cycle, or after 2 days of estrogen treatment in a structure that expresses few estrogen and progesterone receptors (WOOLLEY and MCEWEN 1992), one could speculate that estradiol may act at the cellular membrane to induce changes that indirectly require nuclear activation. These morphological effects of estradiol could be the result of a rapid effect on electrical activity of the hippocampal neurons, with secondary activation of transcription events leading to the product of specific proteins required for changes in neural plasticity and excitability.

The estradiol induction of new excitatory spine synapses on CA1 pyramidal neurons, but not in CA3 or dentate gyrus, was blocked by concurrent administration of MK801 (an NMDA receptor antagonist). The fact that estradiol induced NMDA binding sites in this region of the hippocampus (WEILAND 1992) suggests that the induction of dendritic synapses may be mediated by activation of NMDA receptors which, by opening Ca^{2+} channels, increases intracellular Ca^{2+}. This ill-defined mechanism regulates the formation of new spine synapses. The time course of events either induced by estradiol in ovx rats or during the estrus cycle (GOULD et al. 1990; WOOLLEY et al. 1990; WOOLLEY and McEWEN 1992, 1993) suggests that the final biological response is genomically mediated. However, it could be initiated at the level of the membrane, since this region has a very low density of nuclear (n) ER-α (LOY et al. 1988), though it was recently reported to have a sufficient number of the new nER-β receptor (KUIPER et al. 1997; PENNISI 1997). An argument in favor of the estradiol-mediated membrane events is the recent demonstration in my laboratory that P3 preparations, mainly cellular membranes at low protein concentrations, 0.3 μg/tube, from the hippocampus of intact female rats bind E-6-^{125}IBSA with high affinity and stereospecificity (Fig. 1). A further argument is the fact estradiol, but not 17α-estradiol, can induce rapid remodeling of plasma membranes within 1 min (GARCIA SEGURA et al. 1989) in developing rat cerebro-cortical neurons in culture when added in physiological (nanomolar) doses.

3. Fast Estradiol Effects on Other Areas of the Central Nervous System

In the cerebellum, an area also containing low levels of intracellular nER-α, it was found that, 17β-estradiol hemisuccinate, administered either via iontophoresis or systematically by jugular i.v. injection, significantly increased Purkinje cell excitatory responses to glutamate in ovx rats (SMITH et al.

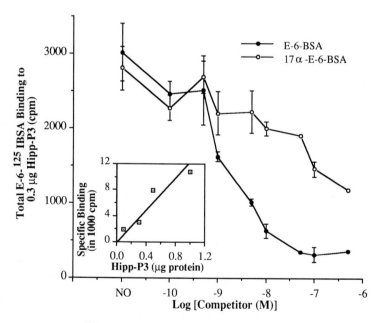

Fig. 1. Binding of E-6-^{125}IBSA to P3 (0.3 μg protein) from hippocampus (*Hipp-P3*) displaced by E-6-BSA (*filled circle*) or its enantiomer 17α-E-6-BSA (*open circle*). The average binding affinity of E-6-^{125}IBSA to Hipp-P3 was 1.3 nM. Mean values of 3–2 experiments in duplicate ±SE; values without bars are $n = 2$ (Liu, Zheng and Ramirez, unpublished observations). *NO* without competitors; *Insert* dose-dependent binding of E-6-^{125}IBSA to Hipp-P3 from 0.1 to 1.0 μg protein

1987a,b; 1988). The effect occurred within 1 min after iontophoretic application and within 10–40 min after i.v. injections; 17α-estradiol was less effective. The antiestrogen tamoxifen was not effective in blocking the estradiol effect. In addition, estradiol could modulate the rapid effect of progesterone on cerebellar responses to GABA and glutamate (SMITH et al. 1987b). Recently, SMITH and CHAPIN (1996) reported rapid effects of estradiol in facilitating the synchronized rhythmic electrical activity of the olivary nucleus in the rat. It was suggested that this rapid estradiol effect may be important in modulating sensorimotor inputs (SMITH 1994).

Rapid estrogen actions were also reported in other brain areas. These include septum (INNES and MICHAL 1970), preoptic septal area (KELLY et al. 1978) and medial amygdala (NABEKURA et al. 1986). In addition, estradiol rapidly attenuated a $GABA_B$ response in hypothalamic neurons (LaGRANGE et al. 1996), but does not seem to affect a glutamate receptor directly in hippocampal neurons (WONG and MOSS 1994).

II. Peripheral Biological Responses

Rapid actions of estradiol have also been shown in the heart, uterus, liver, vessels and bone. The laboratory of SZEGO at UCLA was the first to present convincing evidence that estradiol could act at the cellular membrane of the uterus to initiate biological events. SZEGO and DAVIS (1967) demonstrated a rapid (latency of 15 s) increase in cAMP in the uterus of ovx rats treated with estradiol. Recently, the laboratory of KATZENELLEBOGEN showed that estradiol activates adenylate cyclase, increasing the concentration of cAMP close to 10-fold either in cultured uterine cells or in vivo (ARONICA et al. 1994). This effect was rapid, with a maximal increase of about 30 min in in vitro cell cultures using 1 nM estradiol and a maximal increase of about 3 h in vivo using 5 μg estradiol s.c. in 18-day-old rats. In addition, the effect of estradiol on cAMP was not blocked either by inhibitors of RNA or by protein synthesis, indicating a non-genomic effect of estradiol. Furthermore, they found an increase in cAMP in breast cancer cells, which occurred less than 30 min after application of low doses of estradiol. In human endometrium during the secretory phase, there is an increase in cAMP, and estradiol activates adenylate cyclase (BERGAMINI et al. 1985).

Another important direct effect of estradiol on cell membranes is to regulate Ca^{2+} fluxes. The effect can occur either by directly binding the Ca^{2+} ionophore or by first altering intracellular messengers, such as cAMP or IP_3 and DG, which could indirectly lead to changes in Ca^{2+} permeability. The group of SZEGO was also one of the first to report that physiological doses of estradiol could increase the uptake of radioactive Ca^{2+} by the uterus within 10 min (PIETRAS and SZEGO 1975). Later, several groups reported similar rapid actions of estradiol upon Ca^{2+} metabolism in different peripheral tissues, such as in granulosa cells (MORLEY et al. 1992), cultured human pre-osteoclastic cells (FIORELLI et al. 1996), on maturing human oocytes (TESARIC and MENDOZA

1995) and in prostate cancer cells (Audy et al. 1996). It is important to indicate that in those last three papers, a membrane impermeant complex (E-6-BSA) had similar effects to estradiol on Ca^{2+} fluxes, clearly demonstrating that the initial event leading to increase in intracellular Ca^{2+} occurred at the membrane level.

One peripheral system that has recently received greater attention as a target site of estradiol is the cardiovascular system. Clinical studies indicate that women receiving replacement estrogen treatment have half the risk of dying from myocardial infart than aged-matched women not using estrogen (Sarrell et al. 1994). It is clear that the beneficial effects of estradiol are mediated by indirect actions at the liver, where it probably induces HDL and inhibits LDL production, and by a direct action on smooth muscles of the vascular system and the heart. In general, it seems that high doses of estradiol in the micromolar range are required to produce a rapid vasodilator effect in different vascular preparations from different species (Farhat et al. 1996a).

In part, the complexity of the direct actions of estradiol in the vascular system is due to the separate or complimentary involvement of the vascular endothelium, the smooth muscle and/or the adventitia in the response. As in other tissues, estradiol may mediate its vasomotor control by genomic- and membrane-mediated events. For instance, in cell cultures of human vascular muscle cells, estradiol in micromolar concentrations significantly inhibited calcium influx, whereas estrone, estriol or synthetic estrogens lacked such an effect (Mueck et al. 1996). In cultures of rat pulmonary vascular smooth muscle cells, estradiol stimulated production of cAMP within 5 min of application of the steroid, and the effect was not affected by actinomycin (Farhat et al. 1996b). Comparable doses of testosterone, or 17α-estradiol were ineffective.

The G-protein inhibitor, pertussis toxin (100 ng/ml) did not significantly affect the estradiol-evoked increase in cAMP. However, removal of Ca^{2+} from the incubation medium inhibited the stimulatory effect of estradiol, indicating that the possible membrane-mediated event, induced by estradiol upon cAMP in this tissue, is via a Ca^{2+}-dependent pathway. In another study, low doses of estradiol in the nanomolar range showed potentiation of pressor responses of isolated perfused rat mesenteric vascular bed induced by exogenous NE application, but not by electrically-induced endogenous release of NE. The effect occurred within 2–8 min and was stereospecific for the β-isomer of estradiol. Most interestingly, estradiol conjugated to BSA was as effective or better than free estradiol, indicating that this rapid effect of estradiol is mediated by non-genomic mechanisms (Vargas et al. 1996). A group from Japan (Ogata et al. 1996) reported that estradiol in nanomolar doses induced relaxation of the rabbit basilar artery by inhibition of voltage-dependent Ca^{2+} channels through a pertussis toxin-sensitive GTP-binding protein. Recently, it was reported that estradiol or P, but not T, inhibited smooth-muscle cell proliferation in vitro induced by endothelin-1 or serum within 7 min (Morey et al. 1997). This effect appears to be initiated in the cellular membrane, because E-BSA mimicked the effect of estradiol by also inhibiting MEK- and MAP-K-induced substrate phosphorylation. Aside from these actions, in the systemic vascular system, estradiol has direct effects on myocardial cells (DE BEER and KEIZER 1982; SITZLER et al. 1996; LIU et al. 1997) and coronary arteries (JIANG et al. 1991; GILLIGAN et al. 1994a,b; REIS et al. 1994), which can also be considered non-genomic in nature.

The cells involved in bone formation and bone loss are also target sites for the rapid actions of estradiol. Binding sites for the fluorescent complex E-BSA-FITC were reported in rat osteoclast surfaces cultured in vitro (BRUBAKER and GAY 1994) that reached equilibrium by 50 min, with saturation occurring at $3\,\mu$M. Displacement by free estradiol at $1\,\mu$M represented only 57% of the total. Tamoxifen was also as effective as free estradiol. However, 10 nM estradiol induced a rapid change in the shape of the cells, suggesting alterations in the cytoskeleton following estradiol administration. Interestingly, it was reported that estradiol binds specifically to the cytoskeleton (PUCA and SICA 1981), which could lead to conformational changes in the microfilament/microtubal apparatus of the cell. That these changes in shape may be related to changes in secondary messengers is indicated by the work of LIEBERHERR et al. (1993), who reported that estradiol or E-BSA (E-6-BSA) increased the concentration of intracellular Ca^{2+} within 5 s by mobilizing the cation from the ER following the formation of IP_3 and DAG, most likely due to activation of a PLC linked to a pertussis toxin-sensitive G-protein. Later, this same group (LEMELLAY et al. 1997) presented compelling evidence that the effect of estradiol is mediated by PLC-$\beta2$, one of the isoforms of the family of PLC, since 100 pM estradiol induced a fast increase in cytosolic Ca^{2+} in rat osteoblasts which was specifically blocked by an antibody against PLC-$\beta2$, but not by antibodies against PLC-$\beta1$, PLC-$\beta3$, or PLC-$\gamma1$ or 2.

Recently, FIORELLI et al. (1996) published that in a human pre-osteoclastic cell line, FLG29.1, there are binding sites for E-6-BSA-FITC, since the fluorescent complex labeled approximately 10% of viable cells in culture. The binding was displaced by estradiol, with total disappearance of fluorescent labeling at 100 nM, but not by 17α-estradiol, tamoxifen or progesterone: in these cells, estradiol increased intracellular pH, cAMP, cGMP and intracellular Ca^{2+}. The latter was dependent of extracellular Ca^{2+}, since when EGTA was added to the incubation before estradiol stimulation, this completely abolished the increase in Ca^{2+}. In another study, it was reported that about 50% of osteoblasts responded with a modest mitochondrial membrane potential reduction of 6–13% following estradiol administration (TROYAN et al. 1997), suggesting an action in the mitochondria. Finally, it appears that in avian osteo-clasts, tamoxifen, a drug with agonist and antagonist estrogenic effects, is capable of inhibiting bone resorption, whereas estradiol, DES or a specific antagonist of estradiol, ICI-182780, lacked such an effect (WILLIAMS et al. 1996).

Another peripheral organ known to be a target for estradiol membrane-mediated events is the liver. Studies performed by JENSEN and JACOBSON (1962) nearly 35 years ago showed maximal uptake of ^{3}H-estradiol by the liver within 15 min or less after a subcutaneous injection of 0.1 µg ³H-estradiol in sesame oil. Later, Pietras and Szego (1979) isolated hepatocytes from culture based on the ability of these cells to bind estradiol at the plasma membrane, using estradiol conjugated to BSA, which in turn was linked to nylon fibers anchored to the cultured dish. Those cells responded to estradiol with increased proliferation and cholesterol metabolism. Subsequently, the same authors (Pietras and Szego 1980) reported the existence of specific binding sites for ³H-estradiol in isolated hepatocyte plasma membrane subfractions. Our own current studies indicated that E-6-[125]-IBSA, but not [125]IBSA, was taken up rapidly by the liver in vivo with maximal uptake (45 times higher than blood levels) by 5 min post-intravenous injection. The radioactivity was translocated from the plasmalemmal microsomal fraction to the mitochondria within 1 h post-injection, with little or no change in levels in the nuclear fraction (Moats and Rarmirez 1998). This suggests an active mechanism of transport for estradiol, most likely due to a process of endocytosis. There are few reports of a rapid action of estradiol on the cells of the immune system (Peter et al. 1997).

C. Evidence for Specific Estradiol Binding Sites in Cellular Membranes

I. Central Sites

In spite of the fact that the CNS was well known as a target for rapid actions of hormones, including estradiol, it was much later that specific binding sites were first reported for this lipid-soluble steroid in purified synaptosomal membranes from rat brain (Towle and Sze 1983). Subsequently, other laboratories, including ours, reported specific binding sites in different regions of the CNS (Horvart et al. 1995; Zheng et al. 1996; Ramirez et al. 1996). The Kds reported were in the low (1–20 nM) and high (>100 nM) molar range, with variation depending on the type of neural tissue. This heterogeneity may represent artifacts due to different solubilities of estradiol in different membranes of the CNS or different estrogen-binding proteins. In those studies in which ³H-estradiol was used as a tool, there was the potential risk that estradiol might have diffused to the membrane and specifically intercalated among the lipid–protein interface without binding to a particular protein. In our studies, by using estradiol covalently bound to BSA in position C-6 or C-17 of the steroid molecule, we eliminated this possible artifact. By labeling the BSA molecule with ^{125}I, we can generate a radioactive probe of high specific activity. Furthermore, these probes (E-6-BSA and E-17-BSA) have high affinity for the nuclear ER (Ramirez and Zheng 1999); in a classical uterine cytosolic preparation, in which the receptor was labeled with ³H-estradiol, the Kd calculated from competition studies was about 0.02 nM for the complex E-17-BSA (about 32 moles of estradiol per mole of BSA) with a B_{max} of

1.01 pmol/mg protein. These values are ten times lower than those reported for estradiol by other authors (SUTHERLAND et al. 1982; CLARK and MANI 1994). Interestingly, in this assay, T-3-BSA did not compete, which agrees well with studies using the nuclear ER; however, it is in fact a very good competitor in the membrane preparation.

Using this tool and a crude synaptosomal preparation (P2 fraction), we (RAMIREZ et al. 1996; ZHENG et al. 1996) reported significantly different Kd values of 3 ± 0.7 nM ($n = 3$), 10 ± 1.5 nM ($n = 6$) and 34 ± 7 nM($n = 6$) for the intact female rat hypothalamus, olfactory bulb and cerebellum, respectively. Interestingly, when a plasmalemmal microsomal fraction from the rat brain was used, a 10-fold higher affinity of the conjugate and greater number of sites (31.3 ± 11.8 pmol/mg vs 10.2 ± 4.3 pmol/mg protein) were obtained (RAMIREZ and ZHENG 1999). In the mitochondrial lysosomal fraction (mP2 fraction) the Kd was about 2–4 nM, suggesting that two different protein estrogen binders reside in these two fractions. In a kinetic experiment using a P3 fraction from the corpus striatum, derived from intact female rats, we achieved a rapid association rate ($k_1 = 0.33$ nM^{-1}min^{-1}) with a dissociation rate estimated to have $k_2 = 0.075$ m^{-1}. The Kd calculated from these data revealed a value of 0.23 nM; close to that obtained by the competition assay, thus confirming the validity of the assay (RAMIREZ and ZHENG 1999).

In addition to these ligand studies in neural tissue, the anterior pituitary appears also to be a target for the action of estradiol at the membrane level. The presence of specific sites with an extremely low Kd of 0.04 nM for 17β-estradiol was reported (BRESSION et al. 1986). Interestingly, binding of 2-hydroxy estradiol (a catechol estrogen) to rat anterior pituitary cell membranes has also been reported (SCHAFFER et al. 1980). Two binding sites with Kds of 0.4 nM and 2μM were identified.

II. Peripheral Sites

In peripheral tissue, specific membrane estrogen binding sites were first reported in isolated endometrium and liver cells by SZEGO's group (PIETRAS and SZEGO 1977, 1980). Our own studies (MOATS and RAMIREZ 1998), using a plasmalemmal microsomal fraction from liver of ovx rats confirmed the existence of specific binding sites for estradiol using the complex E-6-^{125}IBSA. The Scatchard plot for homologous competition studies showed a curvilenar plot that was a significant improvement over the linear fit ($P = 0.007$), suggesting two binding sites for this complex, one of high affinity (0.26 nM) and the other of lower affinity (24.6 nM). Binding sites were also reported in isolated membranes from basolateral liver (Kd 26 nM and 26μM, respectively) and canalicular liver (Kd 81 nM and 6.7μM, respectively) (CHANGCHIT et al. 1990).

Estrogen binding sites were also reported (HERNANDEZ-PÉREZ et al. 1979) in the surface of the human sperm (Kd of about 0.66 nM). In cancer cells, such as the breast cancer cell line R75r-1, estradiol binding sites were detected in the surface of the cells by a fluorescent analogue of the steroid, estradiol-BSA-FITC (Nenci et al. 1981). These binding sites on the plasma membrane were distinct from classical cytosolic ERs in their steroid specificity and displayed a temperature-sensitive lateral mobility in the plane of the membrane, leading to the formation of small clusters, larger patches and polar caps. This observation was confirmed by Berthois et al. (1986) using the same technique in two different human breast cancer cell lines; MCF-7 (ER positive) and MDA-MB-231 (ER negative). In addition, they showed that there are two types of estradiol binding sites in the membrane of MCF-7 cells; high-affinity sites (type A, Kd~80nM) and low-affinity sites (type B, Kd estimated from 1/2 saturation curve ~400nM). A high-affinity 2-hydroxyestrone binding site (Kd ~6–10nM) was also found in human breast cancer cell lines (Vandewalle et al. 1988). The number of the binding sites is more in an ER-positive cell line (MCF-7 and VHB1) than in an ER-negative cell line (MDA-MB-231).

Though estradiol appears to play an important role in the immune system (Miller and Hunt 1996), there is little evidence to suggest the existance of surface receptors in the cell of this system (Peter et al. 1997).

D. Evidence for Diverse Protein Estrogen Binders

From the studies described in Sect. CI, it is apparent that E-6-^{125}IBSA could also be used as a specific probe to identify protein estrogen binders in different tissues, using a method called ligand blotting. This method not only identifies specific estrogen-binding proteins in membrane fractions, but also reveals their molecular size. The method is also very useful in determining which particular protein(s) is/are the steroid-binding protein(s) after affinity purification. Furthermore, a single ligand may recognize several proteins of different molecular sizes simultaneously in a mixture, provided they possess ligand-binding activity. First, we will mention our current studies in the CNS, since we are not aware of similar reports from other laboratories. Then, we will make reference to peripheral tissues where estrogen binders have been described.

I. Estrogen Binders in Plasmalemmal Microsomal Fractions

In our initial studies we identified three major proteins corresponding to 23 kDa, 28kDa and 32kDa in the P2 fraction from the hypothalamus, olfactory bulb and cerebellum, respectively, of intact female rats under reducing conditions which were specifically displaced by 1 μM E-6-BSA (Zheng and Ramirez 1997a). Because the P2 fraction represents a rather crude synaptosomal fraction, we examined the subcellular distribution of these estrogen-binding proteins. Subcellular fractions of nuclei (P_1), mitochondrial lysosomal (mP_2) and plasmalemmal microsomal (P_3) fractions were prepared from female rat brain, and the purity of the fractions was checked by electron micrographs (Zheng and Ramirez 1997a). Under those conditions, we showed that the 23-kDa,

28-kDa and 40-kDa proteins were highly concentrated in the mP2 fraction, whereas the 18, 28 and 32 species were richer in the P3 fraction. This suggests that the 23-kDa and 40-kDa proteins are probably mitochondrial-protein estrogen binders, while the 18, 28 and 32 species correspond to the plasmalemmal microsomal fraction. The nature of these latter proteins is unknown and current efforts are directed at their purification. The binding data described above using brain P2 fractions implies that any one of those proteins might have been responsible for the binding. In the following section (see Sect. DI), we will document the identity of one of these proteins (the 23-kDa species) which became a mitochondrial protein, a serendipitous finding which could be of great importance.

Another organ from which specific estrogen-binding proteins were isolated is the rat liver (MOATS and RAMIREZ 1998). These proteins had a molecular size of about 49kDa and 28kDa and were purified with an E-6-BSA affinity column. The P3 fraction was further purified through a sucrose gradient to generate a highly pure plasmalemmal fraction as shown in Fig. 2A. A major doublet protein (Fig. 2B, *line 2*) was highly concentrated by the affinity

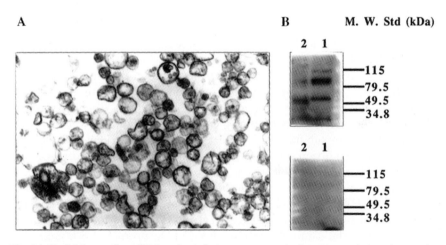

Fig. 2A,B. Male rat liver P3 fractions (microsomes and plasma membranes) were isolated using differential centrifugation (MOATS and RAMIREZ 1998) and plasma membrane subfractions were separated from the P3 fractions following the sucrose gradient method of PIETRAS and SZEGO (1980). The plasma membrane subfractions were centrifuged for 20 min at 10,000 g, 4°C, and the resulting pellet was fixed without resuspension overnight in 2% glutaraldehyde, 2% paraformaldehyde in phosphate-buffered saline (PBS) (pH 7.4). The fixed pellets were stained in uranyl acetate, embedded, and sectioned for electron microscopy. **A** Representative micrograph of the plasmalemmal fraction (magnification, Δ61,250). Plasma membrane subfraction proteins were also purified by E-6-BSA affinity column and separated by SDS-polyacrylamide gel electrophoresis (PAGE) for ligand blotting. **B** Autoradiogram of an E-6-^{125}IBSA ligand blot (*upper panel*) of digitonin-solubilized liver P3 fraction proteins (*Lane 1*) and affinity purified liver plasma membrane subfraction proteins (*Lane 2*). Note the increased concentration of a 50-kDa band in *Lane 2*. Binding was inhibited by an excess (1 μM) of E-6-BSA (*lower panel*). In each lane, 10 μg protein were loaded

column from the solubilized P3 fraction (Fig. 2B, *line 1*) and showed high affinity for E-6-[125]IBSA, since it was completely competed off by 1 μM unlabeled ligand (Moats and Ramirez, unpublished observations).

A photo affinity labeling method was used to identify protein-estrogen binders in mice (Büküsoglu and Krieger 1996). The ligand (progesterone-11-hemisuccinate-2-[125]iodohistamine) identified four proteins, of which a 29-kDa protein was competed off by 17β-estradiol and 17α-estradiol, indicating non-stereospecificity for estradiol.

II. Estrogen Binders in Mitochondrial Lysosomal Fractions

An affinity column linked with E-6-BSA to isolate and purify the 23-kDa protein from a digitonin-solubilized brain P2 fraction was used (for details see Zheng 1996). The protein was specifically recognized by E-6-[125]IBSA with high affinity, but not by [125]IBSA, and was specifically displaced by unlabeled E-6-BSA and 17β-estradiol. Therefore, the estradiol part of the complex, not BSA, was responsible for the interaction between the ligand and the 23-kDa molecule. The N-terminal microsequencing of this protein indicated complete identity to mature rat OSCP (Cretin et al. 1991; Higuti et al. 1993). This is a component of the F0F1 proton ATP-synthase/ATPase of mitochondria required for coupling of proton gradient to ATP synthesis (Pederson and Amzel 1993; Walker and Collinson 1994). The M.W. of this protein, estimated by SDS-PAGE, was close to the calculated M.W. from OSCP cDNA sequence (21 kDa) (Higuti et al. 1993). Furthermore, the N-terminal sequence of the 18-kDa co-purified protein from the affinity column was the same as the delta subunit of rat ATP-synthase/ATPase. The number of major proteins retained by the 17β-E-6-BSA affinity column corresponded well with the expected number of subunits of the F0F1 ATP-synthase/ATPase, which consists of more than a dozen subunits (Hekman and Hatefi 1991; Pederson and Amzel 1993; Walker and Collinson 1994). In addition, this 23-kDa protein was specifically recognized by an antibody against bovine OSCP (Belogrudov et al. 1995; Hekman and Hatefi 1991). Furthermore, a recombinant bovine OSCP, generously supplied to us by Hatefi (Ovchinikov et al. 1984; Walker et al. 1987), had the same mobility as this 23-kDa protein and bound 17β-estradiol-6-[125]IBSA with the same properties (see Fig. 3). All these data indicate that this 23-kDa protein is an estrogen-binding protein and is OSCP. Therefore, OSCP represents a stereospecific membrane estrogen binding site in mitochondria.

III. Estrogen Binders from Other Origins

Some of the estrogen binders could be derived from blood contamination or absorption of serum proteins to the membrane fractions. For instance, SHBG has high affinity (in the nanomolar range) and albumin low affinity (in the micomolar range) for estradiol, respectively. SHBG is known to bind estradiol

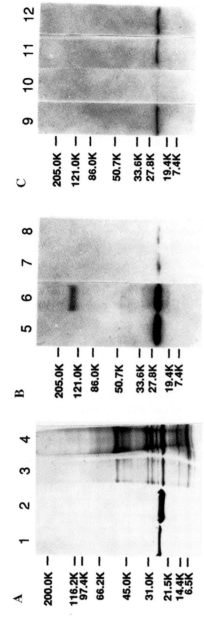

Fig. 3. A SDS-polyacrylamide-gel electrophoresis (PAGE) of recombinant bovine oligomycin sensitivity conferring protein (rbOSCP), (*lanes 1* and *2*, 1 µg and 5 µg, respectively) and 17β-E-6-BSA affinity column-retained proteins from digitonin-solubilized B-P2 (*lanes 3* and *4*, 2 µg and 10 µg, respectively). **B** RbOSCP (5 µg, *lanes 5* and *7*) and purified B-P2 (10 µg, *lanes 6* and *8*) were separated in SDS-PAGE and transferred to a nitrocellulose [^{125}I]BSA (1 × 10^6 cpm/ml, 1.6 nM) in the absence (*lanes 5* and *6*) and presence (*lanes 7* and *8*) of 1 µM unlabeled E-6-BSA. In *lane 6*, not only was the OSCP band specifically identified, but the ligand also labeled a 130-kDa protein. This may represent contamination from plasma membrane. **C** Ligand blotting of rbOSCP (1 µg in each lane) using E-6-^{125}I BSA (1 × 10^6 cpm/ml, 1.6 nM) in the absence of competitors (*lane 9*), and in the presence of 0.5 µM E-6-BSA (*lane 10*), 0.5 µM 17α-E-6-BSA (*lane 11*), and 12 µM BSA (*lane 12*). The broad range *Mr* markers (*a*) and pre-stained broad range *Mr* markers (*b* and *c*) were used

and secondarily binds to SHBG receptors in the cell membrane of several tissues (ROSNER 1996). Indeed, it has been reported that the estradiol-induced proliferation of breast cancer cells were mediated by the SHBG, which led to a rapid increase in cAMP (FISSORE et al. 1994; FORTUNATTI et al. 1996). None of these proteins was part of the final purification step in our procedures described in Sect. D. However, we recently identified two cytosolic proteins that bound E-6-[125]IBSA, one of much smaller size, corresponding to a doublet with a M.W. of 14 kDa and 15 kDa under both reducing or non-reducing conditions, respectively, and another of higher M.W. 210 kDa (ZHENG and RAMIREZ 1997b). The two small proteins were identified by N-terminal sequencing as rat hemoglobin subunits α and β, respectively. However, the presence of an estrogen binder fits a two-site model for the ligand in the brain cytosol and does not correspond to hemoglobin, since the binding was still present after perfusion of the rat brain. Furthermore, in a similar fraction prepared from rat blood cells, i.e., high concentrations of hemoglobin, the binding was only one-third of the total binding detected in the high-speed brain-cytosolic fraction (Zheng and Ramirez, unpublished results).

In addition, the ligand identified another protein of about 37 kDa from a brain P3 fraction (JOE et al. 1997), which by N-terminal sequencing corresponded to a key glycolytic enzyme, G3PD.

Two types of ERs in the plasma membrane could be receptor kinases. MATSUDA et al. (1993) identified estradiol as a ligand for the product of the c-erbB2 (neu or nERα) proto-oncogene, a 185-kDa tyrosine kinase that is expressed normally in fetal tissue and cancer cells and belongs to the epidermal growth factor receptor family. Estradiol bound to this membrane protein with a Kd of 2.7 nM and stimulated its kinase activity. The extracellular domain of this receptor revealed significant (36%) homology to the ligand-binding domain of the nuclear estrogen receptor (nER). The second receptor tyrosine kinase for estrogen could be the non-activated nuclear ER, a 66-kDa glyco-protein identified in the goat uterine plasma membrane which has a high affinity for estradiol (Kd 0.1 nM), but no capacity to bind to DNA on its own (ANURADHA et al. 1994; KARITHIKEYAN and THAMPAN 1996). However, unlike the c-erbB2 proto-oncogene product, estradiol inhibits naER kinase activity (ANURADHA et al. 1994).

E. Diverse Mechanisms in the Fast Actions of Estradiol

It is reasonable to postulate more than one single mechanism for estradiol in its action at the cellular membrane on the basis of the diversity of estradiol target sites and the ubiquitous nature of the steroid. There is compelling evidence indicating that the non-genomic actions of estradiol are mediated by at least three major mechanisms: (1) changes in channel function; (2) activation of membrane transduction systems; and (3) changes in cellular metabolism.

I. Channel Regulator

1. Ca^{2+} Channels

Ca^{2+} channels have received considerable attention and are clearly regulated by estradiol in different cells. For example, estradiol relaxes the rabbit basilar artery by inhibition of voltage-dependent Ca^{2+} channels via a pertussis-sensitive GTP-binding protein (OGATA et al. 1996). In isolated ventricular myocytes, estradiol in micromolar concentrations decreased Ca^{2+}-currents and reduced cytosolic Ca^{2+} concentrations, suggesting a blockade of the channels (JIANG et al. 1992). A similar explanation was put forward by SALAS et al. (1994) for the effect of micromolar concentrations of estradiol and 17α-estradiol on the acute vasorelaxant action on pig coronary arteries. In human lymphocytes, estradiol inhibited the phytohemagglutinin-induced increase in cytosolic free calcium (BELLINI et al. 1990). Using whole-cell patch clamping of A7r5 vascular smooth muscle cell lines, ZHANG et al. (1994) reported that the inhibitory effect of micromolar concentrations of estradiol were due to a reduction in L-type Ba^{2+} and L-type Ca^{2+} currents.

Using whole-cell patch clamp of acutely dissociated and cultured rat neostriatal neurons, physiological concentrations of estradiol reduced Ba^{2+} entry reversibly via Ca^{2+} channels of the L-type (MERMELSTEIN et al. 1996). In other cells, these doses of estradiol had opposite effects, i.e., produced increases in intracellular Ca^{2+}. For instance, estradiol stimulated a rapid Ca^{2+} influx in a human prostate tumor cell line, LNCaP (AUDY et al. 1996), and in granulosa cells (MORLEY et al. 1992), although in this latter case the specificity of the action of estradiol was rather poor. In bone cells, estradiol increased intracellular Ca^{2+} by mobilization from intracellular stores secondary to the formation of IP_3 (LE MELLAY et al. 1997). Previously, however, the same laboratory showed that estradiol can affect Ca^{2+} both by changes in Ca^{2+} influx and by mobilization from intracellular stores (LIEBERHERR et al. 1993). In osteoclast precursors, estradiol induced rapid changes in intracellular Ca^{2+}, which were dependent on extracellular sources of the cation since incubation of the cells in media-containing EGTA abolished the response (FIORELLI et al. 1996). The stimulatory effect of estradiol or the impermeant complex E-6-BSA on intracellular Ca^{2+} at nanomolar doses was observed in maturing human oocytes, an effect also abolished by EGTA (TESARICK and MENDOZA 1995). Thus, it seems that estradiol can either stimulate or inhibit Ca^{2+} influx in cells by regulating Ca^{2+} channels, particularly of L-type, either directly or indirectly. In most of the cases, inhibition of intracellular Ca^{2+} concentrations or influx of Ca^{2+} was achieved by high concentrations of estradiol with a notable exception in striatal neurons. On the other hand, stimulation was induced by much lower concentrations of estradiol (picomolar).

2. Others

Two other channels have been reported to be affected by estradiol; Cl^- channel in NIH3T3 fibroblasts grown in the presence of colchicine (HARDY and

Valverde 1994); and a conductance K⁺ channel in hypothalamic cells, where estradiol rapidly reduced the potency (4-fold) of a μ-opioid agonist, DAMGO (Lagrange et al. 1994 and 1995). Also, a glutamate receptor appears to be the target of direct steroid modulation, but not by estradiol (Wong and Moss 1994). Recently it was reported that estradiol rapidly attenuates a $GABA_B$ receptor response (Lagrange et al. 1996).

II. Transduction Activator

1. Cyclic Adenosine Monophosphate (cAMP)

As in the case of Ca^{2+} channels, early in these studies this classical second intracellular messenger was thought to be responsible for some of the estradiol membrane-mediated events (Szego and Davis 1967). Most of the studies have clearly shown that estradiol increases cAMP in a variety of cells (Aronica et al. 1994) within minutes of exposure to the steroid. This is most likely due to an activation of the adenylate cyclase system (Bergamini et al. 1985). Rather interestingly, pertussis toxin, an inhibitor of G-protein function, led to inhibition of the cAMP response to estradiol and to Ca^{2+} channel activation (see Sect. E.I.1). This suggests that the changes in channel permeability may be, in part, due to a secondary effect of cAMP on channel function. Equally possible is the explanation that the effects of estradiol antagonizing the potency of a μ-opioid agonist in hypothalamic cells appears also to be mediated by increases in cAMP, which then activate PKA, leading to uncoupling of the μ-opioid site from its K⁺ channel (Lagrange et al. 1997).

In another study, estradiol induced phosphorylation of CREB in neurons of the mPOA and BNST of ovx rats within minutes (Zhou et al. 1996). The authors speculated that membrane events initiated by estradiol and leading to an increase in cAMP or in Ca^{2+}-influx could trigger activation of PKA or CaM/kinases with secondary phosphorylation of CREB.

2. Phospholipase C (PLC)

Another membrane enzyme, PLC appears to be a specific target for estradiol action at the membrane level (Graber et al. 1993). Activation of a particular member of the family of this lipase appears to be responsible for the formation of IP_3 and DAG in response to estradiol stimulation (Lieberherr et al. 1993; Le Mellay et al. 1997).

3. Mitogen-Activated Protein Kinase (MAPK)

Recently it was reported by two different laboratories that estradiol can rapidly activate the MAPK cascade, indicating that some of the membrane-mediated events of estradiol could involve this serine–threonine kinase cascade. In MCF-7 cells, 10 nM estradiol led to an increase in phosphorylation of the ERK-1 and 2 proteins (particularly the latter) within 2 min. This was transient, since by 60 min the effect disappeared. Interestingly, the steroid also increased the active form of p21ras-GTP. However, the antiestrogen ICI 812780 abolished the effect of estradiol, and COS cells transfected with the estradiol receptor became responsive in terms of ERK-2 activity following estradiol stimulation. This suggests that estradiol was activating the MAP-kinase cascade secondary to activation of transduction by the classical estradiol receptor (MIGLIACCIO et al. 1996).

More compelling evidence for a membrane-mediated estradiol effect was shown by WATTERS et al. (1997). Using the impermeant E-6-BSA, they showed that this complex activated the MAPK cascade in a human neuroblastoma cell line SK-N-SH, since within 5 min, 10 nM of the complex led to phosphorylation of ERK-1 and 2 kinase proteins. The MAPKK inhibitor PD098059 abolished the effect of E-6-BSA on C-fos gene transcription, an event secondary to activation of this cascade. Equally, in osteoblasts (rat ROS 17/2.8) exposed to physiological doses of estradiol, a rapid and transient MAPK activation was shown (ENDOH et al. 1997). Recall that estradiol at nanomolar concentrations can also inhibit the endothelin-1-stimulated MAPK cascade (MOREY et al. 1997).

III. Metabolic Regulator

1. ATPase/ATP Synthase

Early experiments showed that the mitochondria was a target site for estradiol (SZEGO AND PIETRAS 1984; NOTEBOOM and GORSKI 1965). In the uterus of ovx or immature rats, estradiol decreased ATP concentrations (AARONSON et al. 1965). More circumstantial evidence suggesting an involvement of the ATPase/ATP synthase, was the demonstration that estradiol, administered to immature rats (DEGANI et al. 1984) or directly added to a superfusion bath containing immature uteri (DEGANI et al. 1988), altered high-energy phosphate metabolism. In the in vitro experiments, estradiol 30 nM induced a rapid decrease in β-ATP and γ-ATP, as measured by ^{32}P nuclear magnetic resonance. A direct demonstration that this energy master regulator can be altered by estrogen was later shown in rat liver mitochondria (MCENERY et al. 1986). The authors postulated that the F_0 sector of the ATPase may contain a distinct binding site for DES. In a subsequent paper, the same laboratory used a highly purified F_0 preparation from rat liver mitochondria to show that micromolar concentrations of the synthetic estrogen, DES, modulate proton transport by either inhibiting or uncoupling the ATPase complex (MCENERY et al. 1989). In addition, DES inhibited the H^+-ATPase in bovine chromaffin-granule ghosts also at micromolar concentrations (GRONBERG and FLATMARK 1988).

All these studies strongly suggest an involvement of estradiol on mitochondrial function, but it remains to be shown whether estradiol can bind specifically to one of the subunits of the ATPase complex and, most importantly, affect ATP production. Our recent studies (for details see ZHENG 1996) unquestionably demonstrated that OSCP, a subunit of the multimeric ATPase

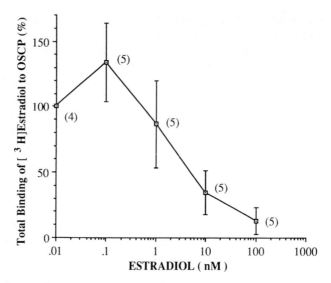

Fig. 4. Binding of [³H] estradiol (1.1×10^6 cpm, 20 nM) to recombinant bovine oligo-mycin sensitivity conferring protein (rbOSCP, 1 μg or 130 nM) in 0.25 ml. The reaction was performed at room temperature for 30 min with constant agitation in phosphate buffer at pH 6.5. The bound was separated from unbound [³H] estradiol by Lipidex-1000 after centrifugation at 7000 rpm for 2 min at 4°C. The data were expressed as mean ± SE. The IC_{50} was about 5 nM. *In parenthesis* number of experiments (ZHENG and RAMIREZ, unpublished observations)

complex and involved in proton translocation (WALKER and COLLINSON 1994), binds with high affinity (nanomolar concentrations) to the E-6-BSA complex (as shown earlier; Fig. 3). Recently, we have shown that it also binds to ³H-E, (Fig. 4). Current studies are addressing the issue of the ability of estradiol to alter ATP production from rat brain mitochondria. Therefore, it is reasonable to postulate that estradiol could be another factor in the regulation of cellular energy by acting on the complex V of the mitochondria, the last step in the oxidative phosphorylation pathway.

2. Glyceraldehyde-3-Phosphate Dehydrogenase (G3PD)

G3PD is an intermediate enzyme in the glycolytic pathway of all cells and plays a key role in glycerol phosphate shuttle across the mitochondrial membrane (DARNELL et al. 1995). In our pursuit to identify estradiol protein binders, we isolated a 37-kDa protein from rat brain that resulted to be G-3PD by *N*-terminal sequencing (JOE et al. 1997). A commercial purified form of this enzyme (rabbit heart muscles, Sigma) binds E-6-¹²⁵BSA with high affinity (IC_{50} ~50 nM) at pH 8.6–6.8. It is intriguing that this enzyme not only appears to be involved in the regulation of the cell's metabolism, but it also seems to play other roles; as a possible activator of transcription (MORGENEGG et al. 1986) or as an actin binding or actin filament network orga-

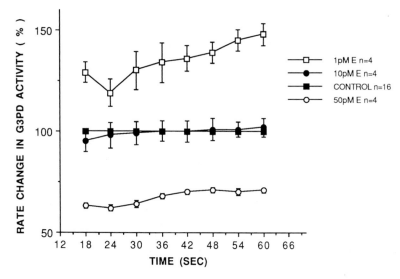

Fig. 5. Rate change (δ/min) in G3PD activity in the presence of estradiol (*1,10pM* or *50pM*) or in its absence (*control*) as measured by a spectrophometric assay (KANT and STECK 1973). The initial rate at 18s and, thereafter, at 6s intervals was considered 100% for the control. All values from the cuvettes containing estradiol were expressed as percentage change from their respective control times. Note that estradiol had a rapid stimulatory effect at low concentrations and an inhibitory effect at higher concentrations ($p < 0.01$). Similar results were obtained with E-6-BSA (data not shown) (JOE, ZHENG, RAMIREZ, unpublished observations)

nizer (ROGALSKI et al. 1997) in neurons. Thus, the fact that E-6-BSA binds to the enzyme and appears to affect its catalytic activity in seconds (unpublished results, Fig. 5) suggests that estradiol may alter the cell metabolism by regulating ATP production, generated either by the glycolytic pathway or by the availability of substrates (NAD^+ NADH) for the electron transport chain at the mitochondria.

3. Others

At the beginning of the second decade of this century, the idea that hormones act by changing the catalytic activity of enzymes was the fashionable research avenue to explain the mechanism of hormone action. An excellent review of those early studies focusing on estradiol was reported by HAGERMAN and VILLEE (1959). For example, micromolar concentrations of estradiol inhibited glutamate dehydrogenase (PONS et al. 1978). Protein disulfide isomerase which is also associated with nERα and shares significant homology in a 40-residue segment was non-competitively inhibited by estradiol with an IC_{50} of about 100nM (TSIBRIS et al. 1989). More recently, it was reported that estradiol can have a biphasic effect on the activity of neuronal constitutive nitric oxide synthase (HAYASHI et al. 1994). Therefore, the fact that estradiol can affect enzyme activity is not without precedent; however, the pharmacological concentrations required to alter enzyme functions reduced the enthusiasm for such an avenue of research. In Chap. 2 of this Handbook, this issue is addressed specifically.

F. Overview and Concluding Remarks

I. The Continuum Theory

The foresight of SZEGO AND COLLABORATORS in the 1970s, regarding the presence of membrane receptors for estradiol in a variety of cells that led her to formulate the "unfolding of a continuum" concept for the mechanism of action of steroids has been fully confirmed. The weight of experimental evidence reviewed here and elsewhere (RAMIREZ and ZHENG 1996) has tilted the balance to re-think the terms of such propositions clearly and elegantly stated by SZEGO in a later review of contributions; hers and others on the subject (SZEGO 1994).

Seemingly, estradiol utilizes different pathways in its continuum of action from the outer to the inner surface of the cells, each one characterized by a particular estrogen molecular interaction. The experimental evidence strongly support the concept that estradiol affects both ionic channels and a variety of key enzymes that regulate a series of different and unique intracellular events. In the former case, the bulk of data indicate that Ca^{2+} channels are a target for estradiol. What is not yet known is whether this is a direct interaction (binding sites in the channel) or is regulated indirectly. In the latter, there is also clear evidence that estradiol modifies: (1) cAMP by an action most likely on adenylate cyclase; (2) Phospholipase C, leading to the formation of IP_3 and DAG; and (3) MAPK leading to phosphorylation of intracellular substrates necessary for transcriptional activation. Our recent work indicates that two other key enzymes in the regulation of the metabolism of the cells are targets for estradiol since OSCP and G3PD binds the conjugated E-6-BSA and the complex alters the catalytic activity of G3PD at picomolar concentrations in seconds (Fig. 5). This diversity of estrogen molecular interactions may explain the so-called protective effects of estradiol to a variety of chemical insults to cells cultured in vitro (BEHL et al. 1995; SINGER et al. 1996; GOODMAN et al. 1996; GREEN et al. 1996, 1997).

The elusive mER might be the missing link between some of the action of estradiol at the membrane level. The mER probably represents a new family of ERs with a selective member present in a particular cell regulating a particular type of function. It is possible that one of the members of the family could be responsible for the translocation of estradiol from the cellular membrane to the interior of cells by a process of endocytosis, as might be the case in the liver (MOATS and RAMIREZ 1998) or the CNS (GARCIA-SEGURA et al. 1989). The cloning of this receptor will clarify some of the issues raised in this review.

Acknowledgements. We thank Ms. LORI HEIL for her skillful help in preparing this manuscript. This work was, in part, support for an NIH grant R01-MH55986–01 to VICTOR D. RAMIREZ.

References

Aaronson SA, Natori Y, Tarver H (1965) Effect of estrogen on uterine ATP levels. Proc Soc Exp Biol Med 120:9–10

Anuradha P, Khan SM, Karthikeyan N, Thampan RV (1994) The nonactivated estrogen receptor (naER) of the goat uterus is a tyrosine kinase. Arch Biochem Biophys 309:195–204

Aronica SM, Kraus WL, Katzenellenbogen BS (1994) Estrogen action via the cAMP signalling pathway-stimulation of adenylate cyclase and cAMP-regulated gene transcription. Proc Natl Acad Sci USA 91:8517–8521

Audy MC, Vacher P, Dufy B (1996) 17β-estradiol stimulates a rapid Ca^{2+} influx in LNCaP human prostate cancer cells. Eur J Endocrinol 135:367–373

Bazzett TJ, Becker JB (1994) Sex differences in the rapid and acute effects of estrogen on striatal D-2 DA receptor binding. Brain Res 637:163–172

Becker JB (1990a) Direct effect of 17β-estradiol on striatum: sex differences in DA release. Synapse 5:157–164

Becker JB (1990b) Estrogen rapidly potentiates AMPH-induced striatal DA release and rotational behavior during microdialysis. Neurosci Lett 118:169–171

Behl C, Widmann M, Trapp T, Holsboer F (1995) 17β-estradiol protects neurons from oxidative stress-induced cell death in vitro. Biochem Biophys Res Commun 216:473–482

Bellini T, Degani D, Matteuzzi M, Dallocchio F (1990) Effect of 17β-estradiol on calcium response to phytohaemagglutnin in human lymphocytes. Biosci Rep 10:73–78

Belogrudov, GI, Tomich JM, Hatefi Y (1995) ATP synthase complex-proximities of subunits in bovine submitochondrial particles. J Biol Chem 270:2053–2060

Bergamini CM, Pansini F, Bettocchi S, Segala V, Dallocchio F, Bagni B, Mollica G. (1985) Hormonal sensitivity of adenylate cyclase from human endometrium: modulation by estradiol. J Steroid Biochem 22:299–303

Berthois Y, Pourreau-Schneider N, Gandilhon P, Mittre H, Tubiana N, Martin PM (1986) Estradiol membrane binding sites on human breast cancer cell lines: use of a fluorescent estradiol conjugate to demonstrate plasma membrane binding systems. J Steroid Biochem 25:963–972

Beyer C, Sawyer CH (1969) Hypothalamic unit activity related to control of the pituitary gland. Frontiers in Neuroendocrinol 1:255–287

Bression D, Michard M, Le Dafniet M, Pagesy P, Peillon F (1986) Evidence for a specific estradiol binding site on rat pituitary membranes. Endocrinology 119:1048–1051

Brinton RD (1993) 17β-estradiol induction of filopodial growth in cultured hippocampal neurons within minutes of exposure. Mol Cell Neurosci 4:36–46

Brubaker KD, Gay CV (1994) Specific binding of estrogen to osteoclast surfaces. Biochem Biophys Res Commun 200:899–907

Büküsoglu C, Krieger NR (1996) Estrogen-specific target site identified progesterone-11α-hemisuccinate-(2-[^{125}I]iodohistamine) in mouse brain membranes. J Steroid Biochem Mol Biol 58:89–94

Changchit A, Durham S, Vore M (1990) Characterization of [^{3}H]estradiol-17β-(β-D-glucuronide) binding sites in basolateral and canalicular liver plasma membranes. Biochem Pharmacol 40:1219–1225

Clark JH, Mani SK (1994) Actions of ovarian steroid hormones. In: Knobil E, Neill JD (eds) The Physiology of Reproduction, 3rd edn. Raven Press, pp 1011–1059

Conklin BR, Bourne HR (1993) Structural elements of Gα subunits that interact with G$\beta\gamma$, receptors, and effectors. Cell 73:631–641

Cretin F, Baggetto LG, Denoroy L, Godinot C (1991) Identification of F0 subunits in the rat liver mitochondrial F0F1-ATP synthase. Biochim Biophys Acta 1058: 141–146

Darnell J, Lodish H, Baltimore D (1995)Molecular Cell Biology Freeman WH and Co., New York, 3rd edn. pp 529–558

DeBeer EL, Keizer HA (1982) Direct action of estradiol-17β on the atrial action potential. Steroids 40:223–231

Degani H, Shaer A, Victor TA, Kaye AM (1984) Estrogen-induced changes in high-energy phosphate metabolism in rat uterus: ^{31}P NMR studies. Biochemistry 23:2572–2577

Degani H, Victor TA, Kaye AM (1988) Effects of 17β-estradiol on high energy phosphate concentrations and the flux catalyzed by creatine kinase in immature rat uteri: ^{31}P nuclear magnetic resonance studies. Endocrinology 122:1631–1638

Di Paolo T, Rouillard C, Bedard P (1985) 17β-estradiol at a physiological dose acutely increases DA turnover in rat brain. Eur J Pharmacol 117:197–203

Di Paolo T (1994) Modulation of brain DA transmission by sex steroids. Rev Neurosci 5:27–42

Endoh H, Sasaki H, Maruyama K, Takeyama K, Waga I, Shimizu T, Kata S, Kawashima H (1997) Rapid activation of MAP kinase by estrogen in the bone cell line. Biochem Biophys Res Commun 235:99–102

Farhat MY, Lavigne MC, Ramwell PW (1996a) The vascular protective effects of estrogen. FASEB J 10:615–624

Farhat MY, Abi-Younes S, Dingaan B, Vargas R, Ramwell PW (1996b) Estradiol increases cAMP in rat pulmonary vascular smooth muscle cells by a nongenomic mechanism. J Pharmacol Exp Therap 276: 652–657

Fiorelli G, Gori F, Frediani U, Franceschelli F, Tanini A, Tosti-Guerra C, Benvenuti S, Gennari L, Becherini L, Brandi ML (1996) Membrane binding sites and non-genomic effects of estrogen in cultured human preosteoclastic cells. J Steroid Bioch Molec Biol 59:233–240.

Fissore F, Fortunati N, Comba A et al. (1996b) The receptor-mediated action of sex steroid-binding protein (SBP, SHBG): Accumulation of cAMP in MCF-7 cells under SBP and estradiol treatment. Steroids 59:661–667

Fortunati N, Fissore F, Fazzari A, Becchis M, Comba A, Catalano MG, Berta L, Frairia R (1996) Sex steroid binding protein exerts a negative control on estradiol action in MCF-7 cells (human breast cancer) through cyclic adenosine 3',5'-monophosphate and protein kinase A. Endocrinology 137:686–692

Foy MR, Teyler TJ (1983) 17α-Estradiol and 17β-estradiol in hippocampus. Brain Res. Bull. 10:735–739

Garcia-Segura LM, Olmos G, Robbins RJ, Hernandez P, Meyer JH, Naftolin F (1989) Estradiol induces rapid remodeling of plasma membranes in developing rat cerebrocortical neurons in culture. Brain Res 498:339–343

Gilligan DM, Quyyumi AA, Cannon RO (1994) Effects of physiological levels of estrogen on coronary vasomotor function in postmenopausal women. Circulation 89:2545–2551

Gilligan DM, Badar DM, Panza JA, Quyyumi AA, Cannon RO (1994) Acute vascular effects of estrogen in postmenopausal women. Circulation 90:786–791

Goodman Y, Bruce AJ, Cheng B, Mattson MP (1996) Estrogens attenuate and corticosterone exacerbates excitotoxicity, oxidative injury, and amyloid β-peptide toxicity in hippocampal neurons. J Neurochem 66:1836–1844

Gould E, Woolley CS, Frankfurt M, McEwen BS (1990) Gonadal steroids regulate dendritic spine density in hippocampal pyramidal cells in adulthood. J Neurosci 10:1286–1291

Graber R, Sumida C, Vallette G, Nuñez EA (1993) Rapid and long-term effects of 17β-estradiol on PIP$_2$-phospholipase C-specific activity of MCF-7 cells. Cell Signalling 5:181–186

Green PS, Gridley KE, Simpkins JW (1996) Estradiol protects against β-amyloid (25–35)-induced toxicity in SK-N-SH human neuroblastoma cells. Neurosci Lett 218:165–168

Green PS, Bishop J, Simpkins JW (1997) 17α-estradiol exerts neuroprotective effects on SK-N-SH cells. J Neurosci 17:511–515

Gronberg M, Flatmark T (1988) Inhibition of the H^+-ATPase in bovine adrenal chromaffin granule ghosts by diethylstilbestrol. FEB 229:40–44

Gu Q, Moss RL (1996) 17β-estradiol potentiate kainate-induced currents via activation of the cAMP cascade. J Neurosci 16:3620–3629

Hagerman DD, Villee CA (1959) Metabolic studies of the mechanism of action of estrogen. Recent Progress in the Endocrinology of Reproduction. LLoyd CW (ed), Academic Press, New York, pp 317–333

Hardy SP, Valverde MA (1994) Novel plasma membrane action of estrogen and antiestrogens revealed by their regulation of a large conductance chloride channel. FASEB J 8:760–765

Hayashi T, Ishikawa T, Yamada K, Kuzuya M, Naito M, Hidaka H, Iguchi A (1994) Biphasic effect of estrogen on neuronal constitutive nitric oxide synthase via Ca^{2+}-calmodulin dependent mechanism. Biochem Biophys Res Commun 203:1013–1019

Hekman C, Hatefi Y (1991) The Fo subunits of bovine mitochondrial ATP synthase complex: purification, antibody production, and interspecies cross-immunoreactivity. Arch Biochem Biophys 284:90–97

Hernández-Pérez O, Ballesteros LM, Rosada A (1979) Binding of 17β-estradiol to the outer surface and nucleus of human spermatozoa. Arch Androl 3:23–29

Higuti T, Kuroiwa K, Kawamura Y, Morimoto K, Tsujita H (1993) Molecular cloning and sequence of cDNAs for the import precursors of oligomycin sensitivity conferring protein, ATPase inhibitor protein, and subunit c of H^+-ATP synthase in rat mitochondria. Biochim et Biophys Acta 1172:311–314

Horvat A, Nikezic G, Martinovic JV (1995) Estradiol binding to synaptosomal plasma membranes of rat brain regions. Experientia 51:11–15

Hruska RE, Silbergeld EK (1980) Estrogen treatment enhances DA receptor sensitivity in the rat striatum. Eur J Pharmacol 61:397–400

Innes D, Michal E (1970) Effects of progesterone and estrogen on the electrical activity of the limbic system. J Exp Zool 175:487–492

Jensen EV, Jacobson H (1962) Basic guides to the mechanism of estrogen action. Recent Prog Horm Res 18:387–414

Jiang C, Poole-Wilson PA, Sarrel PM, Mochizuki S, Collins P, MacLeod KT (1992) Effect of 17β-oestradiol on contraction, Ca^{2+} current and intracellular free Ca^{2+} in guinea-pig isolated cardiac myocytes. Br J Pharmacol 106:39–745

Jiang C, Sarrel PM, Lindsay DC, Poole-Wilson PA, Collins P (1991) Endothelium-independent relaxation of rabbit coronary artery by 17β-estradiol in vitro. Br J Pharmacol 104:1033–1037

Joe I, Zheng J, Ramirez VD (1997) Progesterone-3-BSA and 17β-estradiol-6-BSA bind to glyceraldehyde-3-phosphate dehydrogenase (G3PD). Soc Neurosci Abstr New Orleans

Joels M (1997) Steroid hormones and excitability in the mammalian brain. Frontiers in Neuroendocrinol 182–48

Kant JA, Steck TL (1973) Specificity in the association of glyceraldehyde-3-phosphate dehydrogenase with isolated human erythrocyte membranes. J Biol Chem 248:8457–8464

Karthikeyan N, Thampan RV (1996) Plasma membrane is the primary site of localization of the nonactivated estrogen receptor in the goat uterus: hormone binding causes receptor internalization. Arch Biochem Biophys 325:47–54

Kelly MJ, Moss RL, Dudley CA (1978) The effects of ovariectomy on the responsiveness of preoptic-septal neurons to microelectrophoresed estrogen. Neuroendocrinolology 25:204–211

Kuiper GGJM, Carlsson B, Grandien K, Enmark E, Haggblad J, Nilsson S, Gustafsson J-A (1997) Comparison of the ligand binding specificity and transcript tissue distribution of estrogen receptors α and β. Endocrinology 138:863–870

Lagrange AH, Rønnekleiv OK, Kelly MJ (1997) Modulation of G protein-coupled receptors by an estrogen receptor that activates protein kinase A. Mol Pharmacol 51:605–612

Lagrange AH, Rønnekleiv OK, Kelly MJ (1995) Estradiol-17β and γ-opioid peptides rapidly hyperpolarize GnRH neurons: a cellular mechanism of negative feedback? Endocrinology 136: 2341–2344

Lagrange AH, Rønnekleiv OK, Kelly MJ (1994) The potency of μ-opioid hyperpolarization of hypothalamic arcuate neurons is rapidly attenuated by 17β-estradiol. J Neurosci 14:6196–6204

Lagrange AH, Wagner EJ, Rønnekleiv OK, Kelly MJ (1996) Estrogen rapidly attenuates a GABA$_B$ response in hypothalamic neurons. Neuroendocrinology 64:114–123

Le Mellay V, Grosse B, Lieberherr M (1997) Phospholipase Cβ_2 and membrane action of calcitriol and estradiol. J Biol Chem 272:11902–11907

Levesque D, Di Paolo T (1988) Rapid conversion of high into low striatal D2-DA receptor agonist binding states after an acute physiological dose of 17β-estradiol. Neurosci Lett 88:113–118

Lieberherr M, Grosse B, Kachkache M, Balsan S (1993) Cell signalling and estrogens in female rat osteoblasts: a possible involvement of unconventional nonnuclear receptors. J Bone Mineral Res 8:1365–1376

Liu B, Hu D, Wang J, Liu X-L (1997) Effects of 17β-estradiol on early afterdepolarizations and L-type Ca^{2+} currents induced by endothelin-l in guinea pig papillary muscles and ventricular myocytes. Meth Find Exp Clin Pharmacol 19:19–25

Loy R, Gerlach JL, McEwen BS (1988) Autoradiographic localization of estradiol-binding neurons in the rat hippocampal formation and entorhinal cortex. Developmental Brain Res. 39:245–251

Makriyannis A, Yang D-P, Mavromoustakos T (1990) The molecular features of membrane perturbation by anaesthetic steroids: a study using differential scanning calorimetry, small angle X-ray diffraction and solid state ^{2}H NMR. In: Chadwick D, Widdows K (eds) Ciba Foundation Symposium, Chichester, Wiley, pp 172–189

Matsuda S, Kadowaki Y, Ichino M, Akiyama T, Toyoshima K, Yamamoto T (1993) 17β-estradiol mimics ligand activity of the c-erbB2 protooncogene product. Proc Natl Acad Sci 90:10803–10807.

McEnery MW, Pedersen PL (1986) Diethylstilbestrol: A novel F$_o$-directed probe of the mitochondrial proton ATPase. J Biol Chem 261:1745–1752

McEnery MW, Hullihen J, Pedersen PL (1989) F$_o$ "proton channel" of rat liver mitochondria: rapid purification of a functional complex and a study of its interaction with the unique probe diethylstilbestrol. J Biol Chem 264:12029–12036

Mermelstein PG, Becker JB, Surmeier DJ (1996) Estradiol reduces calcium currents in rat neostriatal neurons via a membrane receptor. J Neurosci 16:595–604

Migliaccio A, Di Domenico M, Castoria G, deFalco A, Bontempo P, Nola E, Auricchio F (1996) Tyrosine kinase/p21ras/MAP-kinase pathway activation by estradiol-receptor complex in MCF-7 cells. EMBO J 15:1292–1300

Miller L, Hunt JS (1996) Sex steroid hormones and macrophage function. Life Sci 59:1–14

Moats RKII, Ramirez VD (1998) Rapid uptake and binding of 17β-estradiol 6-(O-carboxymethyl)oxime: [^{125}I]BSA by female rat liver. Biology of Reproduction 58:531–538

Morey AK, Pedram A, Razandi M, Prins BA, Hu R-M, Biesiada E, Levin ER (1997) Estrogen and progesterone inhibit vascular smooth muscle proliferation. Endocrinology 138:3330–3339

Morgenegg G, Winkler GC, Hubscher U, Heizmann CW, Mous J, Kuenzle CC (1986) Glyceraldehyde-3-phosphate dehydrogenase is a nonhistone protein and a possible activator of transcription in neurons. J Neurochem 47: 54–62

Morley P, Whitfield JF, Vanderhyden BC, Tsang BK, Schwartz J-L (1992) A new, nongenomic estrogen action: the rapid release of intracellular calcium. Endocrinology 131:1305–1312

Moss RL (1997) Estrogen: nontranscriptional signaling pathway. Rec Prog Horm Res 52:33–68

Mueck AO, Seeger H, Lippert TH (1996) Calcium antagonistic effect of natural and synthetic estrogens-investigations on a nongenomic mechanism of direct vascular action. Internat J Clin Pharmacol Therap 34: 424–426

Nabebura J, Oomura Y, Minami T, Mizuno Y, Fukuda A (1986) Mechanism of the rapid effect of 17β-estradiol on medial amygdala neurons. Science 233:226–227

Nenci I, Marchetti E, Marzola A, Fabris G (1981) Affinity cytochemistry visualizes specific estrogen binding sites on the plasma membrane of breast cancer cells. J Steroid Biochem 14:1139–1146

Noteboom WD, Gorski J (1965) Stereospecific binding of estrogens in the rat uterus. Arch Biochem Biophys 111:559–568

Ogata R, Inoue Y, Nakano H, Ito Y, Kitamura K (1996) Oestradiol-induced relaxation of rabbit basilar artery by inhibition of voltage-dependent Ca channels through GTP-binding protein. Bri J Pharmacol 117:351–359

Ovchinnikov YA, Modyanov NN, Grinkevich VA, Aldanova NA, Trubetskaya OE, Nazimov IV, Hundal T, Ernster L (1984) Amino acid sequence of the oligomycin sensitivity-conferring protein (OSCP) of beef-heart mitochondria and its homology with the delta-subunit of the F1-ATPase of Escherichia coli. FEBS Lett 166:19–22

Pasqualini C, Olivier V, Guibert B, Frain O, Leviel V (1995) Acute stimulatory effect of estradiol on striatal DA synthesis by estradiol. J Neurochem 65:1651–1657

Pasqualini C, Olivier V, Guibert B, Frain O, Leviel V (1996) Rapid stimulation of striatal DA synthesis by estradiol. Cell Mol Neurobiol 16:411–415

Pedersen PL, Amzel LM (1993) ATP synthases: structure, reaction center, mechanism, and regulation of one of nature's most unique machines. J Biol Chem 268:9937–9940

Pennisi E (1997) Different roles for Estrogen's two receptors. Science 277:1439

Peter W, Benten M, Lieberherr M, Sekeris CE, Wunderlich F (1997) Testosterone induces Ca^{2+} influx via non-genomic surface receptors in activated T cells. FEBS Lett 407:211–214

Pietras RJ, Szego CM (1975) Endometrial cell calcium and estrogen action. Nature 253:357–359

Pietras RJ, Szego CM (1977) Specific binding sites for estrogen at the outer surfaces of isolated endometrial cells. Nature 265:69–72

Pietras RJ, Szego CM (1979) Metabolic and proliferative responses to estrogen by hepatocytes selected for plasma membrane binding-sites specific for estradiol-17β. J Cell Physiol 98:145–160

Pietras RJ, Szego CM (1980) Partial purification and characterization of estrogen receptors in subfractions of hepatocyte plasma membranes. Biochem J 191:743–760

Pons M, Michel F, Descomps B, de Paulet CA (1978) Structural requirements for maximal inhibitory allosteric effect of estrogens and estrogen analogues on glutamate dehydrogenase. Eur J Biochem 84:257–266

Puca GA, Sica V (1981) Identification of specific high affinity sites for the estradiol receptor in the erythrocyte cytoskeleton. Biochem Biophys Res Comm 103:682–689

Raab H, Pilgrim C, Reisert I (1995) Effects of sex and estrogen on tyrosine hydroxylase mRNA in cultured embryonic rat mesencephalon. Mol Brain Res 33:157–164

Ramirez VD, Zheng J, Khawar MS (1996) Membrane receptors for estrogen, progesterone and testosterone in the rat brain: fantasy or reality? Mol Cell Neurobiol 16:175–197

Ramirez VD, Zheng J (1996) Membrane sex-steroid receptors in the brain. Frontiers in Neuroendocrinol 17:402–439

Ramirez VD, Zheng J (1999) Steroid receptors in membranes of neurons. In: Baulieu EE, Robel P, Schumacher M (eds) Neurosteroids: A New Regulatory Function in the Nervous System. (in press)

Reis SE, Gloth ST, Blumenthal RS et al. (1994) Ethinyl estradiol acutely attentuates abnormal coronary vasomotor responses to acetylcholine in postmenopausal women. Circulation 89:52–60

Rogalski-Wilk AA, Cohen RS (1997) Glyceraldehyde-3-phosphate dehydrogenase activity and F-actin associations in synaptosomes and postsynaptic densities of porcine cerebral cortex. Cell Mol. Neurobiol. 17:51–70

Roselli CE, Horton LE, Resko JA (1985) Distribution and regulation of aromatase activity in the rat hypothalamus and limbic system. Endocrinology 117:2471–2477

Rosner W (1996) Sex steroid transport: binding proteins. In: Adashi EY, Rock JA, Rosenwaks Z (eds) Reproductive Endocrinology, Surgery, and Technology. Pippincott-Raven Publishers, Philadelphia, pp 605–626

Salas E, Lopez MG, Villarroya M, Sanchez-Garcia P, De Pascual R, Dixon WR, Garcia AG (1994) Endothelium-independent relaxation by 17α-estradiol of pig coronary arteries. Eur J Pharmacol 258:47–55

Sarrel PM, Lufkin EG, Oursler MJ, Keefe D (1994) Estrogen actions in arteris, bone, and brain. Scientific American 44–53

Schaeffer JM, Stevens S, Smith RG, Hsueh AJW (1980) Binding of 2-Hydroxyestradiol to rat anterior pituitary cell membranes. J Biol Chem 255:9838–9843

Singer CA, Rogers KL, Strickland TM, Dorsa DM (1996) Estrogen protects primary cortical neurons from glutamate toxicity. Neurosci Lett 212:13–16

Sitzler G, Lenz O, Kilter H, La Rosee K, Bohm M (1996) Investigation of the negative inotropic effects of 17β-oestradiol in human isolated myocardial tissues. Bri J Pharmacol 119:43–48

Smith SS, Waterhouse BD, Woodward DJ (1987a) Sex steroid effects on extrahypo-thalamic CNS. I. Estrogen augments neuronal responsiveness to iontophoretically applied glutamate in the cerebellum. Brain Res 422:40–51

Smith SS, Waterhouse BD, Woodward DJ (1987b) Sex steroid effects on extrahypo-thalamic CNS. II. Progesterone, alone and in combination with estrogen, modulated cerebellar responses to amino acid neurotransmitters. Brain Research 422:52–62

Smith SS, Waterhouse BD, Woodward DJ (1988) Locally applied estrogens potentiate glutamate-evoked excitation of cerebellar Purkinje cells. Brain Res. 475:272–282

Smith SS (1994) Female sex steroid hormones: from receptors to networks to performance-actions on the sensorimotor system. Prog Neurobiol 44:55–86

Smith SS, Chapin JK (1996) The estrous cycle and the olivo-cerebellar circuit. I. Contrast enhancement of sensorimotor-correlated cerebellar discharge. Exp Brain Res 111:371–384

Sutherland RL, Watts CKW, Murphy LC (1982) Binding properties and ligand specificity of an intracellular binding site with specificity for synthetic estrogen antagonists of the triphenylethylene series. In:. Agarwal MK (ed) Hormone Antagonists, Walter de Gruyter & Co., New York, pp 147–161

Szego CM (1994) Cytostructural correlates of hormone action: new common ground in receptor-mediated signal propagation for steroid and peptide agonists. Endocrine 2:1079–1093

Szego CM, Davis JS (1967) Adenosine 3',5'-monophosphate in rat uterus: acute elevation by estrogen. Proc Natl Acad Sci 58:1711–1718

Szego CM, Pietras RJ (1984) Lysosomal functions in cellular activation: propagation of the actions of hormones and other effectors. Int Rev Cytology 88:1–302

Tauboll E, Lindstrom S, Gjerstad L (1994) Acute effects of 17β-estradiol on brain excitability studied in vitro and in vivo. Epilepsy Res 18:107–117

Tesarik J, Mendoza C (1995) Nongenomic effects of 17β-estradiol on maturing human oocytes: relationship to oocyte developmental potential. J Clin Endocrin Metabol 80:1438–1443

Teyler TJ, Vardaris RM, Lewis D, Rawitch AB (1980) Gonadal steroids: effects on excitability of hippocampal pyramidal cells. Science 209:1017–1019

Thompson TL, Moss RL (1994) Estrogen regulation of DA release in the nucleus accumbens: genomic- and nongenomic-mediated effects. J Neurochem 62:1750–1756

Towle AC, Sze PY (1983) Steroid binding to synaptic plasma membrane: differential binding of glucocorticoids and gonadal steroids. J Steroid Biochem 18:135–143

Troyan MB, Gilman VR, Gay CV (1997) Mitochondrial membrane potential changes in osteoblasts treated with parathyroid hormone and estradiol. Exp Cell Res 233:274–280

Tsibris JC, Hunt LT, Ballejo G, Barker WC, Toney LJ, Spellacy WN (1989) Selective inhibition of protein disulfide isomerase by estrogens. J Biol Chem 264:13967–13970

Vandewalle B, Hornez L, Lefebvre J (1988) Characterization of catecholestrogen membrane binding sites in estrogen receptor positive and negative human breast cancer cell-lines. J Receptor Res 8:699–712

Vargas R, Delaney M, Farhat MY, Wolfe R, Rego A, Ramwell PW Effect of estradiol 17β on pressor responses of rat mesenteric bed to norepinephrine, K$^+$, and U-46619. J Cardiovascul Pharmacol 25:200–206

Walker JE, Collinson IR (1994) The role of the stalk in the coupling mechanism of F1Fo-ATPases. FEBS Lett 346:39–43

Walker JE, Gay NJ, Powell SJ, Kostina M, Dyer MR (1987) ATP synthase from bovine mitochondria: sequences of imported precursors of oligomycin sensitivity conferral protein, factor 6, and adenosinetriphosphatase inhibitor protein. Biochemistry 26:8613–8619

Watters JJ, Campbell JS, Cunningham MJ, Krebs EG, Dorsa DM (1997) Rapid membrane effects of steroids in neuroblastoma cells: Effects of estrogen on mitogen activated protein kinase signalling cascade and c-fos immediate early gene transcription. Endocrinology 138:4030–4033

Weiland NG (1992) Estradiol selectively regulates agonist binding sites on the N-methyl-D-aspartate receptor complex in the CA1 region of the hippocampus. Endocrinology 131:662

Williams JP, Blair HC, McKennat MA, Jordan SE, McDonald JM (1996) Regulation of Avian Osteoclastic H$^+$ATPase and Bone Resorption by Tamoxifen and Calmodulin Antagonists. 271:12488–12495

Wong M, Moss RL (1991) Electrophysiological evidence for a rapid membrane action of the gonadal steroid, 17β-estradiol, on CA1 pyramidal neurons of the rat hippocampus. Brain Res 543:148–152

Wong M, Moss RL (1992) Long-term and short-term electrophysiological effects of estrogen on the synaptic properties of hippocampal CA1 neurons. J Neurosci 12:3217–3225

Wong M, Moss RL (1994) Patch-clamp analysis of direct steroidal modulation of glutamate receptor-channels. J Neuroendocrinol 6:347–355

Woolley CS, Gould E, Frankfurt M, McEwen BS (1990) Naturally occuring fluctuation in dendritic spine density on adult hippocampal pyramidal neurons. J Neurosci 10:4035–4039

Woolley CS, McEwen BS (1992) Estradiol mediates fluctuation in hippocampal synapse density during the estrous cycle in the adult rat. J Neurosci 12:2549–2554

Woolley CS, McEwen BS (1993) Roles of estradiol and progesterone in regulation of hippocampal dendritic spine density during the estrous cycle in the rat. J Comp Neurol 336:293–306

Zhang F, Ram JL, Standley PR, Sowers JR (1994) 17β-estradiol attenuates voltage-dependent Ca2+ currents in A7r5 vascular smooth muscle cell line. Am J Physiol 266:C975–C980

Zheng J (1996) Use of estradiol-BSA conjugates to characterize membrane estradiol binding sites in brain cells: biochemical and physiological studies. Ph. D. thesis, University of Illinois at Urbana-Champaign

Zheng J, Ali A, Ramirez VD (1996) The use of steroids conjugated to bovine serum albumin (BSA) as tools to demonstrate the existence of specific steroid neuronal membrane binding sites. J Psychiatry Neurosci 21:187–197

Zheng J, Ramirez VD (1997a) Demonstration of membrane estrogen binding proteins in rat brain by ligand blotting using 17β-estradiol-[^{125}I]bovine serum albumin conjugate. J Steroid Biochem Mol Biol 62:327–336

Zheng J, Ramirez VD (1997b) A brain cytosolic fraction contains specific binding sites for 17β-estradiol-[125I]BSA. Soc Neurosci Abstr 23:713

Zhou Y, Watters JJ, Dorsa DM (1996) Estrogen rapidly induces the phosphorylation of the cAMP response element binding protein in rat brain. Endocrinology 137:2163–2166

CHAPTER 10
Molecular Mechanisms of Antiestrogen Action

L.T. Seery, J.M.W. Gee, O.L. Dewhurst, and R.I. Nicholson

A. Introduction

Antiestrogens are established as compounds which predominantly exert their actions by competing with estrogen for binding to the target steroid receptor. This is evidenced by the observations that their biological effects are most notably recognised in tissues that contain ER; they often structurally resemble estrogens in regions which are important for the binding of the steroid nucleus to the ER and their ER binding, while of differing efficiency, always displaces and/or prevents the association of estrogens (Nicholson, 1993). Simplistically, as a consequence of such binding, antiestrogens subsequently reduce estrogen signalling within responsive cells. In practice, however, they display a bewildering diversity of biological properties, with tissue-specific actions that are not easily reconciled with such a basic model (Furr and Jordan 1984; Nicholson et al. 1986). Thus, while the non-steroidal triphenylethylene compound tamoxifen (the most widely prescribed antiestrogenic drug used in the therapy of breast cancer) promotes objective tumour remissions in approximately 30–50% of women (presumably an antiestrogenic response), it shows many estrogen-like characteristics on endometrium, bone and the cardiovascular system (Powles 1997). Indeed, long-term tamoxifen therapy, while delaying the recurrence of primary breast cancer and reducing the incidence of contralateral cancers (Early Breast Cancer Trialist's Collaborative Group 1992), promotes a significant increase in the development of endometrial cancers (presumably an estrogenic response; Cohan 1997), with possible additional detrimental effects on the liver (Wogan 1997).

Such complex mixed agonist/antagonist properties, now known to be displayed at a species, tissue, cell and gene level, are common to many antiestrogens, especially those which are modified versions of tamoxifen and its metabolites. Compounds in this "tamoxifen-like" category include the antiestrogens TAT-59 (Toko et al. 1990), idoxifene (Coombes et al. 1995) and droloxifene (Ke et al. 1995). Other non-steroidal compounds, however, possess substantially altered properties, with the benzothiophene raloxifene only showing minor agonistic activity on the uterus of ovariectomised rats in comparison with tamoxifen, while demonstrating excellent maintenance of bone density (Draper et al. 1995).

Interestingly, complete loss of all estrogen-like activity of antiestrogens has recently been achieved through the development of steroidal "pure" antiestrogens, including those based on 7α- (Wakeling et al. 1991) and 11β- (van de Velde et al. 1994) substitutions of estradiol, as well as the novel non-steroidal compound EM-800 (Tremblay et al. 1998). However, pure antiestrogens are not equally effective on all estrogen-sensitive tissues. Thus, while in ovariectomised rats the compounds ICI164384 and ICI182780 (now in clinical trials as Faslodex) will fully antagonise the cellular actions of estradiol on the utcrus, vagina and mammary gland, and promote extensive remissions of dimethylbenzanthracene (DMBA)-induced mammary tumours, they have only modest suppressive or no effects on bone density and cholesterol levels, respectively (Wakeling 1993; Dukes et al. 1994; Nicholson et al. 1988).

In light of the extent of responses achievable with antiestrogens, both in normal tissues and neoplasia, the current article aims to summarise the known molecular actions of these compounds, which culminate in their differential transcriptional regulation, not only of classically estrogen-responsive genes, but furthermore of many additional cellular end-points, the gross activities of which will contribute to the response. It will also suggest possible cell- and tissue-specific influences on these events, including those induced by the neoplastic phenotype.

B. Key Elements in Estrogen Receptor Signalling Important for Antiestrogen Action

Estrogen effects on cellular growth and proliferation are mediated primarily through the ER, a ligand-inducible transcription factor that impinges on genes containing estrogen response elements (EREs) in their promoter regions. To date, two genes encoding ERs have been characterised, designated αER and βER, respectively, with the βER protein being homologous to αER in the hormone- (60%) and DNA- (97%) binding domains (Kuiper and Gustafsson 1997). In addition to the wild-type receptor, αER splice variants arising from errors in ER transcription can be co-expressed as a minor species in normal and neoplastic cells (Fuqua and Wolf 1995), while ER point mutations and deletions are occasionally detected (Zhang et al. 1997).

Early investigations showed that the αER is a nuclear transcription factor that is activated by hyperphosphorylation on several serine and tyrosine residues upon binding of estrogens (Kuiper and Brinkmann 1994), most notably estradiol (Arnold et al. 1995). Recent investigations have also demonstrated a parallel binding of estradiol to the βER receptor (Tong et al. 1997). In the absence of hormone, αER resides in a large molecular complex comprising multiple heat-shock proteins (Segnitz and Gehring 1995). On estrogen binding, however, specific conformational changes are induced in the protein that result in the dissociation of the heat-shock proteins, promote receptor dimerisation, maintain the nuclear localisation of the receptor and favour its association with EREs within promoters of target genes. In addition to contacting the basal transcriptional machinery directly, ERs can enhance transcription by recruiting co-activators and/or by overcoming the effects of co-repressor proteins (McDonnell et al. 1992). These proteins appear to be present in limiting amounts in the cell and interpose between the receptor and the basal transcriptional machinery. It appears that co-activators stabilise the pre-initiation complex at the promoter and initiate gene expression, while co-repressors silence genes. To date, an increasing number of co-activators and co-repressors that can interact with ERs have been described, including the co-activators SRC and AIB1 (Smith et al. 1997; Anzick et al. 1997), and the co-repressors Ssn6 and SMRT (McDonnell et al. 1992; Smith et al. 1997). Of further interest is the cell-cycle regulatory protein, cyclin D1, which appears able to behave as an ER co-factor to upregulate ER-mediated transcription (Zwijsen et al. 1997).

Functional analysis of the αER has shown that it is a modular protein (Kumar 1987) with two independent activator functions AF-1 and AF-2 (Kraus et al. 1995). One of the effects of ligand binding is to juxtapose the AF-1 and AF-2 domains through conformational change, thereby generating a productive association (Kraus et al. 1995). In the αER, AF-1 is located at the amino-terminal end of the protein, while AF-2 is positioned towards the carboxy-terminal end, which also contains the well-conserved hormone-binding domain. In contrast, since the βER lacks significant homology with the amino-terminal end of the αER, it probably also lacks comparable AF-1 functions. The activity of the AF-2 domain of the αER is enabled by ligand binding. However, there is increasing evidence that the activity of AF-1 is constitutive, a feature likely to result from considerable ligand-independent influences that also have some bearing on AF-activation. In this light, αER phosphorylation can be mediated by several polypeptide growth factors, such as insulin-like growth factor-1 (IGF-1) (Aronica and Katzenellenbogen 1993), epidermal growth factor (EGF), transforming growth factor alpha (TGFα) (Bunone et al. 1996) and heregulin (Pietras et al. 1995); by dopamine (Mani et al. 1994) via protein kinase C (PKC), mitogen-activated protein (MAP) kinase (Bunone et al. 1996; Kato et al. 1995; Arnold et al. 1995) which is a component that also regulates βER AF-2 (Tremblay et al. 1997), c-src (Arnold et al. 1997, 1995) and protein kinase A (PKA) pathway elements (Le Goff et al. 1994; Aronica and Katzenellenbogen 1993). Additionally, DNA-dependent protein kinase, casein kinase II and cyclinA/cdk2 may also phosphorylate αER (Arnold SF et al. 1994, 1995; Kuiper and Brinkmann 1994; Tzeng and Klinge 1996; Trowbridge et al. 1997). Significantly, these elements appear to induce differential activation of AF-1 and AF-2, with the former being more responsive to EGF and TGFα signalling (Bunone et al. 1996), while IGF-1 and dopamine preferentially activate the latter (Gangolli et al. 1997). While activation by these components occurs most efficiently in the presence of estrogens, their promotion of AF-1 and AF-2 responses is certainly adequate in the absence of hormone.

Interestingly, there is some evidence that, in addition to its transcriptional activation of ERE-containing genes, the αER appears able to activate genes containing AP-1 sites in their promoters (WEBB et al. 1995), an event that is effected by protein–protein interactions (ROCHEFORT 1995) and is enabled by both estrogens (PAECH et al. 1997) and the AP-1 complex components *Fos/Jun* (WEBB et al. 1995). In contrast, the βER is not able to promote AP-1 transcriptional activation in the presence of estrogens (PAECH et al. 1997). Since the AP-1 pathway is normally considered to be growth-factor regulated, its positive co-regulation by the αER provides an alternative mechanism through which αER signalling is diversified. The αER also interacts with the neurotrophic factor (NF)-κB transcription factor, an event that negatively regulates expression of many cell-adhesion molecules, cytokines (such as interleukin-6) and growth factors (SHARMA and NARAYANAN 1996). αER-dependent inhibition of interleukin-6, appears to be mediated via a direct protein–protein interaction with NF-κB (RAY et al. 1997).

C. Molecular Actions of Antiestrogens

The plethora of cellular end-points arising from estrogen-activated signalling pathways are matched by a comparable diversity of effects characteristic of the different classes of antiestrogens, a reflection of their relative agonistic or antagonistic properties. Thus, while they may inhibit expression of many genes and subsequently initiate cell-cycle arrest and suppression of cell proliferation and induce apoptosis, the converse may also be true depending on the cellular and tissue context.

Some of the early molecular effects of antiestrogens on the αER are equivalent to those induced by estrogens, including binding to the hormone binding site of the receptor and dissociation of heat-shock proteins (JORDAN et al. 1997), although some differences in subsequent receptor dimerisation and localisation do exist. Similarly, like estrogens, antiestrogens appear to be able to induce phosphorylation of ERs, albeit inefficiently (ALI et al. 1993; ARONICA and KATZENELLENBOGEN 1993). In contrast, however, marked variations in the resultant transcriptional productivity of antiestrogens are apparent both between the different compounds and versus estradiol. These differences reach an extreme with "pure" antiestrogens, which notably fail to demonstrate inductive effects on gene transcription, a phenomenon probably largely determined by extreme receptor conformational changes and a markedly increased receptor turnover. The following range of molecular actions exhibited by the various classes of antiestrogens are believed to be contributory components to their differing antiestrogenic and estrogen-like effects on transactivation.

I. Effects of Antiestrogens on Binding to the Receptor, Dimerisation and Nuclear Localisation

The classical receptor hormone-binding domain appears to be the primary binding site for both estrogens and antiestrogens. However, there is some evidence suggesting the existence of antagonist-specific (secondary) binding sites within the ER (KUIPER and GUSTAFSSON 1997; HEDDEN et al. 1995). Different classes of antiestrogens have different relative affinities for both primary and secondary sites. It is postulated that if antiestrogens have increased binding for the primary site, they will exhibit increased agonistic activity, while their preferential binding to the secondary site may prevent this event (JENSEN 1996). Such a hypothesis would perhaps explain the bell-shaped dose–response curve observed in vitro on exposure of breast cancer cells to tamoxifen (REDDELL and SUTHERLAND 1984).

Although tamoxifen-like antiestrogens, similar to estrogens, cause receptor dimerisation (METZGER et al. 1995), several studies indicate that some pure antiestrogens may inhibit this event (FAWELL et al. 1990) as a consequence of steric hindrance (BOWLER et al. 1989). Such an effect would subsequently inhibit or reduce DNA binding of the ER (WEATHERILL et al. 1988; WILSON et al. 1990; FAWELL et al. 1990), although this too remains controversial (BERRY et al. 1990; MARTINEZ and WAHLI 1989; WRENN and KATZENELLENBOGEN 1990).

Additionally, in contrast to tamoxifen-like antiestrogens, pure antiestrogens appear to alter the efficiency of receptor–nuclear localisation. In studies of ER distribution in the rat uterus, ER-182,780 complexes were found predominantly in cytosolic and microsomal fractions, suggesting that intracellular trafficking of the receptor is impaired (MENDES et al. 1996).

II. Effects of Antiestrogens on Estrogen Receptor (ER)– Estrogen Response Element (ERE) Binding and Subsequent Transcriptional Activation

1. Altered ER–ERE Binding Efficiency

Recent studies have demonstrated an excellent correlation between the kinetics of ER–ERE interaction induced by estrogens and antiestrogens and their biological effects (CHESKIS et al. 1997). In contrast to the binding of estradiol, which induces the rapid formation of a relatively unstable ER–ERE complex, the binding of antiestrogens leads to the slower for mation of a more stable receptor–DNA complex, a feature most marked with pure antiestrogens. Such events which, to date, have only been observed in a cell-free environment could be hypothesised to influence the frequency of ER–DNA complex formation, hence the level of transcriptional activation.

2. Changes in ER Conformation Influence AF-2 Activity

Considerable evidence exists from the use of monoclonal antibodies to specific antigenic determinants on the αER (HEDDEN et al. 1995; BERTHOIS et al. 1994) that antiestrogens induce different conformations in the ER to those induced by estrogens, by virtue of their bulky, charged side chains (BEEKMAN ET AL. 1993). Such altered receptor behaviour has recently been elegantly extended through a knowledge of the crystal structures of the ligand binding domain of the αER in complex with estradiol and raloxifene (BRZOZOWSKI et al. 1997). While each binds to the same site within the core of the ligand-binding domain, they demonstrate different binding modes and induce distinct conformational changes in the AF-2 transactivation domain. Thus, while estradiol appears to seal the ligand-binding cavity of the ER and generate a competent AF-2 capable of interacting with co-activators (a pre-requisite for efficient transcriptional activation), raloxifene is unable to promote the initial effect and fails to induce a transcriptionally-competent AF-2. Although similar studies with a broader spectrum of antiestrogenic drugs are now necessary to evaluate the generality of the above observations, the data are certainly consistent with earlier investigations, in which activity of an αER AF-2 domain/GAL4 fusion protein could be induced by estradiol, but not by tamoxifen (PHAM et al. 1992). Similar to tamoxifen, pure antiestrogens also appear to effectively block AF-2 activity (McDONNELL et al. 1995).

3. Effects Enabled by AF-1 and the Phenomenon of AF-1/AF-2 Promoter Dependency

Although, as indicated above, tamoxifen-like antiestrogens appear to block AF-2 efficiently, they do not always prevent AF-1 activity and, therefore, function as partial agonists of the ER (TORA et al. 1989; BERRY et al. 1990; McINERNEY and KATZENELLENBOGEN 1996). However, it is notable that different regions of AF-1 appear to be associated with such antioestrogen activation of transcription than with estrogens (McINERNEY and KATZENELLENBOGEN 1996). Since individual gene promoters show a spectrum of dependency on AF-1 and AF-2 (TZUKERMAN et al. 1994), tamoxifen-like antiestrogens appear agonistic whenever they promote binding of the ER to a target ERE from which transcription is normally enabled by constitutive activity of AF-1 (TZUKERMAN et al. 1994). Indeed, a tight correlation exists between the activity of AF-1 in various promoter contexts and the estrogenic effect of 4-hydroxy-tamoxifen (4-OH-T) (TZUKERMAN et al. 1994). In contrast, tamoxifen-like antiestrogens appear to act as pure antagonists whenever the activation of a given promoter is fully dependent on AF-2.

Biologically, AF-1 constitutive activity may be sufficient to explain some of the apparent estrogenicity observed with antiestrogens; however, experiments measuring ER activity in the presence of estradiol, nafoxidine and clomiphene, in HepG2 cells (where AF-2 is not required), showed that the antiestrogens demonstrated only 30% of the effect promoted by estradiol

(McDonnell et al. 1995). This implies the existence of additional modifying factors; perhaps components of the transcriptional machinery, co-activators or co-repressors recognise the estrogen-induced conformation better than that induced by the antiestrogens. Further complexity arises from mutational studies which have demonstrated that additional cellular proteins can substitute for AF-1, or AF-2 function, another feature that may play a role in the phenomenon of antiestrogen agonism (McDonnell et al. 1995).

Studies examining the effects of pure antiestrogens on transactivation seem to indicate that they will permit AF-1-mediated activity to occur in cell-free systems (Berry et al. 1990), However, this event is unlikely in vitro or in vivo, since these compounds are believed to induce a severely perturbed receptor conformation and enhance rapid receptor degradation (Berry et al. 1990; Dauvois et al. 1992; Pink and Jordan 1996).

4. Effects Enabled by the Cellular Levels of Co-Activators/Co-Repressors

Since the efficiency of ER–ERE interactions is likely to be influenced by the presence of co-regulators, the effectiveness of antiestrogens which retain partial agonist activity may be controlled by the ratio of co-activators to co-repressors recruited to the transcription complex by promoter-bound receptors (Jackson et al. 1997; Smith et al. 1997). This is particularly evident in the case of 4-OH-T, whose agonistic activity in HepG2 cells is enhanced by expression of the SRC-1 co-activator, but inhibited by overexpression of the SMRT co-repressor (Smith et al. 1997). A similar pattern was obtained in HeLa cells, where the compound is normally antagonistic to ER-mediated transcription (Smith et al. 1997). Interestingly, deletion of the yeast ER co-repressor Ssn6 confers agonistic activity on ICI 164384 and nafoxidine (McDonnell et al. 1995), suggesting that pure antiestrogenic activity is mediated, at least in part, by interaction with co-repressor proteins (Horwitz et al. 1996). It is believed that since the carboxy-terminal domain (region F) of the αER is involved in the interaction with protein co-factors or transcription factors, this region may have specific modulatory function that affects the agonist/antagonist effectiveness of antiestrogens and the transcriptional activity of the liganded ER in cells (Montano et al. 1995).

5. Effects on Ligand-Independent ERE Transactivation

Growth factors, such as IGF-1, EGF and TGFα, or elements of their signalling pathways have been shown to stimulate the expression of several ERE-containing genes, a feature presumably involving their phosphorylation of the αER (Sect. B). Indeed, these growth factors induce additive or synergistic responses when used in combination with estrogens (Cho et al. 1993; Smith et al. 1993). Although, generally, these actions can be reduced or blocked by antiestrogens (Hafner et al. 1996; Newton CJ et al. 1994; Aronica and Katzenellenbogen 1993; Cho et al. 1993, Smith et al. 1993), in some instances, the effects on particular ERE-containing genes are highly dependent on the class of antiestrogen used. For example, while 4-OH-T is, at best, only moderately effective at reducing the effects of IGF-1, 12-O-tetradecanoylphorbol-13-acetate (TPA) or cyclic adenosine monophosphate (cAMP) on cathepsin-D gene expression, the pure antiestrogen ICI164384 is fully inhibitory (CHALBOS et al. 1993). Conversely, there are instances where the agonistic activity of antioestrogens can be amplified by the presence of excess growth factor signalling components, as exemplified by overexpression of Ras/MAP kinase pathway components (KATO et al. 1995; BUNONE et al. 1996) and also stimulation of PKA (FUJIMOTO and KATZENELLENBOGEN 1994), both increasing tamoxifen agonism. In this light, the recently-developed pure antiestrogenic compound EM-800 may potentially prevent Ras-activated ER transactivation (TREMBLAY et al. 1997).

6. Effects Enabled by Promoter Elements and ERE Sequence

Deletion and mutational analyses have identified a specific *cis* element within the progesterone-receptor distal promoter, which appears to modulate its sensitivity to the inhibitory effects of antiestrogens (MONTANO et al. 1997). Furthermore, modification of the ERE by sequence, variation, number or orientation affects the efficiency with which estrogen or antiestrogen-responsive genes are transcribed (PONGLIKITMONGKOL et al. 1990). Studies in the laboratory of McDONNELL (DANA et al. 1994) have indicated that such changes may permit ER agonism by antiestrogens.

7. Effects Enabled by ER Sub-Type

In contrast to its partial agonistic effects on αER-mediated transactivation, 4-OH-T fails to activate a basal promoter linked to EREs in *Cos*-1 cells in the presence of βERs. This event is believed to be the result of its lack of αER-like AF-1 function (TREMBLAY et al. 1997).

III. Effects on Ap-1- and NF-κB-Mediated Transactivation

The data in this area are sparse, highly controversial and certainly tissue-specific. Initial studies suggested that estrogens increased while antiestrogens antagonised growth factor-induced AP-1 activity, with maximal inhibition by pure antiestrogens (PHILLIPS et al. 1993). However, subsequent data in uterine cells suggested that the tamoxifen–ER complex can act agonistically on promoters regulated by the AP-1 site. The ER DNA-binding domain appears to be required for tamoxifen-dependent AP-1 activation, in contrast to activation by estrogen, which is partially independent of this domain (WEBB et al. 1995). More recent experimental data propose even further complexities: the receptor sub-types αER and βER appear to invoke different AP-1 responses. Thus, while both estrogens and the antiestrogens tamoxifen, raloxifene and ICI164384 bound to the αER do appear to be potent activators of the AP-1 site, only the antiestrogens can activate the βER at this site (PAECH et al. 1997).

With regards to the limited data examining the ER and NF-κB cross-talk, the tamoxifen–ER complex fails to inhibit the expression of genes containing the NF-κB-enhancer element (KUREBAYASHI et al. 1997). In contrast, with the estrogen–ER complex, the effects of pure antiestrogens remain unknown.

IV. Effects on Antiestrogen-Specific Response Elements

There is some evidence that antiestrogens can modulate transcriptional activation from novel (non-ERE) DNA response elements, subsequent to the antiestrogen binding to the ER. Thus, the complex regulation of the human TGFβ_3 gene in bone by raloxifene was shown to involve activation by a poly-purine sequence, termed the raloxifene-response element (RRE), which did not require the DNA-binding domain of the ER. Interaction of the ER with the RRE appears to require a cellular adapter protein (YANG et al. 1996).

V. Effects on ER Degradation

Several reports have established that the treatment of ER-containing cells with the pure antiestrogens ICI164384 or ICI182780 (DEFRIEND et al. 1994; NICHOLSON et al. 1994) and the compound RU58668 (JIN et al. 1995) dramatically increases ER turnover by enhancing the rate of its degradation. This appears to occur in the face of unaltered ER messenger RNA (mRNA) levels (MCCLELLAND et al. 1996; PINK and JORDAN 1996). Thus, the half-life of the receptor protein is reduced from 5h in the presence of estradiol, to less than 1h in the presence of the pure antiestrogens, with a resultant 90–95% decrease in ER levels with such compounds (DAUVOIS et al. 1992; PINK and JORDAN 1996). In contrast, non-steroidal antiestrogens do not appear to have an effect on ER turnover (ECKERT et al. 1984); indeed, both tamoxifen and nafoxidene cause an increase in ER expression (KIANG et al. 1989). It is, thus, likely that the effective removal of ER protein observed with pure antiestrogens is not only central to their lack of transactivation of ERE-containing genes, but furthermore may be highly efficient in their severing ER cross-talk with growth-factor-mitogenic signalling pathways.

VI. Non-ER Actions of Antiestrogens

Although the molecular basis of the antiestrogenic action is primarily attributed to interaction with the ER, tamoxifen has reported direct inhibitory effects on other cellular components, which may influence their biological profile in both ER-positive and -negative cells. As yet, comparable studies have not been performed for other antiestrogens; thus, the generality of their significance can not be assessed.

There are several possible mechanisms by which non-ER mediated antag-onistic and agonistic activities of tamoxifen could occur, depending on the con-centration used, duration of exposure and target cell/tissue type. For example, tamoxifen inhibits calmodulin activation of cAMP-dependent phosphodi-esterase, a key element in the regulation of the cellular levels of cAMP (LAM 1984); it elevates levels of the immunosuppressive cytokine TGFβ1, a mole-cule which, although often inhibitory to the growth of epithelial cells, is pro-tective against bone loss (NOGUCHI et al. 1993; BUTTA et al. 1992). Furthermore, tamoxifen binds to a type-II antiestrogen binding site, potentially elevating the cellular concentrations of the drug and its metabolites (PIANTELLI et al. 1995; SUTHERLAND et al. 1980). Of further interest is its general inhibition of PKC activity (GUNDIMEDA et al. 1996; O'BRAIN et al. 1985), although the PKCe isoform appears to be activated by the antiestrogen (CABOT et al. 1997). Effects on signalling molecules, such as cAMP and PKC, could certainly be envisaged as having wider implications within ER-positive cells, given their potential involvement in receptor phosphorylation.

D. Modifying Effects of the Normal Cellular Phenotype on Antiestrogen Response

Although, in general terms, each somatic cell contains an equivalent number of genes, these are programmed to be differentially expressed, providing a unique phenotypic profile which ascribes highly specialised functions to every cell type. Simplistically, the observed tissue phenotype and functionality is a gross reflection of the concerted actions of its constituent cell types.

ERs are an important component of the cellular phenotype and their expression conferring estrogen sensitivity on many tissues, notably the breast, uterus, bone, liver, brain and cardiovascular system. As outlined above (Sect. C), an increasingly complex picture of the molecular control of estrogen and antiestrogen-occupied ERs and the target gene output is emerging that appears to rely on multiple regulatory parameters. It is likely that this regula-tion is markedly influenced by both the level and activity of many components within the cellular and tissue phenotypes. Additionally, it is envisaged that the biological responses to an antiestrogen further depend on whether the tissue has some reliance on the target gene expression and its resultant networked responses. It is, thus, perhaps not surprising that the antihormonal responses observed are highly tissue-type specific.

Although largely undocumented, potentially important phenotypic influences on ERE-mediated signalling and hence the antiestrogen response (agonism/antagonism) in normal tissues are likely to relate to:

a) The cellular levels of antiestrogen achieved: the cellular uptake or loss of drugs is generally considered to be an important influence on their biological properties, especially those such as antiestrogens which display bell-shaped dose–response curves. In the case of tamoxifen, it has been postulated that agonistic activity is more evident at low cellular levels of the drug, while antagonism is seen at higher concentrations. Variations in the capacity of tissues to accumulate or retain antiestrogens (through the presence of antiestrogen binding sites or metabolising enzymes), therefore, might contribute to their agonistic/antagonistic properties.

b) The availability of different receptor sub-types: for example, the βER was isolated from the prostate gland, where the αER is notably absent, a feature that may determine a different response than with the breast or uterus, where the αsub-type is abundant.

c) The levels and activity of components comprising pathways that participate in ligand-independent phosphorylation of ER: although the details of those components of signal transduction pathways which impinge on ER signalling are largely uncharted between various tissues, their differential activation in an individual tissue could dramatically influence the agonist/antagonist profile of antiestrogens.

d) Availability of co-activators, co-regulators and other transactivation modulators: although the cellular levels of co-activators and co-repressors are largely unknown in most tissue types, variations in their availability to the ER complex has been clearly shown to influence its transcriptional activity.

e) Availability of ER degradation pathways: in light of the apparent importance of receptor degradation to the biological properties of pure antiestrogens, the tissue-selective action of such compounds might be mediated, in part, by variation in the cellular level of the proteins responsible for ER breakdown. Failure to efficiently achieve this process could result in prolonged AF-1-dependent activity.

Additionally, the dependency of the tissue response on genes containing other response elements with which the ER interacts should also be considered, i.e., different ERE sequences, the response elements for AP-1 and NF-κB, and anti-EREs, such as the RRE, should be included with the availability of components of non-ER mechanisms.

Therefore, current important issues include how individual tissues variously interpret cellular signals arising from antiestrogenic drugs and how we can select those properties of antiestrogens that are perceived to be favourable to a clinical setting, while minimising those regarded as detrimental to patient welfare. In this light, selective ER modulators, such as raloxifene, may prove to be of particular value since they comprise a group of structurally diverse compounds that act either as estrogen agonists or antagonists, depending on the target tissue and hormonal milieu, despite their universal interaction with the ER (PALKOWITZ et al. 1997).

E. Modifying Effects of the Cancer Cell Phenotype on Antiestrogen Response

Prior to therapy, it is postulated that at least two classes of tumours utilising ER signalling for their growth exist: (1) those strictly endocrine-dependent tumours that primarily derive their mitogenic stimulus from the influences of the estrogen-activated receptor; and (2) those that have a phenotype that enables a limited or no requirement for estrogens, the receptor being predominantly activated by ligand-independent mechanisms (GEE et al. 1996; NICHOLSON et al. 1996). While the former group of tumours would be predicted to be sensitive to the inhibitory effects of antiestrogens, the second group would, at best, show only a partial response to such therapy and may even be clinically-unresponsive to antiestrogen therapy. Furthermore, tumours with an initial estrogen dependency ultimately undergo adaptive processes triggered by the alterations in the steroid hormone milieu occurring during antihormonal therapy. Given the likelihood of a continued central importance of ER signalling in such tumours (ROBERTSON 1996; NICHOLSON et al. 1995), such events impinging on ER signalling would ultimately enable the circumvention of strict estrogenic control, allowing tumour re-growth to mark the development of acquired antiestrogen resistance. A maintained role for the receptor in resistance is perhaps evidenced by our recent studies demonstrating that a number of tamoxifen-inhibited genes are re-expressed in antiestrogen-resistant tumours (WILLSHER et al. 1998), and also explains why such patients often respond favourably when challenged to alternative forms of antihormone therapy.

While phenotypic changes relating to antiestrogen uptake (JOHNSTON et al. 1993), ER sub-type and variants, ER degradation (MAEDA et al. 1984) and co-activators/co-repressors (SMITH et al. 1997; ANZICK et al. 1997) have all been implicated, their role in modulating antiestrogen response and resistance in tumours remains questionable. Importantly, however, many of the above cellular changes appear to target the expression and/or activity of pathways that initiate ligand-independent ER phosphorylation. For example, our own clinical studies and those of others focusing on the role of the MAP kinase pathway in breast cancer have shown that increased activation arises through alterations in the levels of growth-factor ligands (i.e. TGFα, NICHOLSON et al. 1994b; heregulins, TANG et al. 1996), receptors (EGFR, NICHOLSON et al. 1993; c-erbB2, NICHOLSON et al. 1994c; BORG et al. 1994) and intracellular signalling molecules (e.g. Ras, BLAND et al. 1995; DATI et al. 1991; ARCHER et al. 1995; Raf, EL-ASHRY et al. 1997; PKC, GORDGE et al. 1996; MAP kinase, SIVARAMAN et al. 1997; GEE et al. unpublished observations). It would be envisaged that such events would have dramatic influences on ERE-mediated responses. Additionally, a spectrum of other important mitogenic pathways would also be influenced, since the MAP kinase pathway impinges on many nuclear transcription factors (e.g. Fos AP-1, GEE et al. 1995; AP-2, HURST et al. unpublished observations; Myc, BLAND et al. 1995) which may subsequently have some bearing on both ER signalling and their own specific end-points.

In addition to components of the MAP kinase pathways being elevated in breast cancer, cAMP and PKA levels are also significantly higher than the normal breast (BARTLETT et al. 1996; GORDGE et al. 1996). Moreover, elevated concentrations of cAMP binding proteins are associated with early disease recurrence and poor survival rates. Similarly, elevated expression of pp60c-src (LEHRER et al. 1989; VERBEEK et al. 1996) has been observed in breast cancer and is often associated with a poor prognosis. Since each of these factors can impinge on ER signalling, in addition to the cellular activity initiated by their own specific signalling pathways, it is tempting to conclude that non-ligand activation of the ER is a pivotal element in cancer growth control, a factor that might equally influence antihormone response.

F. Conclusions

It is becoming increasingly apparent that while the molecular actions of antiestrogens, in many ways, mirror the effects of the natural ER ligand, their specific transcriptional regulation of estrogen-responsive genes is markedly differential and far from simplistic. This perhaps reaches an extreme with the recently-developed "pure" antiestrogens, which uniquely lack any resultant inductive effects on gene transcription. The fact that an increasingly complex, multi-tiered picture of ER activation exists, together with the identification of additional target gene families distinct from those classically containing EREs in their promoter regions, is no doubt pivotal to the diversity of molecular effects that can influence gross antiestrogen response. Additionally, it is likely that many of the key elements of ER activation vary according to the phenotype of each estrogen target cell type in normal tissues and, furthermore, can be subverted during the processes of neoplasia as well as on the inevitable development of endocrine resistance.

In summary, it is clear that no single mechanism at the gene, cell or tissue level fully determines the phenomena of antiestrogen antagonism and agonism demonstrated in the clinic. Thankfully, however, the very existence of multiple control points in this process perhaps confers an inherent vulnerability to future pharmacological control.

References

Ali S, Metzger D, Bornert JM, Chambon P (1993) Modulation of transcriptional activation by ligand-dependent phosphorylation of the human estrogen receptor A/B region. EMBO J 12(3):1153–1160

Anzick SL, Kononen J, Walker RL, Azorsa DO, Tanner MM, Guan XY, Sauter G, Kallioniemi OP, Trent JM, Meltzer PS (1997) AIB1, a steroid receptor coactivator amplified in breast and ovarian cancer. Science 277 (5328):965–968

Archer SG, Eliopoulos A, Spandidos D, Barnes D, Ellis IO, Blamey RW, Nicholson RI, Robertson JF (1995) Expression of ras p21, p53 and c-erbB-2 in advanced breast cancer and response to first line hormonal therapy. Br J Cancer 72:1259–1266

Arnold SF, Obourn JD, Jaffe H, Notides AC (1994) Serine 167 is the major estradiol-induced phosphorylation site on the human estrogen receptor. Mol Endocrinol 8(9):1208–1214

Arnold SF, Obourn JD, Jaffe H, Notides AC (1995a) Phosphorylation of the human estrogen receptor ontyrosine 537 in vivo and by src family tyrosine kinases in vitro. Mol Endocrinol 9(1):24–33

Arnold SF, Obourn JD, Jaffe H, Notides AC (1995b) Phosphorylation of the human estrogen receptor bymitogen-activated protein kinase and casein kinase II: consequence on DNA binding. J Steroid Biochem Mol Biol 55(2):163–172

Arnold SF, Melamed M, Vorojeikina DP, Notides AC, Sasson S (1997) Estradiol-binding mechanism and binding capacity of the human estrogen receptor is regulated by tyrosinephosphorylation. Mol Endocrinol 11(1):48–53

Aronica SM, Katzenellenbogen BS (1993) Stimulation of estrogen receptor-mediated transcription and alteration in the phosphorylation state of the ratuterine estrogen receptor by estrogen, cyclic adenosinemonophosphate, and insulin-like growth factor-I. Mol Endocrinol 7(6):743–752

Bartlett JM, Hulme MJ, Miller WR (1996) Analysis of cAMP RI alpha mRNA expression in breast cancer: evaluation of quantitative polymerase chain reaction for routine use. Br J Cancer 73:538–1544

Beekman JM, Allan GF, Tsai SY, Tsai MJ, O'Malley BW (1993) Transcriptional activation by the estrogen receptor requires a conformational change in the ligand binding domain. Mol Endocrinol 7(10):1266–1274

Berry M, Metzger D, Chambon P (1990) Role of the two activating domains of the estrogen receptor in the cell type and promoter-context dependent agonistic activity of the anti-estrogen 4-hydroxytamoxifen. EMBO J 9(9):2811–2818

Berthois Y, Pons M, Dussert C, Crastes de Paulet A, Martin PM (1994) Agonist-antagonist activity of anti-estrogens in the human breast cancer cell line MCF-7: an hypothesis for the interaction with a site distinct from the estrogen binding site. Mol Cell Endocrinol 99(2):259–268

Bland KI, Konstadoulakis MM, Vezeridis MP, Wanebo HJ (1994) Oncogene protein co-expression. Value of Ha-ras, c-myc, c-fos, and p53 as prognostic discriminants for breast carcinoma. Ann Surg 221:706–718

Borg A, Baldetorp B, Ferno M, Killander D, Olsson H, Ryden S, Sigurdsson H (1994) ERBB2 amplification is associated with tamoxifen resistance in steroid-receptor positive breast cancer. Cancer Lett 81:137–144

Bowler J, Lilley TJ, Pittam JD, Wakeling AE (1989) Novel steroidal pure antiestrogens. Steroids 54(1):71–99

Brzozowski AM, Pike AC, Dauter Z, Hubbard RE, Bonn T, Engstrom O, Ohman L, Greene GL, Gustafsson JA, Carlquist M (1997) Molecular basis of agonism and antagonism in the estrogen receptor. Nature 389(6652):753–758

Bunone G, Briand PA, Miksicek RJ, Picard D (1996) Activation of the unliganded estrogen receptor by EGF involves the MAP kinase pathway and direct phosphorylation. EMBO J May 1;15(9):2174–2183

Butta A, MacLennan K, Flanders KC, Sacks NPM, Smith I, McKinna A, Dowsett M, Wakefield LM, Sporon MB, Baum M and Colletta AA (1992) Induction of transforming growth fator βin human breast cancer in vivo following tamoxifen treatment. Cancer Res 52:4261–4264

Cabot MC, Zhang Z-H. Cao H-T, Lavie Y, Giulliano, AE, Han T-Y and Jones RC (1997) Tamoxifen activates cellular phospholipase C and D and elicits protein kinase C translocation. Int J Cancer 7:567–574

Chalbos D, Philips A, Galtier F, Rochefort H (1993) Synthetic antiestrogens modulate induction of pS2 and cathepsin-D messenger ribonucleic acid by growth factors and adenosine 3′,5′-monophosphate in MCF7 cells. Endocrinology 133(2): 571–576

Cheskis BJ, Karathanasis S, Lyttle CR (1997) Estrogen receptor ligands modulate its interaction with DNA. J Biol Chem 272(17):11384–11391

Cho H, Katzenellenbogen BS (1993) Synergistic activation of estrogen receptor-mediated transcription by estradiol and protein kinase activators. Mol Endocrinol 7(3):441–45

Cohan CJ (1997) Tamoxifen and endometrial cancer: Tamoxifen effects on the human female genital tract cancer. Seminars in Oncol 24:S1 55-S1 64

Coombes RC, Haynes BP, Dowsett M, Quigley M, English J, Judson IR, Griggs LJ, Potter GA, McCague R, Jarman M (1995) Idoxifene: report of a phase I study in patients with metastatic breast cancer. Cancer Res 55:1070–1074

Dana SL, Hoener PA, Wheeler DA, Lawrence CB, McDonnell DP (1994) Novel estrogen response elements identified by genetic selection in yeast are differentially responsive to estrogens and antiestrogens in mammalian cells. Mol Endocrinol 8(9):1193–1207

Dati C, Muraca R, Tazartes O, Antoniotti S, Perroteau I, Giai M, Cortese P, Sismondi P, Saglio G, De Bortoli M (1991) c-erbB-2 and ras expression levels in breast cancer are correlated and show a co-operative association with unfavorable clinical outcome

Dauvois S, Danielian PS, White R, Parker MG (1992) Antiestrogen ICI 164,384 reduces cellular estrogen receptor content by increasing its turnover. Proc Natl Acad Sci USA. 89:4037–4041

DeFriend DJ, Howell A, Nicholson RI, Anderson E, Dowsett M, Mansel RE, Blamey RW, Bundred NJ, Robertson JF, Saunders C et al. (1994) Investigation of a new pure antiestrogen (ICI 182780) in women with primary breast cancer. Cancer Res 54(2):408–414

Desai AJ, Luqmani YA, Walters JE, Coope RC, Dagg B, Gomm JJ, Pace PE, Rees CN, Thirunavukkarasu V, Shousha S, Groome NP, Coombes R, Ali S (1997) Presence of exon 5-deleted estrogen receptor in human breast cancer: functional analysis and clinical significance. Br J Cancer 75(8):1173–1184

Draper MW, Flowers DE, Neild JA, Huster WJ, Zerbe RL (1995) Antiestrogenic properties of raloxifene. Pharmacology 50:209–217

Dukes M, Chester R, Yarwood L, Wakeling AE (1994) Effects of a non-steroidal pure antiestrogen, ZM 189,154, on estrogen target organs of the rat including bones. J Endocrinol 141:335–341

Eckert RL, Mullick A, Rorke EA, Katzenellenbogen BS (1984) Estrogen receptor synthesis and turnover in MCF-7 breast cancer cells measured by a density shift technique. Endocrinology 114:629–637

El-Ashry D, Miller DL, Kharbanda S, Lippman ME, Kern FG Constitutive (1997) Raf-1 kinase activity in breast cancer cells induces both estrogen-independent growth and apoptosis. Oncogene 15:423–435

Fawell SE, White R, Hoare S, Sydenham M, Page M, Parker MG (1990) Inhibition of estrogen receptor-DNA binding by the "pure" antiestrogen ICI 164,384 appears to be mediated by impaired receptor dimerization. Proc Natl Acad Sci USA 87(17):6883–6887

Fujimoto N, Katzenellenbogen BS (1994) Alteration in the agonist/antagonist balance of antiestrogens by activation of protein kinase A signaling pathways in breast cancer cells: antiestrogen selectivity and promoter dependence. Mol Endocrinol 8(3):296–304

Fuqua SA, Wolf DM (1995) Molecular aspects of estrogen receptor variants in breast cancer. Breast Cancer Res Treat 35:233–241

Furr BJ, Jordan VC (1984) The pharmacology and clinical uses of tamoxifen. Pharmacol Ther 25(2):127–205

Gangolli EA, Conneely OM, O'Malley BW (1997) Neurotransmitters activate the human estrogen receptor in a neuroblastoma cell line. J Steroid Biochem Mol Biol 61(1–2):1–9

Gee JMW, Ellis IO, Robertson JF, Willsher P, McClelland RA, Hewitt KN, Blamey RW, Nicholson RI (1995) Immunocytochemical localization of Fos protein in human breast cancers and its relationship to a series of prognostic markers and response to endocrine therapy. Int J Cancer 64:269–273

Gee JMW, McClelland RA, Nicholson RI (1996) Growth factors and endocrine sensitivity in breast cancer. In: Pasqualini JR, Katzenellenbogen BS (eds) Molecular and Clinical Endocrinology. Marcel Dekker Publishing, pp 169–197

Gordge PC, Hulme MJ, Clegg RA, Miller WR (1996) Elevation of protein kinase A and protein kinase C activities in malignant as compared with normal human breast tissue. Eur J Cancer. 32 A: 2120–2126

Gundimeda U, Chen Z-H, Gopalakrishna R (1996) Tamoxifen modulates protein kinase C via oxidative stress in estrogen receptor-negative breast cancer cells. J Biol Chem 271(23):13504–13514

Hafner F, Holler E, von Angerer E (1996) Effect of growth factors on estrogen receptor mediated gene expression. J Steroid Biochem Mol Biol 58(4):385–393

Hedden A, Muller V, Jensen EV (1995) A new interpretation of antiestrogen action. Ann N Y Acad Sci 761:109–120

Horwitz KB, Jackson TA, Bain DL, Richer JK, Takimoto GS, Tung L (1996) Nuclear receptor coactivators and corepressors. Mol Endocrinol 10(10):1167–1177

Jackson TA, Richer JK, Bain DL, Takimoto GS, Tung L, Horwitz KB (1997) The partial agonist activity of antagonist-occupied steroid receptors is controlled by a novel hinge domain-binding coactivator L7/SPA and the corepressors N-CoR or SMRT. Mol Endocrinol 11(6):693–705

Jin L, Borras M, Lacroix M, Legros N, Leclercq G (1995) Antiestrogenic activity of two 11 beta-estradiol derivatives on MCF-7 breast cancer cells. Steroids 60(8): 512–518

Johnston SR, Haynes BP, Smith IE, Jarman M, Sacks NP, Ebbs SR, Dowsett M (1993) Acquired tamoxifen resistance in human breast cancer and reduced intra-tumoral drug concentration. Lancet 342 (8886–8887):1521–1522

Jordan VC, Gradishar WJ (1997) Molecular mechanisms and future uses of antiestrogens. Mol Aspects Med 18:167–247

Kato S, Endoh H, Masuhiro Y, Kitamoto T, Uchiyama S, Sasaki H, Masushige S, Gotoh Y, Nishida E, Kawashima H, et al. (1995) Activation of the estrogen receptor through phosphorylation by mitogen-activated protein kinase. Science 270(5241):1491–1494

Ke HZ, Simmons HA, Pirie CM, Crawford DT, Thompson DD (1995) Droloxifene, a new estrogen antagonist/agonist, prevents bone loss in ovariectomized rats. Endocrinology 136:2435–2441

Kiang DT, Kollander RE, Thomas T, Kennedy BJ (1989) Up-regulation of estrogen receptors by nonsteroidal antiestrogens in human breast cancer. Cancer Res 49:5312–5316

Kraus WL, McInerney EM, Katzenellenbogen BS (1995)Ligand-dependent, transcriptionally productiveassociation of the amino- and carboxyl-terminal regions of a steroid hormone nuclear receptor. Proc Natl Acad Sci USA 92(26):12314–12318

Kuiper GG, Brinkmann AO (1994) Steroid hormone receptor phosphorylation: is there a physiological role? Mol Cell Endocrinol 100(1–2):103–107

Kuiper GG, Gustafsson JA (1997) The novel estrogen receptor-beta subtype: potential role in the cell- and promoter-specific actions of estrogens and anti-estrogens. FEBS Lett 410(1):87–90

Kumar V, Green S, Stack G, Berry M, Jin JR, Chambon P (1987) Functional domains of the human estrogen receptor. Cell 51(6):941–951

Kurebayashi S, Miyashita Y, Hirose T, Kasayama S, Akira S, Kishimoto T (1997) Characterization of mechanisms of interleukin-6 gene repression by estrogen receptor. J Steroid Biochem Mol Biol 60(1–2):11–17

Lam, H-YP (1984) Tamoxifen is a calmodulin antagonist in the activation of a cAMP phosphodiesterase. Biochem Biophys Res Comm 118:27–32

Le Goff P, Montano MM, Schodin DJ, Katzenellenbogen BS (1994) Phosphorylation of the human estrogen receptor. Identification of hormone-regulated sites and examination of their influence on transcriptional activity. J Biol Chem 269(6): 4458–4466

Lehrer S, O'Shaughnessy J, Song HK, Levine E, Savoretti P, Dalton J, Lipsztein R, Kalnicki S, Bloomer WD (1989) Activity of pp60c-src protein kinase in human breast cancer. Mt Sinai J Med. 56:83–85

Maeda K, Tsuzimura T, Nomura Y, Sato B, Matsumoto K (1984)Partial characterization of protease(s) in human breast cancer cytosols that can degrade estrogen and progesterone receptors selectively. Cancer Res 44(3):996–1001

Mani SK, Allen JM, Clark JH, Blaustein JD, O'Malley BW (1994) Convergent pathways for steroid hormone- and neurotransmitter-induced rat sexual behavior. Science 265(5176):1246–1249

McClelland RA, Manning DL, Gee JM, Anderson E, Clarke R, Howell A, Dowsett M, Robertson JF, Blamey RW, Wakeling AE, Nicholson RI (1996) Effects of short-term antiestrogen treatment of primary breast cancer on estrogen receptor mRNA and protein expression and on estrogen-regulated genes. Breast Cancer Res Treat 41:31–41

McDonnell DP, Vegeto E, O'Malley BW (1992) Identification of a negative regulatory function for steroid receptors. Proc Natl Acad Sci USA 89(22):10563–10567

McDonnell DP, Dana SL, Hoener PA, Lieberman BA, Imhof MO, Stein RB (1995) Cellular mechanisms which distinguish between hormone- and antihormone-activated estrogen receptor. Ann N Y Acad Sci 761:121–137

McInerney EM, Katzenellenbogen BS (1996) Different regions in activation function-1 of the human estrogen receptor required for antiestrogen- and estradiol-dependent transcription activation. J Biol Chem 271(39):24172–24178

Mendes AF, Caramona MM, Lopes MC (1996) Changes in the subcellular distribution of the rat uterus estrogen receptor as induced by oestradiol, tamoxifen and ZD 182,780. J Pharm Pharmacol 48(3):302–305

Metzger D, Berry M, Ali S, Chambon P (1995) Effect of antagonists on DNA binding properties of the human estrogen receptor in vitro and in vivo. Mol Endocrinol 9(5):579–591

Montano MM, Kraus WL, Katzenellenbogen BS (1997) Identification of a novel transferable cis element in the promoter of an estrogen-responsive gene that modulates sensitivity to hormone and antihormone. Mol Endocrinol 11(3):330–341

Montano MM, Muller V, Trobaugh A, Katzenellenbogen BS (1995) The carboxy-terminal F domain of the human estrogen receptor: role in the transcriptional activity of the receptor and the effectiveness of antiestrogens as estrogen antagonists. Mol Endocrinol 9(7):814–825

Newton CJ, Arzt E, Stalla GK (1994) Involvement of the estrogen receptor in the growth response of pituitary tumor cells to interleukin-2. Biochem Biophys Res Commun 205(3):1930–1937

Nicholson RI, Walker KJ and Davies P (1986) Hormone agonists and antagonists in the treatment of hormone sensitive breast and prostate. Cancer Surv 5:463–486

Nicholson RI, Gotting KE, Gee J, Walker KJ (1998) Actions of estrogens and anti-estrogens on rat mammary gland development: relevance to breast cancer prevention. J Steroid Biochem 30:95–103

Nicholson RI, McClelland RA, Finlay P, Eaton CL, Gullick WJ, Dixon AR, Robertson JF, Ellis IO, Blamey RW (1993) Relationship between EGF-R, c-erbB-2 protein expression and Ki67 immunostaining in breast cancer and hormone sensitivity. Eur J Cancer 29 A:1018–1023

Nicholson RI (1993) Recent advances in the antihormonal therapy of breast cancer. Curr Opin Invest Drugs 2:1259–1268

Nicholson RI, Gee JWM, Eaton CL, and 15 others (1994a) Pure antiestrogens in breast cancer: Experimental and clinical observations. I: Motta M, Serio M (eds) Sex Hormones and Antihormones in Endocrine Dependent Pathology: Basic and Clinical Aspects. Elsevier Science BV, pp 347–360

Nicholson RI, McClelland RA, Gee JM, Manning DL, Cannon P, Robertson JF, Ellis IO, Blamey RW (1994b) Transforming growth factor-alpha and endocrine sensitivity in breast cancer. Cancer Res 54:1684–1689

Nicholson RI, McClelland RA, Gee JM, Manning DL, Cannon P, Robertson JF, Ellis IO, Blamey RW (1994c) Epidermal growth factor receptor expression in breast cancer: association with response to endocrine therapy. Breast Cancer Res Treat 29:117–125

Nicholson RI, Gee JWM, Francis AB, Manning DL, Wakeling AE, Katzenellenbogen BS (1995) Observations arising from the use of pure antiestrogens on estrogen-responsive (MCF-7) and estrogen growth-independent (K3) human breast cancer cells. Endocrine Related Cancer 2:115–121

Nicholson RI, Gee JWM (1996) Growth factors and modulation of endocrine response in breast cancer. In: Vedeckis WV (ed) Hormones and Cancer. Birkhäuser, Boston, pp 227–261

Noguchi S, Motomura K, Inaji H, Imaoka S, Koyama H (1993) Downregulation of transforming growth factor alpha by tamoxifen in human breast cancer. Cancer 72:131–136

O'Brian CA, Liskamp RM, Solomon DH Weinstein IB (1985) Inhibition of protein kinase C by tamoxifen. Cancer Res 45:2462–2465

Paech K, Webb P, Kuiper GG, Nilsson S, Gustafsson J, Kushner PJ, Scanlan TS (1997) Differential ligand activation of estrogen receptors ERalpha and ERbeta at AP1 sites. Science 277(5331):1508–1510

Palkowitz AD, Glasebrook AL, Thrasher KJ, Hauser KL, Short LL, Phillips DL, Muehl BS, Sato M, Shetler PK, Cullinan GJ, Pell TR, Bryant HU (1997) Discovery and synthesis of [6-hydroxy-3-[4-[2-(1-piperidinyl)ethoxy]phenoxy]-2-(4-hydroxy-phenyl)]benzo[b]thiophene: a novel, highly potent, selective estrogen receptor modulator. J Med Chem 40:1407–1416

Pham TA, Hwung YP, Santiso-Mere D, McDonnell DP, O'Malley BW (1992) Ligand-dependent and -independent function of the transactivation regions of the human estrogen receptor in yeast. Mol Endocrinol 6:1043–1050

Piantelli M, Maggiano N, Ricci R, Larocca LM, Capelli A, Scambia G, Isola G, Natali PG, Ranelletti FO (1995) Tamoxifen and quercetin interact with Type II estrogen binding sites and inhibit the growth of human melanoma cells. J Invest Dermatol 105:248–253

Pietras RJ, Arboleda J, Reese DM, Wongvipat N, Pegram MD, Ramos L, Gorman CM, Parker MG, Sliwkowski MX, Slamon DJ (1995) HER-2 tyrosine kinase pathway targets estrogen receptor and promotes hormone-independent growth in human-breast cancer cells. Oncogene 10(12):2435–2446

Pink JJ, Jordan VC (1996) Models of estrogen receptor regulation by estrogens and antiestrogens in breast cancer cell lines. Cancer Res 56:2321–2330

Ponglikitmongkol M, White JH, Chambon P (1990) Synergistic activation of transcription by the human estrogen receptor bound to tandem responsive elements. EMBO J 9(7):2221–2231

Powles TJ (1997) Efficacy of tamoxifen as treatment of breast cancer. Seminars in Oncol 24: S148–S154

Ray P, Ghosh SK, Zhang DH, Ray A (1997) Repression of interleukin-6 gene expression by 17beta-estradiol: inhibition of the DNA-binding activity of the transcription factors NF-IL6 and NF-kappa B by the estrogen receptor. FEBS Lett 409(1):79–85

Reddel RR, Sutherland RL (1984) Tamoxifen stimulation of human breast cancer cell-proliferation in vitro: a possible model for tamoxifen tumour flare. Eur J Cancer Clin Oncol 20(11):1419–1424

Robertson JF (1996) Estrogen receptor: a stable phenotype in breast cancer. Br J Cancer 73:5–12

Rochefort H (1995) Estrogen- and anti-estrogen-regulated genes in human breast cancer. Ciba Found Symp 191:254–265

Segnitz B, Gehring U (1995) Subunit structure of the nonactivated human estrogen receptor. Proc Natl Acad Sci USA 92(6):2179–2183

Sharma HW, Narayanan R (1996) The NF-kappaB transcription factor in oncogenesis. Anticancer Res 16(2):589–596

Sivaraman VS, Wang H, Nuovo GJ, Malbon CC (1997) Hyperexpression of mitogen-activated protein kinase in human breast cancer. J Clin Invest 99:1478–1483

Smith CL, Conneely OM, O'Malley BW (1993) Modulation of the ligand-independent activation of the human estrogen receptor by hormone and antihormone. Proc Natl Acad Sci USA 90(13):6120–6124

Smith CL, Nawaz Z, O'Malley BW (1997) Coactivator and corepressor regulation of the agonist/antagonist activity of the mixed antiestrogen,4-hydroxytamoxifen. Mol Endocrinol 11(6):657–666

Sutherland RL, Murphy LC, Foo MS, Green MD, Whybourne AM (1980) High-affinity antiestrogen binding site distinct from the estrogen receptor. Nature 288: 273–275

Toko T, Sugimoto Y, Matsuo K, Yamasaki R, Takeda S, Wierzba K, Asao T, Yamada Y (1990) TAT-59, a new triphenylethylene derivative with antitumor activity against hormone-dependent tumors. Eur J Cancer 26:397–404

Tong W, Perkins R, Xing L, Welsh WJ, Sheehan DM (1997) QSAR models for binding of estrogenic compounds to estrogen receptor alpha and beta subtypes. Endocrinology 138(9):4022–4025

Tora L, White J, Brou C, Tasset D, Webster N, Scheer E, Chambon P (1989) The human estrogen receptor has two independent nonacidic transcriptional activation functions. Cell 59(3):477–487

Tremblay GB, Tremblay A, Copeland NG, Gilbert DJ, Jenkins NA, Labrie F, Giguere V (1997) Cloning, chromosomal localization, and functional analysis of the murine estrogen receptor beta. Mol Endocrinol Mar;11(3):353–365

Tremblay A, Tremblay GB, Labrie C, Labrie F, Giguere V (1998) EM-800, a novel antiestrogen, acts as a pure antagonist of the transcriptional functions of estrogen receptors alpha and beta. Endocrinology. 139:111–118

Trowbridge JM, Rogatsky I, Garabedian MJ (1997) Regulation of estrogen receptor transcriptional enhancement by the cyclin A/Cdk2 complex. Proc Natl Acad Sci USA 94(19):10132–10137

Tzeng DZ, Klinge CM (1996) Phosphorylation of purified estradiol-liganded estrogen receptor by casein kinase II increases estrogen response element binding but does not alter ligand stability. Biochem Biophys Res Commun Jun 25;223(3):554–560

Tzukerman MT, Esty A, Santiso-Mere D, Danielian P, Parker MG, Stein RB, Pike JW, McDonnell DP (1994) Human estrogen receptor transactivational capacity is determined by both cellular and promoter context and mediated by two functionally distinct intramolecular regions. Mol Endocrinol 8(1):21–30

Van de Velde P, Nique F, Bouchoux F, Bremaud J, Hameau MC, Lucas D, Moratille C, Viet S, Philibert D, Teutsch G (1994) RU 58,668, a new pure antiestrogen inducing a regression of human mammary carcinoma implanted in nude mice. J Steroid Biochem Mol Biol 48:187–196

Verbeek BS, Vroom TM, Adriaansen-Slot SS, Ottenhoff-Kalff AE, Geertzema JG, Hennipman A, Rijksen G (1996) c-Src protein expression is increased in human breast cancer. An immunohistochemical and biochemical analysis. J Pathol 180:383–388

Wakeling AE (1993) The future of new pure antiestrogens in clinical breast cancer. Breast Cancer Res Treat 25:1–9

Weatherill PJ, Wilson AP, Nicholson RI, Davies P, Wakeling AE (1988) Interaction of the antiestrogen ICI 164,384 with the estrogen receptor. J Steroid Biochem 30(1–6):263–266

Webb P, Lopez GN, Uht RM, Kushner PJ (1995) Tamoxifen activation of the estrogen receptor/AP-1 pathway: potential origin for the cell-specific estrogen-like effects of antiestrogens. Mol Endocrinol 9(4):443–456

Willsher PC, Gee JWM, Nicholson RI, Blamey RW and Robertson JF Changes in estrogen receptor and estrogen regulated gene expression in primary breast cancer during long-term tamoxifen therapy. Br J Cancer (Submitted)

Wilson AP, Weatherill PJ, Nicholson RI, Davies P, Wakeling AE (1990) A comparative study of the interaction of oestradiol and the steroidal pure antiestrogen, ICI 164,384, with the molybdate-stabilized estrogen receptor. J Steroid Biochem 35(3–4):421–428

Wogan GN (1997) Review of the toxicology of tamoxifen. Seminars in Oncol 24:S187–S197

Wrenn CK, Katzenellenbogen BS (1990) Cross-linking of estrogen receptor to chromatin in intact MCF-7 human breast cancer cells: optimization and effect of ligand. Mol Endocrinol 4(11):1647–1654

Yang NN, Venugopalan M, Hardikar S, Glasebrook A (1996) Identification of an estrogen response element activated by metabolites of 17beta-estradiol and raloxifene. Science 273(5279):1222–1225

Zhang QX, Borg A, Wolf DM, Oesterreich S, Fuqua SA (1997) An estrogen receptor mutant with strong hormone-independent activity from a metastatic breast cancer. Cancer Res 57(7):1244–1249

Zwijsen RM, Wientjens E, Klompmaker R, van der Sman J, Bernards R, Michalides RJ (1997) CDK-independent activation of estrogen receptor by cyclin D1. Cell 88(3):405–415

Part 3
Biosynthesis and Metabolism of
Endogenous Estrogens

CHAPTER 11

Estrogen Transforming Enzymes

M.J. Reed and A. Purohit

A. Introduction

Estrogens are synthesized in glandular and extraglandular tissues and exert their biological effects by interacting with the estrogen receptor (ER) which is located mainly within the cell nucleus. Estradiol, the estrogen that binds to the ER with highest affinity, therefore, has to cross the cell membrane and cytoplasm in order to reach the nucleus. The biochemical and molecular events that govern this process remain obscure. However, there is a growing awareness that the enzymes involved in the cellular synthesis and metabolism of estrogens may have a crucial role in regulating the availability of estradiol to interact with the ER (Stewart and Sheppard 1992). Thus, estrogen metabolism, rather than just being viewed as a process which ultimately renders estrogens water soluble prior to their excretion, is now considered to be an important mechanism which contributes to the regulation of cellular estrogen concentrations. It has been postulated that the cytochrome P_{450} monooxygenases and other steroid metabolizing enzymes act to regulate the steady-state concentrations of ligands, such as estrogens, which are important in growth and development (Nebert 1991). In keeping with this concept, it is now apparent that the formation of estrogen sulphates and lipoidal estrogens, rather than being end products of metabolism, results in the formation of storage forms of estrogens, which can be activated by the appropriate enzymes. The major pathways of estrogen metabolism are shown in Fig. 1.

B. Metabolism of Estradiol by 17β-Hydroxysteroid Dehydrogenase Type 2 (HSD2)

I. 17βHSD Superfamily

17β-Hydroxysteroid dehydrogenase type 2 (17βHSD2) has a pivotal role in the metabolism of estradiol. The 17βHSDs belong to a superfamily of short-chain alcohol dehydrogenases (Korzowski 1992). Members of this super family of proteins are phylogenetically related groups of enzymes that can act on a range of substrates, including steroids and prostaglandins. They have in common the ability to convert molecules that are involved in cell signalling to either the active or inactive form (Korzowski 1994). For steroids, this

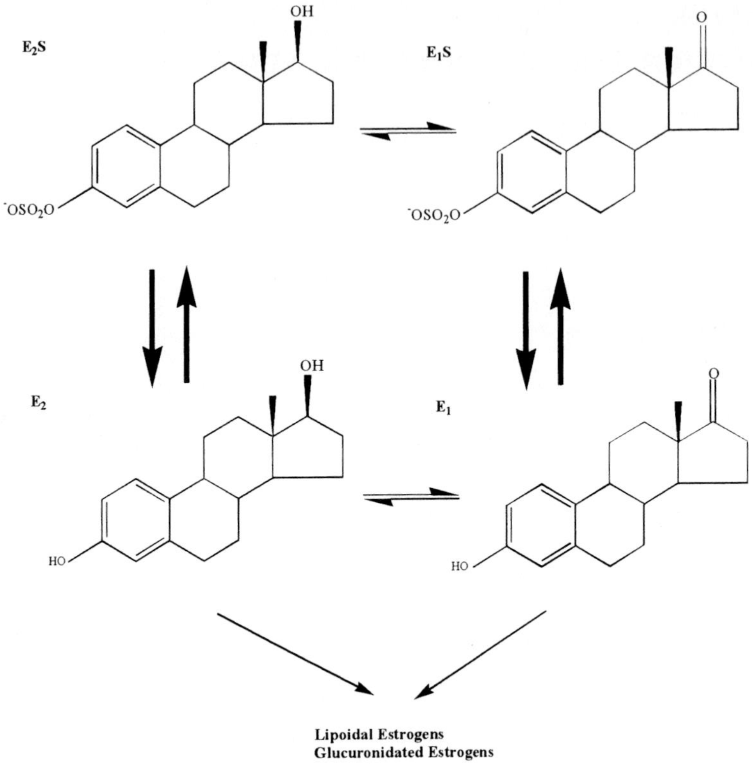

Fig. 1. Main pathways of oestrogen metabolism. *E2* estradiol; *E2-S* E2-sulphate; *E1* estrone; *E1-S* E1-Sulphate; *2-OH-E1* 2-hydroxy-estrone; *2-MeO-E1* 2-methoxy-estrone; *4-OH-E1* 4-hydroxy-estrone; *4-MeO-E1* 4-methoxy-estrone; *16α-OH-E1* 16α-hydroxy-estrone; *E3* estriol

mechanism regulates the availability of the active, i.e., 17β-hydroxysteroid, to interact with the ER. The affinity of 17β-hydroxy estradiol for the ER is about 100-fold greater than for the corresponding keto form.

The complementary deoxyribonucleic acids (cDNAs) for five isoenzymes, designated 17βHSD1–5, have now been cloned and their biochemical properties determined (reviewed by ANDERSSON 1995). 17βHSD1 and -2 have major roles in the activation and inactivation of estrogens, respectively. 17βHSD1 catalyses the reduced nicotinamide adenine dinucleotide phosphate (NADPH)/reduced nicotinamide adenine dinucleotide (NADH)-dependent conversion of estrone to estradiol and has a high specificity for estrogens, i.e. its affinity for estrogens is 100-fold higher than for androgens (ANDERSSON 1995). 17βHSD2 preferentially catalyses the nicotinamide adenine dinucleotide (NAD⁺)-dependent oxidation of C18 and C19 steroids and has an equal affinity for both substrates.

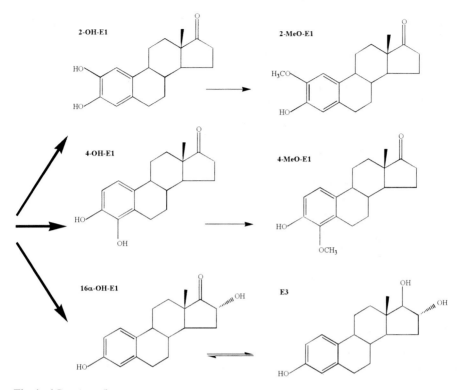

Fig. 1. (*Continued*)

II. Tissue Distribution

Many tissues possess the ability to interconvert estrone and estradiol. Using 4-^{14}C-labelled estradiol to examine its inactivation in the rat, 17βHSD activity was detected in all tissues examined (MARTEL et al. 1992). Apart from the placenta, the highest rate of inactivation of estradiol was detected in the liver. In most tissues, the oxidation of estradiol was 10- to 1000-fold higher than the reduction of estrone. A similar study, employing human tissues, revealed that 17βHSD activity is also widely distributed throughout the body with the highest rates of estradiol inactivation detected in placenta and liver (Fig. 2) (MARTEL et al. 1992).

The cloning of the 17βHSD2 enzyme (WU et al. 1993) has enabled the expression of this gene in different tissues to be examined by Northern analysis. 17βHSD2 mRNA was detectable in RNA prepared from human endometrial tissue (CASEY et al. 1994). The highest levels of messenger ribonucleic acid (mRNA) for 17βHSD2 were found in samples of endometrial tissue obtained during the mid-to-late secretory phase of the menstrual cycle. This is at the time when progesterone levels peak, and the finding is consistent with earlier reports that progesterone can increase the oxidation of estradiol during the

HUMAN TISSUE   **ESTROGENIC 17ß-HSD ACTIVITY**

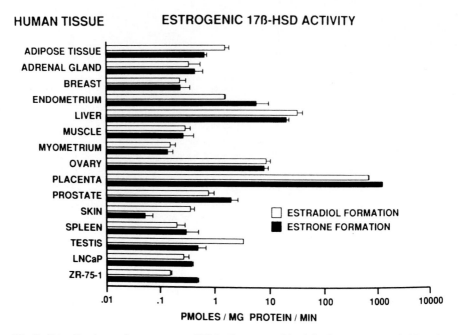

Fig. 2. Distribution of oestrogen 17β-hydroxysteroid dehydrogenase activities in human tissues. Reprinted from MARTEL et al. (1992) J Steroid Biochem Mol Biol, with kind permission from Elsevier Sciences Ltd., Oxford

secretory phase of the menstrual cycle (GURPIDE et al. 1977). 17βHSD2 mRNA expression was also detected in placental tissue where the levels were 10-fold higher than detected in mid-secretory-phase endometrial tissue. 17βHSD2 mRNA was also detected in the pancreas, kidney and small intestine. The role of 17βHSD2 in the placenta remains to be clarified. It has been postulated that it may serve to maintain androgens in their inactive 17-keto form, in order to reduce the risk of virilization of female fetuses (CASEY et al. 1994).

A rat model has been employed to examine the developmental regulation of 17βHSD2 expression. The gene for this enzyme is constitutively expressed in the liver and small intestine, with no variation in the level of expression being detected from late foetal life up to 6 weeks of age (AKINOLA et al. 1997). The constitutive expression of 17βHSD2 in the liver and small intestine throughout the life of these animals presumably reflects the important role that this enzyme has for the inactivation of biologically active sex steroids.

III. Regulation of 17βHSD2

Apart from the effect of progesterone and synthetic progestins in stimulating the expression and activity of 17βHSD2 (TENG and GURPIDE 1979), little is known about other factors regulating this enzyme. Oxidation of estradiol is higher in ER– than in ER+ breast cancer cells (SINGH and REED 1991).

C. Cytochrome P$_{450}$ Mono-Oxygenase Hydroxylation of Estrogens

I. Cytochrome P$_{450}$s

In addition to the oxidation of estradiol to estrone by 17βHSD2, hydroxylation can occur at aromatic positions 2 and 4, to form the catechol estrogens, and aliphatic carbon atoms 6, 7, 11, 14, 16 and 18 of the estrogen nucleus (previously reviewed by REED and MURRAY 1979; BALL and KNUPPEN 1990). The liver is a major site for the hydroxylation of estrogens by cytochrome P$_{450}$ mono-oxygenases, but these enzymes are also present in other body tissues. Although cytochrome P$_{450}$-mediated hydroxylation can occur at a number of positions of the estrogen nucleus (recently reviewed by MARTUCCI and FISHMAN 1993), hydroxylation at C2, C4 and C16 are the major pathways of metabolism by these enzymes.

II. C2-and C16α-Hydroxylation of Estrogens

In humans, the extent of hydroxylation at C2 is greater than at the C16 position (MARTUCCI and FISHMAN 1993). Hydroxylation at C2 is carried out mainly in the liver by cytochrome P$_{450}$IIIA and to a lesser extent by the P$_{450}$IA family of enzymes (KERLAW et al. 1992; MARTUCCI and FISHMAN 1993). The mono-oxygenase responsible for oxidation at C16 remains to be identified.

1. Effect of 2- or 16α-Hydroxylation on Metabolite Estrogenicity

Catechol estrogens, such as 2-hydroxy-estradiol or 2-hydroxy-estrone, and their methoxy derivatives do not possess uterotrophic activity (MARTUCCI and FISHMAN 1979). In contrast, both 16α-hydroxy-estrone and estriol are estrogenic in the uterotrophic assay (MARTUCCI and FISHMAN 1977). The formation of 2-hydroxy or 16α-hydroxy estrogen metabolites, therefore, has a crucial role in regulating the formation of inactive or bioactive estrogen metabolites. Furthermore, there is evidence that 16α-hydroxy-estrone can bind covalently to proteins, including the ER (SWANECK and FISHMAN 1988). Such covalent binding may increase the duration of estrogen action. In view of the different biological properties associated with 2-hydroxy- and 16α-hydroxy-estrogens, metabolism via these competing pathways is being investigated in a number of conditions associated with increased or reduced tissue exposure to estrogen.

2. Modulation by Thyroid Hormones, Body Weight and Diet

Thyroid-hormone status was originally reported to influence the relative extent of formation of 2- and 16α-hydroxyoestrogen metabolites (FISHMAN 1965). In hypothyroid subjects, formation of estriol was increased, while in hyperthyroidism conversion of estradiol to estriol was reduced, with a concomitant increase in the formation of 2-hydroxylated estrogen metabolites.

Nutritional factors and body weight can also modulate the extent of hydroxylation of estrogens at C2/C16α-positions. Ingestion of a low-fat diet decreases urinary excretion of 16α-hydroxyestrone and estriol, and increases production of catechol estrogens (LONGCOPE et al. 1987). Diets with a high protein or carbohydrate content can, respectively, increase or decrease the 2-hydroxylation of estrogens (ANDERSON et al. 1984). In women with anorexia nervosa, conversion of estradiol to estriol was reduced by 44% compared with a 26% increase from control values detected in obese women (FISHMAN et al. 1975). In contrast, the metabolism of estradiol to 2-hydroxyestrone increased 232% in anorexia, while it decreased 59% in obese women. Intense athletic training can also stimulate the 2-hydroxylation pathway of estrogen metabolism (SNOW et al. 1989). The ability of factors, such as thyroid hormones, body weight and dietary fat intake, to influence the extent of formation of 2- and 16α-hydroxyoestrogens in a reciprocal manner suggested that competition for estrogen substrate by the different cytochrome P_{450} mono-oxygenases occurs.

3. Effect of Smoking on C2/C16α-Hydroxylation

Female smokers are reported to have up to a 50% reduction in the relative risk of endometrial cancer (LESKO et al. 1985), but an increased risk of osteoporosis. There is some evidence that the risk of breast cancer may also be reduced in female smokers (LONGCOPE 1990). As estrogens are implicated in the development of endocrine-dependent tumours and bone homeostasis, such evidence suggests that smoking has an anti-estrogenic effect in women.

Using a specific radiometric technique to measure the formation of 2-hydroxy estrone in female smokers, MICHNOVICZ et al. (1986) originally reported a 50% increase in the extent of estrogen-2-hydroxylation in women smoking at least 15 cigarettes a day. Excretion of estriol, the end product of 16α-hydroxylation, was decreased. In a subsequent study, in which urinary estrogen metabolites were measured, excretion of 2-hydroxy estrogen metabolites was considerably higher by smokers (52.2%) than by non-smokers (13.6%) (MICHNOVICZ et al. 1988). Plasma concentrations of estrone and estradiol do not appear to differ in smokers and non-smokers (LONGCOPE and JOHNSTON 1988). However, the concentration of unbound, biologically active estradiol has been reported to be lower in smokers (CASSIDENTI et al. 1990). A lower metabolic clearance rate of estrone was detected in female smokers, but this was not significant when adjusted for weight (LONGCOPE and JOHNSTON 1988).

Epidemiological studies have indicated a protective role for estrogens in the development of heart disease (BUSH and BARRETT-CONNER 1985). The incidence of ischaemic heart disease is higher in men than in aged-matched females, and women are at increased risk after menopause. In male cigarette smokers, a modest increase in peripheral estrogen levels was not found to be associated with an anticipated increase in (protective) high-density lipoprotein (HDL)-cholesterol levels. Excretion of 2-hydroxy estrogens was 70% higher in male smokers, suggesting that increased metabolism via this pathway may provide an explanation for the lack of an increase in HDL cholesterol in the presence of increased plasma estrogen levels (MICHNOVICZ et al. 1989).

4. C2/C16α-Hydroxylation and Breast Cancer

Estrogens have an important role in promoting the growth of breast tumours (JAMES and REED 1980). 16α-Hydroxylation of estrogen is increased, not only in women with breast cancer, but also in those with an increased risk of developing the disease (BRADLOW et al. 1995). Supporting evidence has also been obtained for an alteration of 16α-hydroxylation of estrogens by factors that may increase the risk of tumour development. In cultured mammary epithelial cells derived from 6- to 8-week-old mice, the carcinogen dimethyl benz (a) anthracene (DMBA) acted to increase metabolism of estradiol via the 16α-hydroxylation pathway while, at the same time, reducing the formation of 2-hydroxy estrogen metabolites (TELANG et al. 1992). 16α-Hydroxy estrone also induced the growth of more colonies in cells grown on soft agar than either estrone or estradiol. Results from this study, therefore, provide evidence that, in addition to having a promotional role, 16α-hydroxy estrone might also function as an initiator of the transformation of mammary epithelial cells. An increase in the 16α-hydroxylation of estradiol also occurred when mouse epithelial cells were transfected with the c-Ha-ras oncogene (TELANG et al. 1991a). In terminal lobular ductal units derived from human breast tumours, a 4- to 5-fold increase in 16α-hydroxylation was detected, whereas no increase in activity was detected in breast adipose tissue (TELANG et al. 1991b; OSBORNE et al. 1993).

5. Induction of C2-Hydroxylation by Indole-3-Carbinol

The evidence, as previously discussed, that 2- and 16α-hydroxylation of estrogen is regulated in a reciprocal manner led BRADLOW and his colleagues (1991) to postulate that stimulation of the 2-hydroxylation pathway should result in a concomitant reduction in 16α-hydroxylation. A reduction in the formation of 16α-hydroxy estrone should, therefore, result in a decrease in tissue exposure to bioactive estrogens. A safe means of stimulating the 2-hydroxylation of estrogen in women could provide a novel strategy for the chemoprevention of breast cancer.

Two lines of reasoning appear to have led BRADLOW and his colleagues (1991) to identify the indole, indole-3-carbinol (I3C), as a potent inducer of 2-hydroxylase activity. First, ingestion of cruciferous vegetables, such as broccoli or cabbage, is associated with a lower incidence of some cancers (YOUNG and WOLF 1988). I3C, which is present in such vegetables, can reduce the incidence of mammary tumours in rodents (BRADLOW et al. 1991). Second, 2-,3-,7- and 8-tetrachlorobenzo-p-dioxin (TCDD), which binds to the same receptor as

I3C, was shown to increase estrogen 2-hydroxylase activity 4-fold, while only minimally increasing 16α-hydroxylase activity (NIWA et al. 1994). I3C has, therefore, been examined for its ability to differentially modulate estradiol metabolism as a possible approach to reducing tissue exposure to active estrogens.

In rats fed 250mg/Kg/day I3C, the formation of 2-hydroxy estrone was induced (MICHNOVICZ and BRADLOW 1990). In humans given 5–7mg/Kg/day for 7days, metabolism via the 2-hydroxylase pathway increased from 29.8% to 44.6%. This finding demonstrates that I3C is a potent inducer of 2-hydroxylase activity in humans (MICHNOVICZ and BRADLOW 1991). A distinct shift in the urinary profile of increased excretion of catechol estrogens was also detected.

In a further study to obtain evidence that ingestion of I3C alters estrogen metabolism, 2- and 16α-hydroxy estrogens were assayed using a gas chromatography–mass spectrometry (GC–MS) technique, rather than the radiometric method often employed for such investigations. Results from the analysis by GC–MS confirmed previous radiometric studies, showing that conversion of estradiol to 2-hydroxy estrogens was strongly stimulated in men and women by I3C ingestion (MICHNOVICZ et al. 1997).

When taken orally, I3C undergoes an acid-catalysed rearrangement in the stomach to metabolites such as di-indoylene methane. This dimer may stimulate cytochrome P_{450} activity by binding to the dioxin receptor. In vitro, I3C undergoes spontaneous dimerization in culture medium to stimulate 2-hydroxylase activity in MCF-7 breast cancer cells (NIWA et al. 1994). The presence of ERs appears to be required for the induction of 2-hydroxylase activity by I3C, as no effect of this indole on enzyme activity was detected in ER-MDA-MB-231 cells. Somewhat surprisingly, I3C had no effect on the activity of this enzyme in T47D cells. While NIWA et al (1994) state that these cells only have low ER levels, most investigations consider the cells to be strongly ER+ (POUTANEN et al. 1992). Co-treatment of MCF-7 cells with estradiol, which stimulates their proliferation, and with I3C (50μM) reduced cell growth by 55% (TIWARI et al. 1994). The ability of I3C to reduce the proliferative effect of estradiol on MCF-7 cells provides strong support for the concept that an increase in 2-hydroxylase activity reduces tissue estrogen exposure.

6. C2/C16α-Hydroxylation in Different Ethnic Groups

In an attempt to further examine the hypothesis that there may be a link between increased 16α-hydroxylation of estrogen and the risk of breast cancer, the excretion of urinary estrogen metabolites by Oriental and Caucasian (Finnish) females was measured (ADLERCREUTZ et al. 1994). Caucasian women have a higher risk of breast cancer than Oriental women. In addition, fat consumption, which can increase 16α-hydroxylation of estrogen, was higher in the Caucasian women studied. Although Oriental women excreted less total estrogen, excretion of total and individual 16α-hydroxy estrogens did not differ significantly in the two groups of women. Excretion of catechol estrogens by Finnish smokers has been reported to increase 2-hydroxylase activity and did not differ from excretion by non-smokers.

Support for the concept that a change in the 16α-hydroxy/2-hydroxy estrogen metabolite ratio may be associated with an increased risk of breast cancer has been obtained by measuring estrogen urinary metabolites from Afro-Americans and other ethnic minorities at increased risk of developing breast cancer (TAIOLI et al. 1996; COKER et al. 1997).

7. Effect of Pesticides on C2/C16α-Estrogen Hydroxylation

Exposure to a number of foreign compounds, including pesticides, may increase the risk of developing breast cancer. This may, in part, be due to their intrinsic estrogenicity resulting from their phenolic ring structure. In addition, organochlorine pesticides may increase the risk of breast cancer by stimulating estradiol metabolism via the 16α-hydroxylase pathway (BRADLOW et al. 1995). Pesticides, such as dichlorodiphenyl-trichloroethane (DDT) and 1,1-dichloro-2,2-dichlorophenylethylene (DDE) decreased the formation of 2-hydroxy estrogens whilst significantly increasing 16α-hydroxylase activity.

8. Inhibition of C2-Hydroxylation

An increase in the excretion of 2-hydroxy and decrease in 16α-hydroxy estrogen metabolites has been detected in Korean postmenopausal women with osteopenia (LIMET et al. 1997). Inhibition of the 2-hydroxylation could therefore be used to increase estrogen metabolism via the 16α-hydroxy pathway to generate metabolites with greater estrogenicity. Cimetidine, the drug used for the treatment of peptic ulcers, inhibits cytochrome P_{450}-dependent hydroxylation of estrogens at the C2, but not the C16 position (MICHNOVICZ and GALBRAITH 1991). This drug reduced C2-hydroxylation by 30% and 40%, respectively, in pre-and postmenopausal women. The antimycotic drug, ketoconazole, is also a potent inhibitor of C2-hydroxylation of estrogens (BACK et al. 1989). Such drugs could, therefore, be of benefit as an alternative to hormone-replacement therapy for women suffering from hypo-estrogenic conditions such as osteoporosis.

9. Estrogen-2-Hydroxylation by Placental Aromatase

There is now evidence that, in placental tissues, the cytochrome P_{450} aromatase is also an estrogen 2-hydroxylase (OSAWA et al. 1993). An examination of the ability of three aromatase inhibitors, aminoglutethimide (AG), 4-hydroxyandrostenedione (4-OHA) and CGS 16949A, to inhibit estrogen 2-hydroxylation in vitro in rat liver microsomes, revealed they were only weak inhibitors (PURBA et al. 1994). In vivo, in rats, 2-hydroxylase activity was induced by AG and 4-OHA and may, therefore, enhance the efficiency of these drugs in reducing circulating estrogens.

D. Catechol Estrogens and Carcinogenesis

In addition to the role that estrogens have in carcinogenesis via induction of specific genes through interaction with ERs, the synthesis of catechol estrogen metabolites, such as 2- and 4-hydroxy estradiol, has also been implicated in tumour induction. Catechol estrogens possess the ability to undergo redox cycling through the formation of their semi-quinone and quinone forms (LIEHR 1990). This can give rise to the generation of mutagenic-free radicals. In estrogen target tissues, this may result in damage to DNA and proteins. However, both the 2- and 4-hydroxycatechol estrogens can be rapidly metabolised to their methoxy derivatives by catechol-*O*-methyltransferase (COMT), which is present in liver and red blood cells.

Much of the evidence implicating catechol estrogens in tumour induction has been obtained from experiments using the Syrian hamster kidney model (WEIZ 1992). This organ contains cytochromes P_{450}s that can generate the 2- and 4-hydroxy catechol estrogens. It has been suggested that 4-hydroxy catechol estrogens are of greater importance in tumour development, as the inactivation of 4-hydroxy estradiol by COMT can be inhibited by 2-hydroxy estradiol and catecholamines (ROY et al. 1990; ZHU and LIEHR 1993). This could give rise to high tissue concentrations of 4-hydroxy catechol estrogens which may result in tumour induction. Reduced inactivation of 2-hydroxy catechol estrogens in tumour tissues has also been postulated to have a role in tumour induction (ZHU et al. 1993).

The potential role of catechol estrogen in tumour induction in humans has not been extensively investigated. In human uterine myoma, 4-hydroxylation of estradiol is the predominant form of catechol estrogen present (LIEHR et al. 1995). Its rate of formation in myoma tissue is five times that in the surrounding myometrial tissue. The cytochrome P_{450} specifically catalysing the hydroxylation of estrogen at C4 has now been identified as P_{450} 1B1 (HAYES et al. 1996). Kinetic enzyme analysis of an expressed form of this enzyme revealed that it catalysed the NADP-dependent hydroxylation of estradiol at C4, although some hydroxylation of C2 was also detected.

In view of the potential carcinogenic role of catechol estrogens, some reservations have been expressed as to the wisdom of 2-hydroxylase induction being used for the chemoprevention of breast cancer (ADLERCREUTZ et al. 1994; HAYES et al. 1996). In particular, it has been suggested that factors inducing 2-hydroxylation of estrogens will, at the same time, enhance the formation of 4-hydroxy catechol estrogen (HAYES 1996).

The potential carcinogenic role of catechol estrogen by I3C has been addressed by BRADLOW et al. (1996). As previously noted, cruciferous vegetables contain I3C and their ingestion is associated with a reduced risk of some cancers. It is also possible that the rapid inactivation of catechol estrogen by COMT, in vivo, to the inactive methyl ethers reduces any risk arising from 2- and/or 4-hydroxylase induction.

The role of catechol estrogens in the carcinogenic process, with the exception of the Syrian hamster model, remains controversial. The synthetic estrogens, 17α-ethinyl estradiol (EE) also undergoes hydroxylation at C2 and C4, yet is relatively ineffective in inducing renal carcinomas in the Syrian hamster. EE is a potent inhibitor of the 16α hydroxylation of estrogens (NAMAZAWA and SATOH 1989). It has been postulated that the failure of catechol metabolites of EE to induce tumours in this model, to the same extent as estradiol metabolites, might be due to its ability to inhibit 16α-hydroxylation. However, in one study, no 16α-hydroxylation of estrogens by Syrian hamster kidney tissues was detected (ZHU et al. 1994).

E. Estrogen Sulphates

I. Estrogen Sulphate Formation

Sulphation is a major metabolic pathway for estrogen conjugate formation. Sulpho conjugation of estrogens involves the transfer of a sulphonate radical (SO_3^-) from the universal donor, 3-phosphoadenosine 5-phosphosulphate, to the phenolic group of estrogen. The biochemistry and molecular biology of steroid sulphotransferases (STs) have been the subject of three recent reviews (YAMAZOE et al. 1994; MATSUI and HOMMA 1994; STROTT 1996). Three cytosolic STs that can sulphate estrogens have been identified from human tissues (FALANY and FALANY 1996). These are: a specific estrogen ST (EST) which has a high specificity for estrogens; a hydroxy – or DHA-ST; and a phenol-sulphating form of phenol ST (P-PST). Both DHA-ST and P-PST can sulphate estrogens, although they have a preference for other substrates. These STs have been purified from a number of tissues and their cDNAs cloned (reviewed by STROTT 1996). While the liver is a major site of estrogen sulphation, ST activity is detectable in other body tissues, including the adrenal, kidney, endometrium, platelets and normal and malignant breast tissues.

II. Biological Role of Estrogen Sulphates

Sulphation of estrogens changes their nature from a hydrophobic to hydrophilic form. This renders them water soluble and unable to bind to the ER. It was originally thought that steroid sulphation was the end step in a metabolic process leading to their elimination from the body. However, there is now a considerable body of evidence suggesting that steroid sulphates have an important role in modulating the action of steroid hormones. Concentrations of estrogen sulphates in blood (NOEL et al. 1981) and some tissues (PASQUALINI et al. 1989) are much higher than their unconjugated forms. Furthermore, because estrogen sulphates bind to plasma proteins, they are cleared from the blood at a much slower rate than unconjugated estrogen (RUDER et al. 1972). It is, therefore, likely that the high blood and tissue concentrations of estrogen sulphates act as a storage form of estrogen (REED and PUROHIT 1993). The hydrolysis of estrogen sulphates, by estrone sulphatase, can make the free estrogens available to target tissues. The realisation that estrogen sulphates are an important source of estrogens has been a major stimulus to the development of specific estrone sulphatase inhibitors (see Chap. 31)

Further evidence that EST may have an important role in estrogen action has been provided by immunocytochemical studies demonstrating that it is located in the nuclei of rat hepatocytes (MANCINI et al. 1992) and guinea-pig adrenocorticol cells (WHITNALL et al. 1993). Within the nuclear compartment, EST could function to inactivate estradiol after its interaction with the ER (ROY 1992).

In view of the importance of estrogens in regulating the growth of breast tumours, the abilities of normal and malignant breast epithelial cells to form sulphate conjugates has received particular attention. Although there were early reports that breast cancer cells possess EST activity, it is only recently that advances in molecular biology have allowed the characterization of the different STs in normal and malignant breast epithelial cells. In normal breast epithelial cells, there is evidence of higher ST activity than in breast cancer cell lines (WILD et al. 1991; ANDERSON and HOWELL 1995). Normal breast epithelial cells appear to express the specific hEST (FALANY and FALANY 1996; ANDERSON et al. 1996). In contrast, using an immunoblot technique and employing specific substrates, FALANY and FALANY (1996) were unable to obtain evidence of hEST expression in MCF-7 cells. It was concluded from their investigations that, in MCF-7 cells, the majority of sulphate formation results from P-PST expression with DHA-ST also being able to sulphate estrogens in these cells. However, in a recent preliminary report, hEST mRNA was detected in RNA extracted from MCF-7 cells using a reverse-transcriptase polymerase chain reaction (RT-PCR) technique (ANDERSON et al. 1996).

The observation that estrogen sulphation activity is higher in normal breast epithelial cells than in malignant cells has led to the hypothesis that the role of EST may be to impede estrogen action (ANDERSON and HOWELL 1995). High EST activity in normal breast epithelial cells would protect them from the proliferative effects of estrogens. Part of the malignant transformation may, thus, involve the loss of EST activity.

III. Regulation of Estrogen Sulphation

Relatively little is known about the regulation of estrogen sulphation. It was originally discovered that the highest levels of EST activity in human endometrial tissue was detectable during the luteal phase of the menstrual cycle (BUIRCHELL and HAHNEL 1975). Progesterone levels are maximal during this phase of the menstrual cycle, and it was subsequently shown that progesterone can induce EST activity in cultured endometrial tissue (CLARKE et al. 1982).

In MCF-7 breast cancer cells estradiol has been reported to stimulate estrogen sulphation (ADAMS et al. 1988). In a recent investigation, both progesterone and dexamethasone, but not estradiol, were reported to stimulate estrone sulphate formation in MCF-7 cells (PUROHIT et al. 1999). In this study, tumour necrosis factor alpha (TNFα) was found to be a potent inducer of estrone sulphate formation, with IGF-I and interleukin-1 also enhancing the formation of estrone sulphate.

F. Estrogen Glucuronides

Glucuronyl transferase, which is located in the liver and mucosa of the small intestines, is responsible for the formation of estrogen glucuronide conjugates. The process involves the transfer of a glucuronide moiety from uridine diphosphoglucuronic acid to the steroid. Incubation of estrogen with adult liver tissue revealed that sulphates rather than glucuronides were the major conjugate found in this organ (HOBKIRK et al. 1975). Human kidney homogenates are also capable of estrogen glucuronide formation (HOBKIRK et al. 1974). Furthermore, kidney homogenates are also capable of directly oxidising estradiol-3-glucuronide to estrone-3-glucuronide. The ability of the estrogen glucuronide conjugates to be interconverted in vivo, without removal of the glucuronic acid moiety has also been demonstrated (HOBKIRK and NILSIN 1970).

Estrogens and their conjugates undergo an enterohepatic circulation to a much greater extent than neutral steroids, although the physiological significance of this process remains obscure (ADLERCREUTZ and LUUKAINEN 1967). Estrogen glucuronides appear to be cleared from the circulation at a faster rate than estrogen sulphates (YOUNG et al. 1976). Differences in the rates of clearance of these conjugates may be partly explained in terms of their relative binding to serum proteins (GOEBELSMANN et al. 1973).

G. Lipoidal Estrogens

In1981, SCHATZ and HOCHBERG discovered that estradiol was converted to a mixture of estradiol long-chain fatty acids, when incubated in vitro with various rat tissues. The greatest extent of formation of these steroid derivatives was detected in estrogen-sensitive tissues. Steroid esters were first identified in the adrenal gland as endogenous, non-polar saponofiable derivatives of pregnenolone, 17α-hydroxypregnenolone and DHA (HOCHBERG et al. 1979).

Steroid esters are not excreted in urine and are only present at low concentration in blood. It was, therefore, difficult to obtain sufficient amounts for their full characterization. The structure of the steroid esters remained uncertain and they were named lipoidal derivatives. Lipoidal estradiol (LE2) is present in human ovarian follicular fluid. Its presence in this fluid in high concentration, when obtained from women undergoing ovarian stimulation, finally allowed the structure of LE2 to be determined (LARNER et al. 1993). Five specific estradiol-17-esters were identified in follicular fluid, i.e. arachidonate, linoleate, oleate, palmitate and stearate.

Research into the possible physiological roles of LE2 has now been pursued for almost 20 years, but there is still some uncertainty about their role in vivo. In view of their ester structure, it has been suggested that they may be the endogenous counterparts of the synthetic esters of estrogens. Such derivatives are used therapeutically because they can provoke prolonged estrogenic stimulation. This property is also shared by the endogenous LE2 derivatives. In a comparison of the ability of estradiol or estradiol-17-stearate (E2-17-St) to stimulate uterine growth, the steroids were administered s.c. in an oil vehicle (LARNER et al. 1985a). After three daily injections, estradiol stimulated uterine growth, but uterine weights rapidly reverted to control values with the cessation of injections. In contrast, for animals receiving E2-17-St, it was only 3–4 days after the end of steroid administration that uterine weights peaked. Using an aqueous medium and an i.v. route of administration, E2-17-St produced a 2- to 3-fold greater stimulation of uterine weights than estradiol.

Similar experiments with ^{3}H-labelled steroids revealed a delay in the nuclear uptake of the ester, although it was retained within tissues for a prolonged period of time compared with that of estradiol. The prolonged action of E2-17-esters is also influenced by their resistance to metabolism and decreased clearance from the body. Clearance rates of E2-17-St in rats was nine times slower than that of estradiol (LARNER et al. 1985b). Estradiol-17-esters can undergo hydrolytic cleavage by esterases that are present in human mammary tissues (ABUL-HAJJ and NURIEDDIN 1983). They do not bind to the ER as the ester, indicating that LE2 derivatives act as long-term storage forms of estradiol.

Esters of testosterone also exist in blood (BORG et al. 1995), and it is possible that LE2 derivatives could be formed via the aromatisation of testosterone esters. When testosterone stearate was tested as both an inhibitor and substrate of placental aromatase, no inhibition or conversion to LE2 was detected (LARNER et al. 1992). The most likely route of LE2 production is, therefore, through the esterification of estradiol by an estradiol-acyl COA-acyl transferase (LARNER et al. 1993).

LE2 derivatives can also be formed by incubating estradiol with DMBA-induced mammary tumours (SCHATZ and HOCHBERG 1981). Both ER+ and ER– breast cancer cells can synthesize LE2. A higher rate of LE2 formation was detected in ER– MDA-MB-231 cells than in ER+ MCF-7 breast cancer cells (ADAMS et al. 1986). It was postulated that the lower rate of LE2 formation in MCF-7 cells might reflect a higher turnover rate. This may be required to maintain a higher concentration of estradiol in ER+ cells. It was also proposed that the synthesis of LE2 derivatives could provide a means of concentrating estradiol in an intracellular hydrophobic form. This could enhance its storage or passage through membranes. Such a store would, after subsequent hydrolysis, make free estradiol available to interact with the ER.

Elucidation of the physiological role of LE2 derivatives has been hampered by the difficult, lengthy procedures required for their assay. Using a GC–MS technique, low but detectable levels of LE2 were found in human plasma, but concentrations in adipose tissue were much higher (LARNER et al. 1992). A more robust assay for LE2 derivatives was recently developed based on the separation of lipids from plasma, followed by saponification, with the measurement of free estrogen by radioimmunoassay (RIA) (ARDEVOL et al. 1997).

While it has been almost 20 years since LE2 derivatives were first discovered, research into their physiological role is still in progress. An important clue as to the function of these derivatives has been provided by the observation that estrone-17-esters from adipose tissue may be involved in the regulation of body weight (SANCHIS et al. 1996).

H. Conclusions

Like other steroids, estrogens are transformed by a number of different enzyme systems. Rather than being a process of inactivation, it is now apparent that metabolites with differing degrees of estrogenicity can be generated as well as storage forms of estrogens. While specific inhibitors of most of the cytochrome P_{450} enzymes that are involved in estrogen metabolism still await development, the ability to induce or inhibit 2-hydroxylation of estrogens offers considerable potential for the prevention of diseases such as breast cancer or osteoporosis. Understanding the regulation of cytochrome P_{450} enzymes is still at an early stage. Since cytokines such as TNFα and interleukin (IL)-6 can regulate cytochrome P_{450} aromatase activity (REED and PUROHIT 1997), it will be important to examine whether cytokines are involved in regulating the other P_{450} enzymes involved in estrogen metabolism.

References

Abul-Hajj Y, Nurieddin A (1983) Significance of lipoidal estradiol in human mammary tumours. Steroids 42:417–426

Adams JB, Hall RT, Nott S (1986) Esterification-deesterification of estradiol by human mammary cancer cells in culture. J Steroid Biochem 24:1159–1162

Adams JB, Phillips NS, Hall RS (1988) Metabolic fate of estradiol in human mammary cancer cells in culture: estrogen sulfate formation and cooperativity exhibited by estrogen sulfotransferase Molec Cell Endocrinol 58:231–242

Adlercreutz H, Gorbach SL, Goldin BR, Woods MN, Dwyer JT, Hamalainen E (1994) Estrogen metabolism and excretion in Oriental and Caucasian women. J Natl Cancer Inst 86:1076–1082

Adlercreutz H, Luukainen T (1967) Biochemical and clinical aspects of the enterohepatic circulation of estrogens. Acta Endocrinol Copenh {Suppl 124]:101–140

Akinola LA, Poutanen M, Vihko R, Vihko P (1997) Expression of 17β-hydroxysteroid dehydrogenase type 1 and Type 2, P450 aromatase, and 20α-hydroxysteroid dehydrogenase enzymes in immature, mature and pregnant rats. Endocrinology 138:2886–2892

Anderson EA, Howell A (1995) Estrogen sulphotransferases in malignant and normal human breast tissue. Endocr-Rel Cancer 2:227–233

Anderson E, Roebuck Q, Oojageer A, Howell A (1996) Estrogen inactivation by human breast epithelial cells. Breast Cancer Res Treat 41:{Abstr 323}

Anderson KE, Kappas A, Conney AH, Bradlow HL, Fishman J (1984) The influence of dietary protein and carbohydrate on the principal oxidative biotransformations of estradiol in normal subjects. J Clin Endocrinol Metab 59:103–107

Andersson S (1995) 17β-Hydroxysteroid dehydrogenase: isoenzymes and mutations. J Endocrinol 146:197–200

Ardevol A, Virili J, Sanchis D, Adan C, Fernandez-Real JM, Fernandez-Lopez JA, Remesar X, Alemany M (1997) A method for the measurement of plasma estrone fatty ester levels. Analyt Biochem 249:247–250

Back DJ, Stevenson P, Tjia JF (1989) Comparative effects of two antimycotic agents ketoconazole and terbinafine on metabolism of tolbutamide, ethinyloestradiol and ethoxycoumarin by human liver microsomes in vitro. Br J Clin Pharmac 28:166–170

Ball P, Knuppen R (1990) Formation, metabolism and physiologic importance of catechol estrogens. Am J Obstet Gynec 163:2163–2170

Borg W, Shackleton CHL, Pahuja SL, Hochberg RB (1995) Long-lived testosterone esters in the rat. Proc Natl Acad Sci USA 92:1545–1549

Bradlow HL, Davis DL, Lin G, Sepkovic D, Tiwari R (1995) Effects of pesticides on the ratio of 16α/2-hydroxyestrone: a biological marker of breast cancer risk. Environ Health Perspect {Suppl 103}:147–150

Bradlow HL, Michnovicz JJ, Telang NT, Osborne MP (1991) Effect of dietary indole-3-carbinol on estradiol metabolism and spontaneous tumours in mice. Carcinogenesis 12:1571–1574

Bradlow HL, Spkovic DW, Telang NT, Osborne MP (1995) Indole-3-carbinol: a novel approach to breast cancer prevention. Ann NY Acad Sci 728:180–200

Bradlow HL, Telang NT, Sepkovic DW, Osborne MP (1996) 2-Hydroxyestrone: the 'good' estrogen. J Endocrinol 150:S259–S265

Buirchell BJ, Hahnel R (1975) Metabolism of estradiol-17β in human endometrium during the menstrual cycle. J Steroid Biochem 6:1489–1494

Bush TL, Barrett-Conner E (1985) Noncontraceptive estrogen use and cardiovascular disease. Epidemiol Rev 7:80–104

Casey ML, MacDonald PC, Andersson S (1994) 17β-Hydroxysteroid dehydrogenase type 2: chromosomal assignment and progestin regulation of gene expression in human endometrium. J Clin Invest 94:2135–2141

Cassidenti DL, Vijod AG, Vijod MA, Stanczyk FZ, Lobo RA (1990) Short-term effects of smoking on the pharmacokinetic profiles of micronized estradiol in postmenopausal women. Am J Obstet Gynecol 163:1953–1960

Clarke CL, Adams JB, Wren BG (1982) Induction of estrogen sulphotransferase activity in the human endometrium by progesterone in organ culture. J Clin Endocrinol Metab 55:70–75

Coker AL, Coker MM, Sticca RP, Sepkovic DW (1997) Re: Ethnic differences in estrogen metabolism in healthy women. J Natl Cancer Inst 89:89

Falany JL, Falany CN (1996) Expression of cytosolic sulfotransferases in normal mammary epithelial cells and breast cancer cell lines. Cancer Res 56:1551–1555

Fishman J, Boyar RM, Hellman L (1975) Influence of body weight on estradiol metabolism in young women. J Clin Endocrinol Metab 41:989–991

Fishman J, Hellman L, Zumoff B, Gallagher TF (1965) Effect of thyroid on hydroxylation of estrogen in man. J Clin Endocrinol Metab 25:365–368

Goebelsmann U, Chen L-C, Saga M, Nakamura RM, Jaffe RB (1973) Plasma concentrations and protein binding of oestriol and its conjugates in pregnancy. Acta Endocrinol Copenh 74:592–604

Gurpide E, Tseng L, Gusberg SB (1977) Estrogen metabolism in normal and neoplastic endometrium. Am J Obstet Gynecol 129:809–816

Hayes CL, Spink DC, Spink BC, Cao JQ, Walker NJ, Sutter TR (1996) 17β-Estradiol hydroxylation catalysed by human cytochrome P450 1B1. Proc Natl Acad Sci USA 93: 9776–9781

Hobkirk R, Green RN, Nilsen M, Jennings BA (1974) Direct conversion of 17β-estradiol-3-glucosiduronate and 17β-estradiol-3-sulfate to their keto forms by human kidney homogenates. Can J Biocem 52: 15–20

Hobkirk R, Nilsen M (1970) Metabolism of estrone-3-glucosiduronate and 17β-estradiol-3-glucosiduronate on the human female. Steroids 15:649–667

Hobkirk R, Mellor JD, Nilsen M (1975) In vitro metabolism of 17β-estradiol by human liver tissue. Can J Biochem 53:903–906

Hochberg RB, Bandy L, Ponticorvo L, Welch M, Leiberman S (1979) Naturally occurring lipoidal derivatives of 3β-hydroxy-5-pregnen-20-one; $3\beta,17\beta$-dihydroxy-5-pregnen-20-one and 3β-hydroxy-5-androsten-17-one. J Steroid Biochem 11:333–340

James VHT, Reed MJ (1980) Steroid hormones and human cancer. Prog Cancer Res Ther 14:471–487

Kerlan V, Dreano Y, Bercovici JP, Beaunne PH, Floch HH, Berthou F (1992) Nature of cytochromes P450 involved in the 2-/4-hydroxylations of estradiol in human liver microsomes. Biochem Pharmacol 44:1745–1756

Krozowski Z (1992) 11β-Hydroxysteroid dehydrogenase and the short-chain alcohol dehydrogenase (SCAD) superfamily. Molec Cell Endocrinol 84:C25–C31

Krozowski Z (1994) The short-chain alcohol dehydrogenase superfamily: variations on a common theme. J Steroid Biochem Molec Biol 51:125–130

Lesko SM, Rosenberg L, Kaufman DW, Helmrich SP, Miller DR, Strom B, Schottenfeld D, Rosenshein NB, Knapp RC, Lewis J, et al. (1985) Cigarette smoking and the risk of endometrial cancer. N Engl J Med 313:593–596

Liehr JG (1990) Genotoxic effects of estrogens. Mutat Res 238:269–276

Liehr JC, Ricci MJ, Jefcoate CR, Hannigan EV, Hokanson JA, Zhu BT (1995) 4-Hydroxylation of estradiol by human uterine myometrium and myoma microsomes: implications for the mechanism of uterine tumourigenesis. Proc Natl Acad Sci USA 92:9220–9224

Lim SK, Won YJ, Lee JH, Kwon SH, Lee EJ, Kim KR, Lee HC, Huh KB, Chung BC (1997) Altered hydroxylation of estrogen in patients with postmenopausal osteoporosis. J Clin Endocrinol Metab 82:1001–1006

Larner JM, Hochberg RB (1985b) The clearance and metabolism of estradiol-17-esters in the rat. Endocrinology 117:1209–1214

Larner JM, Pahuja SL, Brown VM, Hochberg RB (1992) Aromatase and testosterone fatty acid esters: the search for a cryptic biosynthetic pathway of estradiol esters. Steroids 57:475–479

Larner JM, Pahuja SL, Shackleton CH, McMurray WJ, Giordano G, Hochberg RB (1993) The isolation and characterization of estradiol – fatty acid esters in human ovarian follicular fluid. J Biol Chem 268:13893–13899

Larner JM, MacLusky NJ, Hochberg RB (1985a) The naturally occurring C-17 fatty acid esters of estradiol are long-acting estrogens. J Steroid Biochem 22:407–413

Larner JM, Shackleton CHL, Roitman E, Schwartz PE, Hochberg RB (1992) Measurement of estradiol-17-fatty acid esters in human tissues. J Clin Endocrinol Metab 75:195–200

Longcope C (1990) Relationship of estrogen to breast cancer, of diet to breast cancer, and of diet to estradiol metabolism. J Natl Cancer Inst 82:896–897

Longcope C, Johnston CC (1988) Androgen and estrogen dynamics in pre- and post-menopausal women: a comparison between smokers and non-smokers. J Clin Endocrinol Metab 67:379–383

Longcope C, Gorbach S, Goldin B, Woods M, Dwyer J, Morrill A, Warram J (1987) The effect of a low fat diet on estrogen metabolism. J Clin Endocrinol Metab 64:1246–1250

Mancini MA, Song CS, Rao TR, Chatterjee B, Roy AK (1992) Spatio-temporal expression of estrogen sulfotransferase within the hepatic lobule of male rats: implications for in situ inactivation. Endocrinology 131:1541–1546

Martel C, Rheaume E, Takahashi M, Trudel C, Couet J, Luu-The V, Simard J Labrie F (1992) Distribution of 17β-hydroxysteroid dehydrogenase gene expression and activity in rat and human tissue. J Steroid Biochem Molec Biol 41:597–603

Martucci C, Fishman J (1977) Direction of estradiol metabolism as a control of its hormonal action: uterotropic activity of estradiol metabolites. Endocrinology 101: 1709–1715

Martucci C, Fishman J (1979) Impact of continuously administered catechol estrogens in uterine growth and luteinizing hormone secretion. Endocrinology 105:1288–1292

Martucci CP, Fishman J (1993) P450 enzymes of estrogen metabolism. Pharmac Ther 57:237–257

Matsui M, Homma H (1994) Biochemistry and molecular biology of drug-metabolizing sulfotransferase. Int J Biochem 26:1237–1247

Michnovicz JJ, Adlercreutz H, Bradlow HL (1997) Changes in levels of urinary estrogen metabolites after oral indole-3-carbinol treatment in humans. J Natl Cancer Inst 89:718–723

Michnovicz JJ, Bradlow HL (1990) Induction of estradiol metabolism by dietary indole-3-carbinol in humans. J Natl Cancer Inst 82:947–949

Michnovicz JJ, Bradlow HL (1991) Altered estrogen metabolism and excretion in humans following consumption of indole-3-carbinol. Nutr Cancer 16:59–66

Michnovicz JJ, Galbraith RA (1991) Cimetidine inhibits catechol estrogen metabolism in women. Metab 40:170–174

Michnovicz JJ, Herscopf RJ, Haley NJ, Bradlow HL, Fishman J (1989) Cigarette smoking alters hepatic estrogen metabolism in men: implications for atherosclerosis. Metab 38:537–541

Michnovicz JJ, Hershcopf RJ, Naganuma H, Bradlow HL, Fishman J (1986) Increased 2-hydroxylation of estradiol as a possible mechanism for the anti-estrogenic effect of cigarette smoking. N Engl J Med 315:1305–1309

Michnovicz JJ, Naganuma H, Hershcopf RJ, Bradlow HL, Fishman J (1988) Increased urinary catechol estrogen excretion in female smokers. Steroids 52:69–83

Nebert DW (1991) Proposed role of drug-metabolizing enzymes: regulation of steady state levels of the ligands that effect growth, homeostasis, differentiation and neuroendocrine function. Molec Endocrinol 5:1203–1214

Niwa T, Swaneck G, Bradlow HL (1994) Alterations in estradiol metabolism in MCF-7 cells by treatment with indole-3-carbinol and related compounds. Steroids 59:523–527

Noel CT, Reed MJ, Jacobs HS, James VHT (1981) The plasma concentrations of oestrone sulphate in postmenopausal women: lack of diurnal variation, effect of ovariectomy, age and weight. J Steroid Biochem 14:1101–1105

Numazawa M, Satoh S (1989) Kinetic studies of inhibition of estradiol 2- and 16α-hydroxylases with 17α-ethynyl and 17β-cyanosteroids: preferential inhibition of the 16α-hydroxylase. J Steroid Biochem 32:85–90

Osawa Y, Higashiyama T, Shimizu Y, Yarborough C (1993) Multiple functions of aromatase and the active site structure; aromatase is the placental 2-hydroxylase. J Steroid Biochem Molec Biol 44:469–480

Osborne MP, Bradlow HL, Wong GYC, Telang NT (1993) Upregulation of estradiol C16α-hydroxylation in human breast tissue: a potential marker of breast cancer risk. J Natl Cancer Inst 85:1917–1920

Pasqualini JR, Gelly C, Nguyen B-L, Vella C (1989) Importance of estrogen sulfates in breast cancer. J Steroid Biochem 34:155–163

Poutanen M, Moncharmont B, Vihko R (1992) 17β-Hydroxysteroid dehydrogenase gene expression in human breast cancer: regulation of expression by a progestin. Cancer Res 45:897–900

Purba HS, King EJ, Richert P, Bhatnagar AS (1994) Effect of aromatase inhibitors on estrogen 2-hydroxylase in rat liver. J Steroid Biochem Molec Biol 48:215–219

Purohit A, De Giovanni CV, Reed MJ (1999) The regulation of estrone sulfate formation in breast cancer cells. J Steroid Biochem Molec Biol 68:129–135

Reed MJ, Murray MAF (1979) The estrogens. In: Gray CH, James VHT (eds) Hormones in Blood Vol 3, Academic Press, London, pp 263–353

Reed MJ, Purohit A (1993) Sulphatase inhibitors: the rationale for the development of a new endocrine therapy. Rev Endocr-Rel Cancer 45:51–62

Reed MJ, Purohit A (1997) Breast cancer and the role of cytokines in regulating estrogen synthesis: an emerging hypothesis. Endocrine Reviews 18:701–715

Roy AK (1992) Regulation of steroid hormone action in target cells by specific hormone-inactivating enzymes. Proc Soc Exp Biol Med 199:265–272

Roy D, Weisz J, Liehr JG (1990) The O-methylation of 4-hydroxy estradiol is inhibited by 2-hydroxy estradiol: implications for estrogen-induced carcinogenesis. Carcinogenesis 11:459–462

Ruder HJ, Loriaux DL, Lipsett MB (1972) Estrone sulphate: production rates and metabolism in man. J Clin Invest 51:1020–1023

Sanchis D, Balada F, Del Mar Grasa M, Virgili J, Peinado J, Monserrat C, Fernandez-Lopez J-A, Remesar X, Alemany M (1996) Oleoyl-estrone induces the loss of body fat in rats. Int J Obesity 20:588–594

Schatz F, Hochberg RB (1981) Lipoidal derivatives of estradiol: the biosynthesis of a non-polar estrogen metabolite. Endocrinology 109:697–703

Singh A, Reed MJ (1991) Insulin-like growth factor I and insulin-like growth factor II stimulate oestradiol 17β-hydroxysteroid dehydrogenase (reductive) activity in breast cancer cells. J Endocrinol 129:R5–R8

Snow RC, Barbieri RL, Frisch RE (1989) Estrogen 2-hydroxylase oxidation and menstrual function among elite oarswomen. J Clin Endocrinol Metab 69:369–376

Stewart PM, Sheppard MC (1992) Novel aspects of hormone action: intracellular ligand supply and its control by a series of tissue specific enzymes. Molec Cell Endocrinol 83:C13-C18

Strott CA (1996) Steroid sulfotransferases. Endocrine Reviews 17:670–697

Swaneck GE, Fishman J (1988) Covalent binding of endogenous 16α-hydroxyestrone to estradiol receptor in human breast cancer cells: characterization and intra nuclear location. Proc Natl Acad Sci USA 85:7831–7835

Taioli E, Garte SJ, Trachman J, Garbers S, Sepkovic DW, Osborne MP, Mehl S, Bradlow HL (1996) Ethnic differences in estrogen metabolism in healthy women. J Natl Cancer Inst 88:617

Telang NT, Axelrod DM, Wong GY, Bradlow HL, Osborne MP (1991b) Biotransformation of estradiol by explant culture of human mammary tissue. Steroids 56:37–43

Telang NT, Narayanan R, Bradlow HL, Osborne MP (1991a) Coordinated expression of intermediate biomarkers for tumorigenic transformation in RAS transfected mouse mammary epithelial cells. Breast Cancer Res Treat 13:155–163

Telang NT, Suto A, Wong GY, Osborne MP, Bradlow HL (1992) Induction by estrogen metabolite 16α-hydroxyestrone of genotoxic damage and aberrant proliferation in mouse mammary epithelial cells. J Natl Cancer Inst 84:634–638

Tiwari RK, Guo L, Bradlow HL, Telang NT, Osborne MP (1994) Selective responsiveness of human breast cancer cells to indole-3-carbinol, a chemopreventive agent. J Natl Cancer Inst 86:126–131

Tseng L, Gurpide E (1979) Stimulation of various 17β- and 20α-hydroxysteroid dehydrogenase activities by progestins in human endometrium. Endocrinology 104:1745–1748

Weisz J, Bui QD, Roy D, Leihr J (1992) Elevated 4-hydroxylation of estradiol by hamster kidney microsomes: a potential pathway of metabolic activation of estrogens. Endocrinology 131:655–661

Whitnall MH, Driscoll WJ, Lee YC, Strott CA (1993) Estrogen and hydroxysteroid sulfotransferases in guinea pig adrenal cortex: cellular and subcellular distribution. Endocrinology 133:2284–2291

Wild MJ, Rudland PS, Back DJ (1991) Metabolism of the oral contraceptive steroids ethynylestradiol and norgestimate by normal (Huma7) and malignant (MCF-7 and ZR75-1) human breast cancer cells in culture. J Steroid Biochem Molec Biol 39:535–543

Wu L, Einstein M, Geissler WM, Chan HK, Elliston KO, Andersson S (1993) Expression cloning and characterization of human 17β-hydroxysteroid dehydrogenase Type 2, a microsomal enzyme possessing 20α-hydroxysteroid dehydrogenase activity. J Biol Chem 268:12964–12969

Yamazoe Y, Nagata K, Osawa S, Kato R (1994) Structural similarity and diversity of sulfotransferases. Chemico-Biol Interactions 92:107–117

Young BK, Jirku H, Kadner S, Levitz M (1976) Renal clearances of estriol conjugates in normal human pregnancy at term. Am J Obstet Gynecol 126:38–42

Young TB, Wolf DA (1988) Case-control study of proximal and distal colon cancer and diet in Wisconsin. Int J Cancer 42:167–175

Zhu BT, Bui QD, Weisz J, Leihr JG (1994) Conversion of estrone to 2- and 4-hydroxyestrone by hamster kidney and liver microsomes: implications for the mechanisms of estrogen-induced carcinogenesis. Endocrinology 135:1772–1779

Zhu BT, Liehr JG (1993) Inhibition of the catechol-O-methyl-transferase-catalyzed O-methylation of 2- and 4-hydroxyestradiol by catecholamines: implications for the mechanism of estrogen-induced carcinogenesis. Arch Biochem Biophys 304:248–256

CHAPTER 12

Metabolism of Endogenous Estrogens

T.H. Lippert, H. Seeger, and A.O. Mueck

A. Introduction

The study of the metabolism of estradiol began many years ago; it was primarily the chemical structures of the various breakdown products that were closely described. Only later was the effort made to research further into the enzyme systems required for that breakdown and, recently, into the receptors to which the metabolites bind in the tissues.

Interest in estradiol metabolites increased greatly with the discovery that these are not merely inactive breakdown products waiting to be excreted, but they can possess effects of their own that fulfil important functions in the organism, independent of the effect of the parent substance. At present, there are a number of indications that they exert special functions in the physiology and pathophysiology of the human organism and, from this perspective, are also of clinical importance. It may be assumed that estradiol achieves some of its varied effects not alone but with the aid of its metabolites. Assumptions of possible effects of metabolites led, as early as the 1960s, to studies on cholesterol-lowering effects, with the result that in animal experiments some metabolites showed stronger effects than estradiol (Gordon et al. 1964).

I. Estradiol Breakdown Mechanisms

Interest in estrogen metabolism has, so far, concentrated mainly on the female organism. Doubtlessly, estradiol secretion is considerably higher in women during their reproductive phase than in men. After the menopause, however, with cessation of ovarial secretion, it approaches that of men. According to the current state of knowledge, the mechanism of metabolic breakdown of estradiol is, in principle, the same in both men and women. It takes place almost exclusively by oxidation.

The first step is the transformation of estradiol into estrone by oxidation in the C17-position, a process that is reversible. However, the equilibrium tends to favor estrone formation, which is also characterized by the fact that the breakdown of estradiol into estrone takes place quickly, and the back reduction is considerably slower (Fishman et al. 1960). The further breakdown from estrone takes place in two different ways: on the one hand, through hydroxylation of the A-ring and, on the other through hydroxylation of the D-ring. The products of the two paths of metabolism take place via separate enzyme systems (Martucci and Fishman 1993), once formed are no longer reversible, and lead in the A-ring to 2-hydroxyestrone and 4-hydroxyestrone, and in the D-ring to 16-hydroxyestrone and estriol (Fishman and Martucci 1980b). Figure 1 shows the chemical structures of the main metabolites involved in the breakdown.

Breakdown paths leading to other metabolites have been described, but play only a subordinate part in the human organism. Table 1 gives a survey of the numerous known metabolites with the structural ring that is involved in the metabolic breakdown. The metabolites possess a broad spectrum of effects of their own (Zhu and Conney 1998), which very probably are not all known yet. The biological activities of the representatives of the two main paths of breakdown can in some cases antagonize each other. Thus, in the metabolites of the A-ring, the catecholestrogens, estrogenicity largely disappears and can even change into antiestrogenicity; the metabolites of the D-ring, however, still develop strong estrogenic effects that may be stronger than those of the parent substance estradiol.

Recently, an anti-carcinogenic effect (Bradlow et al. 1996) and preventive functions in the cardiovascular system (Lippert et al. 1999) have been ascribed to catechol estrogens, together with not-yet clearly defined functions in the brain. Genotoxic and hence carcinogenic qualities are assigned to the D-ring metabolite 16α-hydroxyestrone (Fishman et al. 1995). In principle,

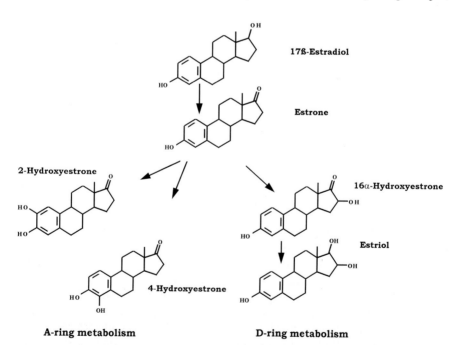

Fig. 1. Chemical structures of the main metabolites of estradiol breakdown

Table 1. Hydroxylated and keto metabolites of 17β-estradiol and estrone according to ZHU and CONNEY (1998)

Position of oxidation	Common name
A-Ring metabolites	1-Hydroxyestrone
	2-Hydroxyestrone
	2-Hydroxyestradiol
	2-Hydroxyestriol
	4-Hydroxyestrone
	4-Hydroxyestradiol
	4-Hydroxyestriol
B-Ring metabolites	6α-Hydroxyestrone
	6β-Hydroxyestrone
	6-Ketoestrone
	6α-Hydroxyestradiol
	6β-Hydroxyestradiol
	6-Ketoestradiol
	6-Ketoestriol
	6α-Hydroxyestriol
	7α-Hydroxyestrone
	7β-Hydroxyestrone
	7α-Hydroxyestradiol
	7β-Hydroxyestradiol
	7-Ketoestradiol
	7α-Hydroxyestriol
C-Ring metabolites	11β-Hydroxyestrone
	11-Ketoestrone
	11β-Hydroxyestradiol
	11-Ketoestradiol
	(11)-Dehydroestradiol-17α
	(9,11)-Dehydroestrone
	14α-Hydroxyestrone
	14α-Hydroxyestradiol
D-Ring metabolites	15α-Hydroxyestrone
	15β-Hydroxyestrone
	15α-Hydroxyestradiol
	15α-Hydroxyestriol (estetrol)
	16α-Hydroxyestrone
	16β-Hydroxyestrone
	16-Ketoestrone
	16α-Hydroxyestradiol (estriol)
	16-Epiestriol
	16-Ketoestradiol
	16,17-Epiestriol
	17α-Estradiol
	17-Epiestriol

the metabolites can all be made more hydrophilic and are, thus, capable of excretion by means of sulfonation or glucuronidation (Musey et al. 1979; Hobkirk 1985; Zhu and Conney 1998), as concerns excretion via the kidney to the urine, and via the liver and gall bladder into the feces. The processes of conjugation are reversible; the enterohepatic circulation in particular depends on this reversibility. In the case of the metabolites of the A-ring, the catechol estrogens, the further breakdown takes place chiefly by methylation at a hydroxy group (Ball and Knuppen 1980). Methylation processes that are not necessarily accompanied by inactivation of the biological features of the breakdown product are reversible.

The cooperation of the two paths of excretion, i.e., urine and feces, has been less studied. As the mode of excretion can be very varied, it is only possible to cover overall metabolism by measuring the excretory products of the two paths. The distribution of excretion – either via urine or feces – appears to depend, among other things, on diet. The intestinal microflora are thought to play a key part in the enterohepatic circulation; the deconjugation of the biliary metabolites renders possible a reabsorption and, thus, keeps the enterohepatic circulation going. By means of a fiber-rich diet, but also by means of certain drugs, such as antibiotics, bacterial activity is restricted and the metabolites are excreted increasingly in the feces and less in the urine.

The hormonal status may also play a part. In a study of human subjects of both genders, in postmenstrual women and in men the excretion of overall estrogens in the feces was similar. However, in women of reproductive age, the fecal excretion was considerably higher (Adlercreutz and Jarvenpaa 1982). A vegetarian regime can raise fecal estrogen excretion several times, with an accompanying fall in urine metabolites and lowering of the estrogen concentration in the blood (Goldin et al. 1986; Gorbach and Goldin 1987).

The human intestine and its content also appear to carry out breakdown processes, such as the transformation of 16α-hydroxyestrone to 15α-hydroxyestrone (Adlercreutz et al. 1976). A radiometric method introduced in the 1960s permits measurement of metabolic processes in the organism as a whole after injection of ^{3}H- and ^{14}C-labeled estrogens into the body, thus avoiding laborious examinations of two-way excretion (Fishman et al. 1980).

II. Sex Differences in Estradiol Metabolism

Differences in estradiol breakdown in men and women were examined using radioactively labeled estradiol. The results did not demonstrate any sex-specific different metabolic pathways, but quantitative differences in breakdown were registered (Fishman et al. 1960, 1972, 1980; Zumoff et al. 1968a). The differences found in breakdown between men and women are shown in Fig. 2. Men and women metabolize estradiol injected into the blood equally quickly into estrone, the half-time of reaction being approximately 10 min for both males and females. There is, however, one difference, in that

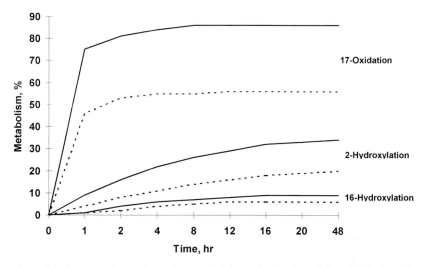

Fig. 2. Oxidative transformation of estradiol in male (----) and female (—) subjects according to FISHMAN et al. 1980. Cumulative percentage curves for the metabolism of estradiol by 17-oxidation, 2-hydroxylation, and 16-hydroxylation. The curves represent average values for male and female subjects

women convert considerably more estradiol into estrone than men, at a percentage rate of about 80% of the administered substrate, while in men it was only 60%: this difference was highly statistically significant. The further breakdown of the A-ring to the 2-hydroxy metabolites takes place more slowly, the half-time for both men and women being about 2 h. Here, too, the metabolism in women was significantly higher than in men, with 35% in women and 20% in men. Slowest was the further metabolism, proceeding from the D-ring by C16-hydroxylation, with a half-time for both genders of about 8 h. There were no longer any sex-specific differences; the figures were between 7% and 10% of the initial value.

The investigations clearly show that the catechol estrogens of the C2-position are the dominant metabolite fraction in both sexes, though to a greater extent in women than in men. In men, a large fraction (40%) of estradiol is neither metabolized nor excreted in the urine. In women, this fraction (20%) was about 50% smaller. Thus, women have available a considerably higher amount of substrate for C2- and C16-hydroxylation. Considering the quantitative difference in estrogen formation in men and women, the difference in the metabolic breakdown process, relating to the ratio of 2- to 16-hydroxymetabolites, is of still greater significance.

FISHMAN and MARTUCCI (1980b) have already compared values for estradiol metabolites in women and men in plasma and urine. As can be seen from Table 2, the differences of plasma concentrations of estradiol are greatest, but become considerably smaller for the metabolites estrone and 2-hydroxyestrone, the value in men approaching that of women in the latter case.

Table 2. Estradiol metabolites in healthy men and women according to Fishman and Martucci 1980b

	Women		Men
	Follicular	Luteal	
Plasma concentrations (pg/ml)			
Estrone	76	81	43
Estradiol	134	136	33
Estriol	20	25	<10
2-Hydroxyestrone	62	71	48
Urinary concentrations (μg/24h)			
Estrone	9.8	10.1	3.1
Estradiol	7.3	6.7	2.7
Estriol	11.1	13.0	2.9
2-Hydroxyestrone	18.2	22.5	9.7
2-Methoxyestrone	14.2	16.4	12.9

The differences between men and women are also apparent in urine excretion. The value of catechol estrogens of men come closest to those of women.

III. Catechol Estrogen Metabolism

Intensive research has dealt with the metabolism of the largest breakdown fraction, the catechol estrogens. However, attention was not directed to gender differences; most of the studies were carried out on men. Kinetic studies showed that the chief metabolite of the catechol estrogens, 2-hydroxyestrone, disappears from the blood very rapidly, faster than any other natural steriod (Merriam et al. 1980; Longcope et al. 1982, 1983). The transformation takes place initially in the blood in the presence of the enzyme catechol-O-methyltransferase of the erythrocytes (Bates et al. 1977), but part is also taken up by the tissue, where it is both methylated and conjugated. Hydroxyestrone, which appears in the urine in the non-methylated form, is thought to have been returned back to its initial form by demethylation during the excretion process. The clearance of 2-methoxyestrone proceeds considerably more slowly than that of 2-hydroxyestrone (Longcope et al. 1982).

Whereas 2-hydroxyestrone makes up the largest catechol estrogen fraction, to a considerably lesser extent the catechol estrogen 2-hydroxyestradiol is formed by direct hydroxylation of estradiol (Kono et al. 1982). Clearance from the blood proceeds about 50% more slowly than in that of 2-hydroxyestrone, but is still considerably faster than in the case of other steroids. It is also methylated principally by the catechol-O-methyltransferase of the erythrocytes. It was noticed early on that 2-methoxyestradiol differs from the other metabolites with respect to its biological activity (Fishman and Martucci 1980b). It shows a very strong affinity to sex-hormone binding globulin (SHBG), which is greater than that of testosterone and estradiol (Dunn et al. 1980). Furthermore, it possesses a strong angiogenesis-inhibiting effect which, in animals, has led to a significant inhibition of tumor growth (FOTSIS et al. 1994).

Further catechol estrogens formed by hydroxylation in the C4-position are 4-hydroxyestradiol and 4-hydroxyestrone. From a quantitative point of view, the C4-hydroxylated catechol estrogens also play a subordinate part in estrogen metabolism (BALL et al. 1977; EMONS et al. 1987). In their further breakdown, they differ from the 2-hydroxyestrogens by a slower clearance, and conjugation is thought more important than methylation. 4-Hydroxyestradiol can be transformed into 4-hydroxyestrone. After administering 4-methoxyestrone, it was largely excreted in the demethylated form.

The 4-hydroxyestrogens have, repeatedly, aroused interest because of their biological properties. They still possess some estrogenicity (FISHMAN and MARTUCCI 1980b) and are associated with the development of tumors in animal experiments (YAGER and LIEHR 1996).

Some research teams have attributed great importance to the relationship between A-ring breakdown with chief representative 2-hydroxyestrone and D-ring breakdown with chief representative 16α-hydroxyestrone. In healthy subjects, the A-ring metabolism should outweigh the D-ring metabolism. The ratio of 2- to 16α-hydroxyestrone in the urine in premenopausal women (CHEN et al. 1996) has shown that the concentration of 2-hydroxyestrone during the whole menstrual cycle was more than twice that of 16α-hydroyxyestrone. The ratio remained fairly constant during the day. It rose during the luteal phase by increased production of the 2-hydroxyestrone by about 50% above the value in the follicular phase. As the number of cases studied was relatively small, no final statement is possible. The ratio was also tested in a larger group of healthy premenopausal women (PASAGIAN-MACAULAY et al. 1996). The variation compass of the results was small. Intervention by means of dietary fat reduction, reduction in weight and increased physical activity showed no significant influence on the ratio. Further studies on healthy post-menopausal women (LIPPERT et al. 1998) showed that despite a considerably lower metabolite excretion, the quotient of A- to D-ring metabolism does not diverge essentially from the values measured in the premenopausal phase (MUECK et al. 1998), i.e., a predominance of 2-hydroxy over 16-hydroxy metabolism was registered. It is to be noted, however, that in the postmenopausal phase, a wider variation of the ratio values predominated.

B. Estrogen Metabolism and Disease

For some time now, interests have been focused on by the breakdown of estradiol under pathophysiological conditions in the organism. Although changes resulting from disease may occur at each step of the breakdown, research has concentrated primarily on the relationship between the two main breakdown paths of the A- and D-rings. The clinical significance of deviations from the normal pattern of breakdown met for a long time with little general attention.

Table 3. Influence of diseases and drugs on estradiol metabolism in human beings

	Result	Reference
Diseases		
Breast cancer	C16-Hydroxylation increased	ZUMOFF et al. (1966); SCHNEIDER et al. (1982); FISHMAN et al. (1984); KABAT et al. (1997)
Endometrial cancer	C16-Hydroxylation increased	FISHMAN et al. (1984)
Cervical cancer	C16-Hydroxylation increased	BRADLOW et al. (1995d); SEPKOVIC et al. (1995)
Prostate cancer	C16-Hydroxylation increased	ZUMOFF et al. (1980)
Larynx papillomas	C16-Hydroxylation increased	NEWFIELD et al. (1993)
Liver cirrhosis	C16-Hydroxylation increased	ZUMOFF et al. (1968b)
Lupus erythematosus	C16-Hydroxylation increased	LAHITA et al. (1979, 1981, 1982, 1984)
Thyroid diseases		
Hyperthyreosis	C2-Hydroxylation increased	FISHMAN et al. (1962b, 1965)
Hypothyreosis	C16-Hydroxylation increased	FISHMAN et al. (1965)
Weight changes		
Obesity	C2-Hydroxylation decreased	SCHNEIDER et al. (1983)
Anorexia nervosa	C2-Hydroxylation increased	FISHMAN et al. (1975)
Osteoporosis	C2-Hydroxylation increased	LIM et al. (1997)
	C16-Hydroxylation decreased	
Drugs		
Cimetidine	C2-Hydroxylation decreased	GALBRAITH and MICHNOVICZ (1989); MICHNOVICZ and GALBRAITH (1991)
Ranitidine	C2-Hydroxylation unchanged	GALBRAITH and MICHNOVICZ (1989)
Omeprazole	C2-Hydroxylation unchanged	GALBRAITH and MICHNOVICZ (1993)
Thyroxine	C2-Hydroxylation increased	MICHNOVICZ and GALBRAITH (1990)
Indole-3-carbinol	C2-Hydroxylation increased	MICHNOVICZ and BRADLOW (1990, 1991); BRADLOW et al. (1995c); MICHNOVICZ et al. (1997)
Omega-3-fatty acids	C16-Hydroxylation decreased	OSBORNE et al. (1988)
Tranylcypromine	C2-Hydroxylation decreased	BANGER et al. (1990)

It was only recently realized that this knowledge might lead to a better under-standing of the etiology of pathological changes in the organism and, thus, also to new aspects for the diagnosis and therapy of diseases. Table 3 summarizes the metabolic changes of the two breakdown paths that have been registered in diseases or in taking drugs. The results were obtained, in most cases, using radiometric methods, and comparison was made with healthy subjects so that deviations could clearly be registered.

The view initially put forward that the two breakdown paths are closely connected and react competitively has not been able to be proved unambigu-ously. However, it has emerged that A-ring breakdown with the formation of catechol estrogens can be induced more rapidly and easily, while D-ring break-down reacts slowly to changes in the organism and is much harder to alter by external influences. Recent studies, however, still permit the assumption of a certain connection between the two systems. Thus, in treating healthy subjects with indole-3-carbinol, a food constituent, it was registered that a rise in the A-ring metabolism was regularly associated with a decline in D-ring metabo-lism (MICHNOVICZ et al. 1997). The proof of a mutual dependence has, however, yet to be made. In Table 3, only those changes in the breakdown path that dom-inated over the others have been cited.

I. Breast Cancer

As is generally known, biosynthesis and metabolism of estradiol take place in neoplastic breast tissue and the surrounding fat tissue. The enzyme activ-ities of aromatases, estradiol dehydrogenase and estrone sulfatase appear to have in many cases a close connection with the growth of breast tumors (CASTAGNETTA et al. 1996; PUROHIT et al. 1996; SIMPSON and ZHAO 1996).

The two breakdown paths proceeding from estrone were examined in cases of breast cancer in women and in men. As can be seen from Table 3, an increase of D-ring metabolism was registered (ZUMOFF et al. 1966; SCHNEIDER et al. 1982; FISHMAN et al. 1984; KABAT et al. 1997). Tissues of the human breast, i.e., the mammary terminal duct lobular units (TDLU) were examined for the biotransformation of estradiol via the C16-hydroxylation pathway. The extent of this metabolism was 4.5-fold higher in TDLU obtained from patients under-going mastectomy for cancer than that observed in TDLU obtained from reduction-mammoplasty patients who did not have cancer (OSBORNE et al. 1993).

The observation is interesting that C16-hydroxylation in the breast corre-lates with the phase of the menstrual cycle in which the tissue was extracted, i.e., the 16-hydroxylated metabolite was higher during the luteal phase than the follicular phase (FISHMAN et al. 1995). It was speculated that the different metabolism represents a tissue reaction to the blood estradiol level, with estra-diol inducing the metabolic reaction as a substrate.

It is surprising that the metabolic changes in cases of breast cancer can also be registered in the organism as a whole, e.g., by urine excretion. This is documented by the results of the already-mentioned studies cited in Table 3, which found a raised concentration of D-ring metabolites in urine.

In one recent study, these results obtained in small groups were checked against a larger number of cases (KABAT et al. 1997). It emerged that the ratio of the estrogen metabolites 2- to 16α-hydroxyestrone was highly significantly changed in favor of the D-ring metabolite in postmenopausal mamma-carcinoma patients. The authors assume that the increase of 16α-hydroxyestrone plays an important role in breast carcinogenesis; however, there is still a lack of large-scale studies to further confirm the association between breast cancer and an increase in D-ring metabolism.

Numerous animal and in vitro studies have been carried out on the biological activities of the D-ring metabolite 16α-hydroxyestrone. The strong estrogenic effect, which has been long known (FISHMAN and MARTUCCI 1980a), was therefore already long associated with the development of tumors of target organs of estradiol. Attention was drawn to the fact that a positive correlation exists between an upregulation of estradiol C16α-hydroxylation and the increase in the risk of breast cancer, the risk being very probably based on genotoxic properties of this metabolite (FISHMAN et al. 1995). Although 16α-hydroxyestrone has a weaker binding affinity than estradiol for the estrogen receptor and SHBG, once bound to the estrogen receptor this binding is irreversible, in contrast to estradiol. 16α-Hydroxyestrone is covalently bound to the nuclear matrix and can cause toxic DNA damage with aberrant hyperproliferations; this process shows resemblances to the effects of chemical carcinogens in cell-culture models. The reaction of 16α-hydroxyestrone with the transcription factor estrogen receptor appears unique insofar as it is irreversible and leads to aberrant gene expression (FISHMAN et al. 1995). However, since 16α-hydroxyestrone is a normal constituent of estradiol metabolism, it has to be found out under what conditions such irreversible bindings come about and whether there are measures to prevent this.

There are some interesting reports on the effect of albumin on the metabolic transformation of estradiol in human cells of the endometrium and the breast. Albumin leads, when given into carcinoma cell cultures, to a significant increase in 16α-hydroxy formation, while the 2-hydroxyestrone formation is significantly lowered (BRADLOW et al. 1995a). As the prognosis for tumor growth is supposed to be dependent in certain cases on the albumin concentration in the tumor tissue (CASTAGNETTA et al. 1996), it may be that the change in the estradiol metabolism plays a decisive part in the stimulation of growth. To sum up, it may be observed that there are indications that estrogen metabolism plays an important part in breast tumor development.

II. Endometrial Cancer

As the endometrium, like the breast, is a target organ of estradiol, patients with endometrial cancer were also included in studies of estradiol metabolism (FISHMAN et al. 1984). As with breast cancer, for this too an increase in estradiol metabolism via the D-ring was found. Although the findings were registered only in a small number of patients, this indicates that there may be similar relationships between endometrial cancer and breast cancer, which hitherto have aroused the greatest interest. The fact that the findings were made in postmenopausal women, with no further monthly shedding of the endometrium, raises the question as to whether the changes in estrogen metabolism proceed from the tissue of the endometrium, and are thus regarded as a result of the disease, or whether the increased D-ring metabolism pre-existed and may have contributed to triggering the cancer formation.

III. Cervical Cancer

The origin and growth of cervical cancer are nowadays connected with the infection of the tissue with human papilloma virus (HPV). Because a correlation was found between the growth of virus-bearing tumors and production of 16α-hydroxyestrone in animal experiments with mice (BRADLOW et al. 1985), interest grew in the reaction of estrogen metabolism in HPV-induced tumors. In vitro studies had shown, on the one hand, that cells of the transformation zone of the cervix are able to form 16α-hydroxyestrone, while on the other the growth of cervical cells immortalized with HPV can be strongly stimulated with 16α-hydroxyestrone (AUBORN et al. 1991).

In women with cervical intraepithelial neoplasma, the urinary excretion of 2- and 16α-hydroxyestrone was examined and compared with that of a group of healthy women (BRADLOW et al. 1995d; SEPKOVIC et al. 1995). In 141 women, carcinomatous changes had been proven by means of biopsies, and 132 women served as controls. The calculated ratio dropped significantly against the control value according to the severity of the degree of disease, i.e., the histopathological changes [cervical intra-epithelial neoplasia (CIN) I to CIN II/III, $P = 0.03$]. Thus, for the first time, it was shown that the cancer process in the epithelial region of the cervix is accompanied by a change in estradiol metabolism in favor of the breakdown of the D-ring.

IV. Prostate Cancer

Although the prostate is regarded as a prototype of an androgenously dependent tissue, there is today proof that estrogens can induce the division and differentiation of epithelial cells in the prostate (CARRUBA et al. 1996). Furthermore, estradiol binding sites have been found in the prostate. In vitro, it was possible to stimulate the prostate epithelium directly, either with estrogens alone or together with androgens, the proliferative activity leading to squamous metaplasia. This is thought to favor the development of prostate hypertrophy and also of prostate cancer (CARRUBA et al. 1996).

There is little known, so far, regarding the metabolism of estradiol in prostate tissue. Of interest is a publication from 1980, which details a study of estradiol metabolism in prostate carcinoma patients using the injection of radioactively labeled estradiol (Zumoff et al. 1980). In cancer patients, a significant increase of the D-ring metabolite estriol was found by comparison with healthy control persons; 16α-hydroxyestrone was not examined in this study. Estradiol, estrone, 2-hydroxyestrone and 2-methoxyestrone were not altered. The authors presume that the prostate carcinoma influences the metabolism of estradiol in a similar way as the rare breast carcinomas in men, which had shown similar results in an earlier study.

V. Papilloma of the Larynx

The human larynx is a target organ for some subgroups of papilloma viruses that cause lesions in the cervix (Steinberg and Abramson 1985). It is known that estrogen binding to cell membranes of papillomas of the larynx is increased (Essman and Abramson 1984). The examination of human laryngeal papillomas obtained by operative removal showed, in comparison with healthy larynx tissue, a significant increase in the metabolite 16α-hydroxyestrone (Newfield et al. 1993). It was remarkable that the C16-hydroxylation of estradiol was comparatively high, even in normal larynx tissue, similar to that in breast cancer cells or the transformation zone of the cervical tissue. There were no differences between the laryngeal tissue of women and men. In the same study, it was demonstrated that both 16α-hydroxyestrone and estradiol strongly stimulated the proliferation of epithelial laryngeal tissue from both normal and papillomatous tissue, while the proliferation could be inhibited by 2-hydroxyestrone. The larynx seems to be, independent of gender, a hormonally very sensitive organ.

Clinical experiments attempting to arrest the growth of laryngeal papillomas by stimulation of 2-hydroxyestrone production by oral administration of 3-indole carbinol, a constituent of crucifer vegetables, have already shown some promising results (Bradlow et al. 1995d; Coll et al. 1997; Rosen et al. 1998).

VI. Liver Disease

It has long been known that diseases of the liver can alter the metabolism of estradiol, the liver being a chief organ of estradiol breakdown. In humans, interruption of the enterohepatic circulation, such as occurs during choleostasis, appears to be of great importance for the metabolic fate of estradiol. Changes in estradiol breakdown had already been diagnosed in the 1960s in both male and female patients (Zumoff et al. 1968b). Together with a not unexpected increase in urinary metabolite excretion, a strongly reduced excretion of the A-ring metabolites 2-hydroxyestrone and 2-methoxyestrone, and an increased excretion of the D-ring metabolites 16α-hydroxyestrone and estriol were registered. There was a remarkable reduction in the breakdown of 16α-hydroxyestrone to estriol. The overall increase in metabolites with still potent estrogenic character is presumably responsible for the symptoms of hyperestrogenicity in liver cirrhosis, which frequently leads to phenomena of feminization in men. Cholestasis was regarded by the authors as the main reason for the change in estradiol metabolism. They had observed similar metabolic changes in the case of extrahepatic biliary obstruction (HELLMAN et al. 1970) and in drug-induced biliary occlusion (ZUMOFF et al. 1970).

VII. Lupus Erythematosus

Lupus erythematosus is a disease of the auto-immune system that occurs predominantly in women. The ratio of women to men is approximately 9:1, with the frequency of the disease in women being lower before puberty and after the menopause. Ovarectomies were found to be associated with an improvement in the symptoms, and normal menstruation, oral contraception and pregnancy with an exacerbation of the symptoms. These facts indicate that the disease is closely connected with the production of sex hormones. Studies of the estradiol metabolism in both female and male patients with lupus erythematosus showed that D-ring metabolism, with production of 16α-hydroxyestrone, was increased (LAHITA et al. 1979, 1981, 1982, 1984). Some first-degree relatives also showed abnormal estradiol hydroxylation (LAHITA et al. 1982).

In the case of rheumatoid arthritis, which is also a disorder of the auto-immune system and occurs predominantly in females, sex hormones appear to play an important pathophysiological role too (CUTOLO and CASTAGNETTA 1996; LAHITA 1996). In individual cases, a predominance of the D-ring metabolism of estradiol was also observed (LAHITA et al. 1981). The mechanism of the disease process appears, however, to differ from that of lupus in that rheumatoid arthritis tends to improve during pregnancy, after hormone-replacement therapy and with the ingestion of the contraceptive pill (VAN VALLENHOVEN and MCGUIRE 1994). To what extent the known effects of estrogen on the immune system, e.g., influence on the B cells by stimulation of antibody formation, T-cell inhibition and stimulation of macrophagic activity (CHAO 1996), may also be triggered by estradiol metabolites is not yet known.

VIII. Disease of the Thyroid

It was recognized at an early stage that thyroid function can also negatively affect the metabolism of estradiol. In the 1960s, it was observed that in the case of hyperthyreosis the A-ring breakdown was intensified (FISHMAN et al. 1962b), while A-ring metabolism was reduced in hypothyreosis (FISHMAN et al. 1965). In the latter case, at the same time, an increase in D-ring breakdown was registered.

Later studies dealt with the effect of the hormone thyroxine (MICHNOVICZ and GALBRAITH 1990). In healthy male subjects, an artificial hyperthyreosis was induced by 2 weeks of treatment with thyroxine, whereupon the estradiol metabolites in the urine were measured. The findings confirmed the results obtained in hyperthyreosis patients, i.e., that C2-hydroxylation rose significantly. Further findings were that C16-hydroxylation remained constant, while the total cholesterol, the low-density lipoprotein (LDL) fraction and the apolipoprotein B-I were lowered. Estradiol, estrone, SHBG, high-density lipoprotein (HDL), very-low-density lipoprotein (VLDL), triglycerides and apolipoprotein A-I remained unaltered. The authors pointed out that, as was known from previous studies, A-ring metabolism is easily influenced, but not that of the D-ring.

IX. Weight Changes

It is well known that pathological changes in body weight accompany endocrinological changes, with disturbances of the reproductive functions. Both overweight and underweight result not only in changes in estrogen biosynthesis, but also in that of estradiol metabolism. Thus, in the case of over-weight, both in premenopausal women and in men, a significant reduction of C2-hydroxylation compared with normal persons was registered (Schneider et al. 1983). In women, the extent of the reduction of A-ring breakdown was considerably larger than in men. Furthermore, in women, a reduction in oxidation in the C17-position of estradiol was registered. C16-hydroxylation was unchanged compared with normal persons. The authors point out that the net effect of the altered estradiol metabolism in adipositas creates a hyperestrogenic status, and an increase in the aromatization of androstendione to estrone, a decrease in the SHBG level, and an increase in plasma estrone sulfate contribute to raising the incidence of reproductive dysfunction and the risk of hormone-dependent neoplasms.

In another study, the estradiol metabolism of overweight women was contrasted with that of female patients with anorexia nervosa (Fishman et al. 1975). Anorexia caused an opposite effect by significantly raising C2-hydroxylation. Since the C2-hydroxylated metabolites, C2-catechol estrogens, can behave like anti-estrogens, in anorexic patients, a hypoestrogenic status prevails, which is accompanied by reproductive dysfunctions such as amenorrhea or infertility. In general, the correction of weight brings about a normalization of the reproductive functions and should, therefore, be accompanied by a normalization of estradiol metabolism (Hershcopf and Bradlow 1987).

X. Depression

It has been known for many years that estrogens may have psychotropic effects (Herrmann and Beach 1978; McEwen et al. 1979). It was therefore speculated whether an abnormal estradiol metabolism might contribute to the development of mental diseases. In a recent study, the metabolism of depressive men was studied and compared with that of a control group of healthy men (Banger et al. 1990). The examinations were concentrated on catechol estrogen metabolism, since it is known that catechol estrogens may have some functions in the brain, although these are not yet clearly defined (Ball and Knuppen 1980).

The authors found significantly lowered values of 4-hydroxyestrogens in the urine of the mentally ill. Estradiol, estrone and 2-hydroxyestrogens showed, initially, only a tendency toward lower concentrations than in the healthy subjects. It was striking, however, that the methylated 2-hydroxyestrogens were twice as high as the non-methylated 2-hydroxyestrogens in the depressive persons.

Furthermore, in this study, the effect of treatment with the monoamine-oxidase (MAO) inhibitor tranylcypromine was tested. After 4 weeks of treatment, an improvement in the mental state of the patients could be noted. However, in all the patients, the state of the estrogen metabolites in the urine remained unchanged, i.e., there remained an increased methylation of the 2-hydroxyestrogens and a reduction in C4-hydroxylation. At the end of the treatment, both the 2-hydroxyestrogens and estradiol were significantly lowered by comparison with the control group. Causal connections of abnormal estrogen metabolism with the disease could not be traced.

XI. Osteoporosis

After the cessation of estrogen production in women with onset of the menopause, the risk of developing osteoporosis rises; this means that estrogens play an important part in the homeostasis of the bones. In recent studies, estradiol metabolism in patients with osteoporosis has been examined (LIM et al. 1997). It was noted that the A-ring metabolites were significantly higher and the D-ring metabolites significantly lower in the urine of these patients. Significant correlations could also be found between the severity of the disorder objectivized by means of measurements of bone density and the metabolite values found in the urine. The serum values of estradiol and estrone, however, did not differ between the patients and healthy control persons.

The findings indicate a close connection between abnormal estrogen metabolism and the symptoms. The considerable rise in the production of catechol estrogens, which can have an anti-estrogenic effect, increased further the hypoestrogenic status of postmenopausal women. The decline in D-ring metabolites, which still possess strong estrogenic properties, contributes further to the estrogen deficiency. This indicates that in postmenopause, a residual estrogenicity is of importance for the functioning of some vital organs. Although a number of other, still unknown, factors may be involved in the development of osteoporosis, the abnormal estrogen metabolism reveals that a marked shift of the ratio of A- to D-ring metabolism induced by some treatments should be aimed with care, as it may lead to pathological changes in the organism.

C. Drugs

Drugs are also able to influence estradiol metabolism; studies in this field are, however, rare. Cimetidine, an H_2-receptor (histamine receptor) antagonist, has been examined for its effect on estradiol breakdown, as there were reports of side effects, such as gynecomastia and sexual dysfunction. It is also known that it can interfere with the metabolism of a number of other medicaments such as theophylline, phenytoin, quinidine and carbamazepine; it is very frequently used for disorders of the stomach. Studies with healthy male subjects showed, after 2 weeks of treatment, a significant fall in the 2-hydroxyestrone content of urine and blood, and a rise in the concentration of estradiol in the blood, while C16-hydroxylation was unaffected (Galbraith and Michnovicz 1989). These findings show that the side effects may be ascribed to changes in estradiol metabolism. In the same study, ranitidine, an H_2-receptor antagonist of the second generation with the same indication was investigated using male subjects. In this case, no changes of estradiol C2-hydroxylation were observed. In a follow-up study, the effect of cimetidine was also analyzed in pre-menopausal and postmenopausal women (Michnovicz and Galbraith 1991). Whereas in both groups C2-hydroxylation dropped significantly after 4 weeks of treatment, only in the postmenopausal women was a significant rise in serum estradiol measured. Yet another drug for gastrointestinal diseases, the proton-pump inhibitor omeprazole, which can also interfere with the metabolism of other drugs, was investigated using male subjects (Galbraith and Michnovicz 1993). The use of this substance led to no changes in estradiol metabolism.

Also of interest is the influence of various hormones of the body since pathophysiological processes in hormonal disorders may be better understood. So far, research findings of this kind have been demonstrated in relation to hyperthyreosis. Thyroxine, as already mentioned, causes a significant rise in A-ring metabolism, while D-ring metabolism remains uninfluenced (Michnovicz and Galbraith 1990).

Research is ongoing in the field of the deliberate influencing of estradiol metabolism under the aspect of reducing the risk of occurrence of hormone-dependent neoplasms. As comprehensive studies have shown, an increased D-ring metabolism is thought, as already mentioned, to favor the risk of the development of breast cancer; increased A-ring metabolism is considered to reduce the risk. A substance from the crucifer plant family, indole-3-carbinol, has hitherto shown properties that promise success in directing the estradiol metabolism positively. In several studies on both men and women, it was demonstrated that indole-3-carbinol is a potent substance for the induction of estradiol C2-hydroxylation (Michnovicz and Bradlow 1990, 1991; Bradlow et al. 1995c; Michnovicz et al. 1997). It must be stressed that even small amounts, for instance 6–7 mg/kg per day – administered per os – lead to significant changes. Measurement of a metabolite profile in the urine (13 metabolites) showed that the total of estrogen metabolites is not significantly changed by indole-3-carbinol, but that raising the A-ring metabolites is accompanied by a lowering of the D-ring metabolites (Michnovicz et al. 1997).

With the use of indole-3-carbinol, promising anti-carcinogenic effects have already been achieved in both animals and humans (Bradlow et al. 1995d).

Indole-3-carbinol has further properties that make it suitable for chemopre-vention, namely an anti-oxidative effect (TABOR et al. 1991). Furthermore, it is able to induce phase-II drug-metabolizing enzymes (BRADFIELD and BJELDANES 1984) and also functions as an anti-estrogen, as has been demon-strated in cell cultures (TIWARI et al. 1994). There are efforts in progress to make this substance available as a pharmacological drug for chemoprevention in medical practice (MICHNOVICZ and BRADLOW 1991).

Another food component, omega-3 fatty acids, possibly also have an influence on estrogen metabolism, as studies on human mamma-carcinoma cell cultures, i.e., MCF-7 cells, have shown (BRADLOW et al. 1995b). Like indole-3-carbinol, they are able to reduce the relationship of the metabolic produc-tion of 16-hydroxyestrone to 2-hydroxyestrone. One brief report gives an account of a reduction of estradiol C16-hydroxylation in humans by omega-3 fatty acids (OSBORNE et al. 1988); however, so far, no extensive studies appear to have been carried out.

As already mentioned, a 4-week treatment with the MAO inhibitor tranyl-cypromine of depressive men, in whom low values of 4-hydroxylated estro-gens were registered before the commencement of treatment, led to a significant lowering of C2 catechol estrogens (BANGER et al. 1990). In this study, no measurement of the D-ring metabolites was carried out.

D. Lifestyle

The influence of lifestyle, such as eating habits, physical activity and smoking, was a further topic for the examination of estrogen metabolism. Eating habits aroused interest at an early stage. When comparing the metabolism of women with vegetarian or omnivorous food consumption, predominantly differences in urine and feces excretion were observed. In vegetarians, a lower excretion of estriol in urine with a simultaneously higher estrogen excretion in feces was striking (GOLDIN et al. 1982). In another study in vegetarian and omnivorous women, it could be observed that the fiber content of food is able to influence estrogen metabolism, by reduction of the estrogen metabolites in the urine (ADLERCREUTZ et al. 1986). The interruption of the enterohepatic estrogen circulation by the fiber-rich food is regarded as the mechanism responsible for this.

Another research team compared the influence of a protein-rich diet with that of a carbohydrate-rich diet in healthy men (ANDERSON et al. 1984). After diets lasting 2 weeks, only in the case of protein-rich food C2-hydroxylation rose significantly. C16-hydroxylation remained the same after both diets. The authors attribute the protein effect to a stimulation of hepatic cytochrome-P450 enzymes, by which not only the oxidative breakdown of estradiol, but also that of various drugs, can be increased.

Recently, changes in the relationship of A- to D-ring metabolites fol-lowing a 12-day broccoli diet (500 g/day) were studied (KALL et al. 1996). Broccoli belongs to the group of crucifers, which contain the substance indole-3-carbinol. It was therefore not a surprising finding that the ratio of metabolites excreted in urine was shifted in favor of A-ring metabolism. Here, the authors also assumed that the enzyme activity of the liver relative to C2-hydroxylation was raised, while that of C16-hydroxylation remained uninfluenced.

According to epidemiological studies a fat-rich diet is supposed to raise the risk of breast cancer, while a low-fat diet is supposed to lower it. To investigate the effects of a low-fat diet on estrogen metabolism, normal young women were studied while eating a Western-style high-fat diet and again after 2 months of consuming a defined low-fat diet (Longcope et al. 1987). The authors found that after consumption of low-fat food, a significantly decreased urinary excretion of C16-hydroxylated metabolites, with a simultaneous increase in C2-hydroxylated metabolites was observed. A study of men arrived at similar conclusions (Dorgan et al. 1996). In the comparison of a 10-week diet with a low-fat and a high-fiber content with a high-fat and low-fiber-content diet, a significant decline in 2-hydroxylated metabolites in urine was registered after the high-fat, low-fiber diet.

A recent study was concerned with a comparison of the estrogen metabolism of women from the Oriental and the Western (Finnish women) cultural spheres (Adlercreutz et al. 1994). The fact that women from the Asian region are less prone to breast cancer was attributed to diet and, dependent on this, to estrogen metabolism. It was expected that Oriental women would produce more A-ring metabolites and less D-ring metabolites (Fishman et al. 1995). However, the authors arrived at a controversial conclusion. In the case of the Finnish women, who ought theoretically to have been subject to a greater risk of breast cancer, an increased C2-hydroxylation metabolism was found, while C16-hydroxylation was the same in both groups. This finding was interpreted by the authors that C2-hydroxy estrogens can also develop carcinogenic properties. Theoretically, this may be possible when the further breakdown of the A-ring metabolites takes place via the formation of semiquinones and quinones, substances that can develop a strong oxidative effect in tissues by means of redox cycling. In a healthy organism, such metabolism seems unlikely (Iverson et al. 1996). It is worth noting, however, that in this study total estrogen production and C4-hydroxy metabolites were higher in Western than in Oriental women. The C4-catechol estrogens appear easier to transform into quinones; in animal experiments they were made responsible for the development of tumors of the kidney (Han and Liehr 1994). To our knowledge there are no reports on the formation of quinones from catechol estrogens in humans. The previously described rapid clearance of the primary catechol estrogens by methylation contributes to their inactivation. A reduced enzymatic methylation activity in the breast tissue, however, is regarded as an increase in the risk of carcinogenesis by A-ring metabolites (Zhu and Liehr 1993).

The results of the cited studies, although not uniform in their conclusions, show that the intake of food, according to its composition, can alter estrogen metabolism and, thus, possibly contribute to the development of diseases.

Investigations into the influence of sporting activity failed to result in a uniform picture. In female rowing athletes, after intensive training, there was no change in A- or D-ring metabolism to be observed in comparison with a normal group (SNOW et al. 1989). One group of women did experience, during intensive training, menstrual disturbances that were thought to be connected with a previously existent increased activity of C2-hydroxy metabolism. The abnormal estradiol metabolism did not correlate with the degree of intensity of training but with the extent of leanness of the sportswomen.

An increase in catechol estrogen production following increased physical activity was registered by another research team investigating sportswomen (DE CREE et al. 1997b, c). The 4-hydroxy estrogen fraction was also affected (DE CREE et al. 1997c). The authors stressed, in particular, the fact that the methylated catechol estrogens were found in a significantly higher percentage than the primary catechol estrogens. They connected the high degree of methylation of the catechol estrogens after muscular training with the epidemiologically demonstrated lower risk of breast cancer disease in sportswomen. In untrained subjects, after a briefly increased physical activity, no significant changes in C2-hydroxylation were found (DE CREE et al. 1997a).

The effect of smoking was also of interest, as epidemological studies have shown that, in female smokers, endometrial cancer occurred less frequently (TYLER et al. 1985), but early menopause and osteoporosis occurred more frequently (BARON 1984). The investigations were carried out first on premenopausal women and showed, in comparison with non-smoking women, an increase in the C2-hydroxylated metabolites, with simultaneous lowering of the estriol values (MICHNOVICZ et al. 1986, 1988). Studies carried out later on men produced similar results (MICHNOVICZ et al. 1989). Because the HDL concentration in the plasma was lowered in the case of the male smokers, the authors presumed that lipoprotein synthesis and estradiol-2 hydroxylation exert a mutual influence in the liver.

E. Pregnancy

Estrogen metabolism occupies a special position in pregnancy; it becomes clearly apparent that the female sex hormones are of great physiological importance not only for the mother but, in particular, for the development of the conceptus. In order to record changes due to pregnancy, first the estrogens in various biological materials from the mother were examined. In contrast to the non-pregnant state, the greatly raised estrogen values are striking, and bear no relationship to the size of the additional compartment of the child. The growth of the conceptus requires a considerable increase in production, with the estrogen concentrations increasing as the pregnancy progresses, reaching the highest concentrations at term (ADLERCREUTZ and LUUKKAINEN

1970). The values rise proportionately higher in the case of multiple pregnancy.

ADLERCREUTZ and MARTIN (1976) made a comparison of various estradiol metabolites measured in maternal plasma, bile, feces and urine in the third trimester of pregnancy. Table 4 summarizes these results. As can be seen, the main estrogen excretion takes place in the urine, while considerably less leaves the organism via feces. In all four compartments, the D-ring metabolites are several times larger than the other breakdown products; D-ring metabolism, with its chief representative estriol, therefore, dominates the endogenous estradiol breakdown in pregnancy.

This is remarkable inasmuch as the biological activity of the D-ring metabolites is characterized by their still-high degree of estrogenicity. In the organism, however, these metabolites are rendered into a neutral state through a considerable degree of conjugation, and appear only in feces and urine in unconjugated forms. The proportion of catechol estrogens, which outside of pregnancy make up about half of the total estradiol metabolism, is much smaller during pregnancy. In contrast to the non-pregnant state, estradiol and estrone are encountered in higher concentrations. Furthermore, during pregnancy, metabolites that are not found in non-pregnant women have been identified. These are 17α-estradiol, 11-dehydro-17α-estradiol and 15α-hydroxyestriol (estetrol) (ADLERCREUTZ and LUUKKAINEN 1970; ADLERCREUTZ et al. 1973; ADLERCREUTZ and MARTIN 1976; FOTSIS et al. 1980b). Their concentrations are small relative to estriol. As in the case of other metabolites, the biological characteristics and clinical significance of these substances have not been clarified.

Table 4. Excretion of various estradiol metabolites expressed as percentages of the total estrogen amount in plasma, bile, feces and urine of women in the third trimester of pregnancy according to ADLERCREUTZ and MARTIN (1976)

Total (%)	Plasma (term) 275 µg/l (%)	Bile (32nd week) 10 µg/l (%)	Feces (33–37th week) 900 µg/24 h (%)	Urine (36–40th week) 32 mg/24 h (%)
E3	41.1	45.9	40.3	77.8
16α-OHE1	18.3	36.6	0.4	7.6
16β-OHE1	3.0	4.9	0.2	1.9
E1	19.1	4.4	11.0	2.1
17β-E2	5.2	0.4	22.8	0.5
17α-E2	0.2	0.2	0.5	0.1
16-epiE3	1.4	1.2	16.0	3.3
17-epiE3	0.2	0.4	4.3	0.5
16-oxo-E2	9.8	3.1	1.1	5.3
15α-OHE1	1.5	1.8	3.4	0.7
2-MeE1	0.2	1.0	0.1	0.3
Total	100%	100%	100%	100%

It is well known that the conceptus contributes essentially to increased estrogen production and metabolism during pregnancy. It was, however, not easy to elucidate the different estrogen metabolisms in the two organisms. FISHMAN et al. (1962a) examined the estradiol metabolism of pregnant and non-pregnant women following intravenous injection of radioactively labeled estradiol. Estriol, which occurs during pregnancy in huge quantities, was not transformed from estradiol to any greater extent in pregnant women than in non-pregnant women. The breakdown of the radioactive estradiol therefore showed no essential differences between pregnant and non-pregnant women.

The conceptus thus caters for its own great need of estrogens; but the way in which it does so differs from that of the maternal organism. Metabolism in the conceptus's compartment does not proceed exclusively from estradiol. The placenta is the chief organ of estrogen production and metabolism for the fetal organism. However, to fulfill these functions, it depends on steroid precursors provided by the fetus. Fetus and placenta therefore form a functional unit in the production and metabolism of sex steriods.

The sequence of the placental–fetal processes was clarified in the 1960s (SIITERI and MacDONALD 1966; DICZFALUSY 1969. Put simply, it consists, as outlined in Fig. 3, according to SPEROFF et al. 1993 of the following steps: progesterone is formed in the placenta using cholesterol of maternal source mainly through uptake of LDL. The placenta is, however, not able to further metabolize progesterone to estradiol. Therefore progesterone is metabolized in the

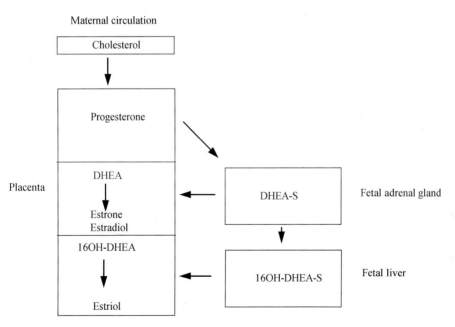

Fig. 3. Simplified scheme of the main steps of estrogen metabolism (steroids and their production places) in the feto–placental unit

fetus mainly by the adrenal glands into dehydroepiandrosterone sulfate (DHEA-S), a central substance for the further estrogen metabolism. DHEA-S, on reaching the placenta, is desulfated and aromatized into estrone. In the placenta, estrone is then either reduced to estradiol or oxidized by A-ring hydroxylation to the catechol estrogens. Estradiol, estrone and the catechol estrogens are partly further broken down or made available to the infant organism, where they have functions that have not yet been completely understood.

The placenta is not able to break down DHEA-S by hydroxylation at the C16-position of the D-ring. Fetal DHEA-S must be transformed in the fetal liver into 16-hydroxy-DHEA-S, in order to be transformed by the placenta into estriol. This takes place by desulfatation and aromatization.

To sum up, the placental estrogen metabolism depends on the fetus providing the precursors, DHEA-S and 16-hydroxy-DHEA-S. This becomes obvious in the case of intrauterine fetal death, when the placenta function is still intact, and progesterone production continues, while estrogen metabolism ceases. Although various other mechanisms of estradiol metabolism were observed in the feto–placental unit – to a small extent fetal organs are also able to metabolize estrogens – the processes described above predominate. The placental estrogen metabolites which reach the maternal organism proceed to final excretion.

I. Changes in Metabolism During Pregnancy

There have been few studies on changes of the estrogen metabolism of the conceptus, i.e., of the feto–placental unit. Interest has been aroused by changes in estriol production observed during the course of pregnancy. Placental insufficiency and retardation of fetal growth are accompanied by a reduced urinary excretion of estriol. Estriol measurements were therefore used for monitoring fetal–placental function (BEISCHER et al. 1991; KOWALCZYK et al. 1998). So far, little is known about the causes of placental insufficiency and any possible connections with the changes in estrogen metabolism.

Treatment of the mother with antibiotics decreases estriol excretion in the urine (WILLMAN and PULKKINEN 1971; PULKKINEN and WILLMAN 1973; ADLERCREUTZ et al. 1975). This does not necessarily seem to be accompanied by a reduction of the infant's wellbeing. Disturbances of the enterohepatic circulation of the mother are thought to contribute to decreased estriol excretion via the kidney.

Catechol estrogen production is increased during pregnancy, but not involved in estrogen metabolism as much as in the case of non-pregnant women. Catechol estrogens are not only formed in the placenta (BARNEA et al. 1988), but may also be produced by various organs of the fetus (CHAO et al. 1981). Not only 2-hydroxyestrone, 2-hydroxyestradiol and 2-hydroxyestriol (GELBKE et al. 1975a, b), but also C4-hydroxylated catechol estrogens have been demonstrated during pregnancy (BALL and KNUPPEN 1978; FOTSIS et al. 1980a). The further breakdown proceeds, in pregnancy too, via methylation by the enzyme catechol-O-methyltransferase (GELBKE and KNUPPEN 1976). The functions of the catechol estrogens during pregnancy have not yet been clarified, nor have the functions of the D-ring metabolites. Attempts to obtain conclusions derived from varying concentrations in maternal blood and urine have so far been unsuccessful (BERG and KUSS 1992).

Because of high concentrations of catechol estrogens in the amniotic fluid of women undergoing spontaneous labor, it has been supposed that they may contribute to the triggering of labor activity (BISWAS et al. 1991). The properties of catechol estrogens in stimulation of prostaglandin synthesis and inhibition of catecholamine breakdown by competition with the breakdown enzymes speak for such a function. However, in vivo experiments in which a uterus-relaxing effect of catechol estrogens was observed (GOYACHE et al. 1995) make this assumption seem rather unlikely.

Whereas in non-pregnant women A-ring metabolism can be influenced by external stimuli relatively easily, during pregnancy there are hardly any investigation concerned with such reaction. There was only one indication to be found that smoking leads to an increased formation of catechol estrogen in the placenta (CHAO et al. 1981), a reaction found also in the organism of smoking non-pregnant women.

So far recording changes in estrogen metabolism either during or outside of pregnancy have not led to sufficient insight into the pathophysiology of hormone dependent diseases to achieve any clinical significance. There are, however, some promising aspects. The under-explored area of the estradiol metabolism certainly needs more attention, not only in order to obtain a better understanding of processes in diseases, but also to discover, among the metabolites, pharmacologically active substances that can be used for the prevention and treatment of diseases.

References

Adlercreutz H, Järvenpää P (1982) Assay of estrogens in human feces. J Steroid Biochem 17:639–645

Adlercreutz H, Luukkainen T (1970) Identification and determination of estrogens in various biological materials in pregnancy. Am Clin Res 2:365–380

Adlercreutz H, Martin F (1976) Estrogen in human pregnancy feces. Acta Endocrinol 83:410–419

Adlercreutz H, Ervast HS, Tenhunen A, Tikkanen MJ (1973) Gas chromatographic and mass spectrometric studies on estrogens in bile. Acta Endocrinol 73:543–554

Adlercreutz H, Martin F, Tikkanen MJ, Pulkinnen M (1975) Effect of ampicillin administration on the excretion of twelve estrogens in pregnancy urine. Acta Endocrinol 80:551–557

Adlercreutz H, Martin F, Pulkkinen M, Dencker H, Rimer U, Sjoberg NO, Tikkanen MJ (1976) Intestinal metabolism of estrogens. J Clin Endocrinol Metab 43:497–509

Adlercreutz H, Fotsis T, Bannwart C, Hamalainen E, Bloigu S, Ollus A (1986) Urinary estrogen profile determination in young Finnish vegetarian and omnivorous women. J Steroid Biochem 24:289–296

Adlercreutz H, Gorbach SL, Goldin BR, Woods MN, Dwyer JT, Hämäläinen E (1994) Estrogen metabolism and excretion in Oriental and Caucasian women. J Natl Cancer Inst 86:1076–1082

Anderson KE, Kappas A, Conney AH, Bradlow HL, Fishman J (1984) The influence of dietary protein and carbohydrate on the principal oxidative biotransformations of estradiol in normal subjects. J Clin Endocrinol Metab 59:103–107

Auborn KJ, Woodworth C, Dipaolo J, Bradlow HL (1991) The interaction between HPV infection and estrogen metabolism in cervical carcinogenesis. Int J Cancer 49:867–869

Ball P, Knuppen R (1978) Formation of 2- and 4-hydroxyestrogens in the brain, pituitary and liver of the human fetus. J Clin Endocrinol Metab 47:732–737

Ball P, Knuppen R (1980) Catecholoestrogens (2- and 4-hydroxyoestrogens). Chemistry, biogenesis, metabolism, occurence and physiological significance. Acta Endocrinol 232[Suppl]:1–127

Ball P, Stubenrauch G, Knuppen R (1977) Metabolism of 2-methoxy and 4-methoxyestrone in man in vivo. J Steroid Biochem 8:989–993

Banger M, Hiemke C, Knuppen R, Ball P, Haupt M, Wiedemann K (1990) Formation and metabolism of catecholestrogens in depressed patients. Biol Psychiatry 28:685–696

Barnea ER, MacLusky NJ, Purdy R, Naftolin F (1988) Estrogen hydroxylase activity in the human placenta at term. J Steroid Biochem 31:253–255

Baron JA (1984) Smoking and estrogen-related disease. Am J Epidemiol 119:9–22

Bates GW, Edman CD, Porter JC, MacDonald PC (1977) Metabolism of catecholestrogen by human erythrocytes. J Clin Endocrinol Metab 45:1120–1123

Beischer N, Brown J, Parkinson P, Walstab J (1991) Urinary estriol assay for monitoring fetoplacental function. Aust N Z J Obstet Gynaecol 31:1–8

Berg FD, Kuss E (1992) Serum concentrations and urinary excretion of classical estrogens, catecholestrogens and 2-methoxyestrogens in normal pregnancy. Arch Gynecol Obstet 251:17–27

Biswas A, Chaudhury A, Chattoraj SC, Dale SL (1991) Do catechol estrogens participate in the initiation of labor? Am J Obstet Gynecol 165:984–987

Bradfield CA, Bjeldanes LF (1984) Effect of dietary indole-3-carbinol on intestinal and hepatic monooxygenase glutathione S-transferase and epoxide hydrolase activities in the rat. Food Chem Toxicol 22:977–982

Bradlow HL, Hershcopf RJ, Martucci CP, Fishman J (1985) Estradiol 16α hydroxylation in the mouse correlates with mammary tumor incidence and presence of mammary tumor virus: a possible model for the hormonal etiology of breast cancer in humans. Proc Natl Acad Sci USA 82:6295–6299

Bradlow HL, Arcuri F, Blasi L, Castagnetta L (1995a) Effect of serum albumin on estrogen metabolism in human cancer cell lines. Mol Cell Endocrinol 115:221–225

Bradlow HL, Davis DL, Lin G, Sepkovic DW, Tiwari R (1995b) Effects of pesticides on the ratio of 16alpha-/2-hydroxyestrone: a biological marker of breast cancer risk. Environ Health Perspect 103:147–150

Bradlow HL, Michnovicz JJ, Wong GYC, Halper MP, Miller D, Osborne MP (1995c) Long term responses of women to indole-3-carbinol or a high fiber diet. Cancer Prev Biomarkers Epidemiol 3:591–595

Bradlow HL, Sepkovic DW, Telang NT, Osborne MP (1995d) Indole-3-carbinol. A novel approach to breast cancer prevention. Ann NY Acad Sci 768:180–200

Bradlow HL, Telang NT, Sepkovic DW, Osborne MP (1996) 2-hydroxyestrone: the good estrogen. J Endocrinol 150:5259–5265

Carruba G, Miceli MD, Camito L, Farruggio R, Sorci CMG, Oliveri G, Amodio R, DiFalco M, D'Amico D, Castagnetta LAM (1996) Multiple estrogen functions in human prostate cancer cells. Ann NY Acad Sci 784:70–84

Castagnetta LA, LoCasto M, Granata OM, Polito L, Calabro M, LoBue A, Bellavia V, Carruba G (1996) Estrogen content and metabolism in human breast tumor tissues and cells. Ann NY Acad Sci 784:314–324

Chao ST, Omiecinski CJ, Namkung MJ, Nelson SD, Dvorchik BH, Juchau MR (1981) Catechol estrogen formation in placental and fetal tissues of humans, macaques, rats and rabbits. Dev Pharmacol Ther 2:1–16

Chao TC (1996) Female sex hormones and the immune system. Chang Keng I Hsueh 19:95–106

Chen C, Malone KE, Prunty JA, Daling JR (1996) Measurement of urinary estrogen metabolites using a monoclonal enzyme-linked immunoassay kit: assay performance and feasibility for epidemiological studies. Cancer Epidemiol Biomarkers Prev 5:727–732

Coll DA, Rosen CA, Auborn K, Potsic WP, Bradlow HL (1997) Treatment of recurrent respiratory papillomatosis with indole-3-carbinol. Am J Otolaryngol 18:283–285

Cutolo M, Castagnetta L (1996) Immunomodulatory mechanisms mediated by sex hormones in rheumatoid arthritis. Ann NY Acad Sci 784:237–251

De Cree C, Ball P, Seidlitz B (1997a) Plasma 2-hydroxycatecholestrogen responses to acute submaximal and maximal exercise. J Appl Physiol 82:364–370

De Cree C, Ball P, Seidlitz B, Van Kranenburg G, Geurten P, Keizer HA (1997b) Effects of a training programm on resting plasma 2-hydroxycatecholestrogen levels in eumenorrheic women. J Appl Physiol 83:1551–1556

De Cree C, Van Kranenburg G, Geurten P, Fujimori Y, Keizer HA (1997c) 4-hydroxycatecholestrogen metabolism responses to exercise and training: possible implications for menstrual cycle irregularities and breast cancer. Fertil Steril 67:505–516

Diczfalusy E (1969) Steroid metabolism in the human feto-placental unit. Acta Endocrinol 61:649–664

Dorgan JF, Judd JT, Longcope C, Brown C, Schatzkin A, Clevidence BA, Campbell WS, Nair PP, Franz C, Kahle L, Taylor PR (1996) Effect of dietary fat and fiber on plasma and urine androgens and estrogens in man: a controlled feeding study. Am J Clin Nutr 64:850–855

Dunn JF, Merriam GR, Eil C, Kono S, Loriaux DL, Nisula BC (1980) Testosterone-estradiol binding globulin binds to 2-hydroxyestradiol with greater affinity than to testosterone. J Clin Endocrinol Metab 51:404–406

Emons G, Merriam GR, Pfeiffer D, Loriaux DL, Ball P, Knuppen R (1987) Metabolism of exogenous 4- and 2-hydroxyestradiol in the human male. J Steroid Biochem 28:499–504

Essman E, Abramson A (1984) Estrogen binding sites on membranes from human laryngeal papilloma. Int J Cancer 33:33–36

Fishman J, Martucci C (1980a) Biological properties of 16α-hydroxyestrone: Implications in estrogen physiology and pathophysiology. J Clin Endocrinol Metabol 51:611–615

Fishman J, Martucci C (1980b) Dissociation of biological activities in metabolites of estradiol. In: McLachlan JA (ed) Estrogens in the environment. Elsevier, North-Holland, pp 131–145

Fishman J, Bradlow HL, Gallagher TF (1960) Oxidative metabolism of estradiol. J. Biol Chem 235:3104–3107

Fishman J, Brown JB, Hellman L, Zumoff B, Gallagher TF (1962a) Estrogen metabolism in normal and pregnant women. J Biol Chem 237:1489–1494

Fishman J, Hellman L, Zumoff B, Gallagher TF (1962b) Influence of thyroid hormone on estrogen metabolism in man. J Clin Endocr 22:389–392

Fishman J, Hellman L, Zumoff B, Gallagher TF (1965) Effect of thyroid on hydroxylation of estrogen in man. J Clin Endocrinol Metab 25:365–368

Fishman J, Boyar RM, Hellman L (1972) The system estradiol – estrone; a sex difference in oxidation of estradiol in man. J Clin Endocrinol Metab 34:989–995

Fishman J, Boyar RM, Hellman L(1975) Influence of body weight on estradiol metabolism in young women. J Clin Endocrinol Metab 41:984–991

Fishman J, Bradlow HL, Schneider J, Anderson KE, Kappas A (1980) Radiometric analysis of biological oxidations in man: sex differences in estradiol metabolism. Proc Natl Acad Sci USA 77:4957–4960

Fishman J, Schneider J, Hershcopf RJ, Bradlow HL (1984) Increased estrogen 16-alpha hydroxylase activity in women with breast and endometrial cancer. J Steroid Biochem 20:1077–1081

Fishman J, Osborne P, Telang NT (1995) The role of estrogen in mammary carcinogenesis. Ann NY Acad Sci 768:91–100

Fotsis T, Järvenpää P, Adlercreutz H (1980a) Identification of 4-hydroxyestriol in pregnancy urine. J Clin Endocrinol Metab 51:148–151

Fotsis T, Järvenpää P, Adlercreutz H (1980b) Purification of urine for quantification of the complete estrogen profile. J Steroid Biochem 12:503–508

Fotsis T, Zhang Y, Pepper MS, Adlercreutz H, Montesano R, Nawroth PP, Schweigerer L (1994) The endogenous estrogen metabolite 2-methoxyoestradiol inhibits angiogenesis and suppresses tumour growth. Nature 368:237–239

Galbraith RA, Michnovicz JJ (1989) The effects of cimetidine on the oxidative metabolism of estradiol. N Engl J Med 321:269–274

Galbraith RA, Michnovicz JJ (1993) Omeprazole fails to alter the cytochrome P450-dependent 2-hydroxylation of estradiol in male volunteers. Pharmacology 47:8–12

Gelbke HP, Knuppen R (1976) The excretion of five different 2-hydroxy estrogen monomethyl ethers in human pregnancy urine. J Steroid Biochem 7:457–463

Gelbke HP, Böttger M, Knuppen R (1975a) Excretion of 2-hydroxyestrone in urine throughout human pregnancies. J Clin Endocrinol Metab 41:744–750

Gelbke HP, Hoogen H, Knuppen R (1975b) Identification of 2-hydroxyestradiol and the pattern of catechol estrogens in human pregnancy urine. J Steroid Biochem 6:1187–1191

Goldin BR, Adlercreutz H, Gorbach SL, Warram JH, Dwyer JT, Swenson L, Woods MN (1982) Estrogen excretion patterns and plasma levels in vegetarian and omnivorous women. N Engl J Med 307:1542–1547

Goldin BR, Adlercreutz H, Gorbach SL, Woods MN, Dwyer JT, Conlon T, Bohn E, Gershoff SN (1986) The relationship between estrogen levels and diets of Caucasian American and Oriental immigrant women. Am J Clin Nutr 44:945–953

Gorbach SL, Goldin BR (1987) Diet and the excretion and enterohepatic cycling of estrogens. Prev Med 16:525–531

Gordon S, Cantrall EW, Cekleniak WP, Albers HJ, Mauer S, Stolar SM, Bernstein S (1964) Steroid and lipid metabolism. The hypocholesteremic effect of estrogen metabolites. Steroids 4:267–271

Goyache FM, Gutierrez M, Hidalgo A, Cantabrana B (1995) Non-genomic effects of catecholestrogens in the in vitro rat uterine contraction. Gen Pharmacol 26:219–223

Han X, Liehr JG (1994) DNA single-strand breaks in kidneys of Syrian hamsters treated with steroidal estrogens: hormone-induced free radical damage preceding renal malignancy. Carcinogenesis 15:997–1000

Hellman L, Zumoff B, Fishman J, Gallagher TF (1970) Estradiol metabolism in total extrahepatic biliary obstruction. J Clin Endocrinol 30:161–165

Herrmann WM, Beach RC (1978) The psychotropic properties of estrogens. Pharmacopsychiatry 11:164–176

Hershcopf RJ, Bradlow HL (1987) Obesity, diet, endogenous estrogens and the risk of hormone-sensitive cancer. Am J Clin Nutr 45:283–289

Hobkirk R (1985) Steroid sulfotransferases and steroid sulfatases. Characteristics and biological roles. Can J Biochem Cell Biol 63:1127–1144

Iverson SL, Shen L, Anlar N, Bolton JL (1996) Bioactivation of estrone and its catechol metabolites to quinoid-glutathione conjugates in rat liver microsomes. Chem Res Toxicol 9:492–499

Kabat GC, Chang CJ, Sparano JA, Sepkovic DW, Hu XP, Khalil A, Rosenblatt R, Bradlow HL (1997) Urinary estrogen metabolites and breast cancer: a case-control study. Cancer Epidemiol Biomarkers Prev 6:505–509

Kall MA, Vang O, Clausen J (1996) Effects of dietary broccoli on human in vivo drug metabolizing enzymes: evaluation of caffeine, oestrone and chlorzoxazone metabolism. Carcinogenesis 17:793–799

Kono S, Merriam GR, Brandon DD, Loriaux DL, Lipsett MB (1982) Radioimmunoassay and metabolism of the catechol estrogen 2-hydroxyestradiol. J Clin Endocrinol Metab 54:150–154

Kowalczyk TD, Cabaniss ML, Cusmano D (1998) Association of low unconjugated estriol and the second trimester and adverse pregnancy outcome. Obstet Gynecol 91:396–400

Lahita RG (1996) The connective tissue diseases and the overall influence of gender. Int J Fertil Menopausal Stud 41:156–165

Lahita RG, Bradlow HL, Kunkel HG, Fishman J (1979) Alterations of estrogen metabolism in systemic lupus erythematosus. Arthritis Rheum 22:1195–1198

Lahita RG, Bradlow HL, Kunkel HG, Fishman J (1981) Increased 16α-hydroxylation of estradiol in systemic lupus erythematosus. J Clin Endocrinol Metab 53:174–178

Lahita RG, Bradlow HL, Fishman J, Kunkel HG (1982) Abnormal estrogen and androgen metabolism in the human with systemic lupus erythematosus. Am J Kidney Dis 2:206–211

Lahita RG, Bucala R, Bradlow HL, Fishman J (1984) Determination of 16α-hydroxyestrone by radioimmunoassay in systemic lupus erythematosus. Arthritis Rheum 28:1122–1127

Lim SK, Won YJ, Lee JH, Kwon SH, Lee EJ, Kim KR, Lee HC, Huh KB, Chung BC (1997) Altered hydroxylation of estrogen in patients with postmenopausal osteopenia. J Clin Endocrinol Metab 82:1001–1006

Lippert TH, Seeger H, Mueck AO (1998) Estradiol metabolism during oral and transdermal estradiol replacement therapy in postmenopausal women. Horm Metab Res 30:598–600

Lippert TH, Seeger H, Mueck AO (1999) Estrogens and prevention of cardiovascular disease: the role of estradiol metabolites. Climacteric 7:296–301

Longcope C, Femino A, Flood C, Williams KIH (1982) Metabolic clearance rate and conversion ratios of ^{3}H-hydroxyestrone in normal men. J Clin Endocrinol Metab 54:374–380

Longcope C, Flood C, Femino A, Williams KIH (1983) Metabolism of 2-methoxyestrone in normal men. J Clin Endocrinol Metab 57:277–282

Longcope C, Gorbach S, Godlin B, Woods M, Dwyer J, Morrill A, Warram J (1987) The effect of a low fat diet on estrogen metabolism. J Clin Endocrinol Metab 64:1246–1250

Martucci CP, Fishman J (1993) P450 enzymes of estrogen metabolism. Pharmacol Ther 57:237–257

McEwen BS, Davis PG, Parsons B, Pfaff DW (1979) The brain as target for steroid hormone action. Ann Rev Neurosci 2:65–112

Merriam GR, Brandon DD, Kono S, Davies SE, Loriaux DL, Lipsett MB (1980) Rapid metabolic clearance of the catechol estrogen 2-hydroxyestrone. J Clin Endocrinol Metab 51:1211–1213

Michnovicz JJ, Bradlow HL (1990) Induction of estradiol metabolism by dietary indole-3-carbinol in humans. J Natl Cancer Inst 82:947–949

Michnovicz JJ, Bradlow HL (1991) Altered estrogen metabolism and excretion in humans following consumption of indole-3-carbinol. Nutr Cancer 16:59–66

Michnovicz JJ, Galbraith RA (1990) Effects of exogenous thyroxine on C-2 and C-16 hydroxylation of estradiol in humans. Steroids 55:22–26

Michnovicz JJ, Galbraith RA (1991) Cimetidine inhibits catechol estrogen metabolism in women. Metabolism 40:170–174

Michnovicz JJ, Hershcopf RJ, Naganuma H, Bradlow HL, Fishman J (1986) Increased 2-hydroxylation of estradiol as a possible mechanism for the anti-estrogenic effect of cigarette smoking. N Engl J Med 315:1305–1309

Michnovicz JJ, Naganuma H, Hershcopf RJ, Bradlow HL, Fishman J (1988) Increased urinary catechol estrogen excretion in female smokers. Steroids 52:69–83

Michnovicz JJ, Hershcopf RJ, Haley NJ, Bradlow HL, Fishman J (1989) Cigarette smoking alters hepatic estrogen metabolism in men: implication for atherosclerosis. Metabolism 38:537–541

Michnovicz JJ, Adlercreutz H, Bradlow HL (1997) Changes in levels of urinary estrogen metabolites after oral indole-3-carbinol treatment in humans. J Natl Cancer Inst 89:718–723

Mueck AO, Seeger H, Gräser T, Oettel M, Lippert TH (1998) Effect of postmenopausal hormone replacement therapy (HRT) and of oral contraceptives (OC) on estradiol metabolism. Naunyn Schmiedebergs Arch Pharmacol 358:R788

Musey PI, Kright K, Preedy JRK, Collins DC (1979) Formation and metabolism of steroid conjugates: effect of conjugation on excretion and tissue distribution. In: Hobkirk R (ed) Steroid biochemistry, vol II. CRC Press Boca Raton, pp 81–132

Newfield L, Goldsmith A, Bradlow HL, Auborn K (1993) Estrogen metabolism and human papilloma virus-induced tumors of the larynx. Anticancer Res 13:227–242

Osborne MP, Karmali RA, Hershcopf RJ, Bradlow HL, Fishman J (1988) Omega-3 fatty acids: modulation of estrogen metabolism and potential for breast cancer prevention. Cancer Invest 8:629–531

Osborne MP, Bradlow HL, Wong GY, Telang NT (1993) Upregulation of estradiol C16alpha-hydroxylation in human breast tissue: a potential biomarker of breast cancer risk. J Natl Cancer Inst 85:1917–1920

Pasagian-Macaulay A, Meilahn EN, Bradlow HL, Sepkovic DW, Buhari AM, Simkin-Silverman L, Wing RR, Kuller LH (1996) Urinary markers of estrogen metabolism 2- and 16alpha-hydroxylation in premenopausal women. Steroids 61:461–467

Pulkkinen MO, Willman K (1973) Reduction of maternal estrogen excretion by neomycin. Am J Obstet Gynecol 115:1153

Purohit A, Reed MJ, Morris NC, Williams GJ, Potter BVL (1996) Regulation and inhibition of steroid sulfatase activity in breast cancer. Ann NY Acad Sci 184:40–49

Rosen CA, Woodson GE, Tompson JW, Hengesteg AP, Bradlow HL (1998) Preliminary results of the use of indole-3-carbinol for recurrent respiratory papillomatosis. Otolaryngol Head Neck Surg 118:810–815

Schneider J, Kinne D, Fracchia A, Pierce V, Anderson KE, Bradlow HL, Fishman J (1982) Abnormal oxidative metabolism of estradiol in women with breast cancer. Proc Natl Acad Sci USA 79:3047–3051

Schneider J, Bradlow HL, Strain G, Levin J, Anderson K, Fishman J (1983) Effects of obesity on estradiol metabolism: decreased formation of non-uterotropic metabolites. J Clin Endocrinol Metab 56:973–978

Sepkovic DW, Bradlow HL, Ho G, Hankinson SE, Gong L, Osborne MP, Fishman J (1995) Estrogen metabolite ratios and risk assessment of hormone-related cancers: assay validation and prediction of cervical cancer risk. Ann NY Acad Sci 768:312–316

Siiteri PK, MacDonald PC (1966) Placental estrogen biosynthesis during human pregnancy. J Clin Endocrinol Metab 26:751–761

Simpson ER, Zhao Y (1996) Estrogen biosynthesis in adipose. Significance in breast cancer development. Ann NY Acad Sci 784:18–26

Snow R, Barbieri R, Frisch R (1989) Estrogen 2-hydroxylase oxidation and menstrual function among elite oarswomen. J Clin Endocrinol Metab 69:369–376

Speroff L, Glass RH, Kase NG, Bohnet HG (1993) Gynäkologische Endokrinologie und steriles Paar. Diesbach Verlag, Berlin, pp 271–301

Steinberg BM, Abramson AL (1985) Laryngeal papillomas. Clin Dermatol 3:130–138

Tabor MW, Coats E, Sainsbury M, Shertzer HG (1991) Antioxidation potential of indole compounds – structure activity studies. Adv Exp Med Biol 283:833–836

Tiwari RK, Guo L, Bradlow HL, Telang NT, Osborne MP (1994) Selective responsiveness of human breast cancer cells to indole-3-carbinol, a chemopreventive agent. J Natl Cancer Inst 86:126–131

Tyler CW, Webster LA, Ory HW, Rubin GL (1985) Endometrial cancer: how does cigarette smoking influence the risk of women under age 55 years having this tumor? Am J Obstet Gynecol 151:899–905

Van Vallenhoven RF, McGuire JL (1994) Estrogen, progesterone and testosterone: can they be used to treat autoimmune diseases? Cleve Clin J Med 61:276–284

Willman K, Pulkkinen MO (1971) Reduced maternal plasma and urinary estriol during ampicillin treatment. Am J Obstet Gynecol 109:893–896

Yager JD, Liehr JG (1996) Molecular mechanism of estrogen carcinogenesis. Annu Rev Pharmacol Toxicol 36:203–232

Zhu BT, Conney AH (1998) Functional role of estrogen metabolism in target cells: review and perspectives. Carcinogenesis 19:1–27

Zhu BT, Liehr JG (1993) Inhibition of the catechol-O-methyltransferase-catalyzed O-methylation of 2- and 4- hydroxyestradiol by catecholamines. Implications for the mechanism of estrogen-induced carcinogenesis. Arch Biochem Biophys 304:248–256

Zumoff B, Fishman J, Cassouto J, Hellman L, Gallagher TF (1966) Estradiol transformation in men with breast cancer. J Clin Endocrinol 26:960–966

Zumoff B, Fishman J, Cassouto J, Gallagher TF, Hellman L (1968a) Influence of age and sex on normal estradiol metabolism. J Clin Endocrinol 28:937–941

Zumoff B, Fishman J, Gallagher TF, Hellman L (1968b) Estradiol metabolism in liver cirrhosis. J Clin Invest 47:20–25

Zumoff B, Fishman J, Levin J, Gallagher TF, Hellman L (1970) Reversible reproduction of the abnormal estradiol metabolism of biliary obstruction by administration of norethandrolone. J Clin Endocrinol 30:598–601

Zumoff B, Freed SZ, Levin J, Whitmore WF, Hellman L, Fishman J, Kukushima DK (1980) Metabolism of ^{3}H-estradiol in men with prostate cancer. Eur J Cancer 16:219–221

Part 4
Physiology and Pathophysiology of Estrogens

Phylogeny of Estrogen Synthesis, Extragenital Distribution of Estrogen Receptors and Their Developmental Role

L. SOBEK and V.K. PATCHEV

A. Phylogeny of Estrogen Synthesis

The biosynthesis of estrogens is phylogenetically well conserved and can be traced back millions of years. The biosynthesis of estrogens appears to occur throughout the entire vertebrate phylum, including mammals, birds, reptiles, amphibians, teleost and elasmobranch fish, and *Agnatha* (hagfish and lampreys) (CALLARD et al. 1978, 1980; CALLARD 1981). It has also been described in the protochordate *Amphioxus* (CALLARD et al. 1984). Estrogen biosynthesis has not been reported in nonchordate animal phyla, even though the CYP19 family appears to be an ancient lineage of *P*-450 gene products, diverging as much as 10^9 years ago (NELSON et al. 1993).

The final step of estrogen biosynthesis, aromatization of androgens, is accomplished by the cytochrome *P*-450aro, an enzyme located in the endoplasmic reticulum. The process involves two hydroxylation steps of the angular C-19 methyl residue, followed by elimination of C-19 and the phenolic rearrangement of ring A (ROBEL 1993). In most vertebrate species that have been examined, aromatase expression occurs predominantly in the gonads and brain. This is true for fish and avian species as well as for laboratory mammals, such as rodents. In addition, estrogen biosynthesis through aromatization has been demonstrated in other tissues, albeit in association with considerable species differences. In humans and a number of higher primates, placenta and adipose tissue produce estrogens. The placenta of numerous ungulate species, such as cattle, pigs and horses, synthesizes estrogens, whereas that of rodents and *Lagomorpha* does not (SIMPSON et al. 1997).

Aromatase expression has been documented in specific brain areas of representatives of each major vertebrate class down to *Cyclostomata*, and the distribution appears to be restricted to the phylogenetically older basal forebrain regions (NEGRI-CESI et al. 1996). Although sexual dimorphism in brain aromatase expression has been documented (HUTCHINSON 1991), and testosterone appears to be the most powerful inducer of aromatase activity (BALTHAZART 1997), as discussed elsewhere in this book, estrogen biosynthesis in the brain should not be solely implicated in the emergence and regulation of gender-specific behavioral repertoires.

B. Extragenital Distribution of Estrogen Receptors and Their Functional Significance

Although organs of the reproductive system are the major targets of estrogens, estrogen receptors (ERs) are expressed in several nonreproductive tissues. These ERs certainly contribute to the exceptional diversity of biological effects of estrogens and underline their importance in several essential aspects of cellular function, e.g., proliferation, growth, synthesis of characteristic proteins, survival, and resistance against environmental and pathological impacts. However, a host of questions related to the physiological relevance of ER-mediated transcription in extragenital tissues awaits clarification. The recent discovery of the β isoform of the ER, which differs from the "classical" ERα in its ligand binding and transactivation domains, and growing evidence for differential anatomical distribution of the two ER subtypes throughout the body may lead to fundamental revisions in our views of the physiological significance of ER signaling in individual organs and systems.

I. Breast

Estrogens are considered as critical hormonal agents in the development and function of the mammary gland, and ERs are constitutively present in breast cells (Bohnet et al. 1977; Shyamala and Haslam 1980; Delouis and Richard 1993). ER-mediated cellular growth and proliferation of parenchymal breast cells initiate the formation of ductal elements, which is ultimately controlled by the orchestrated action of progesterone, prolactin, growth hormone and glucocorticoids (Imagawa et al. 1990). In this context, it is of interest to note that testosterone-mediated restraint of breast growth and differentiation might be based on antagonization of estrogen effects (Santen 1991), whereas, as seen in other estrogen target tissues, biosynthesis of progesterone receptors in breast epithelial cells is under ER-mediated control (Shyamala 1987). Being present in the breast at virtually all functional and developmental stages (Shyamala and Haslam 1980; Shyamala et al. 1992), ERs are believed to belong exclusively to the α isoform; ERα-deficient mice display severe retardation in the development of the mammary gland, despite severalfold increased circulating estrogen levels (Couse and Korach 1998), and ERβ-encoding transcripts have not been identified in breast tissue of these animals (Couse et al. 1997).

II. Bone

The importance of estrogens in skeletal homeostasis was recognized early, and direct effects of estrogens on the bone-forming cells, osteoblasts, are extensively described (Komm et al. 1988; Benz et al. 1991; Takeuchi et al. 1995). Although ER expression in bone is not particularly abundant, ERsα have been demonstrated in osteoblasts, and estrogen responsiveness of these cells apparently correlates with ER expression level (Davis et al. 1994; Huo et al. 1995; Sutherland et al. 1996). The presence of functional ERs was also demonstrated in periosteum (Westerlind et al. 1995). A role for ERs in human bone metabolism is also supported by clinical observations that estrogen resistance emerging from a mutation in the ERα gene is associated with decreased bone mineral density (Smith et al. 1994). In this context, it should be noted that, according to clinical observations (Cutler 1997), in both genders, pubertal bone growth spurt and acceleration of epiphyseal maturation and termination of linear growth appear to be the major estrogen-controlled processes. However, ERα knockout mice display only a modest decrease in bone density (Korach 1994).

The ERβ has been demonstrated in rat bone (Onoe et al. 1997), and recent reports have provided evidence that ERα and ERβ expression have distinct dynamics during human osteoblast differentiation (Arts et al. 1997). Thus, although the functional importance of ERβ in bone metabolism is largely unknown, the presence of both ER subtypes in osteoblasts is suggestive of specific and, probably, different relevance in osteogenesis, mineralization and remodeling. Hoshino et al. (1995) have identified various isoforms of ER mRNA in bone tissue and osteoblasts, and suggested the possibility of a differential functional role. ER-encoding transcripts have been found in mature osteoclasts, and estrogen-mediated inhibition of osteoclasic bone resorption was demonstrated (Mano et al. 1996).

Although ER-mediated transcriptional regulation seems to be the major mechanism of estrogen effects on bone metabolism, evidence for non-genomic effects of estrogens in bone cells is accumulating. Demonstration of specific estradiol binding sites on the surface of osteoblast- and osteoclast-like cells, whose occupation results in altered cytosolic calcium levels and phophospholipid metabolism in female rat osteoblasts (Lieberherr et al. 1993), or changes in the metabolic activity of avian (Gay et al. 1993) and human osteoclasts (Fiorelli et al. 1996), point at alternative pathways of estrogen-mediated regulation of bone metabolism.

III. Gastrointestinal Tract

Attempts to demonstrate ERs in the pancreas have produced inconclusive results. Generally, it can be assumed that ERs are scarcely present in normal pancreatic tissue and, with high probability, virtually absent in primary pancreatic adenocarcinomas (Singh et al. 1997; Ollayos et al. 1994). Some years ago, an estrogen-binding protein was isolated from rat and rabbit pancreas (Grossman and Traish 1992); however, scrutiny of its structure revealed that it differs from the ER by lacking the DNA-binding domain. The significance of this protein in pancreatic physiology remains obscure.

Within the gastrointestinal tract, the liver is the organ which harbors the highest amounts of ERs. As discussed elsewhere in this book, ER-mediated chemical signals have determinant influence on several hepatic functions which also project to physiological and pathological events in other organs and systems. In this part, it is pertinent to note that hepatic ER biosynthesis was recognized early to be under multihormonal control (Norstedt et al. 1981). Gonadectomy and physiological changes in estrogen levels (in association with ovarian cycle, gestation, lactation, etc.) were shown to reflect on ER abundance in the liver (Eriksson 1982; Lax et al. 1983). Gender differences in liver function, ER levels, and responsiveness to estrogens are well known, and gender-specific developmental factors, including differential sex steroid milieu, have been implicated in their emergence (Ignatenko et al. 1992; Mataradze et al. 1992). Previous studies have suggested that a population of high-capacity low-affinity estrogen-binding sites is present in the liver. The latter apparently undergo both perinatal sex-hormone-mediated "imprinting" and postpuberal gender-specific differentiation, and were considered as a possible basis for gender differences in liver function (Sloop et al. 1983; Thompson and Lucier 1983).

IV. Cardiovascular System

Vasculoprotective effects of estrogens have been known for many years, and alterations in ER-mediated influences on the endothelium are increasingly implicated in vascular pathology associated with abnormal ovarian function and potential therapeutic effects of estrogen supplementation. However, the role of ER-dependent transcriptional regulation in the cardiovascular system under normal and pathophysiological conditions certainly requires further clarification.

ERα-encoding transcripts have been detected in the heart and aorta of female and male mice. In both sexes, ERβ mRNA levels were barely detectable in the heart and absent in the aorta (Couse et al. 1997). These findings, as well as demonstrations of impaired nitric oxide synthesis in the aorta and lack of estrogen-induced angiogenesis of ERα-disrupted transgenics (Rubanyi et al. 1997) suggest a major role for ERα in mediating the effects of estrogens on the cardiovascular system. However, vasculoprotective estrogen effects also have been reported in ERα-knockout animals (Iafrati et al. 1997). As discussed below, it is unclear to which extent estrogen effects in excitable tissues (including the cardiovascular system) may be ascribed to cellular actions that do not employ ER-mediated transcriptional regulation.

V. Immune System

A role for estrogens in the dialogue between the endocrine and immune system was postulated almost 15 years ago (Grossman 1984, 1985). Early clinical observations of gender-specific preponderance of immune disorders have persistently nourished research interest in this field. Meanwhile, evidence of an immunomodulatory involvement of estrogens has accumulated (Hall and Goldstein 1984; Berczi 1986); however, elucidation of individual ER-mediated effects using in vivo paradigms has been frequently confounded by estrogen-induced changes in the secretion of gonadotropins and prolactin, which are themselves powerful immunomodulators. Currently, the "cautious" view of estrogen influences on immune regulation is that estrogens suppress cell-mediated immunity while enhancing humoral immune responses (WILDER 1996; JANSSON and HOLMDAHL 1998).

ERs have been demonstrated in reticuloepithelial and T-cell populations of the thymus (MARCHETTI et al. 1984; STIMSON 1988; KAWASHIMA et al. 1992), and pharmacological studies have suggested structural differences between thymic and uterine ERs (BRIDGES et al. 1993); however, the latter assumption has not yet been confirmed. Furthermore, ER expression has been either documented or suggested in B-lymphocyte precursors and supporting stromal cells (SMITHSON et al. 1995), polymorphonuclear leukocytes (ITO et al. 1995), monocytes (BEN-HUR et al. 1995), and synovial macrophages (CUTOLO et al. 1996).

Estrogen-dependent alterations in natural-killer (NK) cells have prompted studies of ER-mediated influences on this important component of the organism's anti-tumor defense (BERRY et al. 1987; GRUBER et al. 1988). However, the mechanisms by which estrogens suppress, and ER antagonists augment, NK cell activity remain elusive, although ERs have been demonstrated in some NK-related cell lineages (CROY et al. 1997). The results of several studies recommend that interpretations of estrogen-induced changes in cytolytic activity exerted by NK cells should take into consideration estrogen effects on NK-targeted tumor cells which, in most experimental paradigms, are ER-positive breast cancer cells. Thus, ER-mediated influences appear to primarily affect the susceptibility of tumor targets toward NK cells (ZIELINSKI et al. 1989; SCREPANTI et al. 1991; BARAL et al. 1995). Finally, estrogen effects in ER-negative immune cells may emerge secondarily as a consequence of estrogen-induced changes in cytokine biosynthesis, which occur in ER-endowed compartments of the immune system (RALSTON et al. 1990; DA-SILVA et al. 1994; DAYAN et al. 1997; ELENKOV et al. 1997).

VI. Pituitary

Significant amounts of ERs are expressed in the adenohypophysis and, as discussed in several chapters of this book, there is no doubt that ER-mediated chemical signals have a primordial importance in the regulation of pituitary gonadotroph and lactotroph secretions. However, little is known about the direct role of estrogens in the control of thyreo-, somato- and corticotroph functions. Although mapping studies of ERs in normal and tumorous human pituitary have produced certain controversies, there is a general consent that prolactin- and gonadotropin-secreting cell populations are the major sites of ER expression (FRIEND et al. 1994; STEFANEANU et al. 1994; ZAFAR et al. 1995). In situ hybridization studies of the distribution of ERs throughout the pituitary lobes (PELLETIER et al. 1988) demonstrated that labeling is confined to the anterior and intermediate lobes, whereas the neurohypophysis was devoid of ERs. Thus, the well-documented effects of estrogens on the biosynthesis and secretion of posterior-pituitary hormones apparently occur at the suprahypophyseal level.

Studies in rats and mice have found that pituitary ERs belong to the α isoform (Couse et al. 1997; Kuiper et al. 1997; Kuiper and Gustafsson 1997), thus suggesting that ERα is the sole mediator of estrogen action in this tissue. This conclusion is supported by the finding that mice with a transgenic disruption of the ERα fail to demonstrate a classical estrogen action, namely, estrogen-mediated negative feedback on the expression of gonadotropin-encoding genes in the pituitary (Scully et al. 1997). However, moderate-to-low expression of ERβ has been found in human fetal pituitaries (Brandenberger et al. 1997). Besides ERα and ERβ, in recent years, various isoforms of the ER have been described in the pituitary gland (Friend et al. 1995, 1997; Demay et al. 1996). These authors suggest the existence of gender-specific distribution of isoforms with respect to defined cell populations in the adenohypophysis, and occurrence of quantitative oscillations in parallel with cyclic changes in estrogen secretion or hormonal treatment. A potential role for such "accessory" estradiol-induced transcriptional regulators has been proposed with respect to gender-specific differences in pituitary responsiveness to estrogens.

VII. Central Nervous System

The distribution of ERs in the rat brain has been described in several excellent reviews (Pfaff and Keiner 1973; Simerly et al. 1990; Shughrue et al. 1996, 1997), the latter of which reveals substantial differences in the topology of ERα and ERβ. Thus, 25 years after the first discoveries, several questions regarding the anatomical localization of estrogen-concentrating neurons in the central nervous system (CNS) can be answered more precisely. It has been demonstrated that ERs are present not only in neurons, but also in astrocytes, ependymal and endothelial cells (Langub and Watson 1992; Santagati et al. 1994). These findings implicate a role for ER-mediated signals also in non-neuronal functions. Co-localization of aromatase and ERs in neurons has been documented in several species (Balthazart et al. 1991; Dellowade et al. 1995; Tsuruo et al. 1995), suggesting possible auto- and paracrine effects of locally formed estrogens.

Much remains to be done concerning the functional significance of ERs in distinct neuronal populations. At this point, it is pertinent to emphasize that ER-mediated chemical signals in the CNS decisively contribute to two major, albeit timely separated, processes: (1) estrogen-dependent organization of neuronal populations with respect to their morphological connections and neurochemical communications, and (2) changes in the activity of hormonally pre-organized neural circuits emerging from physiological fluctuations in estrogen secretions. Thus, expression of ER in the brain as "predictor of function" should be interpreted only under consideration of the fact that ER-mediated functional changes (referred to as "activational" effects; McEwen et al. 1991) occur on a background that has been previously shaped by estrogens. Indeed, high ER densities are present in brain regions that either control the secretory activity of the gonadal endocrine axis or convert the chemical message of estrogen signal into a behavioral pattern which subserves reproduction and infant care. For example, ERs are strongly expressed in a rostro-caudal sequence of structures (olfactory lobe, bed nucleus of stria terminalis, medial preoptic area, ventromedial and arcuate hypothalamic nucleus, amygdala and periaqueductal gray) which are crucially involved (at least, in the female) in the regulation of both the gonadal endocrine axis and reproduction-related behaviors. Less evident is, however, the function-derived necessity for the presence of ERs in brain structures that are not directly involved in neural control of reproduction, e.g., hypothalamic supraoptic, paraventricular and suprachiasmatic nuclei, solitary tract nucleus, raphe nuclei, locus coeruleus, and cerebellum. These findings are indicative of more global aspects of ER importance in the CNS, such as regulation of stress responsiveness or integration of sensory information from intero- and exteroceptive modalities, etc. The recent impressive comparative studies of Shughrue et al. (1997) on the distribution of ERα and ERβ in the brain provide a firm morphological basis for the hypothesis that ER-mediated signals influence almost every aspect of neural function, albeit through distinct receptor populations.

Abundance of ERs in the CNS is subject to substantial changes during ontogenesis (Plapinger and McEwen 1973; Toran-Allerand et al. 1992; Yokosuka and Hayashi 1995). It can be assumed that the "final version" is established at the time when the secretory patterns of estrogen become adult-like. Both the temporary presence of ERs in structures of the developing brain that are considered ER-negative in adulthood and the maturation-associated changes in their abundance are indicative of a role in neural growth and differentiation, and the organization of brain circuitry in the most general sense. Meanwhile, the search for molecular targets and "assistants" of ER-mediated neurotrophic effects is progressing. Interactions between ER and neurotrophin signaling pathways (Miranda et al. 1993) and demonstrations of estrogen-induced changes in the expression of apoptosis-controlling genes in estrogen-sensitive neurons (Arai et al. 1996; Garcia-Segura et al. 1998) suggest novel roles for ER-mediated regulation of neurogenesis and survival. The increasing interest in interactions between steroid hormones and a host of intrinsic factors involved in neural development may contribute to our understanding of the importance of steroids for CNS organization and function.

VIII. Membrane-Bound ERs

For many years, estrogens have been known to exert rapid effects, which remain incompatible with the mainstream view of steroid action on gene transcription (Schumacher 1990; McEwen 1991; Ramirez and Zheng 1996; Joëls 1997). However, although changes in various functional parameters were ascribed to membrane effects of steroids, cell-membrane-located ERs have been neither convincingly demonstrated nor cloned as yet. Also, little is known about the pathways involved in this type of signal transmission by steroid hormones. Indeed, membrane-binding sites for estrogens in brain tissue have been successfully demonstrated (Towle and Sze 1983; Mermelstein et al. 1996; Ramirez et al. 1996; Lagrange et al. 1997). Nonetheless, it is still unclear to what extent these steroid receptors are related to the "classic" intracellular receptors, nor is much known about the regulation of their biosynthesis and ligand-binding characteristics under physiological and pathological conditions. As to the intracellular pathways that contribute to the biological effects of membrane-bound steroid receptors, current views converge to the hypothesis that cellular effects of steroid hormones represent a continuum of events that encompass virtually most of the known signal transduction pathways, from ionic channels to second messengers and nuclear transcription factors (Szego 1994; Revelli et al. 1998).

C. Hormonal Regulation of ER Synthesis

I. Estrogens

The issue of homologous regulation of ER expression by the cognate ligand is of significance for the understanding of changes in tissue sensitivity to estrogens resulting from exposure to the hormone during either individual development or physiological conditions characterized by increased systemic estrogen levels, e.g., the ovarian cycle. However, studies that have employed both in vitro and in vivo paradigms have provided controversial results. For example, ligand-activated ERs in MCF-7 breast cancer cells can transiently suppress ER mRNA levels, presumably by inhibition of transcription, mRNA stability and/or translation (Saceda et al. 1988, 1998; Ree et al. 1989). However, when ERs are expressed in steroid-receptor-negative HeLa cells, promotor-mediated and ligand-dependent upregulation of their own transcription has been documented (Castles et al. 1997). Pre-ovulatory increase of estrogens has been shown to induce synthesis of ERs in the endometrium (Guillomot et al. 1993), while numerous studies have documented homologous downregulation of ERs by their cognate ligand in the liver (Zou and Ing 1998) and brain (DonCarlos et al. 1995; Orikasa et al. 1995; Weiland et al. 1997). Although estrogen-dependent regulation of ER expression in the brain may occur in a site-specific fashion (Orikasa et al. 1996), as discussed below, this process seems to play a determinant role during CNS development with respect to differential estrogen sensitivity of neuroendocrine circuits that operate in a gender-characteristic mode.

II. Thyroid Hormones

There is consistent evidence that thyroid hormones stimulate the biosynthesis of ERs in different tissues. Accumulation of ER mRNA has been shown to occur in parallel with that of thyroid hormone receptors-β (TRβ) mRNA in thyroid hormone-treated hepatocytes (ULISSE and TATA 1994). Similar induction of ERs by thyroid hormones has been seen in pituitary and liver cells (ALTSCHULER et al. 1988; FREYSCHUSS and ERIKSSON 1988; FUJIMOTO et al. 1997), and the CNS (DELLOVADE et al. 1996). Surprisingly, however, these effects of thyroid hormones are not associated with potentiation of estrogen-mediated effects. In several physiological and pathological regulatory aspects, thyroid hormones and estrogens display mutual antagonism (GRAUPNER et al. 1991; ZHOU-LI et al. 1993; ZHU et al. 1997), which is apparently due to interference between these ligand-activated transcription factors at DNA response elements with very close similarity (GLASS et al. 1988).

III. Progesterone

Progesterone is generally considered as the physiological "restrainer" of ER-mediated transcriptional events, although this simplified view requires certain caution. Exemptions from the rule are known, as exemplified by the differential effects of progesterone on epithelial and stromal cells in the endometrium. Thus, during the luteal phase in the rat, progesterone blocks the synthesis of ERs and estrogen-induced cellular proliferation in superficial epithelial cells of the endometrium. However, in the stroma and endometrial glands, progesterone stimulates the synthesis of ERs and promotes pre-decidual cell proliferation through synergism with ER-mediated influences. Nevertheless, in general, physiological increases in progesterone secretion counteract several effects of estrogens in the rat and primate uterus through diminishing ER biosynthesis (BAULIEU et al. 1975; GUILLOMOT et al. 1993). This circumstance is best appreciated by the fact that ER concentrations in the gravid uterus and, consequently, manifestation of estrogen-mediated effects, display a reciprocal relationship to circulating progesterone levels (HADLEY 1984). Furthermore, decrease of ER levels following exposure to progesterone has been well documented in brain and pituitary, thus suggesting that estrogen–progesterone antagonism is not necessarily restricted to the genital tract (BLAUSTEIN and BROWN 1984). However, it is pertinent to note that progesterone-mediated "silencing" of estrogen effects also occurs through several mechanisms which are independent of the regulation of ER expression.

IV. Androgens

Non-aromatizable androgens are known to counteract several neuroendocrine and behavioral effects of estrogens, e.g., estrogen-induced prolactin release, gonadotropin secretion, and female sexual behavior. As already mentioned with regard to the mammary gland, androgens exert their antagonistic action through decreasing estrogen binding by ERs in the brain (Brown et al. 1994), and different types of pituitary cells (Keefer et al. 1987). However, it remains unclear whether androgen–estrogen antagonism depends on binding competition or also involves androgen-mediated alterations in the biosynthesis of ERs. As previously mentioned with regard to progesterone, different classes of steroid hormone receptors may influence the transcription of target genes in opposite modes, and interference emerging from homology in DNA response elements appears to be a quite common phenomenon in steroid-receptor-controlled transcription (von der Ahe et al. 1985; Wahli and Martinez 1991).

V. Glucocorticoids

The ER-mediated transcriptional activation of silent vitellogenin genes in male Xenopus liver proved to be a valuable model for understanding the mechanisms of steroidal regulation of ER expression. Ulisse and Tata (1994) showed that dexamethasone treatment results in prompt upregulation of ER mRNA; the functional significance of the latter phenomenon is manifested by enhanced vitellogenin gene transcription upon subsequent exposure to estrogens. However, administration of glucocorticoids in vivo has been shown to decrease ER concentrations in estrogen target tissues. The latter observation has been implicated in pathophysiological mechanisms of stress- and glucocorticoid-mediated suppression of reproductive functions (Rabin et al. 1990). These examples suggest that glucocorticoids may directly influence ER biosynthesis, albeit in different and, probably, tissue-specific fashion; again, as stated above, net changes in ER-regulated parameters may be "clouded" by coincident cross-talk between ER and other steroid hormone receptors.

VI. Neurotransmitters

Activity-induced changes in the synthesis of specific proteins are a common feature of nerve cells, and steroid-hormone-dependent regulation of neuroendocrine functions provides a good example of convergent interactions of different chemical signals on the output of a defined cell population. More than 10 years ago, several studies provided convincing evidence that pharmacological alterations of monoaminergic neurotransmission can produce significant changes in ER concentrations and, consequently, estrogen responsiveness, in defined estrogen-sensitive hypothalamic neuronal populations and anterior pituitary cells (Blaustein 1986, 1987; Blaustein and Turcotte 1987; Montemayor et al. 1990; Brown et al. 1991; Tetel and Blaustein 1991). Together with reports on ligand-independent activation and/or changes in the binding properties of ERs promoted by monoamine neurotransmitters (Smith et al. 1993; Wooley et al. 1994; Gangolli et al. 1997), the above-mentioned observations indicate that ER-mediated effects in the brain may vary, depending on the neurochemical milieu, and suggest intriguing possibilities for orchestrated interactions of various chemical signals in determining the responsiveness of the CNS to estrogens.

D. Developmental Role of ERs as Exemplified by Hormone-Dependent Brain Differentiation

Estrogen-dependent differentiation of neural circuitry during early brain development is a classic example for ER-mediated structural organization and biochemical "specialization" of defined cell populations. In the physiological context, these estrogen-dependent processes play a determinant role in several aspects of brain functions of adaptive importance, such as control of reproduction, responsiveness to environmental challenges, emotional status, cognitive performance, and "pace" of brain aging. Since these aspects are comprehensively discussed elsewhere in this book, this part will focus on evidence that documents the developmental role of ERs in the CNS.

Extensive research on different aspects of sex-hormone-dependent brain organization in the 1960s and 1970s (PHOENIX et al. 1959; BARRACLOUGH 1967; GORSKI 1971; DÖRNER 1976; GOY and McEWEN 1980) precipitated in a series of postulates which belongs nowadays to the "classics" in neuroendocrinology: (1) tonic or phasic gonadotropin and sex steroid secretions, together with gender-specific patterns of sexual behavior, can serve as common indices of sexual differentiation of the brain; (2) the morpho-functional determination of brain gender results from a brief exposure to gender-specific gonadal steroids during a critical period of perinatal development; (3) a "female-by-default" neural control of gonadal function and sexual behavior is irreversibly "defeminized" by "pure" or testosterone-derived estrogens during this critical developmental period; and (4) in the male rat, this process occurs within the last days of gestation, and the infant brain remains sensitive to estrogens for at least the first 2 weeks of postnatal life. Subsequently, evidence has accumulated that defeminization of the "default" neural circuitry involved in the control of reproduction depends on the aromatization of testosterone to estrogens, which exert their organizing effects through ERs (CHRISTENSEN and GORSKI 1978; MACLUSKY and NAFTOLIN 1981).

The early studies which examined the presence of ERs in the fetal rat brain (MACLUSKY et al. 1979a) were supportive of these postulates; indeed, functionally active ERs could be detected only at the end of gestation (day 21), and in male, but not female fetuses, they were already occupied by circulating estrogens. ER presence was initially confined to the typical estrogen-sensitive areas known in adults, e.g., hypothalamus, amygdala, preoptic area, whereas ERs in the cerebral cortex achieved similar densities only around postnatal day 6. Scrutiny of ER densities in the postnatal period revealed that the male infant brain experiences quite dramatic changes, while, in females, occupancy of ERs, even in estrogen-sensitive areas, barely exceeds 5% of that seen in males (MacLusky et al. 1979b). Experimental methods with higher resolution used in subsequent studies documented the presence of ERs in the fetal rat brain 1 week before term and confirmed the dramatic rise in ER levels during the last week of gestation (Vito and Fox 1981). The presence of functional ERs was demonstrated also in the fetal monkey brain (Pommerantz et al. 1985; Sholl and Kim 1989); although the critical period of sexual differentiation of the primate brain is not precisely defined, increasing ER densities with advancing gestational age were also observed in monkey brain samples.

Substantial support for the (not always unanimous) view that ERs play a crucial role in the process of gender-specific organization of reproduction-related brain functions emerges from the elegant studies of McCarthy (McCarthy et al. 1993; McCarthy 1994) in which antisense "knockout" of the ERs in estrogen-sensitive brain regions of female neonates rendered them insensitive to the "defeminizing" effects of aromatizable androgens. Recent knowledge gathered from ER knockout (ERKO) mice adds weight to the hypothesis that ERs play a primordial role for the sexual differentiation of the brain, as documented by "blunted" or absent sexual dimorphism in several morphological and functional aspects in these animals (Simerly et al. 1997; Ogawa et al. 1997; Rissman et al. 1997).

Notwithstanding this, the postulate that ER-mediated effects are solely responsible for gender-specific brain differentiation has been debated since its emergence. Generally, criticism is based on two major concerns: (1) unlike in the initial studies mentioned above, attempts to "trace" the ontogeny of ERs in the brain have either failed to demonstrate gender-associated differences in ER densities or revealed higher densities of ER in females than in males during early postnatal brain development, and (2) although circulating estrogen levels are similar in male and female rat neonates, females do not become defeminized, albeit still being sensitive to estrogens. Observations in the early 1970s that an α-fetoprotein-resembling protein is produced in significant amounts by the liver of neonates, and specifically binds estradiol with considerable avidity, suggest that the neonatal brain is endowed with an efficient mechanism of protection against estrogens (McEwen et al. 1975; Toran-Allerand 1984). Both arguments deserve special attention in the discussion of the importance of ER-mediated signals for brain differentiation.

Indeed, most recent studies on the developmental profiles of ER expression in rat brain invariantly suggest that, within the first days after birth, females display higher densities of ERs than males (DonCarlos and Handa 1994; Kuhnemann et al. 1994; DonCarlos 1996; Yokosuka et al. 1997). Despite some contradictions to the results obtained in the very early studies on this issue (MacLusky et al. 1979a,b), the latter observations have a far-reaching importance. As summarized by MacLusky et al. (1997), they are an excellent illustration of homologous regulation of ERs in a target organ, whose specific functions during a sizable period of life will depend on the presence or absence of responsiveness to changing estrogen concentrations. In other words, estrogens act on the developing brain during a narrow "window of time" not only to establish a male-specific "hard-wiring", but also to "disable" female-like responsiveness to estrogens in brain areas that control gonadotropin secretion and, probably, sexual behavior. This process apparently involves irreversible downregulation of ER expression by estrogens during early brain development.

As to the protection exerted by α-fetoprotein, this mechanism does not conflict with the developmental role of testosterone-derived estrogens. First, although circulating estradiol levels are similar in both sexes, the gonads of the male rat display a secretory burst of testosterone during late gestation, i.e., add certain amounts of estrogen precursor. Second, the conversion of testosterone to estradiol by aromatase occurs in the intracellular compartment to which the binding protein would not have access. Thus, the "aromatase hypothesis" might serve as a plausible explanation for the occurrence of sexual differentiation, even though males and females may display similar plasma concentrations of testosterone. Pharmacological blockade of this enzyme has been shown to prevent defeminization during early development (BAKKER et al. 1997), and gender differences in aromatase expression and activity have been clearly documented in the juvenile brain, with males displaying higher levels of aromatase-encoding mRNA already in early postnatal life (LAUBER et al. 1997a,b). Further, aromatase expression in the brain is higher in the perinatal period than in adulthood (GEORGE and OJEDA 1982; LEPHART and OJEDA 1990; LEPHART et al. 1992 LAUBER and LICHTENSTEIGER 1994). This appears also to be true in the human, as recently shown by HONDA et al. (1994).

The developmental profile of aromatase expression and activity in the brain supports the hypothesis of its crucial role in gender-specific organization. In the fetal rat, both aromatase activity and mRNA are present in the highly sexually dimorphic preoptic area as early as gestational days 15/16, and peak on gestational days 19/20 before gradually decreasing during the postnatal period to adult-like levels (GEORGE and OJEDA 1982; LEPHART et al. 1992; LAUBER and LICHTENSTEIGER 1994; TSURUO et al. 1994). In contrast, however, aromatase mRNA levels in the estrogen-sensitive structures bed nucleus of the stria terminalis and amygdala remain high throughout development and adulthood (LAUBER and LICHTENSTEIGER 1994; TSURUO et al. 1994). The latter differences might be associated with the functional diversity of these regions and account for the maintenance of estrogen effects in the latter areas, whose importance is not restricted to gender-specific control of reproductive functions.

Thus, while it can be presumed that the male rat brain is able to utilize perinatal testosterone secretions in a way that ensures brain defeminization, it should be mentioned that clinical data obtained from male patients with inborn enzymatic deficiency are less supportive of an indispensable role of aromatization for the gender-specific organization of the human brain (NEGRI-CESI et al. 1996).

The role of ERs in brain areas that are not directly involved in the control of gonadal secretions and reproductive behavior has only begun to be elucidated. It deserves mention that transient expression of ERs has been demonstrated in the neonatal and developing rat hippocampus (O'Keefe and Handa 1990; O'Keefe et al. 1995), thus adding weight to the hypothesis that ERs might play an indispensable role in brain morphogenesis and functional organization. As recently reported, estrogen-mediated brain differentiation may also encompass gender dichotomy in the neuroendocrine responsiveness to stress and glucocorticoids (Patchev et al. 1995; Patchev and Almeida 1998). The questions of whether ER-mediated, gonadal-steroid-dependent organization also involves other aspects of brain function with less obvious sexual dimorphism, and what might be the contribution of distinct isoforms of ER to this process, represent imminent and challenging tasks for future research.

References

Altschuler LR, Ceppi JA, Ritta MN, Calandra RS, Zaninovich AA (1988) Effects of thyroxine on estrogen receptor concentrations in anterior pituitary and hypothalamus of hypothyroid rats. J Endocrinol 119:383–387

Arai Y, Sekine Y, Murakami S (1996) Estrogen and apoptosis in the developing sexually dimorphic preoptic area in female rats. Neurosci Res 25:403–407

Arts J, Kuiper GGJM, Janssen JMMF, Gustafsson JA, Löwik CWGM, Pols HAP, van Leeuwen JPTM (1997) Differential expression of estrogen receptors α and β mRNA during differentiation of human osteoblast SV-HFO cells. Endocrinology 138:5067–5069

Bakker J, Pool CW, Sonnemans M, van Leeuwen FW, Slob AK (1997) Quantitative estimation of estrogen and androgen receptor-immunoreactive cells in the forebrain of neonatally estrogen-deprived male rats. Neuroscience 77:911–919

Balthazart J (1997) Steroid control and sexual differentiation of brain aromatase. J Steroid Biochem Mol Biol 61:323–339

Balthazart J, Foidart A, Surlemont C, Harada N (1991) Distribution of aromatase immunoreactive cells in the mouse forebrain. Cell Tissue Res 263:71–79

Baral E, Nagy E, Berczi I (1995) Modulation of natural killer cell-mediated cytotoxicity by tamoxifen and estradiol. Cancer 75:591–599

Barraclough CA (1967) Modifications in reproductive function after exposure to hormones during the prenatal and early postnatal period. In: Martini L, Ganong WF (eds) Neuroendocrinology, vol 2. Academic Press, New York, pp 61–99

Baulieu EE, Atger M, Best-Belpomme M, Corvol P, Courvalin JC, Mester J, Milgrom E, Robel P, Rochefort H, De Catalogne D (1975) Vitam Horm 33:649–736

Ben-Hur H, Mor G, Insler V, Blickstein I, Amir-Zaltsman Y, Sharp A, Globerson A, Kohen F (1995) Menopause is associated with a significant increase in blood monocyte number and a relative decrease in the expression of estrogen receptors in human peripheral monocytes. Am J Reprod Immunol 34:363–369

Benz DJ, Haussler MR, Komm BS (1991) Estrogen binding and estrogenic responses in normal human osteoblast-like cells. J Bone Miner Res 6:531–541

Berczi I (1986) Gonadotropins and sex hormones. In: Berczi I (ed) Pituitary Function and Immunity, CRC Press, Boca Raton, pp 185–211

Berry J, Green BJ, Matheson DS (1987) Modulation of natural killer cell activity by tamoxifen in stage I post-menopausal breast cancer. Eur J Cancer Clin Oncol 23:517–520

Blaustein JD (1986) Cell nuclear accumulation of estrogen receptors in rat brain and pituitary gland after treatment with a dopamine-beta-hydroxylase inhibitor. Neuroendocrinology 42:44–50

Blaustein JD (1987) The alpha-1-noradrenergic antagonist prazosin decreases the concentration of estrogen receptors in female rat hypothalamus. Brain Res 404:39–50

Blaustein JD, Brown TJ (1984) Progesterone decreases the concentration of hypothalamic and anterior pituitary estrogen receptors in ovariectomized rats. Brain Res 304:225–236

Blaustein JD, Turcotte J (1987) Further evidence of noradrenergic regulation of rat hypothalamic estrogen receptor concentration: possible non-functional increase and functional decrease. Brain Res 436:253–264

Bohnet H, Gomez F, Friesen HG (1977) Prolactin and estrogen binding sites in the mammary gland of the lactating and non-lactating rat. Endocrinology 101:1111–1115

Brandenberger AW, Tee MK, Lee JY, Chao V, Jaffe RB (1997) Tissue distribution of estrogen receptors alpha (ER-alpha) and beta (ER-beta) mRNA in the midgestational human fetus. J Clin Endocrinol Metab 82:3509–3512

Bridges ED, Greenstein BD, Khamashta MA, Hughes GR (1993) Specificity of estrogen receptors in rat thymus. Int J Immunopharmacol 15:927–932

Brown TJ, Blaustein JD, Hochberg RB, MacLusky NJ (1991) Estrogen receptor binding in regions of the rat hypothalamus and preoptic area after inhibition of dopamine-beta-hydroxylase. Brain Res 549:260–267

Brown TJ, Adler GH, Sharma M, Hochberg RB, MacLusky NJ (1994) Androgen treatment decreases estrogen receptor binding in the ventromedial nucleus of the rat brain: a quantitative in vitro autoradiographic analysis. Mol Cell Neurosci 5:549–555

Callard GV (1981) Aromatization is cyclic AMP-dependent in cultured reptilian brain cells. Brain Res 204:451–454

Callard GV, Petro Z, Ryan KJ (1978) Phylogenetic distribution of aromatase and other androgen-converting enzymes in the central nervous system. Endocrinology 103:2283–2290

Callard GV, Petro Z, Ryan KJ (1980) Aromatization and 5α-reduction in brain and non-neural tissues of a cyclostome (*Petromyzan marinus*). Gen Comp Endocrinol 42:155–159

Callard GV, Pudney JA, Kendall SL, Reinboth R (1984) In vitro conversion of androgen to estrogen in amphioxus gonadal tissues. Gen Comp Endocrinol 56:53–58

Castles CG, Oesterreich S, Hansen R, Fuqua SAW (1997) Auto-regulation of the estrogen receptor promoter. J Steroid Biochem Mol Biol 62:155–163

Christensen LW, Gorski RA (1978) Independent masculinization of neuroendocrine systems by intracerebral implants of testosterone or estradiol in the neonatal female rat. Brain Res 146:325–340

Couse JF, Korach KS (1998) Exploring the role of sex steroids through studies of receptor deficient mice. J Mol Med 76:497–511

Couse JF, Lindzey J, Grandien K, Gustafsson JA (1997) Tissue distribution and quantitative analysis of estrogen receptor-α (ERα) and estrogen receptor-β (ERβ) messenger ribonucleic acid in the wild-type and ERα -knockout mouse. Endocrinology 138:4613–4621

Croy BA, McBey BA, Villeneuve LA, Kusakabe K, Kiso Y, van den Heuvel M (1997) Characterization of the cells that migrate from metrial glands of the pregnant mouse uterus during explant culture. J Reprod Immunol 32:241–263

Cutler GB Jr (1997) The role of estrogen in bone growth and maturation during childhood and adolescence. J Steroid Biochem Mol Biol 61:141–144

Cutolo M, Accardo S, Villaggio B, Barona A, Sulli A, Coviello DA, Carabbio C, Felli L, Miceli D, Farruggio R, Carruba G, Castagnetta L (1996) Androgen and estrogen receptors are present in primary cultures of human synovial macrophages. J Clin Endocrinol Metab 81:820–827

Da-Silva JA, Larbre JP, Seed MP, Cutolo M, Villaggio B, Scott DL, Willoughby DA (1994) Sex differences in inflammation-induced cartilage damage in rodents. The influence of sex steroids. J Rheumathol 21:330–337

Dayan M, Zinger H, Kalush F, Mor G, Amir-Zaltzman Y, Kohen F, Sthoeger Z, Mozes E (1997) The beneficial effects of treatment with tamoxifen and anti-oestradiol antibody on experimental systemic lupus erythematosus are associated with cytokine modulators. Immunology 90:101–108

Davis VL, Couse JF, Gray TK, Korach KS (1994) Correlation between low levels of estrogen receptors and estrogen responsiveness in two rat osteoblast-like cell lines. J Bone Miner Res 9:983–991

Dellovade TL, Rissman EF, Thompson N, Harada N, Ottinger MA (1995) Co-localization of aromatase enzyme and estrogen receptor immunoreactivity in the preoptic area during reproductive aging. Brain Res 674:181–187

Dellovade T, Zhu YS, Krey L, Pfaff DW (1996) Thyroid hormone and estrogen inter-act to regulate behavior. Proc Natl Acad Sci U S A 93:12581–12586

Delouis C, Richard P (1993) Lactation. In: Thibault C, Levasseur MC, Hunter RHF (eds) Reproduction in mammals and man. Ellipses, Paris, pp 503–530

Demay F, Tiffoche C, Thieulant ML (1996) Sex- and cell-specific expression of an estro-gen receptor isoform in the pituitary gland. Neuroendocrinology 63:522–529

DonCarlos LL (1996) Developmental profile and regulation of estrogen receptor (ER) mRNA expression in the preoptic area of prenatal rats. Dev Brain Res 94:224–233

DonCarlos LL, Handa RJ (1994) Developmental profile of estrogen receptor mRNA in the preoptic area of male and female neonatal rats. Dev Brain Res 79:283–289

DonCarlos LL, McAbee M, Ramer-Quinn DS, Stancik DM (1995) Estrogen receptor mRNA levels in the preoptic area of neonatal rats are responsive to hormone manipulation. Dev Brain Res 84:253–260

Dörner G (1976) Hormones and brain differentiation. Elsevier, Amsterdam

Elenkov IJ, Hoffman J, Wilder RL (1997) Does differential neuroendocrine control of cytokine production govern the expression of autoimmune disease in pregnancy and the postpartum period. Mol Med Today 3:379–383

Eriksson HA (1982) Different regulation of the concentration of estrogen receptors in the rat liver and uterus following ovariectomy. FEBS Lett 149:91–95

Fiorelli G, Gori F, Frediani U, Franceschelli F, Tanini A, Tosti-Guerra C, Benvenuti S, Gennari L, Becherini L, Brandi ML (1996) Membrane binding sites and non-genomic effects of estrogen in cultured human preosteoclastic cells. J Steroid Biochem Mol Biol 59:233–240

Freyschuss B, Eriksson H (1988) Evidence for a direct effect of thyroid hormones on the hepatic synthesis of estrogen receptors in the rat. J Steroid Biochem 31:247–249

Friend KE, Chiou YK, Lopes MB, Laws ER Jr, Hughes KM, Shupnik MA (1994) Estrogen receptor expression in human pituitary: correlation with immunohisto-chemistry in normal tissue, and immunohistochemistry and morphology in macro-adenomas. J Clin Endocrinol Metab 78:1497–1504

Friend KE, Ang LW, Shupnik MA (1995) Estrogen regulates the expression of several different estrogen receptor mRNA isoforms in rat pituitary. Proc Natl Acad Sci USA 92:4367–4371

Friend KE, Resnik EM, Ang LW, Shupnik MA (1997) Specific modulation of estrogen receptor mRNA isoforms in rat pituitary throughout the estrous cycle and in response to steroid hormones. Mol Cell Endocrinol 131:147–155

Fujimoto N, Watanabe H, Ito A (1997) Up-regulation of estrogen receptor by triiodo-thyronine in rat pituitary cell lines. J Steroid Biochem Mol Biol 61:79–85

Gangolli EA, Conneely OM, O'Malley BW (1997) Neurotransmitters activate the human estrogen receptor in a neuroblastoma cell line. J Steroid Biochem Mol Biol 61:1–9

Garcia-Segura LM, Cardona-Gomez P, Naftolin F, Chowen JA (1998) Estradiol up-regulates Bcl-2 expression in adult brain neurons. Neuroreport 9:593–597

Gay CV, Kief NL, Bekker PJ (1993) Effect of estrogen on acidification in osteoclasts. Biochem Biophys Res Commun 192:1251–1259

Glass CK, Holloway JM, Devary OV, Rosenfeld MG (1988) The thyroid hormone receptor binds with opposite transcriptional effects to a common sequence motif in thyroid hormone and estrogen response elements. Cell 54:313–323

George FW, Ojeda SR (1982) Changes in aromatase activity in the rat brain during embryonic, neonatal, and infantile development. Endocrinology 111:522–529

Gorski RA (1971) Gonadal hormones and perinatal development of endocrine function. In: Ganong WF, Martini L (eds) Frontiers in neuroendocrinology. Oxford University Press, New York, pp 237–290

Goy R, McEwen BS (1980) Sexual differentiation of the brain. MIT Press, Cambridge

Graupner G, Zhang XK, Tzukerman M, Wills K, Hermann T, Pfahl M (1991) Thyroid hormone receptors repress estrogen receptor activation of a TRE. Mol Endocrinol 5:365–372

Grossman CJ (1984) Regulation of the immune system by sex steroids. Endocr Rev 5:435–455

Grossman CJ (1985) Interactions between the gonadal steroids and the immune system. Science 227:257–261

Grossman A, Traish A (1992) Site-specific antibodies to the DNA-binding domain of estrogen receptor distinguish this protein from the ^{3}H-estradiol-binding protein in pancreas. Life Sci 51:859–867

Gruber SA, Hoffman RA, Sothern RB, Lakatua D, Carlson A, Simmons RL, Hrushesky WJ (1988) Splenocyte natural killer cell activity and metastatic potential are inversely dependent on estrous stage. Surgery 104:398–403

Guillomot M, Fléchon JE, Leroy F (1993) Blastocyst development and implantation. In: Thibault C, Levasseur MC, Hunter RHF (eds) Reproduction in mammals and man. Ellipses, Paris, pp 387–410

Hadley ME (1984) Endocrinology. Prentice-Hall, Englewood Cliffs, pp 449

Hall NR, Goldstein AL (1984) Endocrine regulation of host immunity. In: Fenichel RL, Chirigos MA (eds) Immune modulation agents and their mechanisms. Marcel Dekker, New York, pp 533–563

Honda S, Harada N, Takagi Y (1994) Novel exon 1 of the aromatase gene specific for aromatase transcripts in human brain. Biochem Biophys Res Commun 198:1153–1160

Hoshino S, Inoue S, Hosoi T, Saito T, Ikegami A, Kaneki M, Ouchi Y, Orimo H (1995) Demonstration of isoforms of the estrogen receptor in the bone tissues and in osteoblastic cells. Calcif Tissue Int 57:466–468

Huo B, Dossing DA, Dimuzio MT (1995) Generation and characterization of a human osteosarcoma cell line stably transfected with the human estrogen receptor gene. J Bone Miner Res 10:769–781

Hutchinson JB (1991) Hormonal control of behavior: Steroid action in the brain. Curr Opin Neurobiol 1:562–570

Iafrati MD, Karas RH, Aronovitz M, Kim S, Sullican TR Jr, Lubahn DB, O'Donnell TF Jr, Korach KS, Mendelsohn ME (1997) Estrogen inhibits the vascular injury response in estrogen receptor-deficient mice. Nat Med 3:545–548

Ignatenko LL, Mataradze GD, Rozen VB (1992) Endocrine mechanisms for the formation of sex-related differences in hepatic estrogen receptor content and their significance for the realization of an estrogen effect on angiotensin blood levels in rats. Hepatology 15:1092–1098

Imagawa W, Bandyopadhyay GK, Nandi S (1990) Regulation of mammary epithelial cell growth in mice and rats. Endocr Rev 11:494–523

Ito I, Hayashi T, Yamada K, Kuzuya M, Naito M, Iguchi A (1995) Physiological concentration of estradiol inhibits polymorphonuclear chemotaxis via a receptor mediated system. Life Sci 56:2247–2253

Jansson L, Holmdahl R (1998) Estrogen-mediated immunosuppression in autoimmune diseases. Inflamm Res 47:290–301

Joëls M (1997) Steroid hormones and excitability in the mammalian brain. Front Neuroendocrinol 18:2–48

Kawashima I, Seiki K, Sakabe K, Ihara S, Akatsuka A, Katsumata Y (1992) Localization of estrogen receptors and estrogen receptor mRNA in female mouse thymus. Thymus 20:115–121

Keefer DA, Dordai N, Mallonga R, Ziegler K, Shughrue P, Ramirez P (1987) Dihydrotestosterone induces a sexual dimorphism in estrogen uptake by specific anterior pituitary cell types in vivo. Cell Tissue Res 249:477–479

Komm BS, Terpening CM, Benz DJ, Graeme KA, Callegos A, Kore M, Greene GL, O'Malley BW, Haussler MR (1988) Estrogen binding, receptor mRNA, and biologic response in osteoblasts osteosarcoma cells. Science 241:81–83

Korach KS (1994) Insights from the study of animals lacking functional estrogen receptor. Science 266:1524–1527

Kuhnemann S, Brown TJ, Hochberg RB, MacLusky NJ (1994) Sex differences in the development of estrogen receptors in the rat brain. Horm Behav 28:483–491

Kuiper GJM, Gustafsson JA (1997) The novel estrogen receptor-β subtype: potential role in the cell- and promoter-specific actions of estrogens and anti-estrogens. FEBS Lett 410:87–90

Kuiper GJM, Carlsson BO, Grandien K, Enmark E, Häggblad J, Nilsson S, Gustafsson JA (1997) Endocrinology 138:863–870

Langub MC, Watson RE (1992) Estrogen receptor-immunoreactive glia, endothelia and ependyma in guinea pig preoptic area and median eminence: electron microscopy. Endocrinology 130:364–372

Lauber ME, Lichtensteigher W (1994) Pre- and postnatal ontogeny of aromatase cytochrome P450 messenger ribonucleic acid expression in the male rat brain studied by in situ hybridization. Endocrinology 135:1661–1668

Lauber ME, Sarasin A, Lichtensteiger W (1997a) Transient sex differences of aromatase (CYP19) mRNA expression in the developing rat brain. Neuroendocrinology 66:173–180

Lauber ME, Sarasin A, Lichtensteiger W (1997b) Sex differences and androgen-dependent regulation of aromatase (CYP19) mRNA expression in the developing and adult rat brain. J Steroid Biochem Mol Biol 31:359–364

Lagrange AH, Ronnekleiv OK, Kelly MJ (1997) Modulation of G protein-coupled receptors by an estrogen receptor that activates protein kinase A. Mol Pharmacol 51:605–612

Lax ER, Tamulevicius P, Muller A, Schriefers H (1983) Hepatic nuclear receptor concentrations in the rat – influence of age, sex, gestation, lactation and estrous cycle. J Steroid Biochem 19:1083–1088

Lephart ED, Ojeda SR (1990) Hypothalamic aromatase activity in male and female during juvenile-prepubertal development. Neuroendocrinology 51:385–393

Lephart ED, Simpson ER, McPhaul MJ, Kilgore MW, Wilson JD, Ojeda SR (1992) Brain aromatase cytochrome P-450 messenger RNA levels and enzyme activity during prenatal and perinatal development in the rat. Mol Brain Res 16:187–192

Lieberherr M, Grosse B, Kachkace M. Balsan S (1993) Cell signaling and estrogens in female rat osteoblasts: a possible involvement of unconventional non-nuclear receptors. J Bone Miner Res 8:1365–1376

MacLusky NJ, Naftolin F (1981) Sexual differentiation of the central nervous system. Science 211:1294–1302

MacLusky NJ, Lieberburg I, McEwen BS (1979a) The development of estrogen receptor systems in the rat brain: perinatal development. Brain Res 178:129–142

MacLusky NJ, Chaptal C, McEwen BS (1979b) The development of estrogen receptor systems in the rat brain and pituitary: postnatal development. Brain Res 178:143–160

MacLusky NJ, Bowlby DA, Brown TJ, Peterson RE, Hochberg RB (1997) Sex and the developing brain: suppression of neonatal estrogen sensitivity by developmental estrogen exposure. Neurochem Res 22:1395–1414

Mano H, Yuasa T, Kameda T, Miyazawa K, Nakamaru Y, Shiokawa M, Mori Y, Yamada T, Miyata K, Shindo H, Azuma H, Hakeda Y, Kumegawa M (1996) Mammalian mature osteoclasts as estrogen target cells. Biochem Biophys Res Commun 223:637–642

Marchetti P, Scambia G, Reiss N, Kaye AM, Cocchia D, Iacobelli S (1984) Estrogen-responsive creatine-kinase in the reticulo-epithelial cells of rat thymus. J Steroid Biochem 20:835–839

Mataradze GD, Kurabekova RM, Rozen VB (1992) The role of sex steroids in the formation of sex-differentiated concentrations of corticosteroid-binding globulin in rats. J Endocrinol 132:235–240

McCarthy MM (1994) Molecular aspects of sexual differentiation of the rodent brain. Psychoneuroendocrinology 19:415–427

McCarthy MM, Schlenker EH, Pfaff DW (1993) Enduring consequences of neonatal treatment with antisense oligodeoxynucleotides to estrogen receptor messenger ribonucleic acid on sexual differentiation of rat brain. Endocrinology 331:433–439

McEwen BS (1991) Non-genomic and genomic effects of steroids on neural activity. Trends Pharmacol Sci 12:141–147

McEwen BS, Plapinger L, Chaptal C, Gerlach J, Wallach G (1975) The role of feto-neonatal estrogen binding proteins in the association of estrogen with neonatal brain cell nuclear receptors. Brain Res 96:400–407

McEwen BS, Coirini H, Westlund-Danielson A, Frankfurt M, Gould E, Schumacher M, Wooley C (1991) Steroid hormones as mediators of neural plasticity. J Steroid Biochem Mol Biol 39:223–232

Mermelstein PG, Becker JB, Surmeier DJ (1996) Estradiol reduces calcium currents in rat neostriatal neurons via a membrane receptor. J Neurosci 16:595–604

Miranda RC, Sohrabji F, Toran-Allerand CD (1993) Presumptive estrogen target neurons express mRNAs for both neurotrophins and neurotrophin receptors: a basis for potential development interactions of estrogen with neurotrophins. Mol Cell Neurosci 4:510–525

Montemayor ME, Clark AS, Lynn DM, Roy EJ (1990) Modulation by epinephrine of neural responses to estradiol. Neuroendocrinology 52:473–480

Negri-Cesi P, Poletti A, Celotti F (1996) Metabolism of steroids in the brain: a new insight into the role of 5α-reductase and aromatase in brain differentiation and functions. J Steroid Biochem Mol Biol 58:455–466

Nelson DR, Kamataki T, Waxman DJ, Guengerich FP, Estabrook RW, Feyereisen R, Gonzalez FJ, Coon MJ, Gunsalus IC, Gotoh O, Okuda K, Nebert DW (1993) The P450 superfamily: update of new sequences, gene mapping, accession numbers, early trivial names of enzymes and nomenclature. DNA Cell Biol 12:1–51

Norstedt G, Wrange O, Gustafsson JA (1981) Multihormonal regulation of the estrogen receptor in rat liver. Endocrinology 108:1190–1196

Ogawa S, Lubahn DB, Korach KS, Pfaff DW (1997) Behavioral effects of estrogen receptor disruption in male mice. Proc Natl Acad Sci USA 94:1476–1481

O'Keefe JA, Handa RJ (1990) Transient elevation of estrogen receptors in the neonatal rat hippocampus. Dev Brain Res 57:119–127

O'Keefe JA, Li Y, Burgess LH, Handa RJ (1995) Estrogen receptor mRNA alterations in the developing rat hippocampus. Mol Brain Res 30:115–124

Ollayos CW, Riordan GP, Rushin JM (1994) Estrogen receptor detection in paraffin sections of adenocarcinoma of the colon, pancreas, and lung. Arch Pathol Lab Med 118:630–632

Onoe Y, Miyaura C, Ohta H, Nozawa S, Suda T (1997) Expression of estrogen receptor β in rat bone. Endocrinology 138:4509–4512

Orikasa C, Yokosuka M, Hayashi S (1995) Expression of estrogen receptor in the facial nucleus is suppressed by estradiol, but not by testosterone, indicating a lack of requirement for aromatization. Brain Res 693:112–117

Orikasa C, Mizuno K, Sakuma Y, Hayashi S (1996) Exogenous estrogen acts differently on production of estrogen receptor in the preoptic area and the mediobasal hypothalamic nuclei in the newborn rat. Neurosci Res 25:247–254

Patchev VK, Almeida OFX (1998) Gender specificity in the neural regulation of the response to stress: New leads from classical paradigms. Mol Neurobiol 16:63–77

Patchev VK, Hayashi S, Orikasa C, Almeida OFX (1995) Implications of estrogen-dependent brain organization for gender differences in hypothalamo-pituitary-adrenal regulation. FASEB J 9:419–423

Pelletier G, Liao N, Follea N, Govindan MV (1988) Distribution of estrogen receptors in the rat pituitary as studied by in situ hybridization. Mol Cell Endocrinol 56:29–33

Pfaff DW, Keiner M (1973) Atlas of estradiol-concentrating cells in the central nervous system of the female rat. J Comp Neurol 151:121–158

Phoenix CH, Goy RW, Gerall AA, Young WC (1959) Organizing action of prenatally administered testosterone propionate on the tissues mediating mating behavior in the female guinea pig. Endocrinology 65:369–382

Plapinger L, McEwen BS (1973) Ontogeny of estradiol-binding sites in rat brain. Appearance of presumptive adult receptors in cytosol and nuclei. Endocrinology 93:1119–1128

Pommerantz SM, Fox TO, Sholl SA, Vito CC, Goy RW (1985) Androgen and estrogen receptors in fetal rhesus monkey brain and anterior pituitary. Endocrinology 116:83–89

Rabin DS, Johnson EO, Brandon DD, Liapi C, Chrousos GP (1990) Glucocorticoids inhibit estradiol-mediated uterine growth: possible role of the uterine estradiol receptor. Biol Reprod 42:74–80

Ralston SH, Russell RG, Gowen M (1990) Estrogen inhibits release of tumor necrosis factor from peripheral blood mononuclear cells in postmenopausal women. J Bone Miner Res 5:983–988

Ramirez VD, Zheng J (1996) Membrane sex-steroid receptors in the brain. Front Neuroendocrinol 17:402–439

Ramirez VD, Zheng J, Khawar MS (1996) Membrane receptors for estrogen, progesterone and testosterone in rat brain: Fantasy or reality? Cell Mol Neurobiol 16:175–198

Ree AH, Landmark BF, Eskild W, Levy FO, Lahooti H, Jahnsen T, Aakvaag A, Hansson V (1989) Autologous down-regulation of messenger ribonucleic acid and protein levels for estrogen receptors in MCF-7 cells: an inverse correlation to progesterone receptor levels. Endocrinology 124:2577–2583

Rissman EF, Wersinger SR, Taylor JA, Lubahn DB (1997) Estrogen receptor function as revealed by knockout studies: neuroendocrine and behavioral aspects. Horm Behav 31:232–243

Revelli A, Massobrio M, Tesarik J (1998) Nongenomic actions of steroid hormones in reproductive tissues. Endocr Rev 19:3–17

Robel P (1993) Steroidogenesis: the enzymes and regulation of their genomic expression. In: Thibault C, Levasseur MC, Hunter RHF (eds) Reproduction in mammals and man. Ellipses, Paris, pp 135–142

Rubanyi GM, Freay AD, Burton G, Lubahn, DB, Couse JF, Korach KS (1997) Decreased production of endothelium-derived nitric oxide in the aorta of estrogen receptor deficient mice. J Clin Invest 99:2429–2437

Saceda M, Lippman ME, Chambon P, Lindsey RL, Ponglikitmongkol M, Puente M, Martin MB (1988) Regulation of the estrogen receptor in MCF-7 cells by estradiol. Mol Endocrinol 2:1157–1162

Saceda M, Lindsey RK, Solomon H, Angeloni SV, Martin MB (1998) Estradiol regulates estrogen receptor mRNA stability. J Steroid Biochem Mol Biol 66:113–120

Santagati S, Melcangi RC, Celotti F, Martini L, Maggi A (1994) Estrogen receptor is expressed in different types of glial cells in culture. J Neurochem 63:2058–2064

Santen RJ (1991) Male hypogonadism. In: Yen SSC, Jaffe RB (eds) Reproductive endocrinology. Saunders, Philadelphia, pp 739–794

Screpanti I, Felli MP, Toniato E, Meco D, Martinotti S, Frati L, Santoni A, Gulino A (1991) Enhancement of natural killer cell susceptibility of human breast cancer by estradiol and v-Ha-ras oncogene. Int J Cancer 47:445–449

Scully K, Gleiberman AS, Lindzey J, Lubahn DB, Korach KS, Rosenfeld, MG (1997) Role of estrogen receptor α in the anterior pituitary gland. Mol Endocrinol 11:674–681

Schumacher M (1990) Rapid membrane effects of steroid hormones: an emerging concept in neuroendocrinology. Trends Neurosci 13:359–362

Sholl SA, Kim KL (1989) Estrogen receptors in the rhesus monkey brain during fetal development. Dev Brain Res 50:189–196

Shughrue PJ, Komm B, Merchenthaler I (1996) The distribution of estrogen receptor-β mRNA in the rat hypothalamus. Steroids 61:678–681

Shughrue PJ, Lane MV, Merchenthaler I (1997) Comparative distribution of estrogen receptor-α and -β mRNA in the rat central nervous system. J Comp Neurol 388:507–525

Shyamala G (1987) Endocrine and other influences in the normal development of the breast. In: Paterson AHG (ed) Fundamental problems in breast cancer. Martinus Nijhoff, Boston, pp 127–137

Shyamala G, Haslam SZ (1980) Estrogen and progesterone receptors in normal mammary gland during different functional states. In: Bresciani R (ed) Perspectives in steroid receptor research. Raven, New York, pp 193–216

Shyamala G, Schneider W, Guiot MC (1992) Estrogen dependent regulation of estrogen receptor gene expression in normal mammary gland and its relationship to estrogenic sensitivity. Receptor 2:122–128

Simerly RB, Chang C, Maramatsu M, Swanson LW (1990) Distribution of androgen and estrogen receptor mRNA-containing cells in the rat brain: an *in situ* hybridization study. J Comp Neurol 294:76–95

Simerly RB, Zee MC, Pendleton JW, Lubahn DB, Korach KS (1997) Estrogen receptor-dependent sexual differentiation of dopaminergic neurons in the preoptic region of the mouse. Proc Natl Acad Sci USA 94:14077–14082

Simpson ER, Michael MD, Agarval VR, Hinshelwood MM, Bulun SE, Zhao Y (1997) Expression of the CYP19 (aromatase) gene: an unusual case of alternative promoter usage. FASEB J 11:29–36

Singh S, Baker PR, Poulsom R, Wright NA, Sheppard MC, Langman MJ, Neoptolemos JP (1997) Expression of estrogen receptor and estrogen-inducible genes in pancreatic cancer. Br J Surg 84:1085–1089

Sloop TC, Clark JC, Rumbaugh RC, Lucier GW (1983) Imprinting of hepatic estrogen-binding proteins by neonatal androgens. Endocrinology 112:1639–1646

Smith CL, Conneely OM, O'Maley BW (1993) Modulation of the ligand-independent activation of the human estrogen receptor by hormone and antihormone. Proc Natl Acad Sci U S A 90:6120–6124

Smith EP, Boyd J, Frank GR, Takahashi H, Cohen RM, Specker B, Williams TC, Lubahn DB, Korach KS (1994) Estrogen resistance caused by a mutation in the estrogen-receptor gene in a man. N Engl J Med 331:1056–1061

Smithson G, Medina K, Ponting I, Kincade PW (1995) Estrogen suppresses stromal cell-dependent lymphopoiesis in culture. J Immunol 155:3409–3417

Stefaneanu L, Kovacs K, Horvath E, Lloyd RV, Buchfelder M, Fahlbush R, Smyth H (1994) In situ hybridization study of estrogen receptor messenger ribonucleic acid in human adenohypophysial cells and pituitary adenomas. J Clin Endocrinol Metab 78:83–88

Stimson WH (1988) Estrogen and human T lymphocytes: presence of specific receptors in the T-suppressor/cytotoxic subset. Scand J Immunol 28:345–350

Sutherland MK, Hui DU, Rao LG, Wylie JN, Murray TM (1996) Immunohistochemical localization of the estrogen receptor in human osteoblastic SaOS-2 cells: association of receptor levels with alkaline phosphatase activity. Bone 18: 361–369

Szego CM (1994) Cytostructural correlates of hormone action: new common ground in receptor-mediated signal propagation for steroid and peptide agonists. Endocrine 2:1079–1093

Takeuchi M, Tokinn M, Nagata K (1995) Tamoxifen directly stimulates the mineralization of human osteoblast-like osteosarcoma cells through a pathway independent of estrogen response element. Biochem Biophys Res Commun 210:295–301

Tetel MJ, Blaustein JD (1991) Immunocytochemical evidence for noradrenergic regulation of estrogen receptor concentrations in the guinea pig hypothalamus. Brain Res 565:321–329

Thompson C, Lucier GW (1983) Hepatic estrogen receptor responsiveness. Possible mechanisms for sexual dimorphism. Mol Pharmacol 24:69–76

Toran-Allerand CD (1984) On the genesis of sexual differentiation of the central nervous system: morphogenetic consequences of steroidal exposure and possible role of α-fetoprotein. Progr Brain Res 61:93–98

Toran-Allerand CD, Miranda RC, Hochberg RB (1992) Cellular variations in estrogen receptor mRNA translation in the developing brain: evidence from combined [^{125}I] estrogen autoradiography and non-isotopic in situ hybridization histochemistry. Brain Res 576:25–41

Towle AC, Sze PY (1983) Steroid binding to synaptic plasma membranes: differential binding of glucocorticoids and gonadal steroids. J Steroid Biochem 18:135–143

Tsuruo Y, Ishimura K, Fujita H, Osawa Y (1994) Immunocytochemical localization of aromatase containing neurons in the rat brain during pre- and postnatal development. Cell Tissue Res 278:29–39

Tsuruo Y, Ishimura K, Osawa Y (1995) Presence of estrogen receptors in aromatase-immunoreactive neurons in the mouse brain. Neurosci Lett 195:49–52

Ulisse S, Tata JR (1994) Thyroid hormone and glucocorticoid independently regulate the expression of estrogen receptor in male *Xenopus* liver cells. Mol Cell Endocrinol 105:45–53

Vito CC, Fox TO (1981) Androgen and estrogen receptors in embryonic and neonatal rat brain. Brain Res 254:97–110

von der Ahe D, Janich S, Scheidereit C, Renkawitz R, Schulz G, Beato M (1985) Glucocorticoid and progesterone receptors bind to the same sites of two hormonally regulated promoters. Nature 313:706–709

Wahli W, Martinez E (1991) Superfamily of steroid nuclear receptors: positive and negative regulators of gene expression. FASEB J 5:2243–2249

Weiland NG, Orikasa C, Hayashi S, McEwen BS (1997) Distribution and hormone regulation of estrogen receptor immunoreactive cells in the hippocampus of male and female rats. J Comp Neurol 388:603–612

Westerlind KC, Sarkar G, Bolander ME, Turner RT (1995) Estrogen receptor mRNA is expressed in vivo in rat calvarian periosteum. Steroids 60:484–487

Wilder RL (1996) Adrenal and gonadal steroid hormone deficiency in the pathogenesis of rheumatoid arthritis. J Rheumatol 44:10–12

Wooley DE, Hope WG, Thompson-Reece MA, Gietzen DW, Conway SB (1994) Dopaminergic stimulation of estrogen receptor binding in vivo: a reexamination. Recent Prog Horm Res 49:383–392

Yokosuka M, Hayashi S (1995) Transient expression of estrogen receptor-like immunoreactivity (ER-LI) in the facial nucleus of the neonatal rat. Neurosci Res 15:90–95

Yokosuka M, Okamura H, Hayashi S (1997) Postnatal development and sex difference in neurons containing estrogen receptor-alpha immunoreactivity in the preoptic brain, the diencephalon, and the amygdala in the rat. J Comp Neurol 389:81–93

Zafar M, Ezzat S, Ramyar L, Pan N, Smyth HS, Asa SL (1995) Cell-specific expression of estrogen receptor in the human pituitary and its adenomas. J Clin Endocrinol Metab 80:3621–3627

Zhou-Li F, Skalli M, Albaladejo V, Joly-Pharaboz MO, Nicolas B, Andre J (1993) Interference between estradiol and L-triiodothyronine in the control of the proliferation of a pituitary tumor cell line. J Steroid Biochem Mol Biol 45:275–279

Zhu YS, Dellovade T, Pfaff DW (1997) Gender-specific induction of pituitary RNA by estrogen and its modification by thyroid hormone. J Neuroendocrinol 9:395–403

Zielinski CC, Tichatschek E, Muller C, Kalinowski W, Sevelda P, Czerwenka K, Kubista E, Spona J (1989) Association of increased lytic effector cell function with high estrogen levels in tumor-bearing patients with breast cancer. Cancer 63:1985–1989

Zou K, Ing NH (1998) Oestradiol up-regulates estrogen receptor, cyclophilin, and glyceraldehyde phosphate dehydrogenase mRNA concentrations in endometrium, but down-regulates them in liver. J Steroid Biochem Mol Biol 64:231–237

CHAPTER 14

Female Reproductive Tract

C. HEGELE-HARTUNG

A. Embryology of the Reproductive Tract

At about the sixth week of fetal development, invagination of the celomic epithelium to form a furrow creates a groove whose lips later fuse to form the lateral Müllerian (or paramesonephric) ducts. Müllerian ducts first become apparent high on the dorsal wall of the celomic cavity and then progressively grow caudally to enter the pelvis, where they swing medially to fuse. Further caudal growth brings these fused ducts into contact with the urogenital sinus (Fig. 1B). With relatively uncomplicated transformations, the unfused portions mature into the fallopian tubes and the fused caudal portion, into the uterus and the vagina. The upper portion of the vagina is generally held to be of Müllerian origin, and the lower portion is probably derived from the urogenital sinus. It is apparent that the entire lining of the uterus and the tubes is derived from celomic epithelium.

Parallel to the Müllerian ducts are the paired Wolffian (or mesonephric) ducts (Fig. 1A) which, in the male, are destined to form the epididymis and the vas deferens. Normally, the mesonephric duct regresses in the female. Remnants, however, may persist into adult life as epithelial inclusions about the hilus of the ovary and mesosalpinx, designed, respectively, as the epoophoron and paroophoron. If the caudal portions of this mesonephric anlage persist, they may appear as epithelial inclusions within the wall of the lower uterine segment and cervix and as epithelial rests in the lateral walls of the vagina. Sometimes, in the vagina, these rests produce cysts that are designated as Gartner's duct cysts.

The ovary, like the testis, arises from a medial proliferation of the urogenital ridge specified as the genital ridge. At 6 weeks of the fetal development, this sexless gonad has three components: a covering of differentiated celomic lining epithelium, previously incorrectly designated as "germinal epithelium"; an underlying undifferentiated stroma (the mesenchyme); and primitive germ cells. Differentiation of the ovarian mesenchyme provides an origin for theca cells and probably also for the granulosa cell of the follicle. It is generally believed that the germ cells are set aside as totipotent gametes in the earliest stage of formation of the embryonic disk. As the embryo develops, these germ cells migrate or are carried into the primitive mesenchyme destined to become the ovary. It is of interest that they divide during the first

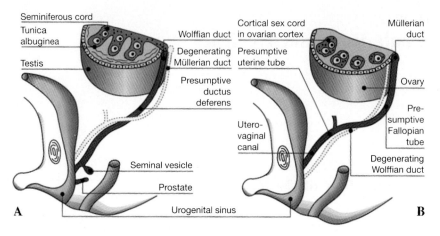

Fig. 1 A,B. Development of gonads and genital ducts in both sexes. **A** Male; **B** Female

half of fetal development to reach a maximum number of 6–7 million, but sub-sequently some regress; thus, at birth approximately 1–2 million oocytes remain. None is formed after the fifth month of fetal development and, indeed, there is a progressive loss so that, at puberty, the number has dwindled into the hundreds of thousands, still more than needed to provide 12 per year during the women's active reproductive life.

B. Sex Differentiation

Sex differentiation can be subdivided into several stages that regulate each other sequentially (Fig. 2):

1. Genetic sex
2. Gonadal sex
3. Somatic sex

Normally, these three steps are coherent with each other. When discrepancy occurs, a clear understanding of the mechanisms of sex differentiation is crucial for diagnosis and treatment.

I. Genetic Sex

Two (XX) or one (X0) X chromosomes lead to a female chromosomal sex, while XY and XXY result in male chromosomal sex. In male embryos, the short arm of the Y chromosome contains a small gene expressed in somatic gonadal cells just prior to testicular differentiation that is responsible for pro-duction of a testis-determining factor (TDF). In the presence of TDF, the

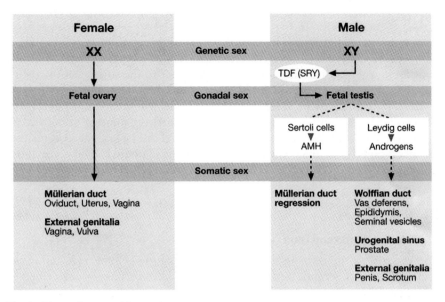

Fig. 2. Flow diagram illustrating process of female and male sex differentiation. *TDF*, testis-determining factor; *SRY*, sex-determining region Y; *AMH*, anti-Müllerian hormone

gonad will become a testis, otherwise the gonad will become an ovary. The search for the TDF on the Y chromosome has had a long history, littered with incorrectly identified candidate genes. However, the identification and cloning of SRY (sex-determining region Y) by SINCLAIR et al. (1990) finally appears to have resolved this issue. It is supposed that SRY contributes to the differentiation of the Sertoli cell, the initial event in testicular differentiation. One would expect that mutations in SRY would result in XY females with either gonadal dysgenesis or hermaphroditism, and this is indeed so; mutations in the SRY gene, all in the HMG (high-mobility group) box, have been documented in 11 individuals (BERTA et al. 1990; JAGER et al. 1992; HAWKINS 1993).

II. Gonadal Sex

1. Male

The initial event of male differentiation is the differentiation of the testes (JIRASEK 1971). The first important step of testicular differentiation is the formation of seminiferous cords, precursors of future seminiferous tubules and the appearance of Sertoli cells (MAGRE and JOST 1980). The second step in testicular development is the differentiation of Leydig cells. After their differentiation, testes in most mammals are displaced from the abdominal cavity to the scrotum through the inguinal canal (GIER and MARION 1969).

2. Female

As for the male, the first evidence of sexual differentiation in the female is observed at the gonadal level. The ovarian anlage keeps a morphologically indifferent aspect. Oogonia move from the center to the surface of the ovary, thereby multiplying very actively, and enter meiotic prophase. Oocytes proceed to the dictyotene stage, in which the cells will remain until ovulation, 12–50 years later. During this process, not only does the number of oocytes no longer increase, but most of them degenerate and disappear. The oocytes that are maintained and that will be present in the definitive ovary are those encompassed by follicular cells to form primordial follicles.

III. Somatic Sex

1. Male Sex Differentiation

The key role of the testis in the differentiation of the genital tract and in the establishment of somatic sex was first shown in experiments performed in rabbit (Zamboni and Upadhyay 1983). Rabbit fetuses that were castrated surgically in utero after the morphological differentiation of the gonads and just before the initiation of genital-tract differentiation (at 19 days of gestation) were permitted to develop until near term (28 days of gestation). Whatever their genetic sex, gonadless fetuses acquired a female genital apparatus – presence of Müllerian ducts, urogenital sinus and external genitalia of female appearance, regression of Wolffian ducts (Jost 1947; Fig. 3). These experimental results were rapidly confirmed by observations in humans. It could be predicted that individuals with gonadal agenesis should present a female morphology, whatever their genetic sex.

It appears, therefore, that fetal testis imposes maleness by exerting two kinds of effects during the differentiation of the genital tract. On the one hand, it provokes the regression of Müllerian ducts, while on the other it induces the development of the male ducts and the masculinization of the urogenital sinus and external genitalia.

There are two different testicular secretions responsible for differentiation of the male genital tract: (1) the anti-Müllerian hormone (AMH) synthesized by the Sertoli cells and (2) the fetal androgen production (testosterone and androstenedione) synthesized by the Leydig cells.

AMH, also known as Müllerian inhibitory substance (MIS), causes regression of the Müllerian ducts, the anlagen of the female internal reproductive structures, in the male fetus (Jost 1947). This glycoprotein is a differentiation factor, which interrupts the development of the uterus, fallopian tube and upper vagina (Josso et al. 1993). Gene-knockout experiments in mice have confirmed that, in the absence of AMH and its receptor AMHRII, Müllerian ducts do not regress (Mishina et al. 1996). There is some evidence that SRY directly regulates transcription of the AMH gene (Haqq et al. 1993). In human, a class of pseudohermaphroditism is characterized by the persistence of uterus

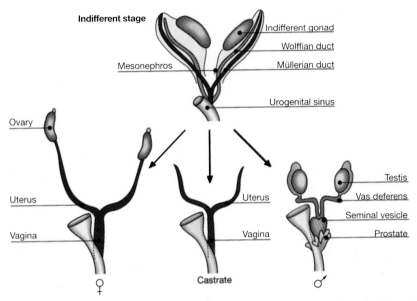

Fig. 3. Schematic representation of the differentiation of the sex ducts in the rabbit fetus. From the indifferent stage, one can differentiate either the female (*left*) or the male (*right*) structure. In castrated fetuses, the genital tract has a gonadless female structure (*middle*)

and oviducts in men with normal external genitalia. The persistent Müllerian-duct syndrome can result either from the absence of the hormone or from a receptor defect (GUERRIER et al. 1989).

The second aspect of male differentiation is the integration of Wolffian ducts in the internal genital system and their subsequent development into epididymides, vasa deferentia and seminal vesicles. This masculinization is produced by testosterone. Cyproterone acetate, an antiandrogenic progestin, given to pregnant rabbits, prevents – in opposing the action of testicular androgen – internal and external masculinization of male fetuses, but does not prevent the inhibition of Müllerian ducts. Therefore, androgens play an essential role in masculinization of the fetus. For masculinizing the external genitalia or urogenital sinus, testosterone must be converted into dihydrotestosterone (DHT) in the target tissue.

2. Female Sex Differentiation

JOST (1947) showed that fetal ovaries play no part in female sex differentiation and that the key factor was, instead, the presence or absence of the fetal testis. In the presence or absence of ovaries, Müllerian ducts differentiate into oviducts and uterine horns. This 'basic femaleness' was shown by castration experiments (Fig. 3), in which the absence of testicular hormone activity induces feminization of the body. In males with either a gonadal agenesis or lack of testosterone production or an androgen insensitivity, Wolffian ducts disappear, Müllerian ducts develop, urogenital sinus and external genitalia feminize.

3. Estrogens and Sex Differentiation

As Müllerian ducts are maintained and Wolffian ducts regress in the gonad-less genital tract of both sexes (Fig. 3), cultured in vitro in a medium devoid of hormones, it may be concluded that the feminization of sexual ducts does not depend on estrogens from either maternal or placental origin.

Although estrogens are not required for normal female differentiation of the reproductive tract, estrogens can interfere with male differentiation. Pre-natal estrogen treatment is associated with reproductive tract abnormalities in human and mouse males (MCLACHLAN et al. 1975; GILL et al. 1979; VISSER et al. 1998) because estrogen blocks the effect of AMH upon Müllerian ducts (NEWBOLD et al. 1984). Of special interest is the fact that regression of the Müllerian ducts in the males is incomplete after prenatal estrogen treatment, leaving remnants that show a female-like differentiation (VISSER et al. 1998). These findings indicate that the Müllerian ducts are targets for estrogen action in male fetuses.

Estradiol formation is initiated in the fetal ovary before definitive sex differentiation occurs and both sexes develop the capacity to respond to estro-gens early in embryogenesis (SMITH et al. 1997). However, the capacity to synthesize estrogens is female specific and occurs later, at the time of sex differentiation, indicating that local estradiol may be important for ovarian differentiation (WILSON et al. 1981).

In the chick, administration of an aromatase inhibitor completely sex-reverses the gonads of genetic females and causes development of a male phe-notype. This suggests that the masculinizing effect of AMH upon mammalian fetal ovaries is linked to the capacity of the hormone to decrease aromatase activity and, therefore, estrogen production (ELBRECHT and SMITH 1992).

C. Estrogens and Ovary

Sex steroids and also estrogens are responsible for the endocrine interplay with the endocrine hypothalamic–pituitary axis that controls ovarian follicu-logenesis. There is little evidence that exogenous estrogens have any action on the morphology of the postmenopausal ovary, but the premenopausal ovary is "suppressed" by the contraceptive use of estrogens with progestins. The estrogenic action of the oral contraceptive, in fact, is to inhibit the release of follicle-stimulation hormone (FSH) and thereby prevent the emergence of a dominant ovarian follicle for ovulation. Therefore, estrogens may be used for the prevention of physiologic ovarian cysts. In contraceptive doses, the estro-gens in oral contraceptives appear to be primarily responsible for the prevention of ovarian cancer, especially by reducing the number of ovulatory events.

D. Fallopian Tubes

The fallopian tubes and the uterus are of fundamental importance to reproduction. They (Fig. 4) conduct the oocyte from the ovary to the uterus and spermatozoa from uterus to the distal part of the tube when they fertilize the oocyte. The fallopian tubes have several roles, but their main function is in the process of natural conception. Gamete transport, fertilization, and early embryo development and transport all take place within the fallopian tube (MAGUINESS et al. 1992). The tubes also provide nutrition in the earliest stages of development and provide a defense against infection (LEESE and DICKENS 1992).

I. Steroidal Regulation

Both the structure and function of the fallopian tubes in mammals are influenced by estradiol and progesterone. The fallopian tubes in humans are paired organs, 7-cm to 14-cm long, enclosed in the mesosalpingeal borders of the broad ligaments. The well-described regions of the fallopian tube are the fimbria, ampulla, ampullary–isthmic region, isthmus and utero-tubal junction (Fig. 4). The isthmus is considered to act as a sperm reservoir and fertilization takes place in the ampullary region. The fallopian tube itself consists of an outer myosalpinx responsible for tubal contraction and an inner endosalpinx which creates the luminal environment (LEESE and DICKENS 1992).

In the fallopian tube, the mucosa is thrown up into numerous high, delicate folds that, on cross section, produce a papillary appearance. The lining

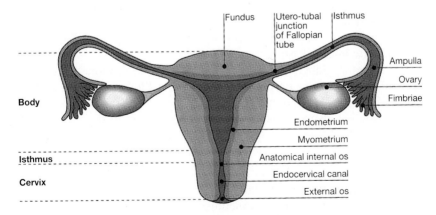

Fig. 4. Coronal section of uterus and fallopian tubes

Table 1. Summary of estrogen receptor (ER) and progesterone receptor (PR) staining in the fallopian tube as determined by immunohistochemistry (AMSO et al. 1994)

	Medial epithelium		Fimbrial epithelium	
	ER	PR	ER	PR
Follicular phase	Low to Moderate	Strong	Strong	Strong
Late luteal phase	Low	Moderate	Moderate	Moderate

epithelium of the tube is made up of three cell types: (1) ciliated columnar cells; (2) non-ciliated, columnar, secretory, mucus cells; and (3) so-called intercalated cells that may simply represent inactive secretory cells. Ciliated epithelium predominates in the fimbrial region and non-ciliated secretory cells in the isthmic portions of the tubes (LEESE and DICKENS 1992). In relation to the ovarian cycle, there are well established, hormone-regulated morphological changes in the tubal epithelium (VERHAGEN et al. 1979). Ciliogenesis is estrogen dependent and takes place during the proliferative phase (JANSEN 1984). In addition, mature differentiated cells are seen at mid-cycle, under the influence of estrogen (JANSEN 1984). Maximal height and ciliation of the luminal epithelium is apparent in the late follicular phase with evidence of atrophy and deciliation at the end of the luteal phase. Progestins increase the number of non-ciliated relative to ciliated cells. The non-ciliated secretory cells show marked cyclic changes, especially in the isthmic region (JANSEN 1984). Throughout the length of the oviduct, there is a circular muscular layer under the mucosa. This circular layer is lined with an external layer and a thin internal layer of longitudinal bands.

The pattern of sex-steroid receptor [estrogen receptor (ER) and progesterone receptor (PR)] in the fallopian tube is shown in Table 1. In general, it can be summarized (AMSO et al. 1994) that the fimbrial region displays a different pattern of ER immunoreactivity from the rest of the tube, i.e., strong staining during the early and mid-follicular phase. ER expression declined during the late luteal phase. Contrary to the ER, the expression of PR protein did not significantly change during the ovarian cycle except for a slight decline during the late luteal phase.

The secretory function of the fallopian tube is hormonally regulated. Estrogen stimulates the production of oviductal fluid (LIPPES et al. 1981). Secretion is maximal at mid-cycle (coincident with the estradiol peak). Protein concentration is, however, inversely related to estrogen levels.

E. Uterus

The human uterus has two distinctive anatomic regions: the cervix, and the corpus (Fig. 4). The three major, functional components of the uterus are the cervix, its glandular living membrane (endometrium) that nourishes the early embryo and the smooth muscle of the uterine body (myometrium), which greatly increases in amount during pregnancy (Fig. 4).

Functionally, the most important tasks of the endometrium are implantation and regeneration, and to assure appropriate maturation and receptivity. Each month, the endometrium must be able to be discarded and regenerate quickly in the absence of a conceptus (CARSON 1992). The myometrium undergoes spontaneous rhythmic contractions that vary with stage of menstrual cycle and stage of pregnancy. Ultimately, at the end of pregnancy, myometrial contractions occur that enable labor and delivery to take place.

I. Estrogen Regulation

Progesterone is essential for the establishment and maintenance of pregnancy in all mammalian species. The first half of the menstrual cycle, the follicular phase, is estrogen dominated and the second half, the luteal phase, is progesterone dominated (NOYES et al. 1950). Progesterone secretion by the corpus luteum converts an estrogen-primed proliferative endometrium into a secretory one, which is receptive to the blastocyst.

Estradiol and progesterone mediate their effects via the ER and PR, respectively. Table 2 summarizes the steroid receptor localization in the endometrium, which has been well described in a number of immunohistochemical studies (SNIJDERS et al. 1992; CRITCHLEY et al. 1993; JONES et al. 1995). In the endometrial epithelium, the PR is induced by estradiol at the transcriptional level, and is downregulated by its own ligand, progesterone, at the transcriptional and post-transcriptional levels (CHAUCHEREAU et al. 1992). In the endometrial stromal cells, however, PR expression is not significantly influenced by progesterone (Table 2). Similarly to PR, the ER is induced by estrogens, and downregulated by progesterone in the endometrial epithelium. In contrast to the changes reported for ER and PR over the menstrual cycle, no changes in glandular or stromal androgen receptor expression have been reported (TAMAYA et al. 1986; HORIE et al. 1992).

At a subcellular level, estrogen induces endometrial glandular cells to produce not only ER and PR, but also alkaline phosphatase and

Table 2. Summary of estrogen receptor (ER) and progesterone receptor (PR) staining in the uterus as determined by immunohistochemistry (SNIJDERS et al. 1992)

| | Endometrium | | | | Myometrium | |
| | ER | | PR | | | |
	Glands	Stroma	Glands	Stroma	ER	PR
Follicular phase	Strong	Strong	Strong	Strong	Strong	Strong
Late luteal phase	Low	Low	Low	Moderate/ strong	Low	Moderate/ strong

prostaglandins $F_{2\alpha}$ and E_2, both important in menstruation and initiation of labor (GURPIDE et al. 1990).

Endometrium and myometrium both proliferate in response to estradiol administration, although the mitogenic response is far more obvious in the endometrium (CHEN et al. 1973). Histologically, endometrial glands demonstrate a clear mitogenic response to estrogen as shown by glandular growth, increased tortuosity, and pseudostratification. Endometrial stromal cells have no clinical response to estrogen, and only proliferate and enzymatically respond to the introduction of progesterone. The addition of estrogen, however, can cause potentiation of the stromal cell response to progesterone (TSENG et al. 1986).

In the human uterine corpus, the clinical response to estrogen closely mirrors findings from the laboratory. When exposed to estrogens alone, endometrial glands demonstrate marked growth stimulation, seen as an increasing ratio of glands to stroma, glandular structures, and pseudostratification of glandular cells (FERENCZY et al. 1979). This estrogenic response appears to be determined by the time of exposure, i.e., the length of time that active estrogens are bound to nuclear estrogen receptors, and not to the dose of estrogen. In the premenopausal patient, this growth proliferation is balanced by progesterone, which inhibits further gland proliferation and induces glandular secretion. Estrogenic stimulation of endometrial glands has been theorized to be that of a continuum. In the absence of estrogen, endometrial glands are atrophic, but with exposure to estrogen in the absence of progesterone (so called "unopposed estrogen"), there are proliferative glandular changes, glandular hyperplasia, and even well-differentiated adenocarcinoma arising from endometrial glands. Endometrial stromal cells, again, appear to have a minimal response to estrogen alone.

Estradiol's actions on endometrial components reflect a composite of circulating estradiol levels and tissue levels of steroid receptors and steroid metabolizing enzymes. Several steroid-metabolizing enzymatic activities have been reported in endometrium (reviewed by STRAUSS and GURPIDE 1991), including aromatase, 17β-hydroxysteroid dehydrogenase (17β-HSD), estrogen (E) sulfate sulfatase, E-sulfotransferase and 20-hydroxysteroid dehydrogenase (20-HSD).

A number of growth factors in the endometrium, including fibroblast growth factor, tumor growth factor and vascular endometrial growth factor (VEGF), has been implicated in angiogenesis (FOLKMAN and KLAGSBRUN 1987; SCHOTT and MORROW 1993). One of the early events caused by 17β-estradiol is the increased angiogenesis and vascular permeability in the endometrium. In the absence of the ER, no uterine water imbibition or uterine hyperemia occur (LUBAHN et al. 1993). Based on the temporal pattern of mRNA expression after the 17β-estradiol administration and capability of increasing vascular permeability as well as the mitogenic activity for vascular endothelial cells, VEGF has been proposed to be a critical factor in the early phase of 17β-estradiol-induced angiogenesis. (CULLINAN-BOVE and KOOS 1993). Similarly, erythropoietin could be shown to be an important factor for the 17β-estradiol-dependent cyclical angiogenesis in the uterus (YASUDA et al. 1998).

II. Estrogens and Uterine Pathology

Endometrial adenocarcinoma occurs during the reproductive and menopausal years. Obesity, nulliparity, and late menopause are classically associated with endometrial cancer, suggestive of a hyperestrogenic state. If a patient is nulliparous, obese, and reaches menopause at an age of 52 years or later, she appears to have a fivefold increase in the risk of endometrial cancer over patients who do not satisfy these criteria.

A number of case-controlled studies in the late 1970s demonstrated the risk of endometrial adenocarcinoma to be markedly elevated in patients taking replacement conjugated estrogens with no progestins. It has since been well accepted by most clinicians that unopposed estrogen use plays an etiologic role in the development of endometrial cancer, with odds ratios as high as 8 for users compared with non-users (WEISS et al. 1979). The use of estrogen and progestins, however, reduces this risk to levels even lower than those seen in postmenopausal women with no hormone replacement (HARLAP 1992).

F. Cervix

The cervix is further divided into the vaginal or anatomic portio and the endocervix. The anatomic portio is that portion of the cervix visible to the eye on vaginal examination. It is covered by a stratified squamous non-keratinizing epithelium reflected off the vaginal vaults onto the cervix. The endocervix is normally not exposed and is lined by columnar mucus-secreting epithelium that dips down into the underlying stroma to produce crypts sometimes designated as endocervical glands. In contrast to the myometrium of the body of the uterus, the normal cervix is a collagenous structure with smooth muscle fibers accounting for only 10–15% of its tissue composition (RORIE and NEWTON 1967). Hence, the cervix has the capacity to undergo profound tissue remodeling during pregnancy and labor, and to rapidly to return to its former state thereafter. The important functions of the cervix are to allow migration and transport of spermatozoa through the cervical canal, to act as a barrier to retain the conceptus and, at an appropriate time, to open and permit the conceptus to be expelled from the uterus (DANFORTH 1983)

I. Estrogen Regulation

A physiologic effect of estrogen is increased endocervical mucus production. This is most evident at the late proliferative phase, when estrogen levels are high, and when cervical glands produce large amounts of clear, thin mucus. This fluid is low in non-migrating mucoid substance that increases 'spinnbarkeit' of the cervical mucus that favors spermatozoa penetration.

Reduction in cervical mucus quantity and quality may signal physiologic evidence of relative estrogen deficiency.

G. Vagina

The vaginal skin is a multilayered, stratified, squamous epithelium that is highly sensitive to estrogen. Climacteric ovarian failure results in a progressive diminution in the layering of the epithelium and concomitant reduction in the glycogen content of individual cells. One of the hallmark effects of estrogen on the genital tract is increased vaginal rugae and vaginal mucosal thickening. Accelerated squamous maturation of the vagina is readily demonstrated after the use of estrogens. This is seen as a shift of desquamated cells on Papanicolaou smears from immature basal cells before estrogen exposure to mature superficial cells after exposure.

I. Estrogens and Vaginal Pathology

In 1971, recognition of a relationship between in utero diethylstilbestrol (DES) exposure and a rare vaginal adenocarcinoma in adolescent female offspring led to the proscription of the use of DES in pregnancy (HERBST et al. 1971). It is currently believed that some of the changes that occur in DES-exposed daughters are exaggerated estrogenic influences on the fetal vagina and cervix. These include vaginal adenosis, cervical ectropion, and structural alterations in the cervix, upper vagina, and endometrial cavity. Vaginal adenosis is the visible encroachment of glandular epithelium beyond the cervix into the upper vaginal fornices and walls that is seen in more than 50% of females exposed to DES in utero; cervical ectropion is the similar process on the exocervix that is commonly obscured by squamous metaplasia. Structural abnormalities include cervical collars and hoods, transverse vaginal ridges, and various irregularities of endometrial cavity, such as T-shaped uteri and cornual constrictions.

Clear cell adenocarcinomas of the upper vagina and cervix apparently develop from a specific type of adenosis derived from tuboendometrial cells at an incidence of 1 per 1000 exposed females. In females found to have extensive vaginal adenosis, who are followed up prospectively, however, clear cell carcinoma is rare. Cervical intraepithelial neoplasia and cervical cancer have been described in the large transformation zones and ectropion of DES-exposed daughters, but it is not thought to be more likely than the occurrence in non-DES-exposed females (ROBBOY et al. 1981).

References

Amso NN, Crow J, Shaw RW (1994) Comparative immunohistochemical study of estrogen and progesterone receptors in the fallopian tube and uterus at different stages of the menstrual cycle and the menopause. Hum Reprod 9:1027–1037

Berta P, Hawkins JR, Sinclair AH, Taylor A, Griffiths BL, Goodfellow PN, Fellous M (1990) Genetic evidence equating SRY and the testis-determining factor. Nature 348:448–450

Carson SA (1992) A summary of normal female reproductive physiology. In: Alexander NJ, d'Areangues (eds) Steroid hormones and uterine bleeding. AAAS Press, Washington

Chauchereau A, Savouret J-F, Milgrom E (1992) Control of biosynthesis and post-transcriptional modification of progesterone receptor. Biol Reprod 46:174–177

Chen L, Lindner HR, Lancet M (1973) Mitogenic action of estradiol-17β on human myometrial and endometrial cells in long-term tissue cultures. J Endocrinol 59:87–97

Critchley HOD, Bailey DA, An CL, Affandi B, Rogers PAW (1993) Immunohistochemical sex steroid receptor distribution in endometrium from long-term subdermal levonorgestrel users and during the normal menstrual cycle. Hum Reprod 8:1632–1637

Cullinan-Bove K, Koos RD (1993) Vascular endothelial growth factor/vascular permeability factor expression in the rat uterus: rapid stimulation by estrogen correlates with estrogen-induced increases in uterine capillary permeability and growth. Endocrinology 133:829–837

Danforth DH (1983) The morphology of the human cervix. Clin Obstet Gynecol 26:7–13

Elbrecht A, Smith RG (1992) Aromatase enzyme activity and sex determination in chickens. Science 255:467–470

Ferenczy A, Bertrand G, Gelfand MM (1979) Proliferation kinetics of human endometrium during the normal menstrual cycle. Am J Obstet Gynecol 133: 859–867

Folkman J, Klagsburn M (1987) Angiogenic factors. Science 235:442–447

Gier HT, Marion GB (1969) Development of the mammalian testis and genital ducts. Biol Reprod 1:1–23

Gill WB, Schumacher GFB, Bibbo M, Straus FH, Schoenberg HW (1979) Association of diethylstilboestrol exposure in utero with cryptorchidism, testicular hypoplasia and semen abnormalities. J Urol 122:36–39

Guerrier D, Tran D, Vanderwinden JM, Hideux S, Van Outryve L, Legeai L, Bouchard M, Van Vliet G, De Laet MH, Picard JY, Kahn A, Josso N (1989) The persistent Müllerian duct syndrome: a molecular approach. J Clin Endcrinol Metab 68:46–52

Gurpide E, Schatz F, Markiewicz L (1990) Steroid effects on endometrial prostaglandin production. In: D'Arcangues C, Fraser IS, Newton JR, Odlind V (eds) Contraception and mechanisms of endometrial bleeding. Cambridge University Press, Cambridge, pp 267–274

Haqq CM, King CY, Donahoe PK, Weiss MA (1993) SRY recognizes conserved DNA sites in sex specific promoters. PNAS 90:1097–1101

Harlap S (1992) The benefits and risks of hormone replacement therapy: an epidemiologic review. Am J Obstet Gynecol 166:1986–1992

Hawkins JR (1993) Mutational analysis of SRY in XY females. Hum Mutat 2:347–350

Herbst AL, Ulfelder H, Poskanzer DC (1971) Adenocarcinoma of the vagina: association of maternal stilbestrol therapy with tumor appearance in young women. N Engl J Med 284:878

Horie K, Takahura K, Imai K, Liao S, Mori T (1992) Immunohistochemical localisation of androgen receptor in the human endometrium, decidua, placenta and pathological conditions of the endometrium. Hum Reprod 7:1461–1466

Jager RJ, Harley VR, Pfeiffer RA, Goodfellow PN Scherer G (1992) A familial mutation in the testis-determining gene SRY shared by both sexes. Hum Genet 90:350–355

Jansen RPS (1984) Endocrine response in the Fallopian tube. Endocr Rev 5:525–551

Jirasek JE (1971) Development of the genital system and male pseudohermaphroditism. John Hopkins Press, Baltimore

Jones RK, Bulner JN, Searle RF (1995) Immunohistochemical characterization of proliferation, estrogen receptor and progesterone receptor expression in endometriosis: comparison of entopic and extopic endomterium with normal cycling endometrium. Hum Reprod 10:3272–3279

Josso N, Cate RL, Picard J-V (1993) Anti-Müllerian hormone: the Jost factor. In: Bardin CW (ed) Recent progress in hormone research, no 48. Academic Press, San Diego, pp 1–59

Jost A (1947) Recherches sur la differenciation sexuelles de l'embryon de lapin. Arch Anat Microsc Morphol Exp 36:271–315

Leese HJ, Dickens CJ (1992) Tubal physiology and function. In: Templeton AA, Drife JO (eds) Infertility. Springer, Berlin Heidelberg New York, pp 157–168

Lippes J, Kasner J, Alfonso LA, Dacalos ED, Lucero R (1981) Human oviductal fluids proteins. Fertil Steril 36:623–629

Lubahn DB, Moyer JS, Golding TS, Couse JF, Korach KS, Smithies O (1993) Alteration of reproductive function but not prenatal sexual development after insertional disruption of the mouse estrogen receptor gene. Proc Natl Acad Sci USA 90:11162–11166

Magre S, Jost A (1980) The initial phases of testicluar organogenesis in the rat. An electron microscopy study. Arch Anat Microsc Morphol Exp 69:297–318

Maguiness SD, Djahanbakhch O, Grudzinskas JG (1992) Assessment of the fallopian tube. Obstet Gynecol Surv 47:587–603

McLachlan JA, Newbold RR, Bullock B (1975) Reproductive tract lesions in male mice exposed prenatally to diethylstilboestrol. Science 190:991–992

Mishina Y, Rey R, Finegold MJ, Matzuk MM, Josso N, Cate RL, Behringer RR (1996) Genetic analysis of the Müllerian-inhibiting substance signal transduction pathway in mammalian sexual differentiation. Genes Dev 10:2577–2587

Newbold RR, Suzuki Y, McLachlan JA (1984) Müllerian duct maintenance in heterotypic organ culture after in vivo exposure to diethylstilboestrol. Endocrinology 115:1863–1868

Noyes RW, Hertig AT, Rock J (1950) Dating the endometrial biopsy. Fertil Steril 1:3–25

Robboy SJ, Szyfelbein WM, Goellner JR, Kaufman RH, Taft PD, Richard RM Gaffey TA, Prat J, Virata R, Hatab PA, McGorray SP, Noller KL, Townsend D, Labarthe D, Barnes AB (1981) Dysplasia and cytologic findings in 4589 young women enrolled in diethyl-stilbestrol adenosis project. Am J Obstet Gynecol 140:579

Rorie DK, Newton M (1967) Histologic and chemical studies of smooth muscle in the human cervix and uterus. Am J Obstet Gynecol 99:466–469

Schott RJ, Morrow LA (1993) Growth factors and angiogenesis. Cardiovasc Res 27:1155–1161

Sinclair AH, Berta P, Palmer MS, Hawkins JR, Griffiths BL, Smith MJ, Foster JW, Frischauf AM, Lovell-Badge R, Godfellow PN (1990) A gene from the human sex-determining region encodes a protein with homology to a conserved DNA-binding motif. Nature 346:240–244

Smith CA, Andrews JE, Sinclair AH (1997) Gonadal sex differentiation in chicken embryos: expression of estrogen receptor and aromatase genes. J Steroid Biochem Mol Biol 60:295–302

Snijders MPML, de Goeij AFPM, Debets-Te Baerts MJC, Rousch MJM, Koudstaal J, Bosman FT (1992) Immunohistochemical analysis of estrogen receptors and progesterone receptors in the human uterus throughout the menstrual cycle and after the menopause J Reprod Fertil 94:363–371

Strauss JF III, Gurpide E (1991) The endometrium: regulation and dysfunction. In: Yen SSC, Jaffe RB (eds) Reproductive endocrinology. Saunders, Philadelphia, pp 309–356

Tamaya T, Murakami T, Okada H (1986) Concentrations of steroid receptors in normal human endometrium in relation to the day of the menstrual cycle. Acta Obstet Gynecol Scand 65:195–198

Tseng L, Mazella J, Sun B (1986) Modulation of aromatase activity in human endometrial stromal cells by steroids, tamoxifen and RU 486. Endocrinology 118:1312–1318

Verhagen HG, Bareither ML, Jaffe RC, Akbar M (1979) Cyclic changes in ciliation secretion and cell height of the oviductal epithelium in women. Am J Anat 156:505–522

Visser JA, McLuskey A, Verhoef-Post M, Kramer P, Grootegoed JA, Themmen APN (1998) Effect of prenatal exposure to diethylstilbestrol on Müllerian duct development in fetal male mice. Endocrinology 139:4244–4251

Weiss NS, Szekely DR, English DR, Schweid AI (1979) Endometrial cancer in relation to patterns of menopausal estrogen use. JAMA 242:261–264

Wilson JD, Griffin JE, George FW, Leshin M (1981) The role of gonadal steroids in sexual differentiation. Recent Prog Horm Res 37:1–39

Yasuda Y, Masuda S, Chikuma M, Inoue K, Nagao M, Sasaki R (1998) Estrogen-dependent production of erythropoietin in uterus and its implication in uterine angiogenesis. J Biol Chem 273:25381–25387

Zamboni L, Upadhyay S (1983) Germ cell differentiation in mouse adrenal glands. J Exp Zool 228:173–193

CHAPTER 15

Estrogen and Brain Function: Implications for Aging and Dementia

S.E. ALVES and B.S. MCEWEN

List of Abbreviations

5-HT	serotonin
apoE	apolipoprotein E
BBB	blood–brain barrier
cAMP	cyclic adenosine monophosphate
ChAT	cholineacetyl transferase
CNS	central nervous system
DA	dopamine
ER	estrogen receptor
ERE	estrogen responsive element
GABA	γ-aminobutyric acid
GAD	glutamic acid decarboxylase
GFAP	glial fibrillary acidic protein
MAPK	mitogen-activated protein kinase
mRNA	messenger ribonucleic acid
NGF	nerve growth factor
NMDA	N-methyl-D-aspartate
OVX	ovariectomy (-ized)
PR	progestin receptor
trkA	tyrosine kinase receptor A

A. Introduction

Estrogen is involved in a multitude of processes throughout the lifespan in the brains of both females and males. For example, this steroid hormone affects critical developmental events such as neuronal differentiation, growth and synaptogenesis, which ultimately leads to the establishment of the sexually differentiated brain (MCEWEN et al. 1977; MACLUSKY et al. 1987; TORAN ALLERLAND 1991). In the adult CNS, estrogen, together with progesterone, orchestrate rapid and continuous neuroplastic events, including the cyclic synaptic turnover which occurs throughout the rat estrus cycle (WOOLEY and MCEWEN, 1993). The aging brain is associated with declines in many neural systems, often resulting in functional deficits; recent evidence suggests that estrogen replacement can decrease the risk of neurodegenerative disorders, such as Alzheimer's disease, and alleviate some of the cognitive deficits in afflicted women (PAGANIN-HILL and HENDERSON 1994; HENDERSON et al. 1994; TANG et al. 1996). Considering these few examples, it becomes apparent that the brain systems targeted by estrogen are many, encompassing both reproductive and non-reproductive circuitries. In this chapter, we will review evidence for estrogen regulation of specific brain-cell phenotypes in an attempt to understand estrogen's influence on brain function. First, we will briefly consider the various means by which estrogen is believed to interact with target cells.

B. Current Views on Estrogen's Mechanism of Action

Estrogen, like other steroid hormones, influences cellular function at the genomic level by acting at specific intracellular receptors. For estrogen, these include the "original" ERα, cloned by GREEN et al. (1986), and the recently identified ERβ (KUIPER et al. 1996); although both isoforms have been identified primarily in the nucleus of target neurons, ERβ immunoreactivity is seen in cell bodies and processes in certain cell populations (Alves, unpublished observations). When bound, the nuclear ligand–receptor complexes act as transcription factors and regulate the expression of responsive genes, that until recently, were believed to be only those genes that contained an ERE. While the precise mechanisms by which the activated-ER dimer/ERE complexes regulate the expression of specific genes remain unclear, very recent evidence indicates that the two ER isoforms can actually form functional heterodimers at an ERE (PETTERSON et al. 1997). In addition, other ways in which estrogen appears capable of altering genomic expression continue to be identified. These include direct interaction in association with a fos/jun heterodimer at an AP-1 enhancer element (UMAYAHARA et al. 1994), as well as indirectly, at the cAMP response element (CRE; ARONICA et al. 1994; GU et al. 1996; ZHOU et al. 1996; MURPHY and SEGAL 1997) or, as recently postulated, via the MAPK cascade (WATTERS et al. 1997). Furthermore, a recent report of estrogen-mediated transcriptional regulation at an AP-1 site indicates that the same ligand, i.e., estradiol 17β, can induce opposite genomic effects, depending upon whether ERβ or ERα is activated (PAECH et al. 1997). Thus, it is now apparent that estrogen regulation of gene expression can be extremely complex and multifaceted.

Estrogen also appears to have rapid, seemingly non-genomic actions, including stimulation of the adenyl cyclase/cAMP or MAPK cascades. Some of these rapid actions appear to involve ion movement. For example, estrogen has been shown to directly potentiate K$^+$-stimulated DA release in the nucleus accumbens (THOMPSON and MOSS 1994). Estrogen also appears capable of inhibiting Ca^{2+} currents in neostriatal neurons of the rat brain, an effect that has been postulated to be mediated at the plasma membrane (MERMELSTEIN et al. 1996). Although all of these so-called rapid actions are suggestive of a plasma membrane-mediated event, to date, no specific membrane binding sites have been identified for estrogen.

C. Non-reproductive Brain Functions Influenced by Estrogen

I. Cognition (Learning and Memory)

Estrogen can influence the brain's capacity for learning and retaining what has been learned as memory. Short-term memory and spatial recognition are two examples of estrogen-sensitive memories. The basal forebrain, hippocampus and cerebral cortex are brain regions involved in the regulation of such cognitive functioning. Thus, we will consider several neural systems within these regions that are likely to mediate estrogen's influence on the brain's ability for learning and memory.

1. Basal Forebrain Cholinergic System

An important example of an estrogen-responsive, non-reproductive circuitry in the lab rat is the basal forebrain cholinergic system, the neurons of which project to the hippocampus and cerebral cortex. This cholinergic neuronal population appears to be sexually dimorphic in cell size and organization and, therefore, probably in projections (WESTLIND-DANIELSSON et al. 1991). It is the female system that appears more responsive to estrogen modulation, a dimorphism that appears to be determined prenatally (LUINE and McEWEN 1983). Physiologically, estrogen is known to increase the activities of the acetylcholine-synthesizing enzyme, ChAT and the degratory enzyme, acetylcholinesterase in OVX rats, suggesting a general stimulatory effect of this hormone on the activity of these neurons (LUINE et al. 1975; LUINE 1985).

More recent work has demonstrated that long-term (5–28 weeks) OVX results in a decline in high-affinity choline uptake and in ChAT activity in the rat frontal cortex and hippocampus (SINGH et al. 1994). While these biochemical alterations were associated with a learning impairment, estrogen replacement ameliorated cholinergic function and cognitive performance in these animals (SINGH et al. 1994). The integrity of this cholinergic system appears to decline with age, and compromised function of these neurons may be, at least in part, responsible for cognitive deficits associated with aging and the age-related dementia of Alzheimer's disease (LUINE et al. 1986; FISHER et al. 1987; DECKER and McGAUGH 1991). Considering the decreased risk of Alzheimer's disease among postmenopausal women who receive estrogen replacement (PAGANNINI and HENDERSON 1994; TANG et al. 1996), as well as the fact that estrogen therapy has been shown to improve cognitive performance of women afflicted by this disorder (HENDERSON et al. 1994), it is likely that this cholinergic circuitry is a primary target for estrogen's beneficial actions on cognition.

The mechanisms by which estrogen mediates these effects are not completely understood; however, neurotrophins, such as NGF may be involved. Exogenous NGF can reverse degeneration of these cholinergic neurons and improve memory (FISHER et al. 1987). Basal forebrain cholinergic neurons contain ERα together with the low-affinity NGF receptor (TORAN-ALLERAND et al. 1992), a member of the tyrosine kinase family, specifically referred to as trkA. NGF is produced by target cells within the hippocampus and is retrogradely transported by forebrain cholinergic neurons. Estrogen has been shown to modulate the expression of NGF, its receptor, as well as ChAT within this circuitry (GIBBS et al. 1994; MCMILLAN et al. 1996). While the results from such studies suggest a complex regulation, the findings by MCMILLAN et al. (1996) are consistent with a positive influence of estrogen on NGF action in this system; OVX-induced decreases in mRNA for trkA, as well as ChAT, were restored by short-term estrogen replacement.

2. Hippocampal γ-Aminobutyric Acid (GABA) Interneurons

The proper functioning of the hippocampus, a primary target of the basal forebrain cholinergic neurons and an integral neural substrate for cognition, requires precise interactions between the excitatory glutamatergic "principle" cells, and the inhibitory GABA-ergic interneurons. Manipulation of hippocampal GABA-ergic activity can influence learning and memory (AMMASSARI-TEULE et al. 1991). Subpopulations of GABA-ergic interneurons within the hippocampus of females and males contain nuclear ERα (WEILAND et al. 1997a, b) and are, thus, likely to be estrogen targets. In fact, the expression of the GABA-synthesizing enzyme, GAD, appears to be positively regulated by estrogen within a distinct population of interneurons associated with the CA1 region (WEILAND 1992b). Collectively, these data indicate that estrogen directly regulates the functioning of specific hippocampal interneurons, which are likely to be involved in the neural transmission for estrogen-sensitive learning and memory.

The GABA-ergic interneurons that contain ERα in the rat hippocampus appear to be of at least two types, based on the Ca^{2+}-binding proteins they express; those containing either calbindin or calretinin (WEILAND et al. 1997b). These binding proteins play a critical role in the buffering/transport of Ca^{2+}, the intra-/extracellular concentrations of which must be precisely monitored to maintain the health and proper functioning of brain cells. Considering that changes in these regulatory proteins have been observed in several neurodegenerative disorders in humans (HEIZMANN and BRAUN 1992) and the presence of nuclear ER within at least a subpopulation of these neurons, it is tempting to speculate a link between age-related decreases in circulating estrogen and a perturbation in Ca^{2+}-sequestering protein expression.

3. Hippocampal CA1 Pyramidal Neurons

These excitatory glutamatergic cells do not appear to contain receptors for estrogen (WEILAND et al. 1997a; WEILAND, unpublished results), yet the synaptic plasticity that occurs on these neurons through the rat estrus cycle is regulated by circulating estrogen (WOOLEY and McEWEN 1993). Although a direct, non-genomic action has not been ruled out, the finding that estrogen induction of CA1-dendritic spines requires functioning NMDA receptors (WOOLEY and McEWEN 1994) suggests an indirect action of estrogen on the CA1 neurons. Interestingly, estrogen regulates agonist-binding sites on the NMDA-receptor complex (WEILAND 1992a), as well as NMDAR1 immunofluorescence (GAZZALEY et al. 1996), specifically in the CA1 region. Furthermore, estrogen apparently increases the sensitivity of CA1 pyramidal cells to NMDA receptor-mediated synaptic input, an action which is correlated with the estrogen-induced increase in dendritic-spine density on the cells (WOOLEY et al. 1997a). Most recent evidence indicates that disinhibition is required for the functional expression of the estrogen-induced excitatory synapses based on long-term potentiation measurements (WOOLEY et al. 1997b). Thus, the ERα-containing GABA-ergic interneurons are likely candidates for such an estrogen-mediated disinhibition in the hippocampus (WEILAND et al. 1997a, b).

The relevance of ovarian steroid-induced fluctuation in NMDA receptor-containing synapses to cognitive function remains unclear. Evolutionarily, such changes in connectivity would likely render the organism better able to perform necessary tasks: mate selection and interaction during proestrus, when levels of circulating estrogen and, therefore, CA1 dendritic synapses are high. DESMOND and LEVY (1997) have dealt with this concept quite poignantly in a recent commentary. They have hypothesized, based on an evaluation of relevant human and rodent cognitive/behavioral data, that *"the optimal number of CA1 synapses fluctuates because different cognitive/behavioral demands come to the fore across the estrus cycle, and different hippocampal-dependent behaviors are optimized by different degrees of connectivity"*. In other words, more synapses do not necessarily mean a general enhancement in all hippocampal functions. For example, while spatial recognition may be hindered by elevated estrogen in rodents (GALEA et al. 1995) and humans (HAMPSON and KIMURA 1988), declarative memory storage appears to benefit from estrogen (ANDERSON et al. 1987; KAMPEN and SHERWIN 1994). However, one must use caution when correlating rodent and human behavioral data, since it has not yet been determined whether the human hippocampus undergoes a similar synaptic fluctuation in response to gonadal steroids.

II. Psychological Function

Gonadal steroids can influence the affective state of an organism. The monoamine 5-HT is believed to play an integral role in the regulation of psychological function. Here, we will consider evidence which suggests that estrogen may influence mood and mental state through this neural system.

1. Midbrain Serotonergic Neurons

Estrogen, as well as progesterone, modulate central 5-HT activity in rodents and primates. Parameters such as 5-HT synthesis, turnover and receptor binding in the dorsal raphe nucleus, where the highest concentration of 5-HT neurons in the brain are found, as well as in numerous forebrain target sites, are sensitive to fluctuations in these hormones. An imbalance in 5-HT neurotransmission is believed to be a crucial element in the etiology of several common psychopathologies, particularly anxiety and depression, both of which occur with higher frequency in women than in men (WEISMANN and KLERMAN 1985). Although the physiological basis of this sex difference in susceptibility is not clear, estrogen therapy significantly reduces symptoms of depression and/or anxiety in women (OPPENHEIM 1983), as well as in a rodent model of depression (BERNARDI et al. 1989). Since symptoms among susceptible women are often amplified during times of low-circulating estrogen levels – the luteal phase of the menstrual cycle, postpartum, and during menopause or when postmenopausal – a regulatory role for this steroid seems evident.

Most of the current therapies used to treat depression and anxiety involve manipulation of synaptic monoamine concentrations, particularly 5-HT, via degradative enzyme inhibitors and re-uptake blockers (MELTZER 1990) or, most recently, with an enhancer of 5-HT re-uptake, tianeptine (WILDE and BENFIELD 1995). In the rat brain, estrogen replacement has been shown to decrease the number of 5-HT re-uptake sites (the 5-HT transporter), specifically in the hippocampus (MENDELSON et al. 1993), but to increase this transporter in other brain regions, including the dorsal raphe, lateral septum and basolateral amygdala (McQUEEN et al. 1997). Furthermore, estrogen replacement to OVX rats significantly increases the density of the 5-HT_{2A} receptor in specific brain regions involved in emotion, such as the cerebral cortex and nucleus accumbens (SUMMER and FINK 1995), suggesting a possible postsynaptic regulation of activity. Collectively, these findings are suggestive of a complex regulation of estrogen on the 5-HT system, and are evidence that the mechanisms of 5-HT-active drugs in treating depression are far from clear. Thus, one should consider how estrogen may affect activity at the cell body, where gene regulation occurs.

Nuclear receptors for estrogen and progesterone exist in the dorsal raphe. However, while these receptors appear to be found in both 5-HT and non-5-HT cells in the dorsal raphe of the macaque (BETHEA 1993), they have been identified only in cells adjacent to, but not in 5-HT neurons in the rat (ALVES et al. 1998). Interestingly, mRNA levels of the 5-HT synthesizing enzyme, tryptophan hydroxylase, appear to be regulated by estrogen replacement in the dorsal raphe of the macaque (PECINS-THOMPSON et al. 1996), but not in the rat (ALVES et al. 1997). Despite these differences, numerous studies have reported increased central 5-HT synthesis and/or activity following estrogen treatment in rodents. Thus, species differences appear to exist in the mechanisms by which ovarian steroids can act upon 5-HT neurons; direct genomic and/or trans-synaptic actions seem likely in the macaque, while trans-synaptic and/or indirect genomic action may occur in the rat.

Taken together, it is not unreasonable to consider that the increased occurrence of depression or decreased mood in menopausal/postmenopausal women may result from a decline and/or loss of "estrogenic tone" on the central 5-HT system.

III. Sensorimotor Performance

Estrogen can affect an animal's motor response to sensory stimuli. In this Section we will review experimental evidence suggesting a regulatory role for this ovarian steroid on the major DA system of the brain, a region which appears to contain very few nuclear ERs.

1. Nigrostriatal Dopaminergic System

Estrogen effects on sensorimotor performance have been well documented in the amphetamine/apomorphine-induced rotational-behavior paradigm in rodents that have had the nigrostriatal DA system unilaterally lesioned with 6-hydroxydopamine (BECKER et al. 1982; KAZANDJIAN et al. 1987; BECKER and CHA 1989). Not surprising then, striatal DA activity which regulates such motor action, is sensitive to natural fluctuations in circulating estrogen (BECKER and RAMIREZ 1980; KAZANDJIAN et al. 1988; BECKER and CHA 1989). Activity and behavior induced by the presynaptic-acting amphetamine are generally increased when estrogen secretion is elevated; during late proestrus/early estrus. Manipulation of estrogen and progesterone levels within the physiological range can mimic estrus cycle-dependent effects on this system. For example, drugs, such as amphetamines, that stimulate DA release induce decreased rotational behavior in OVX rats; amphetamines induce less DA release in these animals as well. Estrogen replacement reinstates rotational behavior, increases striatal DA turnover and increases amphetamine-induced DA release (DI PALOLO et al. 1985; BECKER and BEER. 1986; BECKER 1990). Interestingly, male rats do not exhibit the same extent of amphetamine-induced increases in rotational behavior or locomotor activity as seen in estrus females. In further contrast to females, castration of male rats does not effect amphetamine-stimulated rotational behavior or striatal DA release (BECKER and RAMIREZ 1980; ROBINSON et al. 1981; CAMP et al. 1986). Thus, gonadal steroid regulation of these behaviors appears to occur specifically via estrogen, and the apparent sex differences may be a result of different circuitries established during the sexual differentiation of the brain.

Spontaneous sensorimotor behavior can also be influenced by estrogen. BECKER et al. (1987) demonstrated this by training female rats to walk across a narrow beam, similar to a "tight-rope", suspended several feet above the floor. Sensorimotor function was assessed by recording the accuracy of foot placement while walking; placement of feet directly onto the top surface of the beam was considered correct, as opposed to gripping the sides. Rats in proestrus/estrus, or OVX rats replaced with estrogen directly into the striatum, performed dramatically better than intact, diestrus animals or OVX animals replaced with a control substance. Findings from the hormone replacement study suggest a locally mediated action of estrogen on the activity of these DA neurons (BECKER et al. 1987). Although the exact mechanisms by which estrogen may facilitate striatal DA activity remain unknown, direct, non-genomic action may be involved, considering recent reports of such action on DA neurons (THOMPSON and MOSS 1994; MERMELSTEIN et al. 1996).

Like many neural systems, the nigrostriatal DA pathway shows declines in the aging brain. Parkinson's disease is an extreme example of an age-associated degeneration of this system. Considering the apparently positive effects of estrogen on this neural system, one would suspect that loss of circulating estrogen in postmenopausal women could precipitate a decrease in nigrostriatal DA activity. However, several clinical studies have previously reported an exacerbation of symptoms of several DA disorders, including parkinsonian symptoms in women receiving chronic estrogen treatment, i.e., oral contraceptives or replacement therapy (RIDDOCH et al. 1971; BARBER et al. 1973; BEDARD et al. 1977). Perhaps physiological levels of this hormone act to facilitate DA neurotransmission, but chronic exposure to elevated estrogen levels may actually be disruptive. It has been determined in female and male rats that acute exposure to high doses of estrogen or to low doses for a prolonged period, increases drug-induced behaviors. However, this mechanism contradicts what occurs in intact animals or in those given "physiological" replacements (DI PAULO et al. 1981; HRUSKA and SILDBERGELD 1980). Instead of an enhancement of presynaptic DA release, a decrease is observed, but with a compensatory increase in postsynaptic DA receptors. Thus, prolonged elevation of DA neuronal activity can result in an eventual decline or loss of presynaptic activity. These findings indicate that, in order for an estrogen replacement regimen to be beneficial, it must be carefully determined to mimic, as closely as possible, the natural fluctuations of physiological estrogen levels which occur throughout the ovarian cycle.

IV. Non-Neuronal Functions: "Brain Maintenance"

1. Glial Cells

Glia are involved in numerous aspects of nerve maintenance, and these cells can be targeted by estrogen (LANGUB and WATSON 1992; SANTAGATI et al. 1994). In fact, estrogen has been shown to regulate the expression of specific genes within central glia. For example, recent evidence suggests that estrogen can increase mRNA levels of apoE, a lipophilic protein involved in lipid homeostasis within astrocytes and microglia (STONE et al. 1997). Interestingly, this estrogen effect was observed in glia found within brain regions that undergo cyclic synaptic remodeling through the estrus cycle: the hippocampal CA1, as well as the arcuate nucleus. This study, together with a previous in vitro report of estrogen-mediated changes in hypothalamic astrocytes (TORRES-ALEMAN et al. 1992), have also demonstrated that heterotypic cellular interactions are required for such estrogen regulation. Furthermore, astrocytic morphology within the hippocampus appears to fluctuate with changes in estrogen levels (TRANQUE et al. 1987; LUQUIN et al. 1993; DEL CORRO et al. 1995; KLINTSOVA et al. 1995). Collectively, these findings are relevant not only to estrogen's effects on synaptogenesis in the CA1 and a possible role for this phenomenon in cognition, but also to the putative protective role of this ovarian steroid in Alzheimer's disease. For example, decreases in apoE levels have been reported in the hippocampus and frontal cortex of Alzheimer's patients (BERTRAND et al. 1995).

Gonadal steroids, including estrogen, can also regulate the expression of the astrocyte marker, GFAP. Levels of GFAP mRNA have been shown to increase transiently, specifically within the arcuate nucleus of the female mouse, when circulating estrogen is high during proestrus (KOHAMA et al. 1995b). Similarly, in the male rat hypothalamus, castration inhibits GFAP expression. However, this removal of gonadal steroids actually increases GFAP in the hippocampus (DAY et al. 1993). A similar inhibitory effect of estrogen on astrocytic processes was demonstrated in the CA1 region; during proestrus, when estrogen levels and CA1 synapses are greatest, the volume occupied by astrocytic processes was lowest (KLIMTSOVA et al. 1995). Levels of GFAP mRNA and protein generally increase throughout the aging CNS, regardless of gender or species (mouse, rat, and human) (NICHOLS et al. 1993; KOHAMA et al. 1995a). Thus, gonadal steroids seem to have region-specific effects on GFAP expression. The transient stimulatory effect on GFAP expression, specifically in the hypothalamus, may be relevant to reproductive physiology, i.e., the proestrus gonadotropin surge, sexual behavior. However, in other brain regions, such as the hippocampus, a general inhibition may result, which is eventually lost with the drop in circulating gonadal steroid levels in the aging female and male. For a comprehensive review of the relationship between astrocytes and brain aging, refer to SCHIPPER (1996).

2. Endothelial Cells of the Blood–Brain Barrier (BBB)

Ovarian steroids appear to play a role in the ability of the female brain to utilize glucose, the primary energy source of mammalian brain cells. While OVX rats demonstrate a significantly decreased capacity for glucose utilization, estrogen treatment increases this capacity across the brain by 20–30% (NAMBA and SOKOLOFF 1984). This positive effect of estrogen appears to involve the ability of this hormone to increase glucose-transporter 1 in the endothelial cells of the BBB (SHI and SIMPKINS 1997). At least one study has reported immunoreactivity for ERs in central endothelia at the electron-microscopic level (LANGUB and WATSON 1992). Interestingly, a recent review has suggested a link between age-related increases in blood-glucose levels, due to decreased uptake, and Alzheimer's disease (FINCH and COHEN 1997). Estrogen's ability to enhance glucose transport into the brain, thereby increasing the energy supply for brain cells, may be a means by which estrogen provides beneficial and protective effects in neurodegenerative disorders, including those associated with aging.

D. Summary

The brain is comprised of numerous phenotypically distinct systems that function in concert to collect, integrate, relay and retain information necessary for the survival of the organism. As we have outlined in this chapter, and the list is by no means exhaustive, the functioning of many brain cells' phenotypes, both neuronal and glial, is regulated by estrogen in some capacity. Even endothelial cells of the BBB, which regulate the entrance of the brain's primary energy source, glucose, are influenced by estrogen. Considering that a healthy, working brain requires the precise functioning and integration of all of its counterparts, estrogen is capable of having profound effects on this "chief" regulatory organ and, thus, on the life of the organism, throughout its lifespan. As we have attempted to demonstrate within this chapter, findings from numerous studies, involving various species, suggest that the eventual decline or loss of circulating estrogen which occurs in the aging individual, particularly females, is likely to be involved in age-related decline in brain function. Furthermore, the therapeutic value of estrogen-replacement therapy is likely to depend on the ability to provide a physiological hormone profile.

References

Alves SE, Weiland NG, Hastings NB, Tanapat P, McEwen BS (1997) Estradiol regulation of 5-HT neurons in the dorsal raphe of the rat differs from the monkey: an analysis of tryptophan hydroxylase mRNA levels. Soc Neurosci Abst # 484.1; New Orleans, LA

Alves SE, Weiland NG, Hayashi S, McEwen B (1998) Immunocytochemical localization of nuclear estrogen receptors and progestin receptors within the rat dorsal raphe nucleus. J Comp Neurol 391:322–334

Ammassari-Teule M, Pavone R, Castellano C, McGaugh JL (1991) Amygdala and dorsal hippocampus lesions block the effects of GABAergic drugs on memory and storage. Br Res 551:104–109

Anderson E, Hamburger S, Liu JH, Rebar RW (1987) Characteristics of menopausal women seeking assistance. Am J Obster Gynecol 156:428–433

Aronica SM, Klaus WL, Katzenellenbogen BS (1994) Estrogen action via the cAMP signaling pathway: stimulation of adenylate cyclase and cAMP-regulated gene transcription. Proc Natl Acad Sci USA 91:8517–8521

Barber PV, Arnold AG, Evans G (1973) Recurrent hormone dependent chorea: effects of estrogens and progesterone. Clin Endocrinol 5:291–293

Becker JB, Ramirez VD (1980) Sex differences in the amphetamine stimulated release of catecholamines from rat striatal tissue in vitro. Br Res 204:361–372

Becker JB, Robinson TE, Lorenz KA (1982) Sex differences and estrous cycle variations in amphetamine-elicited rotational behavior. Eur J Pharmacol 80:65–72

Becker JB, Beer ME (1986) The influence of estrogen on nigrostriatal dopamine activity:behavioral and neurochemical evidence for both pre- and postsynaptic components. Behav Br Res 19:27–33

Becker JB, Snyder PJ, Miller MM, Westgate SA, Jenuwine MJ (1987) The influence of estrous cycle and intrastriatal estradiol on sensorimotor performance in the female rat. Pharmacol Biochem Behav 27:53–59

Becker JB, Cha J (1989) Estrous cycle-dependent variation in amphetamine-induced behaviors and striatal dopamine release assessed with microdialysis. Behav Br Res 35:117–125

Becker JB (1990) Direct effect of 17β-estradiol on striatum: sex differences in dopamine release. Synapse 5:57–164

Bedard PJ, Langelier P, Villeneuve A (1977) Estrogens and the extrapyramidal system. Lancet 2:1367–1368

Bernardi M, Vergoni A, Sandrini M, Tagliavini S, and Bertolini S (1989) Influence of ovariectomy, estradiol and progesterone on the behavior of mice in an animal model of depression. Physiol Behav 45:1067–1068

Bertrand P, Poirier J, Oda T, Finch CE, Pasinetti GM (1995) Association of apolipoprotein E genotype with brain levels of apolipoprotein E and apolipoprotein J (clusterin) in Alzheimer's disease. 33:174–178

Bethea C (1993) Colocalization of progestin receptors with serotonin in the raphe neurons of macaque. Neuroendo 57:1–6

Camp DM, Becker JB, Robinson TE (1986) Sex differences in the effects of gonadectomy on amphetamine-induced rotational behavior in rats. Behav Neural Biol 46:491–495

Day JR, Laping NJ, Lampert-Etchells M, Brown SA, O'Callaghan JP, McNeill TH (1993) Gonadal steroids regulate the expression of glial fibrillary acidic protein in the adult male hippocampus. Neurosci 55:435–443

Dekker MW, McGaugh JL (1991) The role of interactions between the cholinergic system and other neuromodulatory systems in learning and memory. Synapse 7:151–168

Del Cerro S, Garcia-Estrada J, Garcia-Segura LM (1995) Neuroactive steroids regulate astroglia morphology in hippocampal cultures from adult rats. Glia 14:65–71

Desmond NL, Levy WB (1997) Ovarian steroidal control of connectivity in the female hippocampus: an overview of recent experimental findings and speculations on its functional consequences. Hippocampus 7:239–245

Di Paolo T, Poyet P, Labrie F (1981) Effect of chronic estradiol and haloperidol treatment on striatal dopamine receptors. Eur J Pharmacol 73:105–106

Di Paolo T, Rouillard C, Bedard P (1985) 17β-estradiol at a physiological dose acutely increases dopamine turnover in rat brain. Eur J Pharmacol 117:197–203

Finch CE, Cohen DM (1997) Aging, metabolism, and Alzheimer disease: review and hypothesis. Exptl Neurol 143:82–102

Fischer W, Wictorin K, Bjorklund A, Williams LR, Varon S, Gage FH (1987) Amelioration of cholinergic neuron atrophy and spatial memory impairment in aged rats by nerve growth factor. Nature 329:65–68

Galea LAM, Kavaliers M, Ossenkopp K-P, Hampson E (1995) Gonadal hormone levels and spatial learning performance in the Morris water maze in male and female meadow voles, *Microtus pennsylvanicus*. Horm Behav 29:106–125

Gazzaley AH, Weiland NG, McEwen BS, Morrison JH (1996) Differential regulation of NMDAR1 mRNA and protein by estradiol in the rat hippocampus. J Neurosci 16:6830–6838

Gibbs RB, Wu D, Hersh LB, Pfaff DW (1994) Effects of estrogen replacement on the relative levels of choline acetyltransferase, trkA, and nerve growth factor messanger RNAs in the basal forebrain and hippocampal formation of adult rats. Expl Neurol 129:70–80

Green S, Walter P, Kumar V, Krust A, Bornet J-M, Argos P, Chambon P (1986) Human estrogen receptor cDNA: sequence, expression and homology to v-erb-A. Nature 320:134–139

Gu G, Rojo AA, Zee MC, Yu J, Simerly RB (1996) Hormonal regulation of CREB phosphorylation in the anteroventral periventricular nucleus. J Neurosci 16:3035–3044

Hampson E, Kimura D (1988) Reciprocal effects of hormonal fluctuations on human motor and perceptual-spatial skills. Behav Neurosci 102:456–459.

Heizmann CW, Braun K (1992) Changes in Ca^{2+}-binding proteins in human neurodegenerative disorders. Trends Neurosci 15:259–264

Henderson VW, Paganini-Hill A, Emanuel CK, Dunn ME, Buckwalter JG (1994) Estrogen replacement therapy in older women: comparisons between Alzheimer's disease cases and nondemented control subjects. Arch Neurol 51:896–900

Hruska RE, Silbergeld EK (1980) Estrogen treatment enhances dopamine receptor sensitivity in the rat striatum. Eur J Pharmacol 61:397–400

Kampen DL, Sherwin BB (1994) Estrogen use and verbal memory in healthy postmenopausal women. Obstet Gynecol 83:979–983

Kazandijian A, Spyraki C, Sfikakis A, Varanos DD (1987) Apomorphine-induced behavior during the oestrus cycle of the rat. Neuropharm 26:1037–1045

Kazandijian A, Spyraki C, Papadopoulou, Sfikakis A, Varanos DD (1988) Behavioral and biochemical effects of haloperidol during the oestrous cycle of the rat. Neuropharm 27:73–78

Klintsova A, Levy WB, Desmond NL (1995) Astrocytic volume fluctuates in the hippocampal CA1 region across the estrous cycle. Br Res 690:269–274

Kohama SG, Goss JR, Finch CE, McNeill TH (1995a) Increases in glial fibrillary acidic protein in the aging female mouse brain. Neurobiol Aging 16:59–67

Kohama SG, Goss JR, McNeill TH, Finch CE (1995b) Glial fibrillary acidic protein mRNA increases at proestrus in the arcuate nucleus of mice. Neurosci Lett 183:164–166

Kuiper GGJM, Enmark E, Pelto-Huikko M, Nilsson S, Gustafsson JA (1996) Cloning of a novel estrogen receptor expressed in rat prostate and ovary. Proc Natl Acad Sci USA 93:5925–5930

Langub MC, Watson RE (1992) Estrogen receptor-immunoreactive glia, endothelia, and ependyma in guinea pig preoptic area and median eminence: electron microscopy. Endocrinol: 130:364–372

Luine VN, Khylchevcskaya RI, McEwen BS (1975) Effect of gonadal steroids on activities of monoamine oxidase and choline acetylase in rat brain. Br Res 86:293–306

Luine VN, McEwen BS (1983) Sex differences in cholinergic enzymes of diagonal band nuclei in the rat preoptic area. Neuroendo 36:475–482

Luine VN (1985) Estradiol increases choline acetyltransferase activity in specific basal forebrain nuclei and projection areas of female rats. Exp Neurol 89:484–490

Luine VN, Renner K, McEwen BS (1986) Sex-dependent differences in estrogen regulation of choline acetyltransferase are altered by neonatal treatments. Endocrinol 119:874–878

Luquin S, Naftolin F, Garcia-Segura LM (1993) Natural fluctuation and gonadal hormone regulation of astrocyte immunoreactivity in dentate gyrus. J Neurobiol 24:913–924

MacLusky NJ, Clark AS, Naftolin F, Goldman-Rakic PC (1987) Estrogen formation in the mammalian brain: possible role of aromatase in sexual differentiation of hippocampus and neocortex. Steroids 50:459–474

McEwen BS, Liederberg I, Chaptal C, Krey LC (1977) Aromatization: important for sexual differentiation of the neonatal rat brain. Horm Behav 9:249–263

McMillan PJ, Singer CS, Dorsa DM (1996) The effects of ovariectomy and estrogen replacement on trkA and choline acetyltransferase mRNA expression in the basal forebrain of the adult female Sprague-Dawley rat. J Neurosci 16:1860–1865

Meltzer HY (1990) Role of serotonin in depression. Ann NY Acad Sci 600:486–500

Mendelson SD, McKittrick CR, McEwen BS (1993) Autoradiographic analyses of the effects of estradiol benzoate on [^{3}H]paroxetine binding in the cerebral cortex and dorsal hippocampus of gonadectomized male and female rats. Br Res 601:299–302

Mermelstein PG, Becker JB, Surmeier DJ (1996) Estradiol reduces calcium currents in rat neostriatal neurons via membrane receptor. J Neurosci 16:595–604

Murphy DD, Segal M (1997) Morphological plasticity of dendritic spines in central neurons is mediated by activation of cAMP response element binding protein. Proc Natl Acad Sci 94:1482–1487

Namba H, Sokoloff L (1984) Acute administration of high doses of estrogen increases glucose utilization throughout the brain. Br Res 291:391–394

Nichols NR, Day JR, Laping NJ, Johnson SA, Finch CE (1993) GFAP mRNA increases with age in rat and human brain. Neurobio Aging 14:421–429

Oppenheim G (1983) Estrogen in the treatment of depression: neuropharmacological mechanisms. Biol Psychiatry 18:721–725

Paech K, Webb P, Kuiper GGJ, Nilsson S, Gustafsson J-A, Kushner PJ, Scanlan TS (1997) Differential ligand activation of estrogen receptors ERαand ERβ at AP1 sites. Science 277:1508–1510

Paganini-Hill A, Henderson VW (1994) Estrogen deficiency and risk of Alzheimer's disease in women. Am J Epidem 140:256–261

Pecins-Thompson M, Brown N, Kohama SG, Bethea C (1996) Ovarian steroid regulation of tryptophan hydroxylase mRNA expression in rhesus macaques. J Neurosci 16:7021–7029

Pettersson K, Grandien K, Kuiper GJM, Gustafsson J-A (1997) Mouse estrogen receptor β forms estrogen response element-binding heterodimers with estrogen receptor α. Mol Endocrinol 11:1486–1496

Riddoch D, Jefferson M, Bickerstaff ER (1971) Chorea and oral contraceptives. Br Med J 4:217–218

Robinson TE, Camp DM, Becker JB (1981) Gonadectomy attenuates turning behavior produced by electrical stimulation of the nigrostriatal dopamine system in female but not male rats. Neurosci Lett 23:203–208

Santagati S, Melcangi RC, Celotti F, Martini L, Maggi A (1994) Estrogen receptor is expressed in different types of glial cells in culture. J Neurochem 63:2058–2064

Schipper HM (1996) Astrocytes, brain aging, and neurodegeneration. Neurobio Aging 17:467–480

Shi J, Simpkins JW (1997) 17β-Estradiol modulation of glucose transporter 1 expression in blood-brain barrier. Am J Physiol 272 (Endocrinol Metab 35):E1016–E1022

Singh M, Meyer EM, Millard WJ, Simpkins JW (1994) Ovarian steroid deprivation results in a reversible learning impairment and compromised cholinergic function in female Sprague-Dawley rats. Br Res 644:305–312

Stone DJ, Rozovsky I, Morgan TE, Anderson CP, Hajian H, Finch CE (1997) Astrocytes and microglia respond to estrogen with increased apoE mRNA in vivo and in vitro. Exptl Neurol 143:313–318

Sumner BE, Fink G (1995) Estrogen increases the density of 5-hydroxytryptamine (2 A) receptors in cerebral cortex and nucleus accumbens in the female rat. J Steroid Biochem Mol Bio 54:15–20

Tang M-X, Jacobs D, Stern Y, Marder K, Schofield P, Gurland B, Andrews H, Mayeux R (1996) Effect of estrogen during menopause on risk and age at onset of Alzheimer's disease. Lancet 348:429–432

Thompson TL, Moss RL (1994) Estrogen regulation of dopamine release in the nucleus accumbens: genomic and nongenomic-mediated effects. J Neurochem 62:1750–1756

Toran-Allerand CD (1991) Organotypic culture of the developing cerebral cortex and hypothalamus: relevance to sexual differentiation. Psychoneuroendo 16:7–24

Toran-Allerand CD, Miranda RC, Bentham WD, Sohrabji F, Brown TJ, Hochberg RB, MacKlusky NJ (1992) Estrogen receptors colocalize with low-affinity nerve growth factor receptors in cholinergic neurons of the basal forebrain. Proc Natl Acad Sci USA 89:4668–4672

Tores-Aleman I, Rejas MT, Pons S, Garcia-Segura LM (1992) Estradiol promotes cell shape changes and glial fibrillary acidic protein redistribution in hypothalamic astrocytes in vitro: a neuronal-mediated effect. Glia 6:180–187

Tranque PA, Suarez I, Olmos G, Fernandez B, Garcia-Segura LM (1987) Estradiol-induced redistribution of glial fibrillary acidic protein immunoreactivity in the rat brain. Br Res 406:348–351

Umayahara Y (1994) Estrogen regulation of the insulin-like growth factor I gene transcription involves an AP-1 enhancer. J Biol Chem 269:16433–16443

Watters JJ, Campbell JS, Cunningham MJ, Krebs EG, Dorsa DM (1997) Rapid membrane effects of steroids in neuroblastoma cells: effects of estrogen on mitogen acitivated protein kinase signalling cascade and c-fos immediate early gene transcription. Endocrinol 138:4030–4033

Weiland NG (1992a) Estradiol selectively regulates agonist binding sites on the *N*-methyl-D- aspartate receptor complex in the CA1 region of the hippocampus. Endocrinol 131:662–668

Weiland NG (1992b) Glutamic acid decarboxylase messanger ribonucleic acid is regulated by estradiol and progesterone in the hippocampus. Endocrinol 131: 2697–2702

Weiland NG, Orikasa C, Hayashi S, McEwen BS (1997a) Distribution and hormone regulation of estrogen receptor immunoreactive cells in the hippocampus of Male and Female rats. J Comp Neurol 388:603–612

Weiland NG, Alves SA and McEwen BS (1997b) Estrogen receptor immunoreactivity colocalizes with calbindin in interneurons in several brain regions including dorsal hippocampus. Soc Neurosci Abst # 453.4;New Orleans, LA

Weissman MM, Klerman GL (1985) Gender and depression. Trends Neurosci 8:416–420

Westlind-Danielsson A, Gould E, McEwen BS (1991) Thyroid hormone causes sexually distinct neurochemical and morphological alterations in rat septal-diagonal band neurons. J Neurochem 56:119–128

Wilde MI, Benfield P (1995) Tianeptine: a review of its pharmacodynamic and pharmacokinetic properties, and therapeutic efficacy in depression and coexisting anxiety and depression. Drugs 49:411–439

Woolley CS, McEwen BS (1993) Roles of estrogen and progesterone in regulation of hippocampal dendritic spine density during the estrous cycle in the rat. J Comp Neurol 336:293–306

Woolley CS, McEwen BS (1994) Estradiol regulates hippocampal dendritic spine density via an N-methyl-D-aspartate receptor-dependent mechanism. J Neurosci 14:7680–7687

Woolley CS, Weiland NG, McEwen BS, Schwartzkroin PA (1997a) Estradiol increases the sensitivity of hippocampal CA1 pyramidal cells to NMDA receptor-mediated synaptic input: correlation with dendritic spine density. J Neurosci 17:1848–1859

Woolley CS, Dlugosch DJ, Schwartzkroin PA (1997b) Functional esxpression of estradiol- induced hippocampal synapses requires disinhibition. Soc Neurosci Abst # 94.18; New Orleans, LA

Zhou Y, Watters JJ, Dorsa DM (1996) Estrogen rapidly induces the phosphorylation of the cAMP response element binding protein in rat brain. Endocrinol 137:2163–216

CHAPTER 16

Cardiovascular System

M. Birkhäuser

A. Introduction

Today, cardiovascular mortality is the leading cause of death in women who have gone through the menopause. In 54%, the cause of death is a coronaropathy. After menopause, the risk of cardiovascular accidents in females approaches that for males (Fig. 1). At the age of 65 years, death is shown to be due to a coronaropathy in one of three women. Ovariectomized women without hormone replacement therapy (HRT) have a higher cardiovascular mortality than non-ovariectomized women of the same age (Johansson et al. 1975). In Europe, the United States and Australia, approximately ten times more women are dying from coronary diseases than from breast cancer, and significantly more women succumb to coronaropathies than to cerebrovascular diseases, infections, gastrointestinal affections, respiratory diseases or diabetes mellitus and its complications (Department of Health, Education and Welfare 1979). The exponential increase of cardiovascular mortality in women after menopause points to an important physiological role of estrogens in the integrity of the cardiovascular system (Johansson et al. 1975; Wuest et al. 1953; Kannel et al. 1976).

Because of the gender-specific secretion of the gonadal steroids, androgens, estrogens and progesterone, the difference in the cardiovascular risk between the two sexes could be linked to the male or female endocrine patterns. In 1952, Pick et al. reported that cocks that were fed estrogens together with a cholesterol-rich diet developed much less atherosclerotic plaques than control animals receiving the same diet, but without the estrogen supplement. The cardioprotective effect of estrogens was later confirmed in rabbits (Haarbo et al. 1991) and primates (Adams et al. 1990; Clarkson et al. 1990). Men suffering from prostate cancer treated with estrogens show an unusually low incidence of atherosclerotic lesions compared with healthy untreated men. Far more than 20 epidemiological studies find that HRT reduces the risk of cardiovascular death by 50% (Oliver and Boyd 1961; McDowell et al. 1967; The Coronary Drug Project 1970; Adam et al. 1981; Stampfer et al. 1985; Bush et al. 1987; Ettinger et al. 1987; Petitti et al. 1987; Avila et al. 1990; Bush 1990; Nabulsi et al. 1993; Rosenberg et al. 1993; Wenger et al. 1993; PEPI Trial Writing Group 1995). All cohort studies, except for the Framingham study (Wilson and Garrison 1985) and the study of La Vecchia

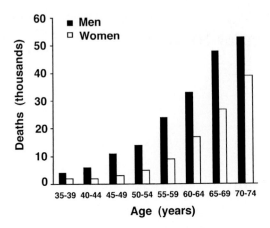

Fig. 1. Cardiovascular death in relation to age and gender (modified from National Institutes of Health 1992)

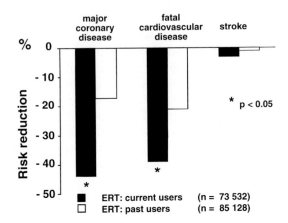

Fig. 2. Reduction of the cardiovascular risk and stroke by estrogen replacement therapy in current and past users (STAMPFER et al. 1991)

et al. (1987), found a significant cardioprotection among estrogen users. However, in the study, there was a clear selection bias for patients and controls. The reanalysis of the Framingham study by EAKER and CASTELLI (1987), using a different definition of risks, also concludes that there is cardioprotection by estrogens. Ross et al. (1989), HUNT et al. (1990), HENDERSON et al. (1991), GRADY et al. (1992) and GRODSTEIN et al. (1997) reported a significant decrease of overall mortality in estrogen users. In HRT users, GRODSTEIN et al. (1997) observed a greater reduction of cardiovascular mortality in women with an high risk of myocardial infarction than in women with a low risk. These epidemiological studies suggest that estrogens used for primary prevention reduce the risk of cardiovascular death by 50–70% (Fig. 2). The situation might be more complex if estrogens are used for secondary cardiovascular prevention, as in the HERS study (HULLEY et al. 1998), but it has to be stressed that the HERS study does not invalidate the data available today on primary prevention. The partially unexpected results of the HERS study might be linked to the progestagen used rather than to the administration of the estrogen.

B. Cardioprotective Mechanism of Estrogens

Today, two mechanisms are known to participate in the cardioprotective effect of estrogens. Estrogens may protect the arteries by an indirect action through a change in the lipid profile or by a direct action at the arterial wall.

I. Indirect Action on the Cardiovascular System by a Change of the Lipid Balance

Sex steroids induce typical changes in the equilibrium of the different lipoproteins, particularly in the balance of the so-called atherogenic (unfavorable) and non-atherogenic (favorable) lipoproteins. In addition, estrogens may reduce the plasma levels of both fibrinogen and the plasminogen-activating inhibitor (PEPI TRIAL WRITING GROUP 1995).

1. Atherogenic Lipoproteins

Very-low-density lipoproteins (VLDLs) are precursors of low-density lipoproteins (LDLs). VLDLs are formed by endogenously synthesized triglycerides, cholesterol and phospholipids, and possess a density below 1.006 g/ml. Peroral estrogens stimulate fasting triglyceride levels through their hepatic first-pass effect in a dose-dependent way by increasing the synthesis of apo-B100 protein. Because the increase of triglycerides is absent if transdermal estradiol is administered (RIJPKEMA et al. 1990), peroral HRT should be avoided in women with elevated basal triglyceride levels. The triglyceride component of VLDL is hydrolyzed in the periphery by lipases. The free surface lipids of the VLDL complexes are transported to the liver by high-density lipoproteins (HDLs). If small VLDL particles remain for a longer period in the circulation, intermediate-density lipoproteins (IDLs, density 1.006–1.019 g/ml) and finally LDLs (density 1.019–1.063 g/ml) are formed by the integration of additional cholesterol particles.

LDL particles are the most important fraction of cholesterol in the peripheral circulation. They are composed mainly by apoprotein B and esterified cholesterol. VLDL, IDL and LDL are atherogenic and consist of several subpopulations of different size, density and composition. Peroral estrogen substitution decreases LDL levels (WUEST et al. 1953; PEPI TRIAL WRITING GROUP 1995). Hepatic LDL receptors bind IDL and LDL and eliminate them from the circulation. In contrast, if oxidized LDL is fixed to the vascular wall and absorbed by macrophages, so-called "Schaumzellen" are formed. The formation of Schaumzellen is the critical step within the arterial wall and initiates the process of atherosclerosis, which progresses from the endothelium to smooth muscular cells. A second mechanism might be mediated through the oxidation of lipids, promoting the proliferation of cellular elements and changing thrombocyte function. Estrogens lower the concentration of LDL and total cholesterol by a reduction of the number of LDL molecules. Because apo-B-100, the main LDL-protein, binds to intra- and extrahepatic apo B100/LDL receptors, LDL particles can be eliminated from the circulation and catabolized. Estrogens regulate the synthesis and clearance of LDL precursors such as VLDL (WALSH et al. 1991).

2. High-Density Lipoprotein

Whereas LDL and triglycerides belong to the so-called unfavorable, atherogenic lipids, HDL belongs to the protective, non-atherogenic lipids. Peroral estrogen substitution induces an increase of HDL (WUEST J et al. 1953; PEPI TRIAL WRITING GROUP 1995). HDL precursors, which are poor in lipids, incorporate lipids during their stay within the circulation and undergo structural changes. As is true for LDLs, HDLs are structurally heterogeneous and are composed of different lipids and proteins. Most HDL particles contain apolipoprotein type A1, originating from the liver and intestinal tract. Other HDL particles contain apolipoprotein type A2, type C and type E. The smaller and denser HDL subclasses are called HDL-3, the bigger HDL-2. Women possess a higher proportion of big HDL-2 molecules than men. In women, plasma HDL-2 concentration correlates well with total HDL cholesterol. The practical clinical relevance of the HDL subfractions is not yet entirely understood. Peroral administration of natural estrogens stimulates HDL cholesterol, especially the bigger HDL particles containing apo-1. The administration of supraphysiological dosages of estrogens induces a marked increase of the production rate of HDL apolipoprotein-A1 and a simultaneous significant decrease of the hepatic lipase and, therefore, a reduction of HDL clearance, resulting in a marked increase of the HDL-2 lipoprotein subclasses.

Because HDL transports cholesterol from the periphery to the liver, HDL is believed to have an anti-atherogenic activity. However, other factors could participate in the beneficial influence of HDL with respect to the risk of atherosclerotic diseases. For example, high HDL levels could point to a shorter presence within the plasma of other, potentially atherogenic particles. Furthermore, HDL could reduce the accumulation of lipid peroxidases within the LDL complexes. Animal data suggest that an increase of the production of apolipoprotein-A1 reduces the amount of LDL complexes.

3. Lipoprotein (a)

Lipoprotein (a) [Lp(a)] is composed of apolipoprotein-B-100 complexes and apoprotein (a), another large protein. Apoprotein (a) has some similarities with plasminogene. Within the normal population, plasma concentrations of LP(a) are highly variable. They are under strict genetic control. High LP(a) levels correlate strongly with an increased risk of coronary heart diseases. Apo(a) has been found in arterial lesions. Via the fibrinolytic system, Lp(a) could participate in the pathogenesis of atherosclerosis. Low LP(a) levels are considered to be a favorable prognostic parameter for a low risk of cardiovascular diseases. HAENGGI et al. (1997) showed that Tibolone, a synthetic steroid with estrogenic, progestagenic and weak androgenic partial activities, has a particularly strong inhibiting effect on Lp(a).

4. Influence of Estrogen Replacement Therapy and HRT on the Lipid Profile

The PEPI TRIAL WRITING GROUP (1995) reports that peroral estrogens result in a significant HDL increase. Unopposed estrogens induce the highest increase with 5.6%. The second highest increase (4.1%) is observed with a combination of estrogens and micronized progesterone. In the PEPI trial, the cumulative difference between the treatment and the placebo groups was highly significant ($P + 0.001$), although the placebo group also had a LDL decrease of 4.1%. If all treatment groups are pooled and compared with the placebo group, the mean HDL increase induced by estrogens is 5–6mg/dl and the mean LDL decrease 15mg/dl. If 17β-estradiol is administered transdermally, a similar pattern of lipid changes is observed (BIRKHÄUSER and HAENGGI 1994; HAENGGI et al. 1997), but in a attenuated form.

II. Direct Effect of Estrogens on the Vascular Wall

For years, the cardioprotective effect of estrogens has been explained by the favorable influence on the lipid profile (OLIVER and BOYD 1961; CLARKSON et al. 1990; BASS et al. 1993; MAUTNER et al. 1993). However, the results of the Lipid-Research-Clinics program (BUSH et al. 1987) and other newer data (STAMPFER and COLDITZ 1991) suggest that the changes of the lipid profile participate about 20–30% in the reduction of the cardiovascular risk, and it has been accepted that the direct estrogenic effect at the arterial wall is responsible for 70–80% of the cardioprotection by estrogen replacement therapy and is the decisive mechanism of action (Fig. 3).

Since the second half of the 18th century, it has been known that estrogens influence tonus and activity of blood vessels. In 1884, McKENZIE reported that menstrual cycle and pregnancy influence blood perfusion, vascularization and hyperemia of the mucosa. In 1940, REYNOLDS and FOSTER observed, in ovariectomized rabbits, that estrogens induce a dilatation of the small vessels of the ear. At the same time, it has been recognized that estrogen deficiency induces a premature and pronounced atherosclerosis in women. UELAND and PARER (1966) published that there is a decrease of systemic vascular resistance in pregnancy. SILVA and MEIRELLE (1977) studied the dilatation of isolated

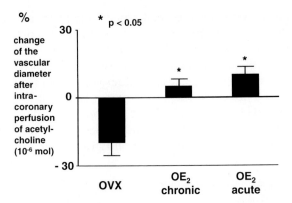

Fig. 3. Change of the vascular diameter after intracoronary perfusion of acetylcholine (10^{-6} M) in ovariectomized animals without (*ovx*) or with (*OE₂*) chronic or acute administration of estradiol (Williams et al. 1990)

estrogen-perfused human umbilical vessels and forwarded the hypothesis of a direct estrogen effect at the vascular wall.

In the earliest studies, high amounts of estrogens were used, resulting in estradiol plasma levels of +300 pg/ml. Less data are available on the effect of low-dose estrogen administration reaching physiological plasma values of 20–100 pg/ml, occurring during the normal menstrual cycle.

The vascular endothelial cells are important for the regulation of vascular motility and the maintenance of the reactivity of the vascular structure (Miller et al. 1988; Herrington et al. 1994; Fig. 3). The well-balanced endothelial secretion of dilating and constricting factors and the equilibrated production of growth promotors and inhibitors control the vascular response. Estrogens influence the vascular reactivity mainly through the stimulation of nitric oxide (NO). NO is secreted by endothelial cells and is probably the most important vasoactive factor produced by the endothelium. In the female rabbit, the endothelium-dependent vascular relaxation is more pronounced than in the male. This difference disappears after gonadectomy (Giscland et al. 1987, 1988). Similar observations have been published from studies in primates and in man (Williams et al. 1990, 1992; Reis et al. 1994) suggesting, again, that the synthesis of NO might be regulated by sex steroids (Weiner et al. 1994).

The synthesis of NO from its precursor l-arginine requires the endothelial enzyme NO synthetase (eNOS). This step is calcium- and calmodulin dependent (Zhang et al. 1994). NO is a potent vasodilator and an inhibitor of platelet function. The action of estrogens on NO is mediated within the endothelium itself by an increase in the production or a decrease in the metabolism of NO (Miller et al. 1988; Ylikorkala et al. 1995). Hayashi et al. (1992) have demonstrated that isolated rings of the aorta with an intact endothelium from healthy, intact females exhibit a higher basal NO secretion than aortic

Estrogen

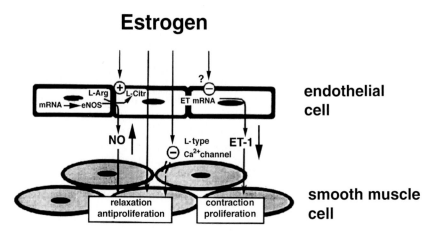

Fig. 4. Estrogen effect on the arterial wall. *eNO*, endothelial NO-synthase; *L-Ar*, L-arginine; *ET*, endothelin; *L-Citr*, L-citruline (adapted from YANG et al. 1996)

rings from ovariectomized females or male animals. In pregnant rats, the secretion of NO is increased (CONRAD et al. 1993). Acetylcholine stimulates the secretion of NO by the intact endothelium within only 20 min of administration (WILLIAMS et al. 1992; REIS et al. 1994). By the stimulation of intracellular cyclic guanosine monophosphate (cGMT) (WILLIAMS et al. 1992; REIS et al. 1994), NO may inhibit indirectly the proliferation and migration of smooth muscle cells. WILLIAMS et al. (1990, 1992, 1994) demonstrated that estrogens delay the progression of a coronary atherosclerosis and increase the endothelium-dependent relaxation in ovariectomized macaques with normal cholesterol levels as well as in monkeys with preexisting atherosclerosis, supporting the hypothesis of an endothelial-mediated estrogen effect directly at the arterial level.

Similar data have been reported in postmenopausal women (ROSANO et al. 1993). Whereas in healthy blood vessels, acetylcholine acts as a vasodilatator, atherosclerotic estrogen-deficient arteries react with a paradoxical vasoconstriction. Estrogens neutralize this paradoxical vasoconstriction and enable an endothelium- and NO-independent relaxation, corresponding to the normal reaction to acetylcholine by healthy arteries (LUDMER et al. 1986; ADAMS et al. 1990; CONRAD et al. 1993; WEINER et al. 1994). Plasma estradiol concenrations greater than 70 pg/ml are able to protect atherosclerotic arteries from the paradoxical vasoconstriction induced by acetylcholine. However, estradiol concentrations lower than 70 pg/ml fail to induce cardioprotection. In postmenopausal women, the administration of estrogens maintains the vasodilating effect of acetylcholine on atheroclerotic coronary arteries (HERRINGTON et al. 1994; REIS et al. 1994). Therefore, estrogens lead to a vasodilation of atherosclerotic arteries and decrease the incidence of angina pectoris.

However, endothelin, produced directly by the endothelial cell, is an extremely powerful local vasoconstrictor (Yanagisawa et al. 1988; Ylikorkala et al. 1995). Estrogens increase NO production and inhibit simultaneously the production of endothelin. Jiang et al. (1992) reported that estrogens modulate the vasoconstrictor response to endothelin-1; estrogens diminish the activity of endothelin-1. This inhibition of endothelin-1 might participate in the favorable effect of estrogens on the cardiovascular system. Men treated with estrogens possess lower endothelin concentrations. In contrast, women treated with testosterone have higher plasma endothelin levels. In male-to-female transsexuals, there is a decrease of endothelin levels after the beginning of estrogen administration (Poldermann et al. 1993). These clinical observations add further evidence to the close relationship between estrogens and endothelin.

A second estrogenic effect is known to act on smooth muscle cells and on coronary microvascularization. Harder and Coulson (1979) reported that smooth muscle cells of the vascular wall react to estrogens. Jiang et al. (1991) demonstrated that estrogens induce an endothelium-independent dilation through a regulatory activity on the ion channels. Estrogens might block a vasoconstriction following the administration of different vasoconstricting agents through a direct relaxing effect on the vascular smooth muscle cells (Jiang et al. 1991). In postmenopausal women with atherosclerosis, a prolonged estrogen-replacement therapy ameliorates the prognosis of a coronaropathy without any specific diet.

Prostacyclin, a potent vasodilator and inhibitor of thrombocytes (Moncada and Vane 1980), is another important physiological factor secreted by the vascular endothelium. Estrogens stimulate basal prostacyclin secretion (Chang et al. 1980) leading again to vascular dilation. In women undergoing estrogen-replacement therapy, prostacyclin is upregulated by an estrogen-mediated induction of apolipoprotein A1, which has a stabilizing effect on prostacyclin (Mikkola et al. 1995).

All these results lead to the conclusion that estrogens influence the vascular tonus directly. This effect must be induced genomically because of its short reaction time. However, it is still open to question whether the estrogen-induced effect on vascular reactivity is mediated exclusively through the endothelial cells, through smooth muscle cells or both. Probably, the mechanism of action does not only depend on the dose of the estrogens and, therefore, on its concentration, but also on the extent of the arterial disease. The second mechanism of action via smooth muscle cells might be preferential in atherosclerotic arteries or in the presence of high pharmacological estrogen concentrations, influencing the function of smooth muscle cells through calcium channels (Zhang et al. 1994), through an endothelin-induced response (Jiang et al. 1992), or through a direct hyperpolarization (Chang et al. 1980). In contrast, physiological estradiol concentrations most likely mediate vasodilation through endothelial cells, both in healthy and in atherosclerotic arteries (Hayashi et al. 1992).

In abnormal atherosclerotic arteries, estrogens most likely act directly on muscular cells because the endothelial function is clearly diminished in atherosclerosis (40). Vascular reactivity decreases in parallel to the deterioration of endothelial function. Although the exact mechanism is not known, estrogens may ameliorate the vasodilation induced by smooth muscle cells in atherosclerosis (WILLIAMS et al. 1992; ROSANO et al. 1993; COLLINS et al. 1995).

New data show that there are two estrogen receptors: estrogen receptors alpha and beta (KATZENELLENBOGEN and KORACH 1997; KUIPER and GUSTAFSSON 1997; KUIPER et al. 1997; PAECH et al. 1997; WATANABE et al. 1997; FOEGH and RAMWELL 1998). The two receptors are not equally distributed within the body. Some tissues and organs possess both receptors, others only one of them. The cardiovascular system expresses beta-subtype receptors. Some antiestrogens, such as tamoxifen (WATANABE et al. 1997), and selective estrogen receptor modulators (SERMs), such as raloxifene (DODGE et al. 1997), can be used in a selective way; because they bind specifically to an estrogen receptor subtype, they can induce a specific estrogen activity only at some selected organs. For example, raloxifene, the first SERM accepted for routine clinical use, has a protective estrogen activity on bone metabolism and on the cardiovascular system, but no activity at the endometrium or on breast tissue.

C. Influence of Progesterone and Progestagens

Preliminary results suggest that progesterone itself, given at a high dose, may induce vasodilation (JIANG et al. 1992). In contrast, some synthetic progestagens, such as medroxyprogesterone-acetate (MPA), may induce the opposite reaction if they are used in a dose corresponding to that used for HRT in healthy postmenopausal women (WILLIAMS et al. 1994). The effects of synthetic progestagens are heterogeneous. Therefore, the adjunction of a synthetic progestagen to an estrogen modulates the estrogen-induced changes of vascular reactivity as a function of the type and dose of progestagen used. Like progesterone (ADAMS et al. 1990; FALKEBORN et al. 1992; PSATY et al. 1994; SAMSIOE and ASTEDT 1996), dydrogesterone does not neutralize the protective effect of estradiol (GODSLAND et al. 1995). The results of the HERS study (HULLEY et al. 1998) might have been determined by the choice of the progestagen, MPA. If this hypothesis is true, it has to be shown by other trials. Although the classical nortestosterone derivative norethisterone-acetate (NETA) results in a less favorable lipid pattern than 17-hydroxy-progesterone derivatives, studies in primates show that NETA does not counteract the protective effect of estrogens on coronary arteries (WILLIAMS et al. 1990, 1992). These observations relate the practical relevance of the presence or absence of androgenic partial effects in progestagens for cardioprotection. The HERS study suggests that other factors and properties of progestagens might be more important.

D. Conclusion

The influence of estrogens and progestagens on the vascular system can be summarized as follows:

1. The increase of triglycerides and VLDL induced by peroral estrogen administration is undesirable. It can be avoided by transdermal estrogen administration. The LDL decrease varies between 11% and 15% if 0.625 mg conjugated equine estrogens or 2 mg estradiol valerate are administered perorally; with transdermal administration it is less. Estrogens increase HDL cholesterol values, particularly HDL2. The increase of HDL varies between 10% and 15% if 0.625 mg conjugated equine estrogen or 2 mg 17β-estradiol is administered perorally. Recent data show that women with total cholesterol levels higher than 240 mg/dl, but combined with HDL levels higher than 50 mg/dl, do not possess an increased cardiovascular risk. In contrast, the risk is increased in women with total cholesterol levels of less than 200 mg/dl combined with HDL levels of less than 50 mg/dl. In women with a HDL level of +50 mg/dl, the cut-off point of 50 mg/dl for HDL levels has the same predictive value as a HDL level of +35 mg/dl in men (Bush et al. 1987; Bass et al. 1993). The analysis of the observed LDL levels in three different groups (A: less than 130, B: 130–159 and C: 160 and more mg/dl) show that in women, higher HDL levels are cardioprotective, independent of the LDL- and triglyceride levels measured. Estrogens modify the serum lipid pattern by decreasing the ratio of LDL cholesterol/HDL cholesterol and the serum levels of lipoprotein(a). Both changes are considered to be cardioprotective.
2. The administration of estrogens leads to an arterial dilation through a direct effect on the vascular wall. This effect is maintained in atherosclerotic coronary arteries. Estradiol prevents the paradoxical vasoconstriction induced by acetylcholine in ovariectomized monkeys and in postmenopausal women. The sublingual administration of estradiol overcomes a coronary vasospasm within 10 min (Sarrel et al. 1992; Rosano et al. 1993). Estrogens decrease directly the risk of vasospasms and angina pectoris in women with a high cardiovascular risk secondary to estrogen deficiency (Grodstein et al. 1997).

References

Adam S, Williams V, Vessey MP (1981) Cardiovascular disease and hormone replacement treatment: a pilot case-control study. BMJ 282:1277–1278

Adams MR, Kaplan JR, Manuck SB, et al. (1990) Inhibition of coronary artery atherosclerosis by 17β-estradiol in ovariectomized monkeys. Lack of an effect of added progesterone. Arteriosclerosis 10:1051–1057

Avila MH, Walker AM, Jick H (1990) Use of replacement estrogens and the risk of myocardial infarction. Epidemiology 1:128–135

Bass KM, Newschaffer CJ, Klag MJ, Bush TL (1993) Plasma lipoprotein levels as predictors of cardiovascular death in women. Arch Intern Med 153:2209–2216

Bush TL (1990) Noncontraceptive estrogen use and risk of cardiovascular disease: an overview and critique of the literature. In: Korenman SG (ed) The menopause Serono symposium, Nowell (NIA)

Bush TL, Barrett-Conner E, Cowan LD et al. (1987) Cardiovascular mortality and non-contraceptive use of estrogen in women: results from the Lipid Research Clinics Program Follow-up Study. Circulation 75:1102–1109

Chang WC, Nakao J, Orimo H, Murota Sl (1980) Stimulation of prostaglandin cyclo-oxygenase and prostacyclin synthetase activities by estradiol in rat aortic smooth muscie cells. Biochim Biophys Acta 620:472–482

Clarkson TB, Adams MR, Williams JK, et al. (1990) Effects of sex steroids on the monkey cardiovascular system: relation to changes in serum lipids and lipoproteins. In: Christiansen C, Overgaard K (eds) Osteoporosis, pp 1798–1805

Collins P, Rosano GMC, Sarrel PM, et al. (1995) 17β-oestradiol attenuates acetylcholine-induced coronary arterial constriction in women but not in men with coronary heart disease. Circulation 92:24–30

Conrad KP, Joffe GM, Kruszyna H, et al. (1993) Identification of increased nitric oxide biosynthesis during pregnancy in rats. FASEB J 7:566–571

Department of Health, Education and Welfare (1979) Vital statistics of the United States, II. Mortality, part B. U.S. Department of Health, Education and Welfare, Washington, DC, publication NIH-79-1102

Eaker ED, Castelli WP (1987) Differential risk for coronary heart disease among women in the Framingham Study. In: Eaker E, Packard B, Wenger N, et al. (eds) Coronary heart disease in women. Haymarket Doyma, New York, pp 122–130

Ettinger B (1987) Overview of the efficacy of hormonal replacement therapy. Am J Obstet Gynecol 156:1298–1303

Falkeborn M, Persson I, Adami H, et al. (1992) The risk of acute myocardial infarction after estrogen and estrogen-progestogen replacement. Br J Obstet Gynaecol 99:821–828

Foegh ML, Ramwell PW (1998) Cardiovascular effects of estrogen: implications of the discovery of the estrogen receptor subtype beta. Curr Opin Nephrol Hypertens 7:83–89

Gisclard V, Flavahan NA, Vanhoutte PM (1987) Alpha adrenergic responses of blood vessels of rabbits after ovariectomy and administration of 17β-estradiol. J Pharmacol Exp Ther 240:466–470

Gisclard V, Miller VM, Vanhoutte PM (1988) Effect of 17β-estradiol on endothelium-dependent responses in the rabbit. J Pharmacol Exp Ther 244:19–22

Godsland, Crook D, Godsland IF, Stevenson JC (1995) The cardiovascular risk profile of hormone replacement therapy containing dydrogesterone: a review. Eur Menopause J 2[Suppl 4]:23–30

Grady D, Rubin SM, Petitti DB, et al. (1992) Hormone therapy to prevent disease and prolong life in postmenopausal women. Ann Intern Med 117:1016–1037

Grodstein F, Stampfer MJ, Colditz GA, et al. (1997) Postmenopausal hormone therapy and mortality. N Engl J Med 336:1769–1775

Haarbo J, Spensen PL, Stender S, et al. (1991) Estrogen monotherapy and combined estrogen–progestogen replacement therapy attenuate aortic accumulation of cholesterol in overectomized cholesterol-fed rabbits. J Clin Invest 87:1274–1279

Haenggi W, Lippuner K, Riesen W, Jaeger Ph, Birkhäuser MH (1997) Long-term influence of different hormone replacement regimens on serum lipids and lipoprotein(a): a randomised study. Br J Obstet Gynaecol 104:708–717

Harder DR, Coulson PB (1979) Estrogen receptors and effects of estrogen on membrane electrical properties of coronary vascular smooth muscle. J Cell Physiol 100:375–382

Hayashi T, Fukuto JM, Ignarro LJ, Chaudhuri G (1992) Basal release of nitric oxide from aortic rings is greater in female rabbits than in male rabbits: implications for atherosclerosis. Proc Natl Acad Sci USA 89:11259–11263

Henderson BE, Paganini-Hill A, Ross RK (1991) Decreased mortality in users of estrogen replacement therapy. Arch Intern Med 151:75–78

Herrington DM, Braden GA, Williams JK, Morgan TM (1994) Endothelial-dependent vasomotor responsiveness in postmenopausal women with and without estrogen replacement therapy. Am J Cardiol 73:951–952

Hulley St, Grady D, Bush T, Furberg C, Herrington D, Riggs B, Vittinghoff E, for the Heart and Estrogen/Progestin Replacement Study (HERS) Research Group (1998) Randomized trial of estrogen plus progestin for secondary prevention of coronary heart disease in postmenopausal women. JAMA 280:605–613

Hunt K, Vessey M, McPherson K (1990) Mortality in a cohort of long-term users of hormone replacement therapy: an updated analysis. Br J Obstet Gynaecol 97:1080–1086

Jiang CW, Sarrel PM, Lindsay DC, Poole-Wilson PA, Collins P (1991) Endothelium-independent relaxation of rabbit coronary artery by 17β-oestradiol in vitro. Br J Pharmacol 104:1033–1037

Jiang CW, Sarrel PM, Lindsay DC, Poole-Wilson PA, Collins P (1992a) Progesterone induces endothelium-independent relaxation of rabbit coronary artery in vitro. Eur J Pharmacol 211:163–167

Jiang CW, Sarrel PM, Poole-Wilson PA, Collins P (1992b) Acute effect of 17β-estradiol on rabbit coronary artery contractile responses to endothelin-l. Am J Physiol 263:H271–H275

Johansson BW, Kaü L, Kullander S, Lenner H-C, Svanberg L, Astedt B (1975) On some late effects of bilateral oophorectomy in the age range 15–30 years. Acta Obstet Gynecol Scand 54:449–461

Kannel WV, Hjortland M, McNamara PM, et al. (1976) Menopause and risk of cardiovascular disease: the Framingham study. Ann Intern Med 85:447–552

Katzenellenbogen BS, Korach KS (1997) A new actor in the estrogen receptor drama – enter ER-beta (editorial). Endocrinology 138:861–862

Kuiper CG, Gustafsson JA (1997) The novel estrogen receptor-beta subtype: potential role in the cell- and promoter-specific actions of estrogens and anti-estrogens. FEBS Lett 410:87–90

Kuiper CG, Carlsson B, Grandien K, Enmark E, Haggblad J, Nilsson S, Gustafsson JA (1997) Comparison of the ligand binding specificity and transcript tissue distribution of estrogen receptors alpha and beta. Endocrinology 138:863–870

La Vecchia C, Franceschi S, Decarli AS, et al. (1987) Risk factors for myocardial infarction in young women. Am J Epidemiol 125:832–843

Ludmer PL, Selwyn AP, Shook TL, et al. (1986) Paradoxical vasoconstriction induced by acetylcholine in atherosclerotic coronary arteries. N Engl J Med 315:1046–1051

MacKenzie JN (1884) Irritation of the sexual apparatus. Am J Med Sci 87:360–365

Mautner SL, Lin F, Mautner GC, Roberts WC (1993) Comparison in women versus men of composition of atherosclerotic plaques in native coronary arteries and in saphenous veins used as aortocoronary conduits. J Am Coll Cardiol 71:1312–1318

McDowell F, Louis S, McDevitt E (1967) A chemical trial of Premarin in cerebrovascular disease. J Chronic Disease 2:679–684

Mikkola T, Turunen P, Avela K, Orpana A, Viinikka L, Ylikorkala O (1995) 17β-estradiol stimulates prostacyclin, but not endothelin-l, production in human vascular endothelial cells. J Clin Endocrinol Metab 80:1832–1836

Miller VM, Gisclard V, Vanhoutte PM (1980) Modulation of endothelium-dependent and vascular smooth muscle responses by estrogens. Phlebology (1988) 224:19–22

Moncada S, Vane JR (1980) Prostacyclin in the cardiovascular systems. Adv Prostaglandin Thromboxane Leukot Res 6:43–60

Nabulsi AA, Aaron B, Folsom R, et al. (1993) Association of hormone-replacement therapy with various cardiovascular risk factors in postmenopausal women. N Engl J Med 328:1070–1075

National Institutes of Health (1992) Chartbook on cardiovascular, lung and blood diseases. US Department of Health and Human Services, Washington DC

Oliver MF, Boyd GS (1961) Influence of reduction of serum lipids on prognosis of coronary heart disease: a five-year study using estrogens. Lancet 2:499–505

Paech K, Webb P, Kuiper CG, Nilsson S, Gustafsson J, Kushner PJ, Scanlan TS (1997) Differential ligand activation of estrogen receptors ERalpha and ERbeta at AP1 sites. Science 277:1508–1510. (see comment on p 1439

PEPI Trial Writing Group (1995) Effects of estrogen or estrogen/ progestin regimes on heart disease risk factors in postmenopausal women: the postmenopausal Estrogen/Progestin Interventions (PEPI) Trial. JAMA 273:199–208

Petitti DB, Perlman JA, Sidney S (1987) Noncontraceptive estrogens and mortality: long-term follow-up of women in the Walnut Creek study. Obstet Gynecol 70:289–293

Pick R, Stamler J, Rodbard S, Katz LN (1952) Estrogen-induced regression of coronary atherosclerosis in cholesterol-fed chicks. Circulation 6:276–280

Polderman KH, Stehouwer CD, van Kamp GJ, Dekker GA, Verheugt FW, Gooren LJ (1993) Influence of sex hormones on plasma endothelin levels. Ann Intern Med 118:429–432

Psaty BM, Heckbert SR, Atkins D, et al. (1994) The risk of myocardial infarction associated with the combined use of estrogens and progestins in postmenopausal women. Arch Intern Med 154:1333–1339

Reis SE, Gloth ST, Blumenthal RS, et al. (1994) Ethinyl estradiol acutely attenuates abnormal coronary vasomotor responses to acetylcholine in postmenopausal women. Circulation 89:5260

Reynolds SRM, Foster Fl (1940) Peripheral vascular action of estrogen observed in the ear of the rabbit. J Pharmacol Exp Ther 68:173–177

Rijpkema AHM, van der Sanden AA, Ruys AHC (1990) Effects of postmenopausal estrogen-progestogen replacement therapy on serum lipids and lipoproteins: a review. Maturitas 12:259–285

Rosano GM, Sarrel PM, Poole-Wilson PA, Collins P (1993) Beneficial effect of estrogen on exercise-induced myocardial ischaemia in women with coronary artery disease. Lancet 342:133–136

Rosenberg L, Palmer JR, Shapiro S (1993) A case-control study of myocardial infarction in relation to use of estrogen supplements. Am J Epidemiol 137:54–63

Ross RK, Paganini-Hill A, Mack TM, Henderson BE (1989) Cardiovascular benefits of estrogen replacement therapy. Am J Obstet Gynecol 160:1301–1306

Samsioe G, Astedt B (1996) Does progesterone co-medication attenuate the cardiovascular benefits of estrogens? In: Birkhäuser MH, Rozenbaum H (eds) Menopause. Proceedings of the European Consensus Development Conference. èditions ESKA, Paris pp 175–183

Sarrel PM, Lindsay D, Rosano GMC (1992) Angina and normal coronary arteries: gynecologic findings. Am J Obstet Gynecol 167:467–472

Silva de Sa MF, Meirelles RS (1977) Vasodilating effect of estrogen on the human umbilical artery. Gynecol Invest 8:307–313

Stampfer MJ, Colditz GA (1991) Estrogen replacement therapy and coronary heart disease: a quantitative assessment of the epidemiologic evidence. Prev Med 20:47–63

Stampfer MJ, Willett WC, Colditz GA, et al. (1985) A prospective study of postmenopausal estrogen therapy and coronary heart disease. N Engl J Med 313:1044–1049

Szklo M, Tonascia J, Gordis L, et al. (1984) Estrogen use and myocardial infarction risk: a case-control study. Prev Med 13:510–516

The Coronary Drug Project (1970) Initial findings leading to modifications of its research protocol. JAMA 714:1303–1313

Ueland K, Parer JT (1966) Effects of estrogens on the cardiovascular system of the ewe. Am J Obstet Gynecol 96:400–406

Walsh BW, Schiff I, Rosner B, Greenberg L, Ravnikar V, Sacks FM (1991) Effects of postmenopausal estrogen replacement therapy on the concentrations and metabolism of plasma lipoproteins. N Engl J Med 325:1196–203

Watanabe T, Inoue S, Ishii Y, Hiroi H, Ikeda K, Orimo A, Muramatsu M (1997) Agonistic effect of tamoxifen is dependent on cell type. ERE-promoter context, and estrogen receptor subtype: functional difference between estrogen receptors alpha and beta. Biochem Biophys Res Commun 236:140–145

Weiner CP, Lizasoain I, Baylis SA, Knowles RG, Charles IG, Moncada S (1994) Induction of calcium-dependent nitric oxide synthetases by sex hormones. Proc Natl Acad Sci U S A 91:5212–5216

Wenger NK, Speroff L, Packard B (1993) Cardiovascular health and disease in women. N Engl J Med 329:247–256

Williams JK, Adams MR, Klopfenstein HS (1990) Estrogen modulates responses of atherosclerotic coronary arteries. Circulation 81:1680–1687

Williams JK, Adams MR, Herrington DM, Clarkson TB (1992) Shor-term administration of estrogen and vascular responses of atherosclerotic coronary arteries. J Am Coll Cardiol 20:452–457

Williams JK, Adams MR, Herrington DM, Clarkson TB (1994a) Short-term administration of estrogen and vascular responses of atherosclerotic coronaries. J Am Coll Cardiol 20:452–457

Williams JK, HonorÇ EK, Washburn SA, Clarkson TB (1994b) Effects of hormone replacement therapy on reactivity of atherosclerotic coronary arteries in cynomolgus monkeys. J Am Coll Cardiol 24:1757–1761

Wilson PW, Garrison RJ (1985) Postmenopausal estrogen use, cigarette smoking, and cardio-vascular morbidity in women over 50. The Framingham Study. N Engl J Med 313:1038–1043

Wuest J, Dry T, Edwards J (1953) The degree of coronary atherosclerosis in bilaterally oophorectomized women. Circulation 7:801–809

Yanagisawa M, Kurihara H, Kimura S, et al. (1988) A novel potent vasoconstrictor peptide produced by vascular endothelial cells. Nature 332:411–415

Yang Z, Do Do-Dai, Spinosa E, Barton M, Arnet U, Lüscher ThF (1996) 17β-Oestradiol inhibits growth of human vascular smooth muscle: similar effects in cells from females and males. J Cardiovasc Pharmacol 28[Suppl 5]:S34–S39

Ylikorkala O, Orpana A, Puolakka J, Pyorala T, Viinikka L (1995) Postmenopausal hormonal replacement decreases plasma levels of endothelin-l. J Clin Endocrinol Metab 80:3384–3387

Zhang F, Ram JL, Standley PR, Sowers JR (1994) 17β-oestradiol attenuates voltage-dependent Ca^{2+} currents in A7r5 vascular smooth muscle cell line. Am J Physiol 266:C975–C980

CHAPTER 17
Immune System

E. Nagy, E. Baral, and I. Berczi

A. Introduction

Selye (1942) was the first to point out that all hormonally active steroids will produce thymic involution. Other studies indicated that gonadal secretions exert a constant moderating effect on the growth of lymphatic organs (Dougherty 1952). Later, estrogens were found to increase host resistance to parasites (Thorson 1970). Space limitations do not allow an exhaustive coverage of the subject here and, therefore, the interested reader is referred to reviews and monographs for additional details (Berczi 1986; Berczi and Nagy 1997; Berczi et al. 1993; Baral et al. 1994, 1996; Cutolo et al. 1993; Kincade et al. 1994; Lahita 1996; Vamvakopoulos and Chrousos 1994; Wilder 1996).

B. Effect On Primary Lymphoid Tissue

The bone marrow, thymus and the bursa of Fabricius in birds are designated as primary lymphoid organs concerned with the generation of bone marrow- or, in birds, bursa-derived (B) lymphocytes and of thymus-derived (T) lymphocytes. Monocyte-macrophages, neutrophilic-, eosinophilic- and basophilic granulocytes, which are also integrated into the immune system, are produced and released by the bone marrow (Roitt et al. 1989).

I. The Effect of Estrogens

1. The Bone Marrow

Classical estrogen receptors (ERα) are present in stromal cells and in multi-potential stem cells of the bone marrow. B-cell precursors and the epithelial cells of the bursa of Fabricius also express ERs (Berczi 1986; Bellido et al. 1993; Kincade et al. 1994; Shevde and Pike 1996; Smithson et al. 1995). Estrogens have a suppressive effect on the bone marrow, including the multi-potential stem cells, B-lymphocyte precursors, polymorphonuclear leukocytes and the cells that differentiate into osteoclasts. This inhibitory effect is exerted by estrogens via action on stromal elements of the bone marrow by the alteration of cytokine and other mediator production, e.g., interleukin (IL)-6, IL-1, tumor-necrosis factor alpha (TNF-α) and prostaglandin (PG) E2, as well as

by direct action on target cells, e.g., direct induction of apoptosis in myelopoietic progenitors (PASSERI et al. 1993; KINCADE et al. 1994; KAWAGUCHI et al.
1995; STEIN and YOUNG 1995; SHEVDE and PIKE 1996; BISMAR et al. 1995; SMITH-
SON et al. 1995).

2. The Thymus

In the thymus, ERα and a variant form, ERβ, were detected. Cytosolic ER
from rat thymus had similar steroid-binding specificity to uterine ERs, but
showed a higher affinity to diethylstilbestrol and cortisone (BERCZI 1986;
BRIDGES et al. 1993; MOSSELMAN et al. 1996).

Thymic sensitivity to sex hormones develops postnatally. Estradiol has a
suppressive effect on the thymus that manifests in decreased weight and cel-
lularity, and in the inhibition of the mitogenic response by thymocytes. Acute
treatment of mice with pharmacological doses of 17β-estradiol (E2) causes the
selective disappearance of CD4+ CD8+ thymocytes. Estrogen blocks T-cell
development in the thymus (BERCZI 1986; NAKAYAMA et al. 1996; RIJHSINGHAMI
et al. 1996). In males, testosterone causes thymus involution after conversion to E2 by aromatase, which can be reversed by aromatase inhibitors
(GREENSTEIN et al. 1992). The development of lympho-hematopoietic progenitors in the gut and the extra-thymic differentiation of T cells in the liver is
stimulated by estrogens (KANAMORI et al. 1996; OKUYAMA et al. 1992; NAKAYAMA
et al. 1996).

C. Effect on Immune Function

I. The Effect of Estrogens

1. Receptors

ERα is present in macrophages, CD8+T lymphocytes and B lymphocytes.
ERβ is also expressed in lymphoid tissue, and membrane-bound receptors may
also be present. Anti-estrogen binding sites (AEBS), which do not bind estrogens, were also detected in lymphoid cells (BARAL et al. 1996; MOSSELMAN
et al. 1996).

2. Cytokines

In mice, endotoxin-induced TNF was significantly increased by estrogens,
while serum IL-6 levels were decreased by 17α ethinyl estradiol (ZUCKERMAN
et al. 1995, 1996). Activated ERs repress the IL-6 promoter by acting on the
nuclear transcription factors, NF-IL6, NF-κB and C/EBPβ (STEIN and YANG
1995; RAY et al. 1997).

In cultured human monocytes, IL-1 receptor antagonist messenger RNA
(mRNA) expression was significantly elevated by physiological levels
(10^{-11} M) of estradiol, while expression was suppressed by higher concen-

trations (LEE et al. 1995). Estradiol 17β induced IL-5 mRNA in a murine T-cell line in vitro (WANG et al. 1993). Estradiol and glucocorticoids both strongly inhibited the activation of the human leukemia inhibitory factor promoter by phorbol ester and ionomycin (BAMBERGER et al. 1997).

3. Humoral Immunity

In general, estrogens enhance the humoral immune response, although antibody formation to certain antigens may actually be suppressed. Estradiol stimulates the secretion of immunoglobulin (IgA and IgG) from the uterus in rats. Estrogens interact synergistically with prolactin (PRL) in the initiation of Ig secretion from the mammary gland (BERCZI 1986; BERCZI and NAGY 1997).

Antigen presentation by uterine epithelial cells was stimulated, and by stromal uterine cells was inhibited, by E2 (WIRA and ROSOLL 1995). Estradiol treatment of mice of various strains for 31 weeks, but not for 4 weeks, significantly increased the frequency of cells producing IgG and IgM (NILSSON and CARESTEN 1994).

4. Cell-Mediated Immunity

Estrogens prolonged skin-graft survival in mice, reduced DNA synthesis markedly, and blocked the generation of cytotoxic lymphocytes in mixed-lymphocyte reactions. Estradiol reduced natural killer (NK) cell activity in various mouse strains that was mediated by suppressor cells of the monocyte-macrophage lineage (BERCZI 1986; NILSSON and CARLSTEN 1994). In contrast, NK activity of human large granular lymphocytes was enhanced after treatment in vitro with 17β-estradiol (SORACHI et al. 1993). Estradiol enhanced the NK cell susceptibility of human breast cancer cells, which was dependent on the presence of classical ERs (SCREPANTI et al. 1991; ALBERTINI et al. 1992). The cytotoxic activity of murine CD4+T cells was dependent on estrogen (MULLER et al. 1995).

Estradiol and dexamethasone showed an additive effect on suppression of delayed-type hypersensitivity reactions in mice. Similar additive effects were not observed in granulocyte-mediated inflammation (CARLSTEN et al. 1996).

5. Inflammation

The chemotactic response of human-polymorphonuclear leukocytes was significantly reduced by physiological concentrations of estradiol (10^{-10}M) (ITO et al. 1995). Granulocyte-mediated inflammation was not influenced in mice by estradiol treatment (CARLSTEN et al. 1996). The degradation of cartilage implants by inflammatory tissue was higher in ovariectomized female mice than in orchidectomized male mice. Degradation was inhibited by estrogen replacement in females and androgen replacement in males (DA-SILVA et al. 1994).

II. The Effect of Anti-Estrogens

1. Lymphocytes

At therapeutic concentrations, tamoxifen (TX) exerted an anti-proliferative effect on lymphocytes (Baral et al. 1989). TX inhibited the expression of C'3 complement receptors by human peripheral B lymphocytes (Baral et al. 1985). Anti-estrogens [TX, toremifene (TO) and ICI 164 384] were reported to enhance or inhibit Ig secretion by human lymphocytes in vitro (Paavonen and Anderson 1993; Teodorczyk-Injeyan et al. 1993).

In vitro treatment of human peripheral-blood lymphocytes with TX, TO or ICI 164 384 suppressed mitogen-induced proliferation, IL-2 production and IL-2 receptor-α expression (Teodorczyk-Injeyan et al. 1993). TX and TO both stimulated the cytokine production [IL-1β, IL-4, -6, -10, TNF and interferon (IFN)] in a human B cell line (Ball) after induction by phorbol myristate acetate (PMA). In contrast, TX stimulated the production of IL-1β, IL-6 and IFN-γ by PMA-induced Molt-4 cells (a T-cell line), whereas TO was inhibitory (Jarvinen et al. 1996).

2. Humoral Immunity

Estradiol significantly augmented the antigen-specific antibody response in vitro that was due to the inhibition of CD8+T suppressor cells. Pre-incubation of the cells with TX prevented this augmentation (Clerici et al. 1991).

Treatment of rats with TX (6 mg/kg s.c.) during immunization with sheep red blood cells significantly inhibited the antibody response, which could be reversed by additional treatment with either PRL or growth hormone (GH) (Nagy and Berczi 1986). Treatment with TX (2–800 μg per mouse, twice per week) had no effect on autoantibody production in female Balb/c mice (Sthoeger et al. 1994).

3. Cell-Mediated Immunity

Contact-sensitivity skin reactions were inhibited in rats by TX treatment, and could be restored by additional treatment with either GH or PRL (Nagy and Berczi 1986). In vitro TX has a suppressive effect on the mixed lymphocyte reaction performed with rat spleen cells (Baral et al. 1991).

Treatment of mice with TX or TO for 6 weeks did not influence NK cell activity (Warri and Kangas 1990). In vitro TX antagonized the enhancing effect of estradiol on the NK activity of human large granular lymphocytes (Sorachi et al. 1993). In breast cancer patients, TX treatment was found to be associated with decreased (Rotstein et al. 1988) and increased NK activity (Berry et al. 1987; Robinson et al. 1993).

TX and TO sensitized ERα-positive and -negative target cells for lysis by NK, lymphokine-activated killer (LAK) and by cytotoxic T lymphocytes (CTL). Current evidence indicates that anti-estrogens facilitate the apoptotic pathway of cytolysis, which requires metabolic participation by the target cell.

A significant proportion (up to 60%) of human ovarian carcinomas and lung carcinomas could also be sensitized by TX and TO towards lysis by autologous killer cells. Moreover, in various murine syngeneic tumor-host systems, TX and TO treatment potentiated the immunotherapeutic effect of NK, LAK and CTL-effector cells and of IL-2 against lethal tumor inocula (LD_{100}) (MANDENVILLE et al. 1984; BARAL and VANKY 1987; BARAL et al. 1994, 1996, 1997a, b, c). TX treatment was also found to potentiate the anti-metastatic effect of IL-2 on the MCA-106 fibrosarcoma in C57Bl/6 mice (KIM et al. 1990).

4. Inflammation

Retroperitoneal fibrosis, which results from fibroblast proliferation and collagen deposition may be treated successfully with TX (CLARK et al. 1991; BENSON and BAUM 1993; SPILLANE and WHITMANN 1995).TX inhibited the carrageenan-induced inflammatory response in female LEW/N rats (MISIEWICZ et al. 1996).

D. Conclusions

Estrogens and anti-estrogens affect immune reactions directly, through specific receptors of lymphoid cells and, indirectly, by acting on the neuroendocrine system and on target cells or tissues of the immune system. Because the neuroendocrine and immune systems form a feedback regulatory network, immune alterations will lead to neuroendocrine changes and vice-versa (Fig. 1). This interaction is facilitated by shared mediators. For instance, it is now apparent that the thymus is a steroid hormone-producing organ. In addition, the immune system is capable of converting precursor steroids into active hormones (BERCZI and NAGY 1997).

There is little doubt that steroid hormones are amongst the most powerful immunoregulators, as they are capable of signal modulation at the level of nuclear transcription factors. Presently, however, it is unknown whether estrogens are required for normal immune function or there is a qualitative difference between the effect of basal and elevated levels of estrogen with regards to immunoregulation. We have little information regarding the possible differences in the immunoregulatory roles of various estrogenic steroids and of the different receptors. Estrogens appear to regulate immune function differently in females than in males.

It seems certain that estrogens are capable of shifting the immune response toward humoral immunity. According to current understanding, this should be advantageous for the survival of the fetus. It is emerging that estrogens and anti-estrogens regulate killer cell-induced apoptosis. This has relevance to immunoregulation and to the pathogenesis of autoimmune disease, chronic inflammatory disease and possibly to age-related immune abnormalities (BERCZI 1986, BERCZI et al. 1993; BERCZI and NAGY 1997; CUTOLO 1993; LAHITA 1996; WILDER 1996).

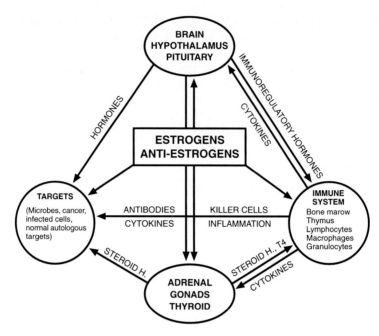

Fig. 1. Pathways of immunomodulation by estrogens and anti-estrogens

Likewise, the effect of anti-estrogens is complex (Fig. 1). Triphenylethylene-derived anti-estrogens show both anti-estrogenic and estrogenic properties and exert biological activities via a receptor other than the ER. There is also evidence to indicate that there are significant differences among the immune effects of various anti-estrogenic drugs (BARAL et al. 1996; MOSSELMAN et al. 1996). Much remains to be elucidated about the immunomodulatory function of estrogens, their agonists and antagonists.

References

Albertini MR, Gibson DFC, Robinson SP, Howard SP, Tans KJ, Lindstrom MJ, Robinson RR, Tormey DC, Jordan VC, Sondel PM (1992) Influence of estradiol and tamoxifen on susceptibility of human breast cancer cell lines to lysis by lymphokine activated killer cells. J Immunother 11:30–39

Bamberger AM, Erdmann I, Bamberger CM, Jenatschke SS, Schulte HM (1997) Transcriptional regulation of the human 'leukemia inhibitory factor' gene: modulation by glucocorticoids and estradiol. Mol Cell Endocrinol 127:71–79

Baral E, Blomgren H, Rotstein S, Virving L (1985) Antiestrogen effects on human lymphocyte subpopulations in vitro. J Clin Lab Immunol 17:33–35

Baral E, Kwok S, Berczi I (1989) Suppression of lymphocyte mitogenesis by tamoxifen. Immunophrmacology 18:57–62

Baral E, Kwok S, Berczi I (1991) The influence of estradiol and tamoxifen on the mixed lymphocyte reaction of rats. Immunopharmacology 21:191–198

Baral E, Nagy E, Berczi I (1994) Tamoxifen as an immunomodulatory agent. In: Berczi I, Szelényi J (eds) Advances in Psychoneuroimmunology, Hans Selye Symposia on Neuroendocrinology and Stress, vol. III. Plenum Press, New York, NY, USA, p 233

Baral E, Nagy E, Berczi I (1996) The effect of tamoxifen on the immune response. In: Kellen JA (ed) Tamoxifen. Beyond the antiestrogen. Birkhäuser, Boston, MA, USA, p 137

Baral E, Nagy E, Kangas L, Berczi I (1997a) Immunotherapy of the SL2–5 murine lymphoma with natural killer cells and tamoxifen or toremifene. Anticancer Res 17:77–84

Baral E, Nagy E, Kangas L, Berczi I (1997b) Combination immunotherapy of the H2712 murine mammary carcinoma with cytotoxic T lymphocytes and anti-estrogens. Anticancer Res 17:3647–3652

Baral E, Nagy E, Kangas L, Berczi I (1997c) Combination immunotherapy of the P815 murine mastocytoma with killer cells, IL-2 and anti-estrogens. Anticancer Res 17:3653–3658

Baral E, Vanki F (1987) Effect of tamoxifen on the cell-mediated auto-tumor lysis. J Clin Lab Immunol 22:97–100

Bellido T, Girasole G, Passeri G, Yu XP, Mocharla H, Jilka RL, Notides A, Manolagas SC (1993) Demonstration of estrogen and vitamin D receptors in bone marrow-derived stromal cells: up-regulation of estrogen receptors by 1,25-dihydroxyvitamin-D3. Endocrinology 133:553–562

Benson JR, Baum M (1993) Tamoxifen for retroperitoneal fibrosis (letter) Lancet 341:836

Berczi I (1986) Gonadotropins and sex hormones. In: Berczi I (ed) Pituitary function and immunity. CRC Press, Boca Raton, FL, USA, p 185

Berczi I, Nagy E (1997) Hormones as immune modulating agents. In: Krezina TF (ed) Immune modulating agents. Marcel Dekker, Inc. New York, NY, USA, p 75

Berczi I, Baragar FD, Chalmers IM, Keystone EC, Nagy E, Warrington RJ (1993) Hormones in self tolerance and autoimmunity: a role in the pathogenesis of rheumatoid arthritis? Autoimmunity 16:45–56

Berry J, Green BJ, Matheson DS (1987) Modulation of natural killer cell activity by tamoxifen in stage I postmenopausal breast cancer. Eur J Cancer Clin Oncol 23:517–520

Bismar H, Diel I, Ziegler R, Pfeilschifter J (1995) Increased cytokine secretion by human bone marrow cells after menopause or discontinuation of estrogen replacement. J Clin Endocrinol Metab 80:3351–3355

Bridges EF de, Greenstein BD, Khamastha MA, Hoghes GR (1993) Specificity of estrogen receptors in rat thymus. Int J Immunopharmacol 15:927–932

Carlsten H, Verdrengh M, Taube M (1996) Additive effects of suboptimal doses of estrogen and cortisone on the suppression of T lymphocyte dependent inflammatory response in mice. Inflamm Res 45:26–30

Clark CP, Vanderpool D, Preskitt JT (1991) The response of retroperitoneal fibrosis to tamoxifen. Surgery 109:502–506

Clerici E, Bergamasco E, Ferrario E, Villa ML (1991) Influence of sex hormones on the antigen-specific primary antibody response in vitro. J Clin Lab Immun 34:71–78

Cutolo M, Sulli A, Barone A, Seriolo B, Accardo S (1993) Macrophages, synovial tissue and rheumatoid arthritis. Clin Exp Rheumatol 11:331–339

Ito I, Hayashi T, Yamada K, Kuzuya M, Naito M, Iguchi A (1995) Physiological concentrations of estradiol inhibits polymorphonuclear leucocyte chemotaxis via a receptor mediated mechanism. Life Sci 56:2247–2253

Dougherty TF, (1952) Effects of hormones on lymphatic tissue. Physiol Rev 32:379–401

Greenstein BD, Bridges EF de, Fitzpatrick FT (1992) Aromatase inhibitors regenerate the thymus in aging rats. Int J Immunopharmacol 14:541–553

Jarvinen LS, Pyrhonen S, Kariemo KJA, Paavonen T (1996) The effect of antiestrogens on cytokine production in vitro. Scand J Immunol 44:15–20

Kanamori Y, Ishimaru K, Nanno M, Maki K, Ikuta K, Nariuchi H, Ishikawa H (1996) Identification of novel lymphoid tissue in murine intestinal mucosa where clusters of c-kit+IL-7R+Thy1+lympho-hemopoietic progenitors develop. J Exp Med 184:1449–1459

Kawaguchi H, Pilbeam CC, Vargas SJ, Morse EE, Lorenzo JA, Raisz LG (1995) Ovariectomy enhances estrogen replacement inhibits the activity of bone marrow factors that stimulate prostaglandins production in mouse calvariae. J Clin Invest 96:539–548

Kim B, Warnaka P, Konrad C (1990) Tamoxifen potentiates in vivo antitumor activity of interleukin-2. Surgery 108:139–145

Kincade PW, Medina KL, Smithson G (1994) Sex hormones as negative regulators of lymphopoiesis. Immunol Rev 137:114–134

Lahita RG (1996) The basis of gender effects in the connective tissue diseases. Ann Med Interne Paris 147:241–247

Lee BY, Huynh T, Prichard LE, McGuire J, Polan ML (1995) Gonadal steroids modulate interleukin-1 receptor antagonist mRNA expression in cultured human monocytes. Biochem Biophys Res Commun 209:279–285

Mandenville R, Ghalli SS, Chausseau J-P (1984) In vitro stimulation of human NK activity by an estrogen antagonist (tamoxifen). Eur J Cancer Clin Oncol 20: 983–985

Misiewicz B, Grieber C, Gomez M, Raybourne R, Zelazowska E, Gold PW, Sternberg EM (1996) The estrogen antagonist tamoxifen inhibits carrageenin induced inflammation in LEW/N female rats. Life Sci 58:PL281–286

Mosselman S, Polman J, Dijkema R (1996) ER beta: identification and characterization of a novel human estrogen receptor. FEBS Lett 392:49–53

Muller D, Chen M, Vikingsson A, Hildeman D, Pederson K (1995) Estrogen influences CD4+T-lymphocyte activity in vitro in β_2-microglobulin-deficient mice. Immunology 86:162–167

Nagy E, Baral E, Kangas L, Berczi I (1997) Anti-estrogens potentiate the immunotherapy of the P815 murine mastocytoma with cytotoxic T lymphocytes. Anticancer Res 17:1083–88

Nagy E, Berczi I (1986) Immunomodulation by tamoxifen and pergolide. Immunopharmacology 12:145–153

Nakayama M, Otsuka K, Sato K, Kasegawa K, Osman Y, Kawamura T, Abo T (1996) Activation by estrogen of the number and function of forbidden T cell clones in intermediate T cell receptor cells. Cell Immunol 172:163–171

Nilsson N, Carlsten H (1994) Estrogen induces suppression of natural killer cell cytotoxicity and augmentation of polyclonal B cell activation. Cell Immunol 158:131–139

Okuyama R, Abo T, Seki S, Ohteki T, Sugiura K, Kusumi A, Kumagai K (1992) Estrogen administration activates extrathymic T cell differentiation in the liver. J Exp Med 175:661–669

Passeri G, Girasole G, Jilka RL, Manolagas SC (1993) Increased interleukin-6 production by murine bone marrow and bone cells after estrogen withdrawal. Endocrinology 133:822–828

Paavonen T, Anderson LC (1985) The estrogen antagonists, Tamoxifen and FC-1157a, display estrogen like effects on human lymphocyte functions in vitro. Clin Exp Immunol 61:467–474

Ray P, Ghosh SK, Zhang DH, Ray A (1997) Repression of interleukin-6 gene expression by 17β-estradiol: inhibition of the DNA-binding activity of the transcription factors NF-IL6 and NF-κB by the estrogen receptor. FEBS Lett 409:78–85

Rijhsinghani AG, Thompson K, Bhatia SK, Waldsmith TJ (1996) Estrogen blocks early T cell development in the thymus. Am J Reprod Immunol 36:269–277

Robinson E, Rubin D, Mekori T, Segal R, Pollack S (1993) In vivo modulation of natural killer cell activity by tamoxifen in patients with bilateral primary breast cancer. Cancer Immunol Immunother 37:209–212

Roitt I, Brostoff J, Male D (1989) Immunology. 2nd edn. Gower Medical Publishing, London, England

Rotstein S, Blomgren H, Petrini B, Wasserman J, Von Steding LV (1988) Influence of adjuvant Tamoxifen on blood lymphocytes. Breast Cancer Res Treat 12:75–79

Screpanti I, Felli MP, Toniato E, Meco D, Martinotti S, Frati L, Santoni A, Gulio A (1991) Enhancement of natural killer-cell susceptibility of human breast cancer cells by estradiol and v-Ha-ras oncogene. Int J Cancer 47:445–449

Selye H (1942) The pharmacology of steroid hormones and their derivatives. Rev Can Biol 1:577–632

Shevde NK, Pike JW (1996) Estrogen modulates the recruitment of myelopoietic cell progenitors in rat through a stromal cell-independent mechanism involving apoptosis. Blood 87:2683–2692

Silva JAP da, Larbre JP, Seed MP, Cutolo M, Villaggio B, Scott DL, Willoughby DA (1994) Sex differences in inflammation induced cartilage damage in rodents. The influence of sex steroids. J Rheumatol 21:330–337

Smithson G, Medina K, Ponting I, Kincade PW (1995) Estrogen suppresses stromal cell- dependent lymphopoiesis in culture. J Immunol 155:3409–3417

Sorachi K, Kumagai S, Sugita M, Yodoi J, Imura H (1993) Enhancing effect of 17β-estradiol on human NK cell activity. Immunol Lett 36:31–36

Spillane RM, Whitman GJ (1995) Treatment of retroperitoneal fibrosis with tamoxifen (letter) AJR 164:515–516

Stein B, Young MX (1995) Repression of the interleukin-6 promoter by estrogen receptor is mediated by NF-κB and C/EBPβ. Mol Cell Biol 15:4971–4979

Sthoeger ZM, Bentwich ZVI, Zinger H, Mozes E (1994) The beneficial effect of the estrogen antagonist, tamoxifen, on experimental systemic lupus erythematosus. J Rheumatol 21:2231–2238

Teodorczyk-Injeyan J, Cembrzynska-Nowak M, Lalani S, Kellen JA (1993) Modulation of biological responses of normal human mononuclear cells by antiestrogens. Anticancer Res 13:279–283

Thorson RE (1970) Direct infection nematodes. In: Jackson GJ, Herman R, Singer I (eds) Immunity to parasitic animals. Appleton-Century-Crofts, New York, p 913

Vamvakopoulos NC, Chrousos GP (1994) Hormonal regulation of human corticotropin releasing hormone gene expression: implications for the stress response and immune/inflammatory reaction. Endocr Rev 15:409–420

Wang Y, Campbell HD, Young IG (1993) Sex hormones and dexamethasone modulate interleukin-5 gene expression in T lymphocytes. J Steroid Biochem Mol Biol 44:203–210

Warri A, Kangas L (1990) Effect of toremifene on the activity of NK cells in NZB/NZW mice. J Steroid Biochem 36:207–209

Wilder RL (1996) Adrenal and gonadal steroid hormone deficiency in the pathogenesis of rheumatoid arthritis. J Rheumatol Suppl 44:10–12

Wira CR, Rossoll RM (1995) Antigen presenting cells in the female reproductive tract: influence of the estrus cycle and antigen presentation by uterine epithelial and stromal cells. Endocrinology 136:4526–4534

Zuckerman SH, Bryan-Poole N, Evans GF, Short L, Glasebrook AL (1995) In vivo modulation of murine tumor necrosis factor and interleukin-6 during endotoxemia by estrogen agonists and antagonists. Immunology 86:18–24

Zuckerman SH, Ahmari SE, Bryan-Poole N, Evans GF, Short L, Glasebrook AL (1996) Estriol: a potent regulator of TNF and IL-6 expression in murine model of endotoxemia. Inflammation 20:581–597

CHAPTER 18

Male Reproductive Function

M. Simoni and E. Nieschlag

A. Introduction

Male reproductive function is essentially androgen-dependent. Androgens are required for the male phenotype, for the maturation of sex characteristics at puberty, for sexual behavior and, together with follicle-stimulating hormone (FSH), for qualitatively and quantitatively normal spermatogenesis (Weinbauer and Nieschlag 1996; Weinbauer et al. 1997). There are no known cases of impaired reproductive function due to estrogen deficiency in man. Nevertheless, the production of transgenic mice, in which the estrogen receptor (ERα) has been knocked out (ERKO mice) (Lubhahn et al. 1993), and the discovery of male patients lacking a functional ER or the aromatase gene (Smith et al. 1994; Morishima et al. 1995; Carani et al. 1997) has shed new light on the role of estrogen in man. This chapter summarizes the actual knowledge of the role of estrogen in male reproductive function.

B. Source of Estrogen in the Male

Serum estrogen concentrations in normal adult men are rather low (<250 pmol/l) and originate mainly from the peripheral conversion of androgens (Macdonald et al. 1979). This reaction is catalyzed by the aromatase cytochrome P450, the product of the CYP19 gene (Simpson et al. 1994), and the same enzyme is active in both sexes. In the male, aromatase expression and estrogen production occur particularly in adipose tissue (Ackermann et al. 1981; Simpson et al. 1989) and, to a lesser extent, in the testes (Kelch et al. 1972), brain (Ryan et al. 1972), liver (Siiteri et al. 1978) and muscle (Longcope et al. 1978). In the rodent testes, aromatase activity is found in the Sertoli cells before puberty and in the Leydig cells and the seminiferous epithelium thereafter (Nitta et al. 1993 and references therein). In adult human testes, aromatase is located in Leydig cells and is absent from Sertoli cells (Inkster et al. 1995).

C. Localization of the Estrogen Receptor in the Male Genital Tract

That the male genital tract expresses ERs has been known for a long time (reviewed by Ciocca and Vargas Roig 1995). The receptor has been classically localized by immunocytochemistry and binding studies and, more recently, by the use of molecular biology techniques. Tissues, such as testes, efferent ducts, epididymis, seminal vesicles and the prostate have been reported to possess ERs (Ciocca and Vargas Roig 1995). The very recent description of a second form of ER, designated as ERβ (Kuiper et al. 1996; Mosselman et al. 1996), now permits a more accurate study of distribution and function of the two receptor forms in several tissues, including the male genital tract. Specific probes and antibodies for ERα and ERβ are now being used, and the old, sometimes contrasting, data on ER localization and function should now be reconsidered.

The ERα has been immunolocalized in Leydig cells and efferent ducts in the rat and the marmoset monkey, whereas Sertoli cells and rete testis do not express it (Fisher et al. 1997). In the rat, efferent ducts express even higher amounts of ERα than uterine tissue, and both ERα and ERβ are found in the efferent ducts and the epididymis (Hess et al. 1997a). In the cynomolgus monkey, the efferent ducts possess large quantities of ERs, localized by immunocytochemistry in the non-ciliated, absorptive cells (West and Brenner 1990). It has been observed that estrogen administration to transsexual men causes atrophy of the seminiferous tubules, an increase in Leydig cells and hyperplasia of the rete testis. In these men, ERs were localized to the efferent ducts by means of immunocytochemistry and, much less, to the epididymis and Leydig cells (Sapino et al. 1987). The localization of ERs in the human efferent ducts has recently been confirmed by both immunocytochemistry and reverse-transcriptase polymerase chain reaction (RT-PCR), while the epididymal duct was negative (Ergün et al. 1997). In some species, estrogen concentration in rete testis fluid is higher than serum estrogen in females (Hess et al. 1997b). Human Leydig cell tumors express abundant ERs (Due et al. 1989).

The presence and role of ERs in the human prostate have long been a matter of controversy. Estrogen receptors have been repeatedly described in the normal and pathological human prostate, especially in prostatic hyperplasia (reviewed by Ciocca and Vargas Roig 1995). The expression, however, is low and mainly limited to stromal cells (Habenicht et al. 1993) and is upregulated by androgen deprivation (Kruithof-Dekker et al. 1996). Using a rat prostate cDNA library, the group of Gustafsson recently isolated the ERβ (Kuiper et al. 1996), which was then specifically localized to epithelial and smooth muscle cells of seminal vesicles and prostate and in Sertoli cells (Saunders et al. 1997). However, in the human, Northern-blot analysis revealed high expression of ERβ in the testes, but not in the prostate (Mosselman et al. 1996).

In summary, the existing data suggest that the ERα is present in Leydig cells, in efferent ducts and the epididymis, whereas the ERβ is found in Sertoli cells, efferent ducts, epididymis, seminal vesicles and prostate. Further insights into localization and function of the two types of ER in the male genital tract are to be expected in the near future.

D. Function of Estrogens in the Male

The function of estrogens in the male has been illustrated by the discovery of the phenotypes associated with naturally occurring inactivating mutations of the aromatase gene (MORISHIMA et al. 1995; CARANI et al. 1997) and of the ER gene (SMITH et al. 1994). The production of transgenic mice lacking a functional ERα (ERKO mice) preceded these case reports (LUBAHN et al. 1993) and definitely demonstrated that the absence of estrogenic function is fully compatible with life, dispelling any doubt of possible lethality related to a supposed irreplaceable role of estrogen in embryo implantation and in the maintenance of pregnancy (KORACH 1994). ERKO mice were phenotypically normal, but infertile (LUBAHN et al. 1993), with a bone density 20–25% lower than in wild-type animals (KORACH 1994). Likewise, men with aromatase or an ER deficit show persistent linear growth, osteoporosis and, although inconsistently, testicular impairment. To what extent functions such as skeletal maturation, feedback control of gonadotropin secretion and lipid metabolism are estrogen-dependent has been elegantly shown by the effects of testosterone and estradiol treatment in a man with aromatase deficiency (CARANI et al. 1997).

I. Bone Maturation

Important functions of the androgens secreted at puberty are the induction of the growth spurt, bone maturation and ossification of the epiphyseal cartilages. Part of these actions are clearly mediated by and require estrogens derived from androgen aromatization. In fact, in a patient with aromatase deficiency, testosterone treatment had no effects on bone density. In contrast, the treatment with estradiol induced a remarkable increase in bone mineral density, epiphyseal closure and changes in biochemical parameters of bone metabolism similar to those occurring during normal skeletal maturation at puberty (CARANI et al. 1997). However, patients with complete androgen insensitivity demonstrated no bone problems (GRIFFIN 1992), whereas patients with 17α-hydroxylase/17,20-lyase deficiency showed tall stature, retardation of bone age, osteoporosis and eunuchoid skeleton (YANASE et al. 1991). These findings led to the concept that the skeletal abnormalities commonly seen in male hypogonadism are largely due to estrogen deficiency. A possible role of estrogen deficiency in the osteoporosis occurring during male senescence is presently an object of investigation (SLEMENDA et al. 1997).

II. Testicular Function

Estrogen affects testicular function by operating in the regulation of gonadotropin secretion from the pituitary gland and by playing an important role in fluid resorption in the efferent ducts.

The secretion of gonadotropins is stimulated by the hypothalamic hormone gonadotropin releasing hormone (GnRH) and is inhibited by testosterone, which acts both directly at the pituitary level and indirectly by suppressing the secretion of GnRH. Part of the testosterone action is mediated by conversion into estrogens (WEINBAUER and NIESCHLAG 1996). The three men discovered so far who either lack ERs or have complete inactivation of aromatase demonstrate normal or high serum testosterone levels and (slightly) elevated gonadotropin concentrations (SMITH et al. 1994; MORISHIMA et al. 1995; CARANI et al. 1997), supporting the concept that aromatization of androgen is physiologically required for the complete feedback control of gonadotropin secretion at the hypothalamus–pituitary level (FINKELSTEIN et al. 1991a, b; BAGATELL et al. 1994). In fact, testosterone administration did not affect serum gonadotropin levels in men with aromatase deficiency, whereas luteinizing hormone (LH) and FSH became completely suppressed during estradiol treatment (CARANI et al. 1997). These findings indicate that, in the male, the sex-steroid gonadotropin feedback is mainly mediated by testosterone, but some testosterone must be converted to estrogen. The action of estrogen at the central level, however, is unclear. At the pituitary level, estradiol might modulate GnRH receptor number and function (MCARDLE et al. 1992), but no ERs, at least of the α type, can be found in GnRH secreting neurons (SULLIVAN et al. 1995), leaving the issue regarding the effects of steroids on hypothalamic neurons unresolved.

A previously unsuspected physiological action of estrogen in testicular function was revealed by ERKO mice. Adult, sexually mature, male ERKO mice were infertile in the presence of anatomically normal male accessory sex organs (KORACH 1996); testicular volume was reduced and spermatogenesis was profoundly altered. Testicular histology in these mice shows atrophic and degenerating seminiferous tubules, together with dilated tubules with a thin layer of Sertoli cells and tubules with disorganized spermatogenesis or lacking the lumen and containing only Sertoli cells. Some germ cells, however, can develop to produce spermatocytes (EDDY et al. 1996). The degeneration of spermatogenesis is progressive: testicular histology is normal at 10 days of age and starts degenerating at 20 days; tubules are completely dilated at 40–60 days. The degenerative process starts at the caudal pole of the testis and progresses cranially. When tested in vitro, sperm from ERKO mice demonstrated a significantly reduced capacity to fertilize eggs compared with wild-type sperm (EDDY et al. 1996).

In ERKO mice, tubular fluid is not absorbed in the efferent ducts, which are the anatomical structures in the male genital tract that have the highest concentration of ERs. As a consequence, the seminiferous tubules upstream of the rete testis become swollen, first causing an increase in testicular volume up to the age of 2 months and then a progressive degeneration with drastic reduction of spermatogenesis and shrinkage of the testes. Studies in vitro have shown that efferent ducts of ERKO mice are unable to reabsorb fluid and that this defect can be partially mimicked by an antiestrogen in wild-type efferent ducts, suggesting a role for estrogens in fluid resorption. Therefore, the knock-out of the ERα in the male mouse causes infertility because the lack of fluid resorption in the efferent ducts induces a progressive swelling and degeneration of the seminiferous tubules (HESS et al. 1997b).

It is not possible at the moment to say whether the same mechanism operates in the human testes. The presence of abundant ERs in the human efferent ducts would speak in favor of such an idea. However, the existing examples of men with estrogen deficiency are not really consistent with this. The only man with estrogen resistance discovered so far, a human equivalent of the ERKO mice, had normal testis volume and normal sperm count with slightly reduced motility (SMITH et al. 1994). Of the two men with aromatase deficiency reported in the literature, one had normal testis volume (MORISHIMA et al. 1995), the other showed reduced testicular volume, severe oligozoospermia and infertility (CARANI et al. 1997). The physiological role of estrogens in human testicular function, therefore, remains to be ascertained.

The crucial role of estrogens in determination of male fertility in the mouse induces a reconsideration of the supposed adverse effect of compounds with estrogenic activity for human fertility. The unproven claim that human sperm quality and quantity has deteriorated over the past decades, probably one of the major mystifications of the nineties (LERCHL and NIESCHLAG 1996; NIESCHLAG and LERCHL 1996), and the concept that estrogens might be involved (SHARPE and SKAKKEBAEK 1993) continue to preoccupy the media. However, prenatal exposure to high doses of diethylstilbestrol did not cause any impairment of fertility in men despite a higher rate of hypospadias (WILCOX et al. 1995). There is no evidence that exposure to estrogen in utero plays any adverse effect on male fertility; one should consider that pregnancy is physiologically characterized by large estrogen concentrations.

III. Prostate

The effects of estrogens on the prostate are complex and not fully understood (THOMAS and KENAN 1994). Prostate development is androgen-dependent, but prostate diseases develop in elderly men when serum androgen levels decrease and estrogens rise. Aromatase is expressed similarly in stromal cells surrounding the hyperplastic tissue in patients with benign prostatic hyperplasia and in the carcinomatous glands of men with prostatic carcinoma (HIRAMATSU et al. 1997). The local aromatization of androgens has been proposed to play a causal role in benign prostatic hyperplasia (EL ETREBY 1993) and the administration of an aromatizable androgen to cynomolgus monkeys induced microscopic, estrogen-related hyperplastic changes that could be prevented by the simultaneous administration of an aromatase inhibitor (HABENICHT et al. 1987).

In a recent study, long-term androgen abuse in body builders led to a 20% increase of the volume of the central prostate, the region of origin of benign prostatic hypeplasia, whereas administration of large estrogen doses to male-to-female transsexuals reduced total prostate volume by 30% (Jin et al. 1996). However, the reduction in prostate size in the latter group was lower than expected and much higher than the prostate volume in untreated hypogonadal males of similar age (Behre et al. 1994). These data are consistent with the hypothesis that estrogens exert a direct, stimulatory effect on prostate growth, as suggested by numerous in vitro and in vivo studies (Thomas and Keenan 1994). However, the molecular mechanism of estrogen's action is still unclear. The recent description of the ERβ, isolated from a prostate cDNA library, is expected to induce stimulating research in this field.

IV. Behavior

The reproductive behavior of male ERKO mice is characterized by normal motivation to mount females, but reduced mating activity, with less intromissions and practically no ejaculation (Ogawa et al. 1997; Eddy et al. 1996). Moreover, male aggressive behavior is profoundly reduced in ERKO mice and the open-field behavior is of female-type (Ogawa et al. 1997). These data suggest that, in mice, a functional ERα is necessary during development in order to achieve the normal male neuronal organization of the brain. Recent data in human male and female hypogonadal adolescents, treated with testosterone or estrogen, indicate strong effects of low estrogen doses on aggressive behavior in girls. Since aggressive behavior in boys was induced by testosterone only at mild doses, it was speculated that the conversion to estrogen could be involved in steroid-dependent aggressive behavior in males as well (Finkelstein et al. 1997). However, in the three men with an ER defect or aromatase deficiency, reported so far, abnormalities of sexual orientation and behavior were not noticed (Smith et al. 1994; Morishima et al. 1995, Carani et al. 1997), indicating that estrogens in humans might not be so critical for the sexual differentiation of the brain.

E. Conclusions

While, in mice, estrogen is necessary for male fertility, the data in men are still inconclusive. Since ERα is present in the human testes and epididymis and estrogens are produced locally, a physiological function is highly probable, but still not characterized. Instead, estrogens seem to play a very important role in the feedback regulation of gonadotropin secretion in man, an issue that should be considered during the pharmacological manipulation of the hypothalamo–pituitary testis axis, oriented to complete gonadotropin suppression, e.g., for contraceptive purposes. Finally, estrogens might have a stimulatory effect on the inner prostate. The characterization of ERα and ERβ in the human male genital tract is necessary for further elucidation of estrogen function in male reproduction.

References

Ackerman GE, Smith ME, Mendelson CR, MacDonald PC, Simpson ER (1981) Aromatization of androstenedione by human adipose tissue stromal cells in monolayer culture. J Clin Endocrinol Metab 53:412–417

Bagatell CJ, Dahl KD, Bremner WJ (1994) The direct pituitary effect of testosterone to inhibit gonadotropin secretion in men is partially mediated by aromatization to estradiol. J Androl 15:15–21

Behre HM, Bohmeyer J, Nieschlag E (1994) Prostate volume in testosterone treated and untreated hypogonadal men in comparison to age-matched normal controls. Clin Endocrinol (Oxf) 40:341–349

Carani C, Qin K, Simoni M, Faustini-Fustini M, Serpente S, Boyd J, Korach KS, Simpson ER (1997) Effect of testosterone and estradiol in a man with aromatase deficiency. New Engl J Med 337:91–95

Ciocca DR, Vargas Roig LM (1995) Estrogen receptors in human nontarget tissues: biological and clinical implications. Endocr Rev 16:35–62

Due W, Dieckmann KP, Loy V, Stein H (1989) Immunohistological determination of estrogen receptor, progesterone receptor, and intermediate filaments in Leydig cell tumors, Leydig cell hyperplasia and normal Leydig cells of the human testis. J Pathol 157:225–234

Eddy EM, Washburn TF, Bunch DO, Goulding EH, Gladen BC, Lubahn DB, Korach KS (1996) Targeted disruption of the estrogen receptor gene in male mice causes alteration of spermatogenesis and infertility. Endocrinology 137:4796–4805

El Etreby MF (1993) Atamestane: an aromatase inhibitor for the treatment of benign prostatic hyperplasia. A short review. J Ster Biochem Mol Biol 44:565–572

Ergün S, Ungefroren H, Holstein AF, Davidoff MS (1997) Estrogen and progesterone receptors and estrogen receptor-related antigen (ER-D5) in human epididymis. Mol Reprod Devel 47:448–455

Finkelstein JS, O'Dea LS, Whitcomb RW, Crowley WF (1991a) Sex steroid control of gonadotropin secretion in the human male. II. Effects of estradiol administration in normal and GnRH deficient men. J Clin Endocrinol Metab 73:621–628

Finkelstein JS, Whitcomb RW, O'Dea LS, Longcope C, Schoenfeld DA, Crowley WF (1991b) Sex steroid control of gonadotropin secretion in the human male. I. Effects of testosterone administration in normal and GnRH deficient men. J Clin Endocrinol Metab 73:609–620

Finkelstein JW, Susman EJ, Chinchilli VM, Kunselman SJ, D'Arcangelo MR, Schwab J, Demers LM, Liben LS, Lookingbill G, Kulin HE (1997) Estrogen or testosterone increases self-reported aggressive behaviors in hypogonadal adolescents. J Clin Endocrinol Metab 82:2423–2438

Fisher JS, Millar MR, Majdic G, Saunders PTK, Fraser HM, Sharpe RM (1997) Immunolocalisation of estrogen receptor-α within the testis and excurrent ducts of the rat and marmoset monkey from perinatal life to adulthood. J Endocrinol 153:485–495

Griffin JE (1992) Androgen resistance – the clinical and molecular spectrum. N Engl J Med 326:611–618

Habenicht UF, Schwarz K, Neumann F, El Etreby MF (1987) Induction of estrogen-related hyperplastic changes in the prostate of the cynomolgus monkey (*Macaca fascicularis*) by androstenedione and its antagonization by the aromatase inhibitor 1-methyl-androsta-1,4-diene3,17-dione. Prostate 11:313–326

Habenicht UF, Tunn UW, Senge T, Schröder FH, Schweikert HU, Bartsch G, El Etreby MF (1993) Management of benign prostatic hypeplasia with particular emphasis on aromatase inhibitors. J Ster Biochem Mol Biol 44:557–563

Hess RA, Gist DH, Bunick D, Lubahn DB, Farrel A, Bahr J, Cooke PS, Greene GL (1997a) Estrogen receptor (α and β) expression in the excurrent ducts of the adult male rat reproductive tract. J. Androl 18:602–611

Hess RA, Bunick D, Lee K-H, Bahr J, Taylor JA, Korach KS, Lubahn DB (1997b) A role for estrogens in the male reproductive system. Nature 390:509–512

Hiramatsu M, Meehara I, Ozaki M, Harada N, Orikasa S, Sasano H (1997) Aromatase in hyperplasia and carcinoma of the human prostate. Prostate 31:118–124

Inkster S, Yue W, Brodie A (1995) Human testicular aromatase: immunocytochemical and biochemical studies. J Clin Endocrinol Metab 80:1941–1947

Jin B, Turner L, Walters WAW, Handelsman DJ (1996) Androgen or estrogen effects on human prostate. J Clin Endocrinol Metab 81:4290–4295

Kelch RP, Jenner MR, Weinstein RL, Kaplan SL, Grumbach MM (1972) Estradiol and testosterone secretion by human, simian and canine testes, in males with hypogonadism and in male pseudohermaphrodites with the feminizing testes syndrome. J Clin Invest 51:824–830

Korach KS (1994) Insights from the study of animals lacking functional estrogen receptor. Science 266:1524–1527

Korach KS, Couse JF, Curtis SW, et al. (1996) Estrogen receptor gene disruption: molecular characterization and experimental and clinical phenotypes. Recent Prog Horm Res 51:159–186

Kruithof-Dekker IG, Tetu B, Janssen PJA, Van der Kwast TH (1996) Elevated estrogen receptor expression in human prostatic stromal cells by androgen ablation therapy. J Urol 156:1194–1197

Kuiper GGJM, Enmark E, Pelto-Huikko M, Nilsson S, Gustafsson JA (1996) Cloning of a novel estrogen receptor expressed in rat prostate and ovary. Proc Natl Acad Sci U S A 93:5925–5930

Lerchl A, Nieschlag E (1996) Decreasing sperm count? a critical (re)view. Exp Clin Endocrinol Diabetes 104:301–307

Longcope C, Pratt JH, Schneider SH, Fineberg SE (1978) Aromatization of androgens by muscle and adipose tissue in vivo. J Clin Endocrinol Metab 46:146–152

Lubahn DB, Moyer JS, Golding TS, Couse JF, Korach KS, Smithies O (1993) Alteration of reproductive function but not prenatal sexual development after insertional disruption of the mouse estrogen receptor gene. Proc Natl Acad Sci USA 90:11162–11166

MacDonald PC, Madden JD, Brenner PF, Wilson JD, Siiteri PK (1979) Origin of estrogen in normal men and in women with testicular feminization. J Clin Endocrinol Metab 49:905–916

McArdle CA, Schomerus E, Gröner I, Poch A (1992) Estradiol regulates gonadotropin-releasing hormone receptor number, growth and inositol phosphate production in αT3-1 cells. Mol Cell Endocrinol 87:95–103

Morishima A, Grombach MM, Simpson ER, Fisher C, Qin K (1995) Aromatase deficiency in male and female siblings caused by a novel mutation and the physiological role of estrogens. J Clin Endocrinol Metab 80:3689–3698

Mosselman S, Polman J, Dijkema R (1996) ERβ: identification and characterization of a novel human estrogen receptor. FEBS Lett 392:49–53

Nieschlag E, Lerchl A (1996) Declining sperm counts in European men–fact or fiction? Andrologia 28:305–306

Nitta H, Bunick D, Hess RA, Janulis L, Newton SC, Millette CF, Osawa Y, Shizuta Y, Toda K, Bahr JM (1993) Germ cell of the mouse testis express P450 aromatase. Endocrinology 132:1396–1401

Ogawa S, Lubahn DB, Korach KS, Pfaff DW (1997) Behavioral effects of estrogen receptor gene disruption in male mice. Proc Natl Acad Sci USA 94:1476–1481

Ryan KJ, Naftolin F, Reggy V, Flores F, Petro Z (1972) Estrogen formation in the brain. Am J Obstet Gynecol 114:454–460

Saunders PTK, Maguire SM, Gaughan J, Millar MR (1997) Expression of estrogen receptor beta (ERβ) in multiple rat tissues visualized by immunohistochemistry. J Endocrinol 154:R13–R16

Sapino A, Pagani A, Godano A, Bussolati G (1987) Effects of estrogen on the testis of transexuals: a pathological and immunocytochemical study. Virchows Arch 411:409–414

Sharpe RM, Skakkebaek NE (1993) Are estrogens involved in falling sperm counts and disorders of the male reproductive tract? Lancet 341:1392–1395

Siiteri PK, Seron-Ferre M (1978) Secretion and metabolism of adrenal androgens to estrogens. In: James VHT, Serio M, Giusti G, Martini L (eds) The endocrine function of the human adrenal cortex. New York, Academic Press, pp 251–264

Simpson ER, Mahendroo MS, Means GD, Kilgore MW, Hinshelwood MM, Graham-Lorence S, Amarneh B, Ito Y, Fisher CR, Michael MD, Mendelson CR, Bulun SE (1994) Aromatase cytochrome P450, the enzyme responsible for estrogen biosynthesis. Endocr Rev 15:342–355

Simpson ER, Merril JC, Hollub AJ, Graham-Lorence S, Mendelson CR (1989) Regulation of estrogen biosynthesis by human adipose cells. Endocr Rev 10:136–148

Slemenda CW, Longcope C, Zhou L, Hui SL, Peacock M, Johnston CC (1997) Sex steroids and bone mass in older men. Positive associations with serum estrogen and negative associations with androgens. J Clin Invest 100:1755–1759

Smith EP, Boyd J, Frank GR, Takahashi H, Cohen RM, Specker B, Williams TC, Lubahn DB, Korach KS (1994) Estrogen resistance caused by a mutation in the estrogen receptor gene in a man. New Engl J Med 331:1056–1061

Sullivan KA, Witkin JW, Ferin M, Silverman (1995) Gonadotropin-releasing hormone neurons in the rhesus macaque are not immunoreactive for the estrogen receptor. Brain Res 685:198–200

Thomas JA, Keenan EJ (1994) Effects of estrogens on the prostate. J Androl 15:97–99

Yanase T, Simpson ER, Waterman MR (1991) 17α-hydroxylase/17,20-lyase deficiency: from clinical investigation to molecular definition. Endocr Rev 12:91–108

Weinbauer GF, Gromoll J, Simoni M, Nieschlag E (1997) Physiology of testicular function. In: Nieschlag E, Behre HM (eds) Andrology. Male reproductive health and dysfunction. Springer, Berlin, pp 25–57

Weinbauer GFW, Nieschlag E (1996) The Leydig cell as a target for male contraception. In: Payne AH, Hardy MP, Russell LD (eds) The Leydig Cell. Cache River Press, Clearwater, pp 629–662

West NB, Brenner RM (1990) Estrogen receptor in the ductuli efferentes, epididymis, and testis of rhesus and cynomolgus macaques. Biol Reprod 42:533–538

Wilcox AJ, Baird DD, Weinberg CR, Hornsby PP, Herbst AL (1995) Fertility in men exposed prenatally to diethylstilbestrol. N Engl J Med 332:1411–1416

CHAPTER 19

The Effect of Estrogens and Antiestrogens on the Urogenital Tract

A. HEXTALL and L. CARDOZO

A. Introduction

The female lower genital and urinary tracts start to develop from the primitive urogenital sinus as early as the fourth week of embryological life and remain anatomically closely related. Estrogen receptors have been identified in the tissues of both systems and, therefore, fluctuations in the circulating level of sex steroids may lead to symptomatic, cytological and functional changes in the bladder, urethra and vagina. Urogenital symptoms often fluctuate during the menstrual cycle and frequently develop during pregnancy. In addition, following menopause, when there is generalised atrophy of all Estrogen-sensitive tissues, lower urinary and vaginal symptoms are common. However, these changes may also be related to the effects of ageing, so a cause-and-effect relationship has been difficult to prove. This chapter reviews the effects of estrogens and antiestrogens on the urogenital tract. Data are mainly derived from a number of epidemiological studies and also clinical trials evaluating estrogen replacement therapy for postmenopausal women.

B. Pathophysiology

I. Hormonal Influences on the Urogenital Tract

Estrogen receptors are consistently expressed in the squamous epithelium of the urethra and vagina, and also in the trigone of the bladder in areas that have undergone squamous metaplasia (IOSIF 1981; BLAKEMAN 1996a). However, they are not present in the transitional epithelium of the bladder dome, reflecting the different embryological origin of this tissue. The pubococcygeous muscle of the pelvic floor is also a target site for estrogens (INGELMAN-SUNDBERG 1981; SMITH 1993). Estrogens increase cell-cycle activity in the female lower urinary tract (BLAKEMAN 1996b) and this is demonstrated by an increase in the number of intermediate and superficial cells in the urethra and bladder (SAMSIOE 1985), with similar changes occurring in the vagina of postmenopausal women (SMITH 1976; SEMMENS 1985). Alterations in urinary cytology during the menstrual cycle are comparable with those seen in vaginal cytology (McCALLIN 1950), changes which also occur in the urinary sediment following treatment with estrogens (SOLOMAN 1958).

Cyclical variations in urinary symptoms occurring during the menstrual cycle can be measured objectively using urethral pressure profilometry (UPP) (van Geelen 1981). The functional and anatomical lengths of the urethra increase midcycle and early in the luteal phase, reflecting serum estrogen concentrations. Changes also occur during pregnancy (Stanton 1980; Cutner 1992), which can partially be explained by an increase in urine output and pressure effects from the gravid uterus. However, the prevalence of detrusor instability antenatally is significantly greater than that found postpartum (Cutner 1993), suggesting a possible hormonal effect on the bladder thought to be mediated through progesterone (Cutner 1991).

Progesterone receptors are expressed inconsistently in the lower urinary tract and may be dependant on the estrogen status of the woman (Blakeman 1996a). Androgen receptors are found in both the female bladder and urethra, but their role is at present unclear (Blakeman 1997).

II. The Effect of Ageing on the Bladder

Many women consider the development of urinary symptoms as they get older to be a normal phenomenon rather than the manifestation of a disease (Svanberg 1997). Indeed, in a study by Gjorup et al. (1987), over 50% of women aged over 75 years thought that their symptoms were normal for elderly people. Symptomatic and functional changes which are difficult to differentiate from those due to estrogen deficiency certainly do occur in the lower urinary tract as a result of the ageing process. Younger women tend to excrete the bulk of their fluid intake before they go to bed, whereas in the elderly this pattern may be reversed. Postural effects lead to daytime pooling of extracellular fluid, especially in the ankles, and when this returns to the vasculature during the night there is a consequential increase in urine output. This, combined with an alteration in sleep patterns, may lead to nocturia.

Urodynamic studies have shown that the urethra and bladder become less efficient with age (Rud 1980b; Rud et al. 1980; Malone-Lee 1993). Elderly women have a reduced urine flow rate, increased urinary residual, higher first sensation of a desire to void and increased bladder capacity, although the latter may fall in the eighth and ninth decades (Collas and Malone-Lee 1996). In addition, detrusor pressures at urethral opening and closure during voiding fall in absolute terms as women become older (Wagg 1996). Histologically, there is an age-related increase in fibrosis of the bladder neck (Brocklehurst 1972) and in collagen content in the human female bladder (Sussett 1978). The number and diameter of muscle fibres in the pelvic floor also decrease with age (Kolbl 1989), although neuronal damage secondary to childbirth may be a confounding factor (Smith 1989; Allen and Warrell 1992).

The ageing population are at risk of developing a number of illnesses, including diabetes mellitus, congestive cardiac failure and renal disease, which can present with lower urinary tract symptoms. An individual with impaired mobility may also be incontinent of urine if suitable access to a toilet is not available, a situation which may be exacerbated if medications such as diuretics or hypnotics are being taken.

C. Estrogen and Urinary Incontinence

Postmenopausal women with lower urinary tract dysfunction often present with a combination of symptoms. The most common are frequency of micturition, nocturia, stress incontinence, urgency and urge incontinence. They may also complain of urinary-tract infections and vaginal symptoms of soreness, itching or dryness. An accurate diagnosis of the underlying pathology is made by careful clinical assessment and urodynamic studies.

There are a number of causes of urinary incontinence (Table 1), with transient problems particularly affecting the elderly (Table 2). The two most common presentations are stress and urge incontinence. Genuine stress incontinence is the involuntary loss of urine when the intravesical pressure exceeds the maximal urethral closure pressure in the absence of a detrusor contraction. This is commonly precipitated by coughing, sneezing or exercise. Urge incontinence is the involuntary loss of urine, associated with a strong desire to void, and may be secondary to unstable detrusor contractions which cause a rise in the intravesical pressure.

I. Epidemiological Studies

Urinary symptoms secondary to estrogen deficiency may only develop many years after menopause and, therefore, may be underreported by both patient

Table 1. Common causes of female urinary incontinence

Genuine stress incontinence
Detrusor instability (detrusor hyperreflexia)
Overflow incontinence
Fistulae (vesicovaginal, ureterovaginal, urethrovaginal)
Urethral diverticulum
Functional (immobility)
Transient problems (Table 2)

Table 2. Transient causes of urinary incontinence in the elderly

Urinary-tract infection
Faecal impaction
Oestrogen deficiency
Restricted mobility
Drug therapy
Depression
Confusional state

and doctor. Epidemiological studies have shown that the incidence of urogenital problems increases with age, with many women delaying seeking treatment for many years. In a study of 2045 women aged between 55 years and 85 years, BARLOW and colleagues (1997) showed that urogenital symptoms had affected 48.5% of postmenopausal women at some time, but only 11% were affected by individual symptoms at the time of the study. At least two-thirds of women did not relate their vaginal or urinary complaint to menopause. IOSIF and BEKASSY (1984) studied 2200 women aged 61 years and also found that the incidence of lower genital tract disorders was high, with 49% of women having some symptoms. Urinary symptoms are certainly common after menopause, with one in five women attending a menopause clinic complaining of severe urgency and nearly 50% of stress incontinence (CARDOZO 1987).

The prevalence of postmenopausal incontinence in the community is thought to be 16–29% (THOMAS 1980; VETTER 1981; BARLOW 1997). While the ageing process is clearly a significant aetiological factor in the pathogenesis of urinary incontinence, there is conflicting evidence as to whether menopause and estrogen deficiency are also important. JOLLEYS (1988) surveyed 937 women registered with a rural general practice and found the prevalence of incontinence was most common in the 45-year to 55-year age group; a period which includes menopause in most cases (Fig. 1). HILTON (1981) also found a similar pattern in hospital practice; 40% of women referred to a urogynaecology unit were aged between 40 years and 60 years, with a mean comparable with the average age of menopause. In addition, 70% of the incontinent

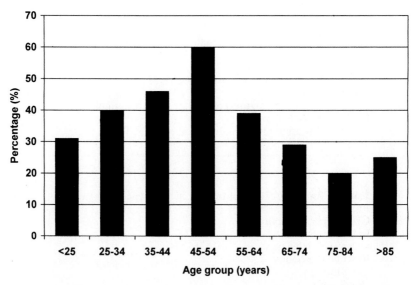

Fig. 1. The changing prevalence of stress incontinence with age. Adapted from JOLLEYS (1988)

women studied by IOSIF and BEKASSY (1984) related the onset of their urinary leakage to their final menstrual period. Urge incontinence in particular is found more commonly after menopause, although the prevalence of stress incontinence starts to fall (KONDO 1990) (Fig. 2). Most studies, however, show that many women develop incontinence at least 10 years before becoming climateric, with JOLLEYS (1988) finding that significantly more premenopausal women were affected than postmenopausal women.

II. Estrogen and the Continence Mechanism

The estrogen-sensitive tissues of the bladder, urethra and pelvic floor all play an important role in the continence mechanism. For a woman to remain continent, the urethral pressure must exceed the intravesical pressure at all times except during micturition (ABRAMS 1990). The urethra has four estrogen-sensitive functional layers, all of which play a part in the maintenance of a positive urethral pressure:

1. Epithelium
2. Vasculature
3. Muscle
4. Connective tissue

The cellular changes in the epithelial layer of the urethra that occur in response to estrogen have already been described. There is evidence to suggest that the vasodilatory effect of estrogens which occurs in the systemic

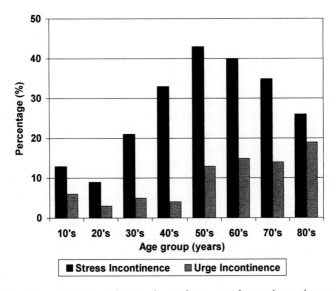

Fig. 2. Changes in prevalence of stress incontinence and urge incontinence with age. Adapted from KONDO (1990)

Table 3. Mechanisms by which oestrogens may treat female urinary incontinence

---

Increased urethral closure pressure:

1. Improved urethral cell maturation
2. Increased urethral blood flow
3. Increased alpha-adrenergic receptor sensitivity in urethral smooth muscle
4. Stimulation of periurethral collagen production

Increased sensory threshold of the bladder
Improved abdominal pressure transmission to the proximal urethra
Reduced incidence of urinary-tract infection
Improved mood and quality of life

---

circulation (GANGER 1991; JACKSON and VYAS 1998) may also take place in the urogenital tract. VERSI and CARDOZO (1986) have shown that the vascular pulsations seen on the UPP due to blood flow in the urethral submucosa and urethral sphincter increase in size in response to estrogen. Connective-tissue metabolism is stimulated by estrogens, increasing the production of collagen in periurethral tissues and, therefore, possibly reversing the changes that occur as a result of ageing (JACKSON 1996a). Finally, the alpha receptors in the urethral sphincter are sensitised by estrogens (SCREITER 1976), helping to maintain muscular tone. The estrogen status of a woman can therefore have a significant effect on urethral pressure (RUD 1980a), and this may be particularly important when there is already a degree of weakness.

There are a number of reasons why estrogens may be useful in the treatment of women with urinary incontinence (Table 3). In addition to improving the "maturation index" of urethral squamous epithelium (BERGMAN 1990), estrogens increase urethral closure pressure and improve abdominal pressure transmission to the proximal urethra (HILTON and STANTON 1983; BHATIA 1989; KARRAM 1989). The sensory threshold of the bladder may also be raised (FANTL 1988).

SALMON (1941) was the first to report the successful use of estrogens to treat urinary incontinence over 50 years ago. It is now well recognised that there is a poor correlation between a woman's symptoms and the subsequent diagnosis following appropriate investigation (JARVIS 1980). Unfortunately, initial trials took place before the widespread introduction of urodynamic studies and, therefore, almost certainly included a heterogeneous group of individuals with a number of different pathologies. Lack of objective outcome measurements also limit interpretation of the initial trials.

III. Estrogens for Stress Incontinence

The role of estrogen in the treatment of stress incontinence has been controversial even though there are a number of reported studies. Some have given promising results, but this may be because they were observational, not randomised, blinded or controlled. The situation is further complicated by the fact that a number of different types of estrogen have been used with varying doses, routes of administration and durations of treatment. There have now been two meta analyses performed which have helped to clarify the situation further. In the first, a report by the Hormones and Urogenital Therapy (HUT) committee, the use of estrogens to treat all causes of incontinence in post-menopausal women was examined (FANTL 1994). Of 166 articles identified which were published in English between 1969 and 1992, only 6 were controlled trials and 17 were uncontrolled series. The results showed that there was a significant subjective improvement for all patients and those with genuine stress incontinence. However, assessment of the objective parameters revealed that there was no change in the volume of urine lost. Maximum urethral closure pressure did increase significantly, but this result was influenced by only one study which showed a large effect. In the second meta analysis, SULTANA and WALTERS (1995) reviewed 8 controlled and 14 uncontrolled prospective trials and included all types of estrogen treatment. They also found that estrogen therapy was not an efficacious treatment of stress incontinence but may be useful for the often-associated symptoms of urgency and frequency.

Two studies using oral estrogen, which were not included in the meta analyses, have recently been reported. FANTL et al. (1996) treated 83 hypo-estrogenic women who presented with urodynamic evidence of genuine stress incontinence and/or detrusor instability with conjugated equine estrogens (0.625 mg) and medroxyprogesterone (10 mg) cyclically for 3 months. Controls received placebo tablets. At the end of the study period, the clinical and quality-of-life variables had not changed significantly in either group. JACKSON et al. (1996b) treated 57 postmenopausal women with genuine stress incontinence or mixed incontinence with estradiol valerate (2 mg) or placebo daily for 6 months. There was no significant change in objective outcome measures, although both the active and placebo group reported subjective benefits, possibly because estrogens improve quality of life.

Estrogen, when given alone, therefore, does not appear to be an effective treatment for stress incontinence. However, several studies have shown that it may have a role in combination with other therapies. BEISLAND et al. (1984) treated 24 women with genuine stress incontinence using phenyl-propanolamine (50 mg twice daily) and oestriol (1 mg/day vaginally), separately and in combination. They found that the combination cured eight women and improved a further nine and was more effective than either drug given alone. HILTON and colleagues (1990) used estrogen (vaginal or oral) alone or in combination with phenylpropanolamine to treat 60 post-menopausal women with genuine stress incontinence in a double-blind, placebo-controlled study. Subjectively, the symptoms of stress incontinence improved in all groups, but objectively only in the women given combination therapy. This type of treatment may be particularly useful for women with mild stress incontinence or for those not suitable for surgery.

IV. Estrogens for Urge Incontinence

Estrogen has been used for many years to treat postmenopausal urgency and urge incontinence, but there have been very few controlled trials performed to confirm that it is of benefit. WALTER and colleagues (1978) found that a combination of oestradiol (2 mg) and oestriol (1 mg daily) cured the symptoms of urge incontinence in 7 of 11 women, whereas placebo cured only 1 of 10 patients. SAMSIOE (1985) also used oral oestriol (3 mg daily) to treat 34 women aged 75 years in a double-blind placebo-controlled crossover study. Overall, a substantial subjective improvement was found in the 12 women with urge incontinence and 8 women with mixed incontinence. However, these reports need to be interpreted with caution because of the small patient numbers and lack of objective outcome measures despite the known large placebo effect which occurs in treatment of this condition.

A double-blind multi-centre study of 64 postmenopausal women with the "urge syndrome" has failed to confirm these results (CARDOZO 1993). All women underwent pretreatment urodynamic investigation to establish that they either had sensory urgency or detrusor instability. They were then randomised to treatment with oral oestriol (3 mg daily) or placebo for 3 months. Compliance was confirmed by a significant improvement in the maturation index of vaginal epithelial cells in the active group but not the placebo group. Oestriol produced subjective and objective improvements in urinary symptoms, but it was not significantly better than placebo. In a further study, sustained-release 17β-oestradiol vaginal tablets (25 μg, Vagifem, Novo Nordsk) or placebo were used to treat 110 postmenopausal women (BENNESS et al. unpublished data). Urodynamic investigations confirmed that the women had sensory urgency, detrusor instability or no objective abnormality. At the end of the 6-month treatment period the only significant differences between the active and placebo groups was an improvement in the symptoms of urgency in the women who had a diagnosis of sensory urgency. It is possible that this low dose of local estrogen was reversing atrophic changes in the lower urinary/genital tract rather than treating the underlying pathology.

These studies may not have shown any benefit, possibly because the wrong type of estrogen was used for too short a time period, or it may have been given via the wrong route. Oestriol, although a naturally occurring estrogen, has little effect on the endometrium and does not prevent osteoporosis. It is therefore also questionable whether the low dose used in these studies is sufficient to treat urinary symptoms. Sustained-release 17β-oestradiol vaginal tablets are well absorbed and have been shown to induce maturation of the vaginal epithelium within 14 days (NILSSON 1992), but higher systemic levels may be needed for therapy to be effective.

D. Estrogen for Recurrent Urinary-Tract Infections

Alterations in the vaginal flora following the menopause place women at an increased risk of urinary-tract infections, particularly if they are sexually active. There is a rise in vaginal pH and a fall in the number of lactobacilli, allowing colonisation of gram-negative bacteria which act as uropathogens. Estrogen reverses these changes, an effect which enables it to be used for either treatment or prophylaxis.

BRANDBERG et al. (1985) treated 41 elderly women with recurrent urinary-tract infections using oral oestriol and showed that their vaginal flora was restored to the premenopausal type and that they required fewer antibiotics. KIRKENGEN and colleagues (1992) randomised 40 elderly women with recurrent urinary-tract infections to receive either oral oestriol (3 mg/day for 4 weeks followed by 1 mg/day for 8 weeks) or matched placebo. After the first treatment period, no difference was found between oestriol and placebo. However, following the second treatment period, oestriol was significantly better than placebo in reducing the incidence of urinary-tract infections.

A randomised, double blind, placebo-controlled study of 93 post-menopausal women has also shown that intravaginal oestriol cream prevents recurrent urinary-tract infections in those women presenting with this problem (RAZ and STAMM 1993). Midstream urine cultures were obtained at enrolment, monthly for 8 months and whenever urinary symptoms occurred. Changes in the vaginal pH and colonisation with lactobacilli were present within the estriol group only within 1 month of the start of treatment. The incidence of urinary-tract infection in the group given oestriol was significantly reduced compared with that in the group given placebo (0.5 vs 5.9 episodes per patient per year). Unfortunately, we have been unable to reproduce these results in a double-blind, placebo-controlled study of oral oestriol in the prevention of recurrent urinary-tract infections in elderly women (CARDOZO et al. 1998). Although both oestriol and placebo improved urinary symptoms during the trial, the incidence of urinary-tract infection did not differ significantly between the two groups.

E. Estrogen for Vaginal Atrophy

Estrogen deficiency results in atrophy of all steroid-hormone-sensitive tissues, including the vagina. Cytology is the most frequently used method of objective assessment and is based on the examination of desquamated cells collected from vaginal smears. Cells are classified as parabasal, intermediate or superficial, depending on their degree of differentiation. Several different quota, such as the KPI (karopycnotic index), MV (maturation value) and MI (maturation index), can be calculated to express the degree of atrophy as a numeric value (HAMMOND 1977). To standardise the evaluation of cytological assessment, morphometric analyses of the characteristics of cell nuclei have been developed, as have immunohistochemical studies of Ki-67 antigen, found in normal proliferating cells (McCORMICK 1993).

Estrogen-replacement therapy induces maturation of postmenopausal atrophic vaginal epithelium. This is measured cytologically by a disappearance of parabasal cells and a significant increase in superficial cells on vaginal smear tests. In addition, an increased thickness and number of cell layers in vaginal epithelium can be seen on vaginal biopsy specimens, as can the presence of Ki-67 antigen binding sites (NILSSON 1995). Estrogen therapy has also been shown to increase vaginal blood flow (SEMMENS 1985), reduce vaginal dryness and lower vaginal pH (NILSSON 1995).

Genital atrophy may be successfully treated with low-dose topical estrogen regimens (creams, tablets, pessaries and rings) without the potential side effects and endometrial proliferation associated with systemic preparations. However, it does appear that prolonged administration, sometimes of more than a year, may be necessary for maximum benefit and that symptoms recur on discontinuation of treatment (IOSIF 1992). A dose of 0.5mg oestriol given twice weekly is usually an effective treatment of vaginal atrophy. The messiness and inconvenience associated with vaginal creams or pessaries can be overcome by the use of a sustained-release silicone intravaginal ring. SMITH and co-workers (1993) evaluated a 55-mm-diameter silicone vaginal ring releasing 5–10μg of oestradiol per 24h for a period of at least 90days. Cure or improvement of atrophic vaginitis was recorded in more than 90% of the subjects, with the majority of women finding this form of treatment acceptable even during sexual intercourse. Unfortunately, no placebo-controlled trials of this device have been reported yet and it is, therefore, possible that the ring has an effect on the vaginal epithelium merely by acting as a foreign body in the vagina.

Tibolone has also been shown to cause vaginal maturation and improve the symptoms of genital atrophy (RYMER 1994). This synthetic compound has estrogenic, progestogenic, and androgenic effects, with its estrogenic potency approximately 1/50 that of ethinyl estradiol. The main benefit of this medication is that it causes very little endometrial stimulation (GENAZZANI 1991; TREVOUX 1983), eliminating the need for women to have a progestogen-induced withdrawal bleed.

F. Antiestrogens and the Urogenital Tract

Tamoxifen, perhaps the best-known antiestrogen, has an established role in the adjuvant treatment of breast carcinoma in women of all age groups and all stages of the disease (ZIEGLER and BUZDAR 1991; Early Breast Cancer Trialists' Collaborative Group 1998). The recurrence and mortality rates are significantly reduced in women taking this therapy compared with controls, as is the risk of developing a new primary cancer in the contralateral breast. There is at present conflicting evidence as to its role in the prevention of disease in apparently healthy women with a strong family history of breast cancer (PRITCHARD 1998).

In view of tamoxifen's important benefits in the treatment of breast disease, its effect on the urogenital tract may be of little practical importance and therefore be underreported. However, there is an association between the use of tamoxifen and development of endometrial pathology, including endometrial polyps, different types of endometrial hyperplasia and endometrial carcinoma (FORANDER 1989; DE MUYLDER 1991; CORLEY 1992; HULKA and HALL 1993). Endometrial abnormalities have been detected in 29–39% of endometrial biopsies in women on long-term tamoxifen treatment (COHEN 1993; KEDAR 1994), although increased surveillance may account for some of the apparent increase in risk.

Several studies have shown that tamoxifen also has an estrogen-like action on the vagina, maturing the epithelium in postmenopausal women (FERRAZZI 1977; BOCCARDO 1981; EELLS 1990). There is unfortunately a paucity of information regarding the effects of antiestrogens on the lower urinary tract, although it is likely that they have a beneficial rather than an adverse effect. This area requires further investigation before the mechanism by which tamoxifen induces agonistic estrogen effects in the urogenital tract, but is antagonistic in the breast, is understood and firm conclusions can be made.

G. Conclusions

The lower urogenital tract is sensitive to the effect of sex steroids, with estrogen deficiency implicated in the pathogenesis of a wide range of urogenital complaints. Urinary incontinence presents most frequently around the time of menopause, but unfortunately estrogen-replacement therapy, when given alone, does not objectively improve stress incontinence; when given in combination with alpha-adrenergic agonists it may be more effective. Estrogen may improve urinary frequency and urgency, but further studies are required to establish the optimal type of estrogen, route of administration and duration of therapy. There is evidence to suggest that estrogen may be useful clinically for prophylaxis against recurrent urinary-tract infections and in the treatment of vaginal atrophy when given systemically or locally. The effect of antiestrogens on the urogenital tract has not been extensively investigated and, therefore, their mechanism of action and clinical significance at this site are still to be determined.

References

Abrams P, Blaivas JG, Stanton SL et al. (1990) The standardisation of terminology of lower urinary tract function. Br J Obstet Gynaecol 97:1–16

Allen RE, Warrell DW (1992) The role of pregnancy and childbirth in partial denervation of the pelvic floor. Neurourol Urodyn 6:183–184

Barlow DH, Cardozo LD, Francis RM, Griffin M, Hart DM, Stephens E, Sturdee DW (1997) Urogenital ageing and its effect on sexual health in older British women. Br J Obstet Gynaecol 104:87–91

Beisland HO, Fossberg E, Moer A, Sander S (1984) Urethral sphincteric insufficiency in postmenopausal females treatment with phenylpropanolamine and estriol separately and in combination. Urol Int 39:211–216

Bergman A, Karram MM, Bhatia NN (1990) Changes in urethral cytology following estrogen administration. Gynecol Obstet Invest 29:211–213

Bhatia NN, Bergman A, Karram MM (1989) Effects of estrogen on urethral function in women with urinary incontinence. Am J Obstet Gynecol 160:176–181

Blakeman PJ, Hilton P, Bulmer JN (1996a) Mapping estrogen and progesterone receptors throughout the female lower urinary tract. Neurourol Urodyn 15:324–325

Blakeman PJ, Hilton P, Bulmer JN (1996b) Estrogen status and cell cycle activity in the female lower urinary tract. Neurourol Urodyn 15:325–326

Blakeman PJ, Hilton P, Bulmer JN (1997) Androgen receptors in the female lower urinary tract. Int Urogynecol J Pelvic Floor Dysfunct 8:S54

Boccardo F, Bruzzi P, Rubagotti A, Nicolo GU, Rosso R (1981) Estrogen like action of tamoxifen on vaginal epithelium in breast cancer patients. Oncology 38:281–285

Brandberg A, Mellstrom D, Samsioe G (1985) Peroral estriol treatment of older women with urogenital infections. Lakartidningen 82:3399–3401

Brocklehurst JC (1972) Ageing of the human bladder. Geriatrics 27:154

Cardozo LD, Tapp A, Versi E (1987) The lower urinary tract in peri and postmenopausal women. In: Samsioe E, Bonne Erickson P (eds) The urogenital estrogen deficiency syndrome. Novo Industri A/S, Bagsverd, Denmark, pp 10–17

Cardozo LD, Rekers H, Tapp A, Barnick C, Shepherd A, Schussler B, Kerr-Wilson R, van Geelan J, Barlebo H, Walter S (1993) Oestriol in the treatment of postmenopausal urgency: a multicentre study. Maturitas 18:47–53

Cardozo L, Benness C, Abbott D (1998) Low dose estrogen prophylaxis for recurrent urinary tract infections in elderly women. Br J Obstet Gynaecol 105:403–407

Cohen I, Rosen D, Tepper R, Cordoba M, Shapira Y, Altaras MM, Yigael D, Beyth Y (1993) Ultrasonographic evaluation of the endometrium and correlation with endometrial sampling in postmenopausal patients treated with tamoxifen. J Ultrasound Med 12:275–280

Collas DM, Malone Lee J (1996) Age-associated changes in detrusor sensory function in women with lower urinary tract symptoms. Int Urogynecol J Pelvic Floor Dysfunct 7:24–29

Corley D, Rowe J, Curtis MT, Hogan WM, Noumoff JS, Livolsi VA (1992) Postmenopausal bleeding for unusual endometrial polyps in women on chronic tamoxifen therapy. Obstet Gynecol 79:111–116

Cutner A (1993) The lower urinary tract in pregnancy. MD thesis, University of London

Cutner A, Burton G, Cardozo LD et al. (1991) Does progesterone cause an irritable bladder? Int Urogynecol J Pelvic Floor Dysfunct 98:1181–1183

Cutner A, Carey A, Cardozo LD (1992) Lower urinary tract symptoms in early pregnancy. J Obstet Gynaecol 12:75–78

de Muylder X, Neven P, de Somer M, van Belle Y, Vanderick G, De Muylder E (1991) Endometrial lesions in patients undergoing tamoxifen therapy. Int J Gynaecol Obstet 36:127–130

Early Breast Cancer Trialists' Collaborative Group (EBCTCG) (1998) Tamoxifen for early breast cancer: an overview of the randomised trials. Lancet 351:1451–1467

Eells TP, Alpern HD, Grzywacz C, MacMillan RW (1990) The effect of tamoxifen on cervical squamous maturation in Papanicolaou stained cervical smears of post menopausal women. Cytopathology 1:263–268

Fantl JA, Wyman JF, Anderson RL, Matt DW, Bump RC (1988) Postmenopausal urinary incontinence: comparison between non-estrogen and estrogen supplemented women. Obstet Gynecol 71:823–828

Fantl JA, Cardozo LD, McClish DK (1994) Estrogen therapy in the management of urinary incontinence in postmenopausal women: a meta-analysis. First report of the Hormones and Urogenital Therapy Committee. Obstet Gynecol 83:12–18

Fantl JA, Bump RC, Robinson D, McClish DK, Wyman JF (1996) Efficacy of estrogen supplementation in the treatment of urinary incontinence. Obstet Gynecol 88:745–749

Ferrazzi E, Cartei G, Mattarazzo R, Fiorentino M (1977) Estrogen like effects of tamoxifen on vaginal epithelium. BMJ 1:1351–1352

Fornander T, Rutqvist LE, Cedermark B, Glas U, Mattsson A, Silfversward C, Skoog L, Somell A, Theve T, Wilking N, et al. (1989) Adjuvant tamoxifen in early breast cancers: occurrence of new primary cancers. Lancet 1:117–119

Gangar KF, Vyas S, Whitehead M, Crook D, Meire H, Campbell S (1991) Pulsatility index in the internal carotid artery in relation to transdermal oestradiol and time since menopause. Lancet 338:839–842

Genazzani AR, Benedek-Jaszman LJ, Hart DM, Andolsek L, Kicovic PM, Tax L (1991) ORG OD 14 and the endometrium. Maturitas 13:243–251

Gjorup T, Hendriksen C, Lund E, Stromgard E (1987) Is growing old a disease? A study of the attitudes of elderly people to physical symptoms. J Chronic Dis 40:1095–1098

Hammond D (1977) Cytological assessment of climacteric patients. Clin Obstet Gynecol 4:49–70

Hilton P (1981) Urethral pressure measurement by microtransducer: observations on the methodology, the pathophysiology of stress incontinence and the effects of treatment in the female. MD Thesis, University of Newcastle-upon-Tyne

Hilton P, Stanton SL (1983) The use of intravaginal estrogen cream in genuine stress incontinence. Br J Obstet Gynaecol 90:940–944

Hilton P, Tweddel AL, Mayne C (1990) Oral and intravaginal estrogens alone and in combination with alpha-adrenergic stimulation in genuine stress incontinence. Int Urogyn J Pelvic Floor Dysfunct 12:80–86

Hulka CA, Hall DA (1993) Endometrial abnormalities associated with tamoxifen therapy for breast cancer: sonographic and pathologic correlation. AJR Am J Roentgenol 160:809–812

Ingelman-Sundberg A, Rosen J, Gustafsson SA, Carlstrom K (1981) Cytosol estrogen receptors in urogenital tissues in stress incontinent women. Acta Obstet Gynecol Scand 60:585–586

Iosif CS (1992) Effects of protracted administration of estriol of the lower genitourinary tract in postmenopausal women. Arch Gynecol Obstet 251:115–120

Iosif CS, Bekassy Z (1984) Prevalence of genito-urinary symptoms in the late menopause. Acta Obstet Gynecol Scand 63:257–260

Iosif CS, Batra S, Ek A, Astedt B (1981) Estrogen receptors in the human female lower urinary tract. Am J Obstet Gynecol 141:817–820

Jackson S, Vyas S (1998) A double-blind, placebo controlled study of postmenopausal estrogen replacement therapy and carotid artery pulsatility index. Br J Obstet Gynaecol 105:408–412

Jackson S, Avery N, Shepard A et al. (1996a) The effect of oestradiol on vaginal collagen in postmenopausal women with stress urinary incontinence. Neurourol Urodyn 15:327–328

Jackson S, Shepherd A, Abrams P (1996b) The effect of oestradiol on objective urinary leakage in postmenopausal stress incontinence; a double blind placebo controlled trial. Neurourol Urodyn 15:322–323

Jarvis GJ, Hall S, Stamp S, Millar DR, Johnson A (1980) An assessment of urodynamic investigation in incontinent women. Br J Obstet Gynaecol 87:893–896

Jolleys JV (1988) Reported prevalence of urinary incontinence in women in a general practice. BMJ 296:1300–1302

Karram MM, Yeko TR, Sauer MV, Bhatia NN (1989) Urodynamic changes following hormone replacement therapy in women with premature ovarian failure. Obstet Gynecol 74:208–211

Kedar RP, Bourne TH, Powles TJ, Collins WP, Ashley SE, Cosgrove DO, Campbell S (1994) Effects of tamoxifen on uterus and ovaries of postmenopausal women in randomised breast cancer prevention trial. Lancet 343:1318–1321

Kirkengen AL, Andersen P, Gjersoe E, Johannessen GR, Johnsen N, Bodd E (1992) Oestriol in the prophylactic treatment of recurrent urinary tract infections in post-menopausal women. Scand J Prim Health Care 10:139–142

Kolbl H, Strassegger H, Riis PA et al. (1989) Morphologic and functional aspects of pelvic floor muscles in patients with pelvic relaxation and genuine stress inconti-nence. Obstet Gynecol 74:789–795

Kondo A, Kato K, Saito M et al. (1990) Prevalence of handwashing incontinence in females in comparison with stress and urge incontinence. Neurourol Urodyn 9:330–331

Malone Lee J, Wahenda I (1993) The characterisation of detrusor contractile function in relation to old age. Br J Urol 72: 873–880

McCallin PF, Taylor ES, Whitehead RW (1950) A study of the changes in the urinary sediment during the menstrual cycle. Am J Obstet Gynecol 60:64–74

McCormick D, Chong H, Hobbs C, Datta C, Hall PA (1993) Detection of Ki-67 antigen in fixed and wax embedded sections with the monoclonal antibody MIB1. Histopathology 22:355–360

Nilsson K, Heimer G (1992) Low dose oestradiol in the treatment of urogenital estro-gen deficiency – a pharmacokinetic and pharmacodynamic study. Maturitas 15:121–127

Nilsson K, Risberg B, Heimer G (1995) The vaginal epithelium in the postmenopause – cytology, histology and pH as methods of assessment. Maturitas 21:51–56

Pritchard KI (1998) Is tamoxifen effective in prevention of breast cancer? Lancet 352:80–81

Raz R, Stamm WE (1993) A controlled trial of intravaginal estriol in postmenopausal women with recurrent urinary tract infections. N Engl J Med 329:753–756

Rud T (1980a) The effects of estrogens and gestagens on the urethral pressure profile in urinary continent and stress incontinent women. Acta Obstet Gynecol Scand 59:265–270

Rud T (1980b) Urethral pressure profile in continent women from childhood to old age. Acta Obstet Gynecol Scand 59:331–335

Rud T, Andersson KE, Asmussen M, Hunting A, Ulmsten U (1980) Factors maintain-ing the urethral pressure in women. Invest Urol 17:343–347

Rymer J, Chapman MG, Fogelman I, Wilson PO (1994) A study of the effect of tibolone on the vagina in postmenopausal women. Maturitas 18:127–133

Salmon UL, Walter RI, Gast SH (1941) The use of estrogens in the treatment of dysuria and incontinence in postmenopausal women. Am J Obstet Gynecol 14:23–31

Samsioe G, Jansson I, Meelstron D, Svanborg A (1985) Occurrence, nature and treat-ment of urinary incontinence in a 70 year old female population. Maturitas 7:335–342

Screiter F, Fuchs P, Stockamp K (1976) Estrogenic sensitivity of alpha receptors in the urethra musculature. Urol Int 31:13–19

Semmens JP, Tsai CC, Semmens EC, Loadholt HB (1985) Effects of estrogen therapy on vaginal physiology during menopause. Obstet Gynecol 66:15–18

Smith ARB, Hosker GL, Warrell DW (1989) The role of partial denervation of the pelvic floor in the aetiology of genito-urinary prolapse and stress incontinence of urine. A neurophysiological study. Br J Obstet Gynaecol 96:24–28

Smith P (1993) Estrogens and the urogenital tract. Acta Obstet Gynecol Scand Suppl 72:1–26

Smith P, Heimer G, Lindskog M, Ulmsten U (1993) Oestradiol releasing vaginal ring for treatment of postmenopausal urogenital atrophy. Maturitas 16:145–154

Smith PJB (1976) The effect of estrogens on bladder function in the female. In: Management of the menopause and post menopausal years. MTP, Lancaster, pp 291–298

Soloman C, Panagotopoulos P, Oppenheim A (1958) The use of urinary sediment as an aid in endocrinological disorders in the female. Am J Obstet Gynecol 76:56–60

Stanton SL, Kerr-Wilson R, Harris VG (1980) The incidence of urological symptoms in normal pregnancy. Br J Obstet Gynaecol 87:897–900

Sultana CJ, Walters MD (1995) Estrogen and urinary incontinence in women. Maturitas 20:129–138

Susset JG, Servot-Viguier D, Lamy F, Madernas P, Black R (1978) Collagen in 155 human bladders. Invest Urol 16:204–206

Svanberg A (1977) The gerontological and geriatric population study in Goteborg, Sweden. Acta Med Scand 611:1–37

Thomas TM, Plymat KR, Blannin J, Meade TW (1980) Prevalence of urinary incontinence. BMJ 281:1243–1245

Trevoux R, Dieulangard P, Blum A (1983) Efficacy and safety of ORG OD 14 in the treatment of climacteric complaints. Maturitas 5:89–96

van Geelen JM, Doesburg WH, Thomas CMG, Martin CB Jr (1981) Urodynamic studies in the normal menstrual cycle: the relationship between hormonal changes during the menstrual cycle and the urethral pressure profile. Am J Obstet Gynecol 141:384–392

Versi E, Cardozo LD (1986) Urethral instability: diagnosis based on variations in the maximum urethral pressure in normal climacteric women. Neurourol Urodyn 5:535–541

Vetter NJ, Jones DA, Victor CR (1981) Urinary incontinence in the elderly at home. Lancet 2:1275–1277

Wagg AS, Lieu PK, Ding YY, Malone-Lee JG (1996) A urodynamic analysis of age associated changes in urethral function in women with lower urinary tract symptoms. J Urol 156:1984–1988

Walter S, Wolf H, Barlebo H et al. (1978) Urinary incontinence in postmenopausal women treated with estrogens. Urol Int 33:135–143

Ziegler LD, Buzdar AU (1991) Current status of adjuvant therapy of early breast cancer. Am J Clin Oncol 14:101–110

CHAPTER 20

Effects of Estrogens on Various Endocrine Regulations

F. GOMEZ

A. The Complexity of Estrogen's Action on Various Tissues

I. Introduction

The steroid-receptor molecule has been defined as a site for crosstalk between the endocrine system and either growth factors or neurotransmitters, since the latter have been shown to drive receptor activation, nuclear translocation and steroid-dependent signal transduction in a ligand-independent manner. In turn, steroid receptors activate growth-factor expression in tissues (reviewed by MOUTSATSOI and SEKERIS 1997). A similar crosstalk has been suggested among subregions of the estrogen receptor (ER), depending on the nature of the ligand. Agonist binding might not be necessary for ER binding to an estrogen-responsive element (ERE) in DNA, but its main role may consist of converting an ER into an active form for transcription. The two transcription activating functions (AF) that have been described in nuclear receptors, AF-1 and AF-2, can be activated or selectively interfered with by the ligand, depending on its agonist, mixed agonist/antagonist, or pure antagonist nature (METZGER et al. 1995).

II. Estrogen Receptor-Mediated Tissue Specificity

Studies with synthetic ligands of the ER are unveiling the complexity of estrogen's action on tissues. This is illustrated by the case of raloxifene, which appears to have agonistic or antagonistic properties, depending on the tissue considered. Raloxifene's antagonistic properties have been demonstrated in rat estrogen-dependent mammary tumors. However, this non-steroidal ER ligand activates the gene encoding transforming growth factor-$\beta 3$ (TGF-$\beta 3$) (YANG et al. 1996a) in bone via pathways that are independent of EREs, but which require the integrity of the AF-2-activating function of ERs. TGF-$\beta 3$ is an important mediator of bone remodeling which inhibits the differentiation and bone-resorptive activities of osteoclasts. Therefore, in this system, raloxifene behaves as an ER agonist (YANG et al. 1996b). Animal and pre-clinical human studies demonstrated that raloxifene decreases the markers of bone turnover, which constitutes a typical estrogenic effect, although it does not cause uterine growth as natural estrogens would do (Black et al. 1994; Draper et al. 1996).

A 2-year interim report of an ongoing long-term clinical trial on raloxifene administered to post-menopausal women has demonstrated a significant dose-dependent increase in trabecular and cortical bone density, as assessed by dual-energy X-ray absorptiometry (DEXA), confirming the ER agonist effect of raloxifene on bone. Furthermore, a substantial number of these patients underwent sequential endovaginal ultrasonography at 6-month intervals, and no endometrial thickening was observed with any dose of raloxifene compared with controls (Delmas et al. 1997). Therefore, in endometrial ERs, raloxifene functioned as an antagonist or an indifferent agent. In the liver-modulated blood lipids, the effects were intermediate between a classical estrogenic agonist [reduction of total and low-density lipoprotein (LDL) cholesterol] and an antagonist or an indifferent agent [no increase in triglycerides and high-density lipoprotein (HDL) cholesterol]. Similarly, hot flashes were not inhibited by raloxifene, suggesting lack of estrogenic action at the cerebral level. Thus, this synthetic analogue has tissue-specific agonistic, indifferent or antagonistic properties on ERs. Rather, this tool permits dissociation of specific properties of ERs, which are different in different tissues and cannot be distinguished easily when the broader effects of the natural endogenous ligands are studied.

III. Sex-Hormone-Binding Globulin-Mediated Estrogenic Action

An additional action of estrogens that may be of importance is that mediated by the sex-hormone-binding globulin (SHBG). This plasma protein is more than a mere steroid carrier, since it is active on specific membrane receptors of its own (Khan et al. 1990). SHBG receptor activation, as assessed by intracellular accumulation of cAMP, only occurs when the steroid ligand of SHBG is secondarily added. In contrast, SHBG that has been previously liganded by a steroid cannot bind to its receptor. Therefore, in addition to their well-known effect on regulating SHBG plasma levels, estrogens exert a control over the SHBG–receptor interaction (Hryb et al. 1990). However, a SHBG-mediated estrogenic action on tissues has long been thought to take place, in terms of facilitated estrogen transport into target tissues (discussed by Rosner 1990). Also in this system, evidence suggests that there may be a different effect depending on the tissue involved. This has been shown in studies of the rat, a species that lacks SHBG. Concomitant bolus administration of human SHBG resulted in an increased accumulation of injected ^{3}H-estradiol in the uterus and the oviduct, compared with the administration of the radiolabeled steroid alone, whereas other tissues, equally putative estrogen targets, such as the pituitary, the liver and the brain, did not show such increased accumulation of ^{3}H-estradiol (Noe et al. 1992).

In conclusion, estrogens exert their systemic actions via mechanisms that may differ from one tissue to another, and this may result in marked differences in the intensity of their actions.

B. Estrogens and the Thyroid

I. Sexual Dimorphism in Thyroid Disease

The prevalence of thyroid disease, in general, has long been recognized to be predominant in women. In a large population-survey study in England, the female:male ratio on the prevalence of visible and/or palpable goiter was 4:1, and if only the larger goiters were taken into account, this ratio was 13:1. After the age of menopause, this high prevalence in women declined sharply (TURNBRIDGE et al. 1977). These and similar observations have led to the concept that estrogens play a facilitating role in the development of abnormal thyroid enlargement.

II. The Effect of Estrogens in Thyroid Enlargement

Various mechanisms may be involved and, among them, a possible direct action of estrogens on thyroid cells has received some attention. However, studies on the presence of ERs in the thyroid resulted in conflicting results, depending on the methodology used and the tissue condition (reviewed by CIOCCA and VARGAS ROIG 1995). Owing to the relative paucity of ERs in the thyroid, binding assays, such as the dextran-coated charcoal assay performed on cytosol preparations from tissue homogenates, may not be sensitive enough, and immunocytochemical studies are interfered with by tissue thyroperoxidase (GIANI et al. 1993). In contrast, cytosol immunoassays constitute a more suitable method to study ERs when they are present at low concentrations. Using an enzyme immunoassay based on a monoclonal antibody against ERs, BONACCI et al. (1996) demonstrated that the normal thyroid tissue surrounding neoplastic lesions contained ERs more often (61–71% of the cases) than the adenomatous (39%) or carcinomatous (35%) contiguous lesions.

ERs in thyroid tissue appear to be functional, since both normal and adenoma thyroid cells in culture display an enhanced expression of the gene encoding thyroglobulin and an enhanced ^{3}H-thymidine incorporation into cultured follicles when estrogen is added, whereas carcinomatous cultured cells fail to respond to estrogen (DEL SENNO et al. 1989). Interestingly, in molecular constructs, the estradiol-liganded ER can induce gene expression, not only by binding to DNA-specific ERE, but also via thyroid hormone response elements (TRE), provided that no thyroid receptor (TR) is present. However, if unliganded TR is present, ER binding to TRE is prevented, as if non-activated TR "protects" its own response element from activation by estrogens, in order to preserve TRE fidelity to T3 (GRAUPNER et al. 1991). Whether these findings are relevant for the understanding of the facilitating role of estrogens on goiter and on thyroid disease, in general, is not known. But it must be admitted that, as far as thyroid cancer is concerned, estrogen action on the thyroid has limited interest, contrary to breast cancer, where ER abundance has independent prognostic significance (VOLLENWEIDNER-ZERARGUI et al. 1986). Researchers are skeptical about any possible clinical application of ER determination in thyroid cancer (JAKLIC et al. 1995).

III. Thyroid Dysfunction

The prevalence of thyroid function abnormalities is also characterized by a marked sexual dimorphism, consistent with a facilitating role of estrogens. In the above-mentioned population survey (TUNBRIDGE et al. 1977), the female:male ratio of hyperthyroidism was between 8 and 14 and that of hypothyroidism between 14 and 19, with 19 of the 1000 women showing overt hyperthyroidism and 14 showing overt hypothyroidism. Although the prevalence of anti-thyroid autoantibodies in that population was much higher than that of overt clinical dysfunction, the sexual dimorphism was still clearly apparent: 10.3% of women presented with thyroid cytoplasmic antibodies compared with 2.7% of men. These differences persist with advancing age.

In a large population study of elderly people in Sweden, SUNDBECK et al. (1995) found that 20% of the women were positive for anti-thyroid peroxidase (TPO) antibodies at a titer greater than 100 kilounits/l and 7% at greater than 1000 kilounits/l; the corresponding figures for men were 9% and 4%, respectively. Owing to the sharp decline of circulating estrogens in women after the menopause, it may be concluded that, if estrogens are implicated in the appearance of autoimmune thyroid disease, they are not necessary for its maintenance. In addition, cross-sectional and longitudinal studies of euthyroid and hypothyroid aging women and of women with post-partum thyroiditis have demonstrated a remarkable constancy of the epitopic recognition by recombinant TPO autoantibodies over time, generated from DNA extracted from plasma cells infiltrating thyroids with autoimmune disease (JAUME et al. 1995a; JAUME et al. 1995b). Since this "epitopic fingerprint" may thus be inherited, the facilitating role of estrogens on the development of autoimmune thyroid disease should be a quantitative one, without influencing the qualitative differences among TPO autoantibodies. However, such a role of estrogens cannot be confirmed in the pregnancy model, where thyroid autoimmunity declines as estrogens markedly increase. The CD4:CD8 lymphocyte ratio, autoantibody titer and disease activity are significantly lower during the third trimester than at the beginning of pregnancy (GLINOER et al. 1994; AMINO et al. 1982; STAGNARO-GREEN et al. 1992; PATTON et al. 1987), whereas autoimmunity rebounds after delivery (ROTI and EMERSON 1992). The state of relative immune suppression that characterizes pregnancy, clearly overcomes any possible facilitating role of estrogens on anti-thyroid immune modulation.

IV. Thyroid Hormone-Binding Globulin

In states of relatively high estrogen impregnation, such as pregnancy and oral contraceptive intake, thyroid hormone-binding globulin (TBG) is increased

2- to 3-fold. The mechanism involved in this increase has been ascribed to a stimulatory effect of estrogens on hepatic TBG synthesis (REFETOFF et al. 1976; GLINOER et al. 1977), but a major contributing factor is the prolongation of the in vivo half-life ($t_{1/2}$) of circulating TBG, induced by estrogens. On isoelectric focusing (IEF), TBG discloses a marked microheterogeneicity, like other glycoproteins. This is due to changes in the carbohydrate moiety (that constitutes as much as 20% of the molecular weight of TBG) and, in particular, to its sialic acid content. Glycoproteins with high sialic acid content show an increased $t_{1/2}$, whereas the desialylated forms are rapidly taken up and cleared by the liver.

The serum of pregnant women contains forms of TBG that are particularly rich in sialic acid and that disclose more acidic isoelectric points. When these forms are purified on IEF and are injected to rats, a species which has no TBG, they display a markedly prolonged $t_{1/2}$ (AIN et al. 1987). These changes in the carbohydrate moiety of TBG in pregnant women are considered to be a post-translational effect of the increased circulating levels of estrogens during pregnancy. Since 70% of thyroid hormones are bound to TBG, increases in circulating TBG levels result in corresponding increases of total circulating T4 and T3. However, the levels of free hormones, which are a determinant for the biological activity, remain unaltered.

V. Thyrotropin Regulation

Regarding the pituitary regulation of thyroid function, apparent effects of estrogens have been described, although their physiological significance is not well established. Classical studies in the human have demonstrated that estrogens potentiate the thyroid-stimulating hormone (TSH) response to exogenously administered thyrotropin-releasing hormone (TRH). This was the case in men treated with estrogens (FAGLIA et al. 1973) and in women taking oral contraceptives (RAMEY et al. 1975). The mechanisms behind these effects have not been elucidated. The anterior pituitary of female rats contains more TRH receptors than that of males, and this may justify an increased TSH response, but it also contains more T3 receptors and more T4-5' deiodinase activity which, on theoretical grounds, could contribute to an increased negative feedback of T3 on the pituitary and, therefore, to the opposite result in terms of TSH response to TRH (DONDA et al. 1990). These studies, however, have not determined whether the observed sexual differences concern specifically the thyrotropic cells or other anterior pituitary cells as well. These studies, together with contradictory results published on the response to TRH (REYMOND and LEMARCHAND-BERAUD 1976), cast doubt on the importance of an estrogen regulation of TSH secretion. Indeed, neither free T4 and T3 nor basal TSH are significantly altered in normal individuals of either sexes as a consequence of an increased exposure to estrogens alone.

C. Estrogens and the Adrenal Glands

I. Sexual Dimorphism in Adrenal Disease?

We know very little about the role of estrogens and ERs in adrenal disease. Adrenal medulla catecholamine-secreting tumors (pheochromocytomas) and extra-adrenal sympathetic paragangliomas occur at an approximately equal incidence in adult males and females. However, tumors in children occur more frequently in males, suggesting that hormonal changes at puberty may influence tumor growth and secretion (reviewed by Tischler 1991). As for adrenal cortex disease, there is an overall moderate female sex predilection, with a female:male ratio of 3–4:1 for Cushing's syndrome, 2:1 for primary aldosteronism and 2.5:1 for autoimmune Addison's disease, with an equal prevalence of tuberculous Addison's disease (Labhart 1986).

II. Regulation of Adrenal Sex-Steroid Precursors

The adrenal cortex of humans and superior primates is unique in that it secretes large amounts of dehydroepiandrosterone (DHEA) and DHEA-sulfate (DHEA-S) which are converted into potent androgens and estrogens in target tissues. Indeed, the key enzymes involved in this peripheral conversion have been shown to be widely distributed in the tissues of rhesus monkeys. This includes sulfatase, 3β-hydroxysteroid dehydrogenase (3β-HSD), 17β-hydroxysteroid dehydrogenase (17β-HSD), 5α-reductase and aromatase activities (Martel et al. 1994). This conversion allows for a substantial extra-gonadal sex steroid production from the steady flow of adrenal precursors in superior primates, in contrast with the exclusively gonadal origin of circulating sex steroids in lower species, such as small rodents. In turn, estrogens are implicated in the regulation of adrenal DHEA and DHEA-S secretion, in a manner that resembles a feedback mechanism.

On the one hand, studies on human fetal adrenal cells in culture have suggested that estrogens contribute to the low 3β-HSD activity observed in fetal adrenals, which in the presence of corticotropin (ACTH) results in a typically low cortisol and high DHEA secretion by these cells (Fujieda et al. 1982). On the other hand, the markedly elevated placental estrogens that characterize the second half of gestation exert a regulatory control on adrenal precursor production. Pregnant baboons in which estrogens were prematurely elevated twofold at midgestation by maternal administration of androstenedione, showed an attenuated or suppressed DHEA response to ACTH (Pepe and Albrecht 1990). It is not certain, however, that this effect of estrogens is exerted mainly on the fetal adrenals, since comparable suppressions of DHEA and DHEA-S secretions were observed when high doses of estradiol benzoate were administered to pregnant baboons, to pregnant baboons after fetectomy, and to non-pregnant baboon females (Albrecht and Pepe 1995).

From these studies, it can be inferred that estrogens control steroid-precursor secretion, certainly in the adult adrenal cortex, possibly also in the fetal adrenals. These actions could be exerted by the activation of specific ERs. The presence of ERs has been demonstrated for the first time in primate adrenals by HIRST et al. (1992), using a careful technique that prevents receptor loss from the nucleus by stabilizing the ERs with previous incubation of fresh tissue with estrogen ligand to increase the nuclear DNA affinity of ERs. Interestingly, these authors found abundant ERs in the nuclear extracts of adrenal tissue from adult animals of either sex, but very limited amounts of ER in fetal adrenals.

III. Anti-toxic Properties of the Adrenal Glands

Adrenals are involved in protecting the organism from the detrimental effects of free radicals and in the metabolizing of a variety of foreign compounds, and there is evidence that these properties are influenced by the sex-steroid hormonal milieu, including a marked effect of estrogens. α-Tocopherol is one of the major antioxidants that protects cell membranes from free radicals. Free radical-initiated processes in membranes, such as lipid peroxidation (LP), are inhibited by α-tocopherol and, after scavenging, α-tocopherol is regenerated by ascorbic acid. Adrenal cortex in rodents contains strikingly high concentrations of α-tocopherol, which are approximately tenfold greater than those of liver, and the concentrations of ascorbic acid are also particularly high. The highest α-tocopherol concentrations are found in the outer zone (zona glomerulosa and zona fascicul(aris), which contains low LP activity, whereas the inner zone (zona reticularis) contains little α-tocopherol. Experimental reduction of α-tocopherol results in increased LP activity (COLBY et al. 1995).

It has been shown that the concentrations of α-tocopherol in adrenals and liver are far greater in female rats than in male rats, and experiments on castration and sex-steroid replacement have demonstrated that estrogens increase and androgens decrease α-tocopherol content in those tissues (FEINGOLD et al. 1993). These effects may be exerted directly on the adrenals through ERs or indirectly via changes in lipid metabolism, since androgens and estrogens can modify the lipoprotein uptake by tissues and α-tocopherol is largely associated with lipoproteins in blood.

In contrast, the adrenal cortex possesses the ability to demethylate and hydroxylate a variety of xenobiotics. This property appears to be associated with the activity of a distinctive cytochrome P450 that is not involved in steroidogenesis and that is suppressed, not stimulated, by ACTH. This 52-kDa protein is mainly concentrated in the inner zone (zona reticularis) of the adrenal cortex, and estrogens appear to suppress both protein concentration and enzymatic activity (BLACK 1994). This negative effect of estrogens explains the lesser ability of females to metabolize xenobiotics, compared with males, and contrasts with the above-mentioned scavenging ability for free radicals that is located in the outer zones and is stimulated by estrogens.

IV. Corticosteroid-Binding Globulin

The 50- to 60-kDa glycoprotein corticosteroid-binding globulin (CBG) binds over 90% of circulating cortisol, but it is more than a simple buffer or carrier for glucocorticoid. It plays a major role in delivering cortisol into inflammatory tissues, where the serine-protease elastase system cleaves the sequence close to the carboxy terminus to release bound cortisol in situ. Moreover, CBG binds to specific membrane receptors eliciting a biological action in cells (reviewed by Seralini 1996). Therefore, any factors that influence the levels of CBG or its binding properties, may result in changes in glucocorticoid activity. Estrogens figure among these factors. As is the case for other circulating glycoproteins, CBG presents a molecular-size microheterogeneity due to differences in glycosylation (Avvakumov et al. 1993). The carbohydrate moieties confer conformational stability and solubility to CBG, while possibly protecting it from proteolysis. Sialic acid content influences both the isoelectric point and the $t_{1/2}$ of CBG, similarly to TBG (see Sect. B.IV), and desialylated forms have an increased liver clearance.

Although CBG secretion and glycosylation depend on the preservation of an intact peptidic structure, as demonstrated in human CBG-transfected Chinese hamster ovary (CHO) cells by Avvakumov and Hammond (1994), estrogens may directly influence the glycosylation of CBG. They may also directly enhance the expression of the human CBG gene. The combination of these two mechanisms results in the well-known increase of plasma CBG during pregnancy and during the course of oral contraceptive intake. The existence of a natural variant of CBG with modified carbohydrate moieties has indeed been demonstrated in human pregnancy serum (Avvakumov and Hammond 1994; for review, see also Hammond 1990).

V. Corticotropin Regulation

It has been known for many years that there is a sex difference in the resting and stress-stimulated activity of the hypothalamo–pituitary–adrenal (HPA) axis in lower species, with females displaying higher corticosterone levels than males. A model for the gender-specific organization of HPA early in life has been proposed, in which testicular testosterone determines the male pattern after its aromatization to estradiol in the brain and then acting via ERs (Patchev et al. 1995). In the adult female rat, estrogen treatment to ovariectomized animals resulted in an increased HPA activity, as assessed by the ACTH and/or the corticosterone response to footshock or ether stress, possibly by impairing the function of the glucocorticoid receptors in the hippocampus and in other structures involved in the negative feedback on corticotropin secretion (Burgess and Handa 1992).

Intact adult female mice display a greater corticosterone response than males to bacterial endotoxin administration (Spinedi et al. 1992). Since estradiol was less effective than testosterone in controlling endotoxin stimulation of both the HPA axis and the secretion of tumor necrosis factor (TNF) in gonadectomized animals, these authors concluded that there was neuroendocrine-immunological sexual dimorphism in mice. The basis for this may reside, at least in part, in gender differences in the expression of the corticotropin-releasing hormone (CRH) gene. VAMVAKOPOULOS and CHROUSOS (1993) found estrogen, but not androgen or glucocorticoid-responsive elements, in the 5' flanking region of the human (h) CRH gene. Estradiol activated the transcription of hCRH constructs, indicating that ERs may bind the ERE sequences in the promoter region of the hCRH gene.

Whether these observations are of relevance for the understanding of human physiology or pathology is not clear. The in vivo evidence of a sexual dimorphism in the human species, in terms of basal or stress-induced HPA activity, is much more difficult to ascertain than in small rodents, and the ACTH and cortisol responses to CRH in young individuals are unaffected by gender (MARINI et al. 1991). The blunted response observed during the third trimester of pregnancy is probably related to the existence of large amounts of endogenous CRH of placental origin as well as of circulating CRH-binding proteins, rather than to a decreased sensitivity of the pituitary corticotrophs. These data indicate that estrogens do not participate in a significant manner in the regulation of the HPA axis activity in the human.

D. Estrogens and the Phosphocalcic Metabolism

I. Bone Remodeling and Calcium Kinetics

No clear gender difference has been documented for experimental animals or humans with respect to the architecture of cancellous bone, where androgens and estrogens seem to exert analogous effects. In contrast, males have greater cortical bone width than females, and this sex difference continues to increase well after the pubertal growth spurt and closure of the epiphyses. In case of gonadal failure in both sexes, before and after puberty, osteopenia develops (reviewed by TURNER et al. 1994). Menopause and estrogen deprivation in women result in a reduction of cancellous bone due to the removal of entire trabeculae, and a similar observation can be made in the ovariectomized rat, with no significant decrease in trabecular width. In cortical bone, an age-dependent continuous loss obscures the demonstration of a role for estrogens in the human, but histomorphometric studies performed in the ovariectomized rat model have clearly shown that long-term (1 year) estrogen deprivation induces a loss in both cancellous and cortical bone (LI and WRONSKI 1995).

Estrogens are recognized as the hormones with the most profound effect on the female skeleton, and estrogen replacement therapy (ERT) is the single most effective way to prevent post-menopausal osteoporosis and reduce the incidence of fractures of the vertebrae, distal forearm and hip. Estrogens suppress bone remodeling by decreasing bone resorption and increasing bone formation, but the variety of mechanisms involved is not completely known. The effects of estrogens could be exerted either directly on bone cells to modulate bone remodeling, or after the induction of blood-borne hormones or locally elicited mediators in the bone. Some of the metabolic effects observed could be the result of one or more of these mechanisms. For example, studies on calcium kinetics with oral ^{44}Ca-stable calcium isotope in young girls with Turner's syndrome have shown that short-term (4-week) ERT with ethinyl estradiol results in a significant reduction of whole-body calcium turnover and a significant increase of calcium absorption and of (urinary) calcium retention (MAURAS et al. 1997). However, in this study a significant increase was also observed in the serum concentration of 1,25-dihydroxy vitamin D (1,25(OH)$_2$D or calcitriol), a common observation on ERT (see Sect. D.III. below). Therefore, the observed results on calcium kinetics could be the consequence of a direct stimulation of bone formation, an effect mediated by increasing calcitriol levels, or both.

The complex interaction among the organs involved in calcium handling and regulation, namely bone, kidney and the intestines, and the main calciotropic hormones, namely parathyroid hormone (PTH), calcitriol and calcitonin, has been reviewed extensively by PRINCE (1994). This author suggested that the fall in plasma-ionized calcium that can be observed during ERT may be proportional to the rise in bone mass, and is expected to be greater in individuals with a high bone turnover.

In a double-blind, placebo-controlled, randomized study of postmenopausal women with low forearm bone density, it was observed that combined estrogen and progestogen treatment resulted in a marked and persistent decrease in urinary calcium excretion, to a greater extent than could be accounted for by the decrease in filtered calcium load (PRINCE et al. 1991a). It appears that in the estrogen-deficient state, a renal calcium leak develops that is reverted by hormonal replacement therapy. In addition, estrogen treatment in osteoporotic women results in an apparent increase in the kidney response to PTH since, despite a smaller ability shown by these patients to increase PTH in response to the hypocalcemia induced by edetic acid (EDTA) infusion compared with untreated patients, the urinary phosphate/creatinine and cAMP/creatinine ratios increased significantly and similarly in both groups (COSMAN et al. 1994).

A similar, sensitive mechanism is probably operating in the ovariectomized rat model, since, soon after ovariectomy, these animals disclosed a significant increase in the maximal renal tubular reabsorption of phosphate, even when there was no clear change in renal calcium handling in this experiment (DICK et al. 1996). Some of the effects of estrogens in the calcium kinetics could occur in the kidneys by interaction of estrogens with ERs located in the renal cortex (DICK and PRINCE 1997). However, direct effects of estrogens on bone cells may certainly occur, since the existence of functional ERs has been demonstrated in cultured human osteoblast-like cells, avian osteoclasts, and human osteoclast-like cells from an osteoclastoma (OURSLER et al. 1994; OURSLER et al. 1991; ERIKSEN et al. 1988; KOMM et al. 1988). The anti-resorptive action of estrogens on bone results, at least in part, from control of cytokines and growth factors produced in situ. A major candidate mediator is TGFβ, which is produced by many cells including osteoclasts and osteoblasts, and which strongly inhibits osteoclastic function. Estrogens have been shown to increase the production and secretion of TGFβ by human osteoblast-like cells and by avian osteoclasts in culture. In the bone, TGFβ may act in a paracrine or autocrine manner to suppress osteoclasts (reviewed by TURNER et al. 1994). However ERs in bone cells, which were longtime elusive to researchers, are still difficult to demonstrate, and the in vitro response of these cells to estrogens are not largely reproducible. This contrasts with in vivo studies, in which the trophic effects of estrogens on bone metabolism are readily demonstrable. For this reason, it can be suspected that many of these effects are indirect, and are mediated by calciotropic hormones, mainly PTH and calcitriol.

II. Parathyroid Hormone

Careful observation of the initiation of ERT in post-menopausal women discloses a moderate decrease in serum ionized calcium that is not matched by an appropriate PTH increase (BOUCHER et al. 1989; SELBY and PEACOCK 1986a). Furthermore, the PTH response to an EDTA-induced hypocalcemia, is blunted in such patients (COSMAN et al. 1994). These observations suggest that estrogen treatment resets the threshold for PTH secretion at a lower calcium level. This effect may not be a direct one on ERs in the parathyroids.

Although evidence of the expression of mRNA for ERs was obtained in rat parathyroid after amplification by PCR (NAVEH-MANY et al. 1992), studies on human parathyroids, using cytosol binding assays or immunocytochemical assays of nuclear ERs with monoclonal antibodies to the receptor protein, have demonstrated that this tissue contains little or no ERs. Only 15 of 161 glands assayed showed ERs in small concentrations, in contrast to 107 of 163 that contained glucocorticoid receptors in much larger concentrations, irrespective of the histopathological diagnosis (normal parathyroids, adenoma, hyperplasia or carcinoma) and of the sex or menopausal status (SANDELIN et al. 1992). Other studies on excised glands from patients with primary hyperparathyroidism using similar techniques, failed to obtain any evidence of the expression of ERs (SAXE et al. 1992; PRINCE et al. 1991b). Therefore, the effects of estrogens on PTH secretion may be indirect, possibly mediated by the increased calcitriol levels, since parathyroid cells contain vitamin-D receptors (PRINCE 1994) and estrogens increase calcitriol concentrations in plasma (see Sect. D.III).

In contrast, estrogens modulate the catabolic effects of PTH on bone. PTH, prostaglandin E_2 and isoproterenol induce adenylate cyclase activity in murine osteoblastic cells. Elevation of cAMP in osteoblasts is thought to be linked to increased osteoclastic activity. Pre-treatment with estrogen decreased the cell's response to the above hormones (MAJESKA et al. 1994).

Intravenous infusion over 20 hours of (1–34) human PTH in post-menopausal women with osteoporosis, resulted in acute increases of such biochemical markers of bone resorption as urinary pyridinoline, deoxypyridinoline and hydroxyproline. The increments were significantly less in estrogen-treated than in non-treated women, indicating that the estrogenized post-menopausal osteoporotic skeleton is less sensitive to the bone resorbing effects of acutely administered PTH (Cosman et al. 1993).

Indirect evidence suggests that a similar protective effect of estrogens is operating under chronic exposure to elevated PTH. Grey et al. (1996a) performed sequential measurements of bone mineral density by DEXA at 6-month intervals over 2 years in untreated post-menopausal women with mild primary hyperparathyroidism. They observed an accelerated bone loss in comparison with age-matched post-menopausal controls, also untreated. In another study by the same group (Grey et al. 1996b), estrogen administration (together with medroxyprogesterone) to post-menopausal women with mild hyperparathyroidism in a double-blind, randomized, placebo-controlled protocol, resulted in a clear-cut reversal of bone loss, although both groups of patients, steroid-treated and controls, remained similarly mildly hyperparathyroid over the 2-year observation period.

Besides preventing PTH bone resorption, estrogens potentiate PTH's anabolic properties. PTH and TGFβ stimulate fibronectin (FN) production by cultured primary rat and human bone cells, which are osteoblast-enriched, and by osteoblast-like osteosarcoma cells. The precise role of FN in bone is unknown, but this extracellular matrix glycoprotein contains collagen-binding sequences that enable it to organize framework for collagen. The stimulatory effect of PTH on FN is significantly increased in the presence of estrogen, while the effect of TGFβ is not (Eielson et al. 1994). Finally it has long been known that intermittent PTH treatment in osteoporotic patients increases bone formation more than it increases bone resorption, and PTH administration may prevent bone loss during acute estrogen deprivation (Finkelstein et al. 1994). It remains to be proven whether periodical or continuous estrogen replacement further potentiates the overall bone anabolic effect of intermittent PTH.

III. Vitamin D

Vitamin D is biosynthesized in the skin by the action of ultraviolet light on 7-dehydrocholesterol. In the circulation, vitamin D is conveyed to the sites of hydroxylation and to the target tissues by vitamin D-binding protein (DBP). The 25-hydroxylation occurs in the liver and hydroxylated vitamin D [25(OH)D] has the highest affinity for plasma DBP. The 1-hydroxylation of 25(OH)D occurs in the kidneys and 1,25(OH)$_2$D (calcitriol) has the highest affinity for nuclear vitamin D receptors and is considered the active form of the hormone.

States of relative hyperestrogenemia, such as human pregnancy or oral estrogen intake, are associated with increased levels of plasma calcitriol, which are not ascribed to the increased DBP levels since only 1–2% of sterol binding sites are occupied (RAY 1996). Conversely, calcitriol levels are decreased during medically induced hypoestrogenemia (HARTWELL et al. 1990), while short-term (1 month) ERT, given to young girls with Turner's syndrome (MAURAS et al. 1997) or post-menopausal women (CHEEMA et al. 1989), or a prolonged treatment (2 years) given to post-menopausal women (VAN HOOF et al. 1994), results in sustained increases in calcitriol levels. These in vivo observations are interpreted as indicating stimulation by estrogens of the 25(OH)D 1-hydroxylase activity.

Investigations on the ovariectomized rat model suggest that PTH exerts a permissive role for this action. The ^{3}H-calcitriol production by cortical kidney slices and the ^{3}H-calcitriol in plasma were markedly increased by pre-treatment with estrogen in animals that had received the precursor ^{3}H-25(OH)D intraperitoneally. The stimulatory effect on 1-hydroxylase could not be demonstrated in parathyroidectomized animals, indicating that estrogens and PTH co-operate for 1-hydroxylase activity and suggesting that perhaps the main effect of estrogens, in this respect, is to potentiate the well known PTH action on 1-hydroxylase (ASH and GOLDIN 1988). Therefore, a complex system is operating, where PTH and vitamin D co-operate with estrogens to modulate each other's activity. Finally, oral estrogens but not transdermally administered estrogens, increase circulating DBP (SELBY and PEACOCK 1986b), although this is unrelated to the actual enhancement of 1-hydroxylase and the actual increase of free calcitriol (VAN HOOF et al. 1994; CHEEMA et al. 1989). In addition, owing to the limited binding of calcitriol to DBP, this protein rise is not expected to hamper the hormone bioactivity.

Acknowledgements. The author wishes to thank Dr Rolf C. Gaillard for helpful comments, and Marie-Claude Evraere for editorial assistance.

References

Ain KB, Mori Y, Refetoff S (1987) Reduced clearance rate of thyroxine-binding globulin (TBG) with increased sialylation: a mechanism for estrogen-induced elevation of serum TBG concentration. J Clin Endocrinol Metab 65: 689–696

Albrecht ED, Pepe GJ (1995) Suppression of maternal adrenal dehydroepiandrosterone and dehydroepiandrosterone sulfate production by estrogen during baboon pregnancy. J Clin Endocrinol Metab 80:3201–3208

Amino N, Tanizawa O, Mori H, Iwatani Y, Yamada T, Kurachi K, Kumahara Y, Miyai K (1982) Aggravation of thyrotoxicosis in early pregnancy and after delivery in graves' disease. J Clin Endocrinol Metab 55:108–112

Ash SL, Goldin BR (1988) Effects of age and estrogen on renal vitamin D metabolism in the female rat. Am J Clin Nutr 47:694–699

Avvakumov GV, Warmels-Rodenhiser S, Hammond GL (1993) Glycosylation of human corticosteroid-binding globulin at aspargine 238 is necessary for steroid binding. J Biol Chem 268:862–866

Avvakumov GV, Hammond GL (1994) Glycosylation of human corticosteroid-binding globulin. Differential processing and significance of carbohydrate chains at individual sites. Biochemistry 33:5759–5765

Black LJ, Sato M, Rowley ER, Magee DE, Bekele A, Williams DC, Cullinan GJ, Bendele R, Kauffman RF, Bensch WR et al. (1994) Raloxifene (LY139481 HCl) prevents bone loss and reduces serum cholesterol without causing uterine hypertrophy in ovariectomized rats. J Clin Invest 93:63–69

Black VH (1994) Estrogen, not testosterone, creates male predominance of a P4501-related cytochrome in adult guinea pig adrenals. Endocrinology 135:299–306

Bonacci R, Pinchera A, Fierabracci P, Gigliotti A, Grasso L, Giani C (1996) Relevance of estrogen and progesterone receptors enzyme immunoassay in malignant, benign and surrounding normal thyroid tissue. J Endocrinol Invest 19:159–164

Boucher A, D'Amour P, Hamel L, Fugère P, Gascon-Barré M, Lepage P, Ste-Marie LG (1989) Estrogen replacement decreases the set point of parathyroid hormone stimulation by calcium in normal postmenopausal women. J Clin Endocrinol Metab 68:831–836

Burgess LH, Handa RJ (1992) Chronic estrogen-induced alterations in adrenocorticotropin and corticosterone secretion, and glucocorticoid receptor-mediated functions in female rats. Endocrinology 131:1261–1269

Cheema C, Grant BF, Marcus R (1989) Effect of estrogen on circulating "free" and total 1,25-dihydroxyvitamin D and on the parathyroid-vitamin D axis in postmenopausal women. J Clin Invest 83:537–542

Ciocca DR, Vargas Roig LM (1995) Estrogen receptors in human nontarget tissues: biological and clinical implications. Endocr Rev 16:35–62

Colby HD, Larson IW, Kowalski C, Levitt M (1995) Zonal differences in adrenocortical lipid peroxidation: role of alpha-tocopherol. Free Radic Biol Med 18:373–376

Cosman F, Shen V, Xie F, Seibel M, Ratcliffe A, Lindsay R (1993) Estrogen protection against bone resorbing effects of parathyroid hormone infusion. Ann Intern Med 118:337–343

Cosman F, Nieves J, Horton J, Shen V, Lindsay R (1994) Effects of estrogen on response to edetic acid infusion in postmenopausal osteoporotic women. J Clin Endocrinol Metab 78:939–943

del Senno L, degli Uberti E, Hanau S, Piva R, Rossi R, Trasforini G (1989) In vitro effects of estrogen on tbg and c-myc gene expression in normal and neoplastic human thyroids. Mol Cell Endocrinol 63:67–74

Delmas PD, Bjarnason NH, Mitlak BH, Ravoux AC, Shah AS, Huster WJ, Draper M, Christiansen C (1997) Effects of raloxifene on bone mineral density, serum cholesterol concentrations, and uterine endometrium in postmenopausal women. N Engl J Med 337:1641–1647

Dick IM, John AS, Heal S, Prince RL (1996) The effect of estrogen deficiency on bone mineral density, renal calcium and phosphorus handling and calcitropic hormones in the rat. Calcif Tissue Int 59:174–178

Dick IM, Prince RL (1997) Estrogen effects on the renal handling of calcium in the ovariectomized perfused rat. Kidney Int 51:1719–1728

Donda A, Reymond F, Rey F, Lemarchand-Béraud T (1990) Sex steroids modulate the pituitary parameters involved in the regulation of TSH secretion in the rat. Acta Endocrinol 122:577–584

Draper MW, Flowers DE, Huster WJ, Neild JA, Harper KD, Arnaud C (1996) A controlled trial of raloxifene (LY139481) HCl: impact on bone turnover and serum lipid profile in healthy postmenopausal women. J Bone Miner Res 11:835–842

Eielson C, Kaplan D, Mitnick MA, Paliwal I, Insogna K (1994) Estrogen modulates parathyroid hormone-induced fibronectin production in human and rat osteoblast-like cells. Endocrinology 135:1639–1644

Eriksen EF, Colvard DS, Berg NJ, Graham ML, Mann KG, Spelsberg TC, Riggs BL (1988) Evidence of estrogen receptors in normal human osteoblast-like cells. Science 241:84–86

Faglia G, Beck-Peccoz P, Ferrari P, Ambrosi B, Spada A, Travaglini P (1973) Enhanced plasma thyrotrophin response to thyrotrophin-releasing hormone following oestradiol administration in man. Clin Endocrinol 2:207–210

Feingold IB, Longhurst PA, Colby HD (1993) Regulation of adrenal and hepatic alpha-tocopherol content by androgens and estrogens. Biochim Biophys Acta 1176:192–196

Finkelstein JS, Klibanski A, Schaefer EH, Hornstein MD, Schiff I, Neer RM (1994) Parathyroid hormone for the prevention of bone loss induced by estrogen deficiency. N Engl J Med 331:1618–1623

Fujieda K, Faiman C, Feyes FI, Winter JSD (1982) The control of steroidogenesis by human fetal adrenal cells in tissue culture. IV. The effects of exposure to placental steroids. J Clin Endocrinol Metab 54:89–94

Giani C, Campani D, De Negri F, Martini L, Fabbri R, Bonacci R, Ciancia EM, Gigliotti A, Fierabracci P, Pinchera A (1993) Interference of thyroperoxidase on immunocytochemical determination of steroid receptors in thyroid tissue. J Endocrinol Invest 16:37–43

Glinoer D, Gershengorn MC, Dubois A, Robbins J (1977) Stimulation of thyroxine-binding globulin synthesis by isolated rhesus monkey hepatocytes after in vivo beta-estradiol administration. Endocrinology 100:807–813

Glinoer D, Riahi M, Grün JP, Kinthaert J (1994) Risk of subclinical hypothyroidism in pregnant women with asymptomatic autoimmune thyroid disorders. J Clin Endocrinol Metab 79:197–204

Graupner G, Zhang XK, Tzukerman M, Wills K, Hermann T, Pfahl M (1991) Thyroid hormone receptors repress estrogen receptor activation of a TRE. Mol Endocrinol 5:365–372

Grey AB, Stapleton JP, Evans MC, Reid IR (1996a) Accelerated bone loss in post-menopausal women with mild primary hyperparathyroidism. Clin Endocrinol 44:697–702

Grey AB, Stapleton JP, Evans MC, Tatnell MA, Reid IR (1996b) Effect of hormone replacement therapy on bone mineral density in postmenopausal women with mild primary hyperparathyroidism. Ann Intern Med 125:360–368

Hammond GL (1990) Molecular properties of corticosteroid binding globulin and the sex-steroid binding proteins. Endocr Rev 11:65–79

Hartwell D, Riis BJ, Christiansen C (1990) Changes in vitamin D metabolism during natural and medical menopause. J Clin Endocrinol Metab 71:127–132

Hirst JJ, West NB, Brenner RM, Novy MJ (1992) Steroid hormone receptors in the adrenal glands of fetal and adult rhesus monkeys. J Clin Endocrinol Metab 75:308–314

Hryb DJ, Khan MS, Romas NA, Rosner W (1990) The control of the interaction of sex hormone-binding globulin with its receptor by steroid hormones. J Biol Chem 265:6048–6054

Jaklic BR, Rushin J, Ghosh BC (1995) Estrogen and progesterone receptors in thyroid lesions. Ann Surg Oncol 2:429–434

Jaume JC, Costante G, Nishikawa T, Phillips DIW, Rapoport B, McLachlan SM (1995a) Thyroid peroxidase autoantibody fingerprints in hypothyroid and euthyroid individuals. I. Cross-sectional study in elderly women. J Clin Endocrinol Metab 80:994–999

Jaume JC, Parkes AB, Lazarus JH, Hall R, Costante G, McLachlan SM, Rapoport B (1995b) Thyroid peroxidase autoantibody fingerprints. II. A longitudinal study in postpartum thyroiditis. J Clin Endocrinol Metab 80:1000–1005

Khan MS, Hryb DJ, Hashim GA, Romas NA, Rosner W (1990) Delineation and synthesis of the membrane receptor-binding domain of sex hormone-binding globulin. J Biol Chem 265:18362–18365

Komm BS, Terpening CM, Benz DJ, Graeme KA, O'Malley BW, Haussler MR (1988) Estrogen binding receptor mRNA, and biologic response in osteoblast-like osteosarcoma cells. Science 241:81–84

Labhart A (1986) Adrenal cortex. In: Labhart A (ed) Clinical Endocrinology, Theory and Practice. Springer, Berlin, pp 349–486

Li M, Wronski TJ (1995) Response of femoral neck to estrogen depletion and parathyroid hormone in aged rats. Bone 16:551–557

Majeska RJ, Ryaby JT, Einhorn TA(1994) Direct modulation of osteoblastic activity with estrogen. J Bone Joint Surg 76-A:713–721

Marini M, Rey F, Gomez F (1991) Test à la corticolibérine ovine (oCRH): application clinique et valeurs de référence chez l'adulte jeune et la femme postménopausée. Schweiz Med Wschr 121:653–659

Martel C, Melner MH, Gagné D, Simard J, Labrie F(1994) Widespread tissue distribution of steroid sulfatase, 3beta-hydroxysteroid dehydrogenase/delta5-delta4 isomerase (3beta-HSD), 17beta-HSD, 5alpha-reductase and aromatase activities in the rhesus monkey. Mol Cell Endocrinol 104:103–111

Mauras N, Vieira NE, Yergey AL (1997) Estrogen therapy enhances calcium absorption and retention and diminishes bone turnover in young girls with Turner's syndrome: a calcium kinetic study. Metabolism 46:908–913

Metzger D, Berry M, Ali S, Chambon P (1995) Effect of antagonists on DNA binding properties of the human estrogen receptor in vitro and in vivo. Mol Endocrinol 9:579–591

Moutsatsou P, Sekeris CE (1997) Estrogen and progesterone receptors in the endometrium. Ann N Y Acad Sci 816:99–115

Naveh-Many T, Almogi G, Livni N, Silver J (1992) Estrogen receptors and biologic response in rat parathyroid tissue and C cells. J Clin Invest 90:2434–2438

Noé G, Cheng YC, Dabiké M, Croxatto HB (1992) Tissue uptake of human sex hormone-binding globulin and its influence on ligand kinetics in the adult female rat. Biol Reprod 47:970–976

Oursler M, Osdoby P, Pyfferoen J, Riggs BL, Spelsberg TC (1991) Avian osteoclasts as estrogen target cells. Proc Natl Acad Sci USA 88:6613–6617

Oursler M, Peterson L, Fitzpatrick L, Riggs BL, Spelsberg TC (1994) Human giant cell tumor of the bone (osteoclastoma) are estrogen target cells. Proc Natl Acad Sci USA 91:5227–5231

Patchev VK, Hayashi S, Orikasa C, Almeida OFX (1995) Implications of estrogen-dependent brain organization for gender differences in hypothalamo-pituitary-adrenal regulation. FASEB J 9:419–423

Patton PE, Coulam CB, Bergstralh E (1987) The prevalence of autoantibodies in pregnant and nonpregnant women. Am J Obstet Gynecol 157:1345–1350

Pepe GJ, Albrecht ED (1990) Regulation of the primate fetal adrenal cortex. Endocr Rev 11:151–176

Prince RL, Smith M, Dick IM, Price RI, Garcia Webb P, Henderson NK, Harris MM (1991a) Prevention of postmenopausal osteoporosis. A comparative study of exercise, calcium supplementation, and hormone-replacement therapy. N Engl J Med 325:1189–1195

Prince RL, MacLaughlin DT, Gaz RD, Neer RM (1991b) Lack of evidence for estrogen receptors in human and bovine parathyroid tissue. J Clin Endocrinol Metab 72:1226–1228

Prince RL (1994) Counterpoint: estrogen effects on calcitropic hormones and calcium homeostasis. Endocr Rev 15:301–309

Ramey JN, Burrow GN, Polackwich RJ, Donabedian RK (1975) The effect of oral contraceptive steroids on the response of thyroid stimulating hormone to thyrotropin-releasing hormone. J Clin Endocrinol Metab 40:712–714

Ray R (1996) Molecular recognition in vitamin D-binding protein. Proc Soc Exp Biol Med 212:305–312

Refetoff S, Fang VS, Marshall JS, Robin NI (1976) Metabolism of thyroxine-binding globulin in man. Abnormal rate of synthesis in inherited thyroxine-binding globulin deficiency and excess. J Clin Invest 57:485–495

Reymond M, Lemarchand-Béraud T (1976) Effects of estrogens on prolactin and thyrotrophin responses to TRH in women during the menstrual cycle and under oral contraceptive treatment. Clin Endocrinol 5:429–437

Rosner W (1990) The functions of corticosteroid-binding globulin and sex hormone-binding globulin: recent advances. Endocr Rev 11:80–91

Roti E, Emerson CH (1992) Clinical review 29. Postpartum thyroiditis. J Clin Endocrinol Metab 74:3–5

Sandelin K, Skoog L, Humla S, Farnebo LO (1992) Estrogen, progesterone, and glucocorticoid receptors in normal and neoplastic parathyroid glands. Eur J Surg 158:467–471

Saxe AW, Gibson GW, Russo IH, Gimotty P (1992) Measurement of estrogen and progesterone receptors in abnormal human parathyroid tissue. Calcif Tissue Int 51:344–347

Selby PL, Peacock M (1986a) Ethinyl estradiol and norethindrone in the treatment of primary hyperparathyroidism in postmenopausal women. N Engl J Med 314:1481–1485

Selby PL, Peacock M (1986b) The effect of transdermal estrogen on bone, calcium-regulating hormones and liver in postmenopausal women. Clin Endocrinol 25:543–547

Séralini GE (1996) Regulation factors of corticosteroid-binding globulin: lesson from ontogenesis. Horm Res 45:92–196

Spinedi E, Suescun MO, Hadid R, Daneva T, Gaillard RC (1992) Effects of gonadectomy and sex hormone therapy on the endotoxin-stimulated hypothalamo-pituitary-adrenal axis: evidence for a neuroendocrine-immunological sexual dimorphism. Endocrinology 131:2430–2436

Stagnaro-Green A, Roman SH, Cobin RH, El-Harazy E Wallenstein S, Davies TF (1992) A prospective study of lymphocyte-initiated immunosuppression in normal pregnancy: evidence of a T-cell etiology for postpartum thyroid dysfunction. J Clin Endocrinol Metab 74:645–653

Sundbeck G, Edén S, Jagenburg R, Lundberg PA, Lindstedt G (1995) Prevalence of serum antithyroid peroxidase antibodies in 85-year old women and men. Clin Chem 41:707–712

Tischler AS (1991) The adrenal medulla and extra-adrenal paraganglia. In: Kovacs K, Asa L (eds) Functional Endocrine Pathology. Blackwell, Boston, pp 509–545

Tunbridge WMG, Evered DC, Hall R, Appleton D, Brewis M, Clark F, Grimley Evans J, Young E, Bird T, Smith PA (1977) The spectrum of thyroid disease in a community: the Whickham survey. Clin Endocrinol 7:481–493

Turner RT, Riggs BL, Spelsberg TC (1994) Skeletal effects of estrogen. Endocr Rev 15:275–300

Vamvakopoulos NC, Chrousos GP (1993) Evidence of direct estrogenic regulation of human corticotropin-releasing hormone gene expression. Potential implications for the sexual dimorphism of the stress response and immune/inflammatory reaction. J Clin Invest 92:1896–1902

van Hoof HJC, van der Mooren MJ, Swinkels LMJW, Rolland R, Benraad TJ (1994) Hormone replacement therapy increases serum 1,25-dihydroxyvitamin D: a 2-year prospective study. Calcif Tissue Int 55:417–419

Vollenweider-Zerargui L, Barrelet L, Wong Y, Lemarchand-Béraud T, Gomez F (1986) The predictive value of estrogen and progesterone receptors' concentrations on the clinical behavior of breast cancer in women. Clinical correlation on 547 patients. Cancer 57:1171–1180

Yang NN, Venugopalan M, Hardikar S, Glasebrook A (1996a) Identification of an estrogen response element activated by metabolites of 17beta-estradiol and raloxifene. Science 273:1222–1225

Yang NN, Bryant HU, Hardikar S, Sato M, Galvin RJS, Glasebrook AL, Termine JD (1996b) Estrogen and raloxifene stimulate transforming growth factor-beta3 gene expression in rat bone: a potential mechanism for estrogen- or raloxifene-mediated bone maintenance. Endocrinology 137:2075–2084

Subject Index

A

abortion, imminent 8
acetylcholine 6
– intracoronary perfusion 334 (fig)
– nitric oxide secretion 335
activator protein 1 transcription factor
  complex (AP1) 118, 155
adenohypophysis 280
adipose tissue 353
adrenal glands 2–3
– anti-toxic properties 385
– cortex, zones 385
– sexual dimorphism in adrenal
  disease 384
adrenal sex-steroid precursors,
  regulation 384–385
Agnatha 275
AIB1 203
albumin, concentration in tumour
  tissues 252
aldicarb 50
alkylphenols 49
α-fetoprotein 287
α-tocopherol 385
Alzheimer's disease 9, 317, 323
amniotic fluid 265
amphetamines 321
Amphioxus 275
androgens 283–284, 301(fig), 353
– metabolism 5
– puberty, function 355
androstendione 3
angina pectoris 335
angiogenesis 308
anordiol 70
anorexia nervosa 228, 256
anti-estrogen binding sites 344
anti-Müllerian hormone 302
antiestrogens
– definition 82, 201
– molecular actions
– – altered ER-ERE binding efficiency
  205
– – cellular levels of Co-Activators/
  Co-Repressors 207
– – effect enabled by AF-1 206–207
– – effects on binding to the receptor,
  dimerisation and nuclear
  localisation 205
– – effects on ER degradation 209
– – ER conformation influence AF-2
  activity, changes 206
– – ligand-independent ERE transactivation 207–208
– – non-ER actions 209–210
– – phenomenon of AF-1/AF-2
  promoter dependency 206–207
– – promoter elements and ERE
  sequence 209
– partial agonists 55-75
– response
– – modifying effects of the cancer cell
  phenotype 212–213
– – modifying effects of the
  normal cellular phenotype
  210–211
– urogenital tract 372–373
antithrombin III 7
antitumour antibiotics 33
apolipoprotein A 336
apolipoprotein B 138, 256
apolipoprotein II 138
apoprotein a 332–333
apoptosis 344
arginine 6
aromatase deficiency 353–358
arylphenols 49
ascorbic acid 385
astrocytes, hypothalamic 322–323
– brain ageing 323
atherogenic lipoproteins 331–332
atherosclerosis 7, 332–333
athletic training 228
ATPase/ATPsynthase 189–190
atrazine 50
azoresorcinol 50

B

basal forebrain cholinergic system
    317–318
bedocarb   50
behaviour, aggressive in boys and girls
    358
benign prostatic hyperplasia   357
benzothiophenes   62
benzothiopyrans   62
17-β estradiol   43–44
– anti-atherosclerotic effect, role
    of increased endothelial nitric oxide
    164–165
11-β-substituted estradiol derivatives
    72–74
β-zeraneol   90
benzofurans   62
biochanin A   47
biphenyl   49
Birch reduction   24
bisphenol A   49, 90
bladder
– effect of ageing   364–365
– female:
– – androgen receptors   364
– – neck, fibrosis   364
– sex steroid effects   363
blood-brain barrier
– endothelial cells   323
– – glycose-transporter-1   323
body weight   228
bone   276–277
bone marrow   343–344
bone mineral density   277
– estradiol   355
– testosterone   355
bone remodelling   387–389
"brain maintenance"   322–323
breakthrough bleedings   8
breast   276
breast cancer   101, 251–252
– Asian women   260
– C2/C16 a hydroxylation   229
– Finnish women   260
– NK cells, effect of estradiol   345
– Oriental women   260
– sports women   261
– Western women   260
broccoli   259-260
bursa of Fabricius   343

C

C57B1/6 mice   347
calcitonin   6, 388
calcitriol   388–389, 391
calcium kinetics   387–389

calmodulin   155
carbamazepine   258
carbaxyl   50
cardiovascular system   278, 329–338
– mortality   329
– – hormone replacement therapy
    329
– – relation to age and gender
    330 (fig)
catechol estrogens   227
– amniotic fluid   265
– anti-carcinogenic effect and preventive
functions   224
– carcinogenesis   232–233
– depression   256–257
– metabolism   248–249
– production in pregnancy   264
– prostaglandin synthesis, stimulation
    265
catecholestrogens   5–6
caveolin   155
CBP/p300   118
cell-mediated immunity   345–346
centchroman   63
cerebral arteries   6
cervical cancer   253, 310
cervical ectropion   310
cervix   309–310
– estrogen regulation   309–310
chlordan   49
chlordecone   49-50
cholestasis   255
cholineacetyl transferase (ChAT)   317
chromium reagents   25
chromosome 6, estrogen receptor
    gene   6
ciliogenesis   306
cimetidine   257–258
climacteric complaints
– estrogens, use   8–9
– women   2
clomiphene   56
cognitive cerebral functions   9
– learning and memory   317–319
comparative molecular field analysis
    (CoMFA)   82
conception   263
coronaropathy   329
coronary arteries   6
corticosteroid-binding globulin   386
corticotropin regulation   386–387
coumestrol   47, 90
cruciferous vegetables   229,
    259–260
cyclic adenosine monophosphate
    (cAMP)   188

cyclic AMP response element-binding
protein (CREB)   118
cyclin D1   203
cyclofenyl   60
Cyclostomata   275
cynomolgus monkey, estrogen receptors
   354
CYP19 gene   353
cyproterone acetate   303
cytochrome P450   223, 227
cytokines   344–345
cytotoxic T lymphocytes   346

D
daidzein   47
dehydroepiandosterone   3
dehydroepiandrosterone sulfate
   (DHEA-S)   7, 264
delta-9-tetrahydrocannabinol   50
depression   256–257
dexamethasone   154, 234
– delayed type hypersensitivity,
   suppression   345
1,2-diarylethanes   44–47
1,1-dichloro-2,2-dichlorophenylethylene
   (DDE)   231
dichlorodiphenyl-trichloroethane (DDT)
   49, 231
dieldren   49
Diels-Alder approach/reaction   19–20,
   31
diet   228
– carbohydrate-rich   259
– fat-rich   260
– fiber-rich   246
– low-fat   228
– protein-rich   259
– vegetarian   246
diethylstilbesterol (DES)   44–47, 56
– derivatives   90
– metabolism   46–47
– prenatal exposure   357
– vaginal adenocarcinoma   310
dimethyl benz   229
dioxin   49
2,3-diphenyl-2H-1-benzopyrans
   98–100
1,1-diphenylethylene-derived agents
   66–67
1,2-diphenylethylene-derived agents
   67–68
diuretics   345
diverse protein estrogen binders,
   evidence   182–186
– plasmalemmal microsomal fractions
   estrogen binders in   182–186

– – mitochondrial lysosomal fractions
   184
– – other origins   184–186
dopamine   6
– disorders   322
dorsal raphe   320
droloxifene   57–58, 60 (fig), 201
drug-induced biliary occlusion   255
dual-energy X-ray absorptiometry
   (DEXA)   380

E
EM-139   71 (fig)
EM-800   66 (fig), 202
enclomiphene   56
endometrial cancer   9, 252–253
endometrium
– adenocarcinoma   201, 309
– – female smokers   261
– proliferation   308
– tasks   307
endosulfan   49
endothelin   6
endothelin-1   336
endothelium-derived relaxing factor
   (EDRF)   6, 155
Ene-reaction   26
enterohepatic circulation   235, 246
– pregnancy   264
epidermal growth factor (EGF)   117,
   203
epididymis   299, 301 (fig)
16-epiestriol   3
epoophoron   299
equilenin   3, 17
equilibrium dissociation constants (Kds)
   127
equlin   3, 17
ERKO mice   122, 162–164, 286,
   353–358
erythropoietin   308–309
"esterified estrogens"   8
estradiol   8, 17, 172
– actions on endometrium   308
– bone density   355
– derivatives   33–35
– EC50, estrogen receptors α and β
   121 (table)
– effect on calcium influx   178
– effects at the target organs
– – breasts   4
– – hypothalamus and pituitary
   gland   4–5
– – vagina and endometrium   4
– fast actions, diverse mechanisms
   186–192

estradiol
– – ATPase/ATP synthase   189–190
– – calcium channels   177
– – cyclic adenosine monophosphate
    (cAMP)   188
– – glyceraldehyde-3-phosphate
    dehydrogenase (G3PD)
    190–191
– – mitogen-activated protein kinase
    (MAPK)   188–189
– – others   187–188
– – phospholipase C (PLC)   188
– isolation   2–3
– labeling   33–35
– levels in pregnancy   7
– metabolism
– – 17-β-hydroxysteroid dehydrogenase
    type   2, 223–226
– – breakdown   243–246
– – drugs   257–259
– – influence of diseases and drugs in
    human beings   250 (table)
– – prostate tissue   253–254
– – sex differences   246–248
– pharmacophore   21, 24
– urge incontinence   370
estradiol valerate   369
estramustin   33
estriol   17
– antibiotic treatment in pregnancy
    264
– effect on calcium influx   178
– isolation   2–3
– levels in pregnancy   7, 263
– measurements, fetal-placental function
    monitoring   264
estrogen
– antiestrogens with a modified steroid
    structure   82–90
– – A ring modification   84
– – B ring modification   85–87
– – C ring modification   87
– – D ring modification   84–85
– behavioural effects   283
– bioconversion   35
– biogenesis   5
– body fluids and tissues   3
– brain function, implications on ageing
    and dementia   315–324
– C2/C16 α-hydroxylation   227–231
– – breasts cancer   229
– – different ethnic groups   230–231
– – effect on metabolite estrogenicity
    227
– – effect of pesticides   231
– – effect of smoking   228–229

– – estrogen-2-hydroxylation by
    placental aromatase   231
– – induction by indole-3-carbinol
    229–230
– – inhibition   231
– – modulation by thyroid hormones,
    body weight and diet   227–228
– carcinogenesis   9
– cardioprotective effect   329, 331–337
– cardiovascular protection   156
– cytochrome P450 mono-oxygenase
hydroxylation   227
– deficiency, male hypogonadism   355
– dependent differentiation, early brain
    development   285–288
– determination   4
– direct effect on:
– – coronary microvascularization   336
– – smooth muscle cell   336
– – vascular wall   333–337
– effects on:
– – coagulation   7
– – lipids   6–7
– endogenous   43
– – metabolism   243–265
– enterohepatic metabolism   5
– fetal endocrinology   7
– food   3
– heart disease, protective roll   228
– hormone replacement therapy see
    estrogen replacement therapy
– immune function, effect   344–347
– immunomodulation, pathways
    348 (fig)
– induced prolactin release   283–284
– ischaemic heart disease   228
– isolation   1–2
– life expectancy   9
– male, function
– – behaviour   358
– – bone maturation   355
– – prostate   357–358
– – testicular function   355–357
– male, source   353
– mechanism of action   6
– metabolism
– – fetoplacental unit   263 (fig),
    264–265
– – main pathways   224–225 (figs)
– metabolism and disease   249–257
– – breast cancer   251–252
– – cervical cancer   253
– – depression   256–257
– – endometrial cancer   252–253
– – liver disease   254–255
– – lupus erythematosus   255

– – osteoporosis   257
– – papilloma of the larynx   254
– – prostate cancer   253–254
– – thyroid disease   255–256
– – weight changes   256
– mood and mental state influence 319–321
– natural/synthetic   7–8
– non-aromatic   49–50
– non-genomic effects   172–192
– non-steroidal see non-steroidal estrogens
– nucleophilic aromatic substitution 22–23
– ovary   304–305
– physiologic effects   5–6
– – genomic actions   5–6
– – non-genomic actions   6
– plant-derived   47
– pregnancy   7
– primary lymphoid tissues   343–344
– production
– – adipose tissue   353
– – localization   2–3
– replacement therapy see estrogen replacement therapy
– research, history   1–9
– sceletal homeostasis   276–277
– sex differentiation   304
– steroidal see steroidal estrogens
– synthesis
– – partial/total   2
– – phylogeny   275–276
– testicular function   355–357
– treatment
– – chronic, parkinsonian symptoms 322
– – climacteric complaints   7–9
– – myocardial infarction risk   178
– – pregnancy   8
– – preparations   7–8
– – side effects   8
– urinary incontinence   365–370
– urinary tract infections, recurrent 370–371
– uterine pathology   309
– vaginal pathology   310
estrogen glucuronides   235
estrogen receptor (ER)   81, 130–132
– α and β 276–277, 337
– a knock-out mouse, functional model   122
– developmental role, hormone-dependent brain differentiation 285–288

– extragenital distribution and functional significance   276–282
– – bone   276–277
– – breast   276
– – cardiovascular system   278
– – central nervous system   280–281
– – gastrointestinal tract   277–278
– – immune system   278–279
– – pituitary gland   279–280
– Fallopian tubes   306 (table)
– functions and transcription   116–119
– – DNA binding properties   116–117
– – ligand-dependent transcriptional activity   117–119
– – ligand-independent activity   117
– isotype diversity generates specificity 120–122
– – α and β differential activity 121–122
– – α and β heterodimerization   120
– – tissue distribution and ligand binding   120–121
– localization, male genital tract 353–354
– male reproductive function 353–358
– mediated tissue specificity   379–380
– membrane bound   281–282
– relative binding affinity   81, 83 (table)
– signalling, key elements for anti-estrogen action   202–204
– structure
– – A/B domain   112
– – co-activators   119 (fig)
– – DNA-binding domain   112–115
– – ligand-binding domain   115
– structure-activity relationship 81–101
– synthesis, hormonal regulations
– – androgens   283–284
– – estrogens   282
– – glucocorticoids   284
– – neurotransmitters   284–285
– – progesterone   283
– – thyroid hormones   283
– – uterus   307 (table)
estrogen replacement therapy   153, 387
– female skeleton   387
– influence on the lipid profile   333
– vaginal atrophy   371–372
estrogen response element   116
estrogen sulphates   233–235
– biological role   233–234
– formation   233
– regulation of sulphation   234–235

estrogen sulphates
estrogen transforming enzymes
    223–237
estrogen-regulated genes   127–139
– extra-genomic effects of estrogens
    139
– non-transcriptional control of gene
    activity by estrogens   137–139
– post-transcriptional effects
    of estrogens   138–139
– transcriptional control of gene activity
    by estrogens   129–137
– – estrogen receptors   130–132
– – estrogen response elements
    132–135
– – target cell environment   135–137
– – target gene promoters   135
estrone   17
– effect on calcium influx   178
– levels in pregnancy   7
ethinylestradiol   7
ethylenes   44–47
7,17 α-ethinylestradiol   18
extrahepatic biliary obstruction   255

F
Fallopian tubes   305–306
– estrogen receptor   306 (table)
– progesterone receptor   306 (table)
– steroidal regulation   305–306
fast estradiol-evoked biological responses
    173–180
– fast estradiol effects on:
– – catecholamine system   173–174
– – central nervous system   176–177
– – hippocampus   174–176
– peripheral biological responses
    177–180
female chromosomal sex   300
female reproductive tract   299–310
– cervix   309–310
– embryology   299–300
– estrogens and ovary   304–305
– Fallopian tubes   305–306
– sex differentiation   300–304
– uterus   306–309
– vagina   310
female sexual behaviour   284
fermentation   47
fetal brain   5
fetal endocrinology   7
fetoplacental unit, estrogen metabolism
    263 (fig)
fetus   263
fibrinolytic system   333
fibroblast growth factor   308

fibronectin   390
flavins   153
flavones   47
follicle-stimulating hormone (FSH)   2,
    304, 353
– estradiol treatment   356
– oral contraceptive pill, estrogens
    304–305

G
G-protein-mediated receptors   172
galanin   6
Gartner's duct cysts   299
gastrointestinal tract   277–278
genetic sex   300–301
genistein   47, 90, 121 (table)
genital atrophy   372
glial cells   322–323
glial fibrillary acidic protein   323
glucocorticoid response element   133
glucocorticoids   284
glucuronyl transferase   235
glutamate receptor   188
glyceraldehyde-3-phosphate dehydro-
    genase (G3PD)   190–191
gonadal sex   300–302
– female   302
– male   301
gonadectomy   278
– endothelium dependent vascular
    relaxation   334
gonadotropin secretion   284
– estrogen dependent   355
gonadotropin-releasing hormone (GnRH)
    6, 356
GW-5638   60 (fig)
gynaecomastia   258

H
Hajos-Weichert-Eder ketone   19
HEK-293 cell line   121
heregulin   203
HERS study   331, 337
hexachlorocyclohexane   49
hexestriol (HES)   44–46
high-density lipoprotein   6–7, 332
hippocampal CA1 pyramidal neurones
    318–319
hippocampal-aminobutyric acid (GABA)
    interneurins   318
Hohlweg effect   4–5
homocystein   7
hormone replacement therapy see
    estrogen replacement therapy   116
hormone replacement therapy
– cardiovascular mortality   329

– influence of the lipid profile   333
– women with elevated basal
    triglycerides   331
Hormones and Urogenital Therapy
    committee (1994)   369
human chorionic gonadotrophin (HCG)-
    stimulated fetal tests   5–6
human papilloma virus   253
humoral immunity   345–346
4,2-hydroxy estradiol   17
16 α-hydroxyestrone   229
16-α-β-hydroxyestrone   3
16,18-hydroxyestrone   3
17 α-hydroxylase/17,20-lyase deficiency
    355
17 β-hydroxysteroid dehydrogenase
    type   2, 223–226
hyperestrogenemia, relative   391
hyperthyreosis   258
hyperthyroidism   227–228
– artificial   255–256
hypnotics   365
hypocalcaemia, edentic acid (EDTA)
    induced   388
hypogonadism   355
hypospadias   357
hypothalamus
– feed-back of estrogens   4–5
– P2 fraction   182
hypothyroidism   227–228

I
ICI-164, 70–71, 85, 120–121, 202,
    346, 384,
– synthesis   72 (fig)
ICI-182,780   71 (fig), 86 (table), 93,
    175, 202
idoxifene   58, 60 (fig), 201
immune system   278–279, 343–348
indenes   62
indole-3-carbinol   258–260
– growth of laryngeal papillomas
    254
– induction of C2-hydroxylation
    229–230, 232
inflammation   345, 347
insulin-like growth factor-1   203
interleukin(s)   6, 343
International Conference of the
    Standardization of Sex Hormones
    (1932)   2
intestine function   246
intravaginal ring, silicone   372
isoflavones   47
isoproterenol   389

K
karyopicnotic index   371
keoxifene   63-64
kepone   49
Ki-67 antigen, immunohistochemical
    studies   371–372
Koeber reaction   4

L
Lagomorpha   275
levormeloxifene   164
Leydig cells   301–302
– aromatase activity   353–354
lipid metabolism, estrogen dependent
    355
lipoidal estradiol   235–237
lipoidal estrogens   235–237
lipoprotein a   332–333
liver disease   254–255
low-density lipoprotein   6–7, 256, 331
lupus erythematosus   255
luteinizing hormone (LH)   2
– estradiol treatment   356
– releasing hormone (LHRH)   5
LY-117018   63
lymphocytes   346
lymphokine-activated killer (LAK)   346

M
McMurry reaction   57, 58 (fig)
macrolactones   47–49
male chromosomal sex   300
male reproductive function   353–358
mammary cancer   9
maturation index   371
maturation value   371
MCA-106 fibrosarcoma   347
MCF-7 cells   63, 82, 84, 230
– breast cancer, estrogen sulphation
    stimulation   235
medroxyprogesterone   369, 390
menopause, atrophy of estrogen sensitive
    tissues   363
menstruation   1, 255
– luteal phase, mood changes   320
– urogenital symptoms   363–364
MER-25   56 (fig)
methamomyl   50
methoxychlor   49
MG63 osteosarcoma cells   129
midbrain serotonergic neurons
    319–321
midstream urine culture   371
mitochondrial lysosomal (mP2)
    fraction   182

mitogen-activated protein kinase
    (MAPK) 117, 188–189
MK801 175
monoamine-oxidase (MAO) inhibitors
    257, 259
Müllerian duct 299–304
Müllerian duct syndrome 303
Müllerian inhibitory substance 302
mycotoxins 48
myelopoietic progenitors 344
myocardial infarction 9
myogenic transcription factor D (Myo D)
    118

N
N-fluorobis 22
N-fluoropyridinium triflate 22
N-methyl-D-aspartate (NMDA)
    receptors 319
nafoxidene 56, 62
natural killer (NK) cells 279
– estradiol effect 345
neonatal brain 286
neurotransmitters 284–285
neutrophic factor (NF)-kB sites 154
NF-1 155
Nf-kB-mediated transactivation
    208–209
nicotinamide adenine dinucleotide
    phosphate (NADPH) 153
nigrostriatal dopaminergic system
    321–322
nitric oxide 6, 153
– arterial wall, effect on 335 (fig)
– bioactivity 162
– increased endothelial, role in the
    antiatherosclerotic effects 164–165
– production effect of estrogen 156
– secretion 334
– – acetylcholine induced 335
nitric oxide synthase (NOS)
– isoenzymes
– – I 154
– – II 154–155
– – III 155
– regulation of isoenzymes by
    17β-estradiol 155–165
– – inhibition of NOS-II activity by
    17β-estradiol 155–156
– – upregulation of endothelial NOS-III
    activity by 17β-estradiol 156–165
nitromifene 61
nocturia 364
Nolvadex 57
non-steroidal estrogens 43–50
– antiestrogens 90–101

– structural classification
– – alkylphenols 49
– – arylphenols 49
– – 1,2-diarylethanes 44–47
– – ethylenes 44-47
– – flavones 47
– – isoflavones 47
– – macrolactones 47
– – non-aromatic estrogens 49–50
nonylphenol 90
norethisterone-acetate (NETA) 337
NOS-I (neuronal NOS) 153–154
NOS-II (inducible NOS) 153–155
NOS-III (endothelial NOS) 153–155
– enzyme activity 161–162
– gene expression 159–161
– regulation, role of the estrogen
    receptor 162–164

O
O-quinodimethane 20
omega-3 fatty acids 259
omeprazole 258
oocytes 302
oogonia 302
oral contraceptive pill 255
– parkinsonian symptoms 322
ornitine-d-aminotransferase gene
    transcript, retinoblastoma cells 138
osteoblasts 276–277
– function 6
osteoclasts 6
osteopenia 387
osteoporosis 101, 257
– estrogens 9
– female smokers 261
– men 355
ovarectomy 255
ovarian cancer 9, 347
ovarian extracts 1
ovary 299
– development 300 (fig)
– endocrine function 1
– estrogens 304–305
oxamyl 50
16-oxoestradiol 3

P
p-53 155
panomifene 61
Papanicolaou smears 310
papilloma of the larynx 254
parathormone 6
parathyroid hormone 388–390
paroophoron 299
partial agonists 55–75

pelvic floor
- muscle fibres   364
- neuronal damage secondary to childbirth   364
PEPI Trial Writing Group   333
peroxisome proliferator-activated receptor   116
pesticides, C2/C16 α-estrogen hydroxylation effect   231
2-phenylbenzo[b]thiophene derivatives   94
2-phenylindole derivatives   68–70, 91–93
phenylpropanolamine   369
phenytoin   258
pherol red   50
phloridzin   47
phosphocalcic metabolism
- bone remodelling   387–389
- calcium kinetics   387–389
phospholipase C (PLC)   188
phthalates   50
phytoestrogens   47, 90
pituitary gland   279–280
- anterior   4–5
- tumour cells, tamoxifen   128
placenta   7, 263
- catechol estrogens   264
- endocrine function   1
- insufficiency   264
plasmalemmal microsomal (P3) fraction   182
plasminogen   332
platelet-derived growth factor AB and BB (PDGF-AB and BB)   154
polychlorinated biphenyls   90
positron emission tomography (PET)   35
postmenopausal women   246
- acetylcholine effect on atherosclerotic coronary arteries   335
- Alzheimer's disease   317
- cimetidine   258
- endometrial cancer   253
- estrogen replacement therapy, clinical trials   363
- hyperthyroidism   390
- ionized calcium levels   389
- loss of "estrogenic tone"   320–321
- lower urinary tract dysfunction   365
- osteoporosis   257
- stress relief   48
postpartum, mood changes   320
pregnancy   255
- systemic vascular resistance   333–334
- thyroid autoimmunity   382
- urogenital symptoms   363–364
Prins reaction   26
progestagens, cardiovascular system influence   337
progesterone   154, 235, 283
- cardiovascular system influence   337
- detrusor instability, antenatally/post partum   364
- endometrium proliferation   308
- metabolism in fetus   263–264
- receptors
- - Fallopian tube   306 (table)
- - uterus   307 (table)
- withdrawal bleeding   8
progestins   306
prolactin   345
propoxur   50
prostacyclin   6, 336
prostaglandin E2   343, 389
prostate   301 (fig)
- aromatase   357–358
- development   300 (fig)
- - androgen dependent   357
- - estrogen receptors, presence and role   354
prostate cancer   253–254, 357
- estrogen treatment   329
protein kinase C   203
prunetin   47
psychological function   319–321
puberty   353
- androgens, function   355
"pure antiestrogens"   55, 202
pyrrolidine dithiocarbamate   155

Q
QSAR models   82, 84
quinidine   258

R
raloxifene   63–64, 128–129, 201, 206, 337, 379–380
- synthesis   65 (fig)
ranitidine   258
receptor tyrosine kinase   172
reimplantation, classical experiments of extirpation   1
9-cis-retinoic acid receptor   116
retinoid-X receptor   116
retinoids   154
retroperitoneal fibrosis   347
reverse-transcriptase polymerase chain reaction (RT-PCR)   234, 354
rheumathoid arthritis   255

RU-58668   93
– synthesis   74 (fig)
RU-45144   71 (fig)
RU58668   73

S
"schaumzellen"   332
secorticoids   50
selective estrogen receptor modulators
   (SERMs)   43
sensorimotor performance   321–322
serotonin   9, 319
Sertoli cells   301–302
– aromatase activity   353–354
sex differentiation   287, 300–304
sex hormone binding globulin (SHBG)
   5, 184–186
– mediated estrogenic action   380
sex-steroid receptor   306
sexual dysfunction   258
"sexual zentrum"   4–5
short term memory   317
sleep pattern, serotonin levels   9
smoking, effect on C2/C16a
   hydroxylation   228–229
SMRT   203
somatic sex   300–304
– female sex differentiation
   303–304
– male sex differentiation   302–303
spatial recognition   317
specific estradiol binding sites, evidence
   in cellular membranes
– central sites   180–181
– peripheral sites   181–182
spermatogenesis   353
sporting activity   261
SRC1/NCoA1   118
SRC   203
Ssn6   203
steroidal estrogens   17–35
– partial synthesis   20–33
– – ring-A substitution   21–24
– – ring-B substitution   24–28
– – ring-C substitution   28–30
– – ring-D substitution   30–33
– total synthesis   18–20
stilbestrol   3, 7
– use in pregnancy   8
stress incontinence   368–370
– age related, prevalence
   366–367 (figs)
subcellular fractions of nuclei (P1)
   182
7 α-substituted estradiol derivatives
   70–72

T
T47D cells   82, 230
tamoxifen   9, 55–57, 337
– chemical structure   94 (fig)
– endometrial carcinoma   372
– fixed-ring analogues   61–63
– long term therapy   201
– lymphocytes, effect on   346
– McMurry reaction   58 (fig)
– production, alternative routes
   59 (fig)
– retroperitoneal fibrosis   347
– rodents   128
– structural analogues   99 (fig)
– triphenylethylene derivatives related
   to   57–61
TAT-59   60 (fig), 201
testes   299
– development   300 (fig)
– estradiol production   2–3
– estrogen production   353–354
testis determining factor   300
testosterone   3, 172, 287, 303
– bone density   355
– gonadotropins inhibition   356
2,3,7 and 8-tetrachlorobenzo-p-dioxin
   (TCDD)   229–230
tetrahydrobiopterin   153
theca folliculi   2
theophylline   258
thromboxane   6
thymus   343–344
– sensitivity to sex hormones   344
– testosterone effect   344
thyroid gland
– cancer   382
– disease   255–256
– – sexual dimorphism   381
– dysfunction   382
– enlargement, estrogen effect
   381–382
thyroid hormone-binding globulin
   382–383
thyroid hormones   283
– status   227
thyroid-stimulating hormone   383
thyroxine   255, 258
tianeptine   320
Tibolone   333, 372
TIF2/GRIP1   118
toremifene   60 (fig), 61, 346
Torgov total synthesis   28
toxaphene   49
transsexual men   354
transforming growth factor a (TGF α)
   203

transforming growth factor ß (TGF β) 154
tranylcypromine 259
1,2,3-triarylpropenone-derived antiestrogens 63–66
tricarbonyl chromiun 26
trimeric G-protein-coupled receptors 172
trioxifene 63–64
triphenylbutene antiestrogens, early clinical trials 60 (fig)
triphenylethylene derivatives 56–66, 70, 94–98
tryptophan levels 9
tumour growth factor (TGF) 308
tumour necrosis factor a (TNF α) 235, 343, 387
Turner's syndrome
– calcium kinetic studies 388
– estrogen replacement treatment 391

U
"unopposed estrogen" 308
urethra
– female, androgen receptors 364
– sex steroids effect 363
urge incontinence 370
„urge syndrome" 370
urinary incontinence 365–370
– continence mechanism 367–368
– elderly, transient causes 365 (table)
– epidemiological studies 365–367
– female, common causes 365 (table)
– – treatment, mechanism 368 (table)
– stress incontinence 368–370
– urge incontinence 370
urinary tract infection
– lower 364–365
– recurrent 370–371, 371
urine output, day time/night time 364
urodynamic studies 364
urogenital tract
– antiestrogens 372–373
– hormonal influences 363–364

uterine hyperemia 308
uterine-weight test 55
uterus 306–309
– cornual 310
– estrogen regulation 307–309
– estrogen and uterine pathology 309
– T-shaped 310

V
vagina 299, 301 (fig), 310
– sex steroid effect 363
vaginal adenocarcinoma 310
vaginal adenosis 310
vaginal atrophy 371–372
vaginal-cornification 55
vas deferens 299, 301 (fig)
vascular endometrial growth factor (VEGF) 308
vascular endothelial cells 334
vegetarians 259
very-low-density lipoprotein 331
vitamin D 6, 390–391
vitellogenin II 138
vulva 301 (fig)

W
weight changes 256
Wolffian (mesonephric) ducts 299–300, 302, 304

X
X chromosome 300
xenoestrogens 49

Y
Y chromosome 301

Z
zearalenone 48
zindoxifene 68 (fig)
– analogues 69 (fig)
ZK164,015 93
ZK-119,010 92
ZR-75 cells 82, 87

Springer
and the
environment

At Springer we firmly believe that an international science publisher has a special obligation to the environment, and our corporate policies consistently reflect this conviction.
We also expect our business partners – paper mills, printers, packaging manufacturers, etc. – to commit themselves to using materials and production processes that do not harm the environment. The paper in this book is made from low- or no-chlorine pulp and is acid free, in conformance with international standards for paper permanency.

Printing: Saladruck, Berlin
Binding: H. Stürtz AG, Würzburg